Applied Exterior Calculus

APPLIED EXTERIOR CALCULUS

DOMINIC G. B. EDELEN

Center for the Application of Mathematics
Lehigh University

A Wiley-Interscience Publication

JOHN WILEY & SONS

New York Chichester Brisbane Toronto Singapore

Library of Congress Cataloging in Publication Data:

Edelen, Dominic G. B.
Applied exterior calculus.

"A Wiley-Interscience publication."
Bibliography: p.
Includes index.
1. Calculus. 2. Exterior forms. 3. Manifolds (Mathematics) I. Title.
QA614.5.E34 1985 515 84-19575
ISBN 0-471-80773-7

Printed in the United States of America

10 9 8 7 6 5 4 3 2 1

APPLIED EXTERIOR CALCULUS

DOMINIC G. B. EDELEN

Center for the Application of Mathematics
Lehigh University

A Wiley-Interscience Publication

JOHN WILEY & SONS

New York Chichester Brisbane Toronto Singapore

Library of Congress Cataloging in Publication Data:

Edelen, Dominic G. B.
Applied exterior calculus.

"A Wiley-Interscience publication."
Bibliography: p.
Includes index.
1. Calculus. 2. Exterior forms. 3. Manifolds (Mathematics) I. Title.
QA614.5.E34 1985 515 84-19575
ISBN 0-471-80773-7

Printed in the United States of America

10 9 8 7 6 5 4 3 2 1

To all my fledglings

Damien
Marybeth
Mittie-ellen
Dominique
Clarissa
John Dominic

PREFACE

The purpose of this book is to provide upper division undergraduate and beginning graduate students with access to the exterior calculus. Such access is essential simply because, much of the modern literature both in mathematics and in the quantified sciences has come to use the exterior calculus as an expository vehicle. Whole segments of the literature and, of greater importance, essential and often simplifying concepts are thus unintelligible to the student who is unfamiliar with the exterior calculus.

Most texts that treat the exterior calculus are oriented toward global results and the needs of the research worker. The reader thus is confronted at the very beginning, and rightly so, with fundamental aspects of topology and the theory of differentiable manifolds. The demands on the student are great, a quantum jump in both sophistication and maturity often being self-evident from the start. A significant portion of the exterior calculus and its attendant concepts can be mastered, however, without recourse to global concepts. In fact, most of the theory can be developed using only local notions, and it can be done in such a way that it places only moderate demands on a student with previous exposure to upper division algebra and analysis courses. These priorities are the basis on which this book has been written.

The analysis and discussions are confined from the outset to local questions. We start with standard n-dimensional number space and restrict attention to what happens in a single neighborhood of a point. It is even possible to restrict attention to what happens in a single neighborhood of a point that carries a single fixed coordinate cover. This, however, is adequate for discussion of tangents to curves and for the attachment of an n-dimensional vector space to each point of the neighborhood in a natural way. Tangent spaces and vector fields then follow as direct consequences, and a change of viewpoint leads to the realization of a vector field as a derivation of the associative algebra of C^∞ functions. Careful consideration of the properties of vector fields under mappings between neighborhoods of points in different spaces then give the more customary results. Exterior forms of degree 1 become immediately available as elements of the dual space of the tangent space. Forms of higher degree are defined operationally by introducing exterior products of basis elements for 1-forms. Again careful study of the mapping properties for exterior forms, which are induced by mappings between neighborhoods of points in different spaces, leads to the pull back map and to the standard results obtained in the more customary approach. With mastery of these facts, the exterior derivative and the Lie derivative follow in a natural way. The notions of closed and exact forms lead to the theory of linear homotopy operators that play the role of inverses of the exterior derivative on an appropriately defined subalgebra of the exterior algebra.

This approach places strong emphasis on the ability to compute at the earliest possible stage. Specific examples are given as soon as the student has been shown how to perform nontrivial calculations. This is further advanced by a list of straightforward but nontrivial exercises at the end of each of the first five chapters. The student is strongly encouraged to try as many of these exercises as possible and to compare answers with those that are given. If you cannot compute with facility, then there is something wrong, something missed along the way.

A second aim of the first five chapters is to provide the student with a sound foundation for modern studies in partial differential equations. While nineteenth-century mathematics was primarily concerned with developing specific methods to solve specific equations, it is twentieth-century mathematics that has developed general techniques for looking at large classes of problems. This is why exterior algebra, vector fields, differential forms, Lie derivatives, and group-theoretic methods were invented. Many of the exotic changes of variables that proliferate in the older literature as special tricks are both immediate and natural from the vantage point of the exterior calculus. Thus vector fields are used directly for a full discussion of the method of characteristics, whereby general solutions are constructed to linear and quasilinear partial differential equations of the first order. This is followed by study of systems of quasilinear partial differential equations with the same principal part. Systems of simultaneous linear partial differential equations are then studied by means of Lie subalgebras and reduction to Jacobi normal form. Partial differential equations are revisited in the context of exterior forms by a systematic exploitation of the Frobenius, Darboux, and Cartan theorems for differential systems. Antiexact forms and the linear homotopy operator, which are the subjects of Chapter Five, have their primary practical significance in answering the question of how to go about solving exterior differential equations in a systematic manner. This leads directly to the study of connection, tension, and curvature forms and to the Cartan equations of structure.

The student with a more applied flair may be put off by the definition, theorem, and proof format adopted throughout the first seven chapters. It is my personal preference because it causes the key ideas to stand out in bold relief and provides a means of rapid perusal and review. It is essential that the student learn to articulate specific points and established facts in a careful and exact manner, for without this ability things rapidly degenerate to utter confusion. Once the student has become accustomed to this mode of exposition, it is often a revelation how a careful statement and summary can point to new and previously unsuspected directions of investigation.

This is a book on applied exterior calculus, as the title states. The text has therefore been divided into three parts. Part One, comprising the first five chapters, covers the essential elements of the exterior calculus together with a number of abbreviated applications. Chapters Six and Seven, which comprise Part Two, deal with specific detailed applications of the exterior calculus to group-theoretic questions in nonlinear second-order partial differential equations and to problems in the calculus of variations. These two chapters are reasonably complete and bring the student to the frontier of modern work. They also provide background material necessary for understanding what follows.

Part Three consists of three chapters that provide in-depth studies of physical disciplines *via* the exterior calculus. Here we relax the definition, theorem, proof format in favor of direct exposition. Chapter Eight is concerned with classical and irreversible thermodynamics. It is based largely on Carathéodory's approach (the Darboux theorem and inaccessibility) with nontrivial extensions to nonconservative mechanical forces and internal degrees of freedom. The fundamental problem of irreversible thermodynamics is then stated and solved through use of the linear homotopy operator introduced in Chapter Five.

Chapter Nine studies electrodynamics with both electric and magnetic charges. General solutions of Maxwell's equations are constructed by converting the field equations to an equivalent system of exterior differential equations, followed by integration through use of the linear homotopy operator. The solutions are obtained without recourse to constitutive relations and are shown to give rise to a four-parameter group of general duality transformations. Restrictions imposed by the vacuum ether relations are then studied. Variational principles for electromagnetic fields in the presence of magnetic charge are obtained through standard techniques of the exterior calculus. This gives rise to closed form evaluations of the 1-forms of forces that act on electric and on magnetic charge distributions that include radiation reaction.

Chapter Ten deals with the modern theory of gauge fields. Here the full scope of the exterior calculus and much of the material in Chapters Six and Seven combine in what is now considered one of the cornerstones in the conceptual containment of physical reality. The simpler situation of a matrix Lie group of internal symmetries is considered first. Derivation of all relevant field equations and geometric structures are obtained even though no direct use is made of the theory of fiber bundles. This is followed by a general theory of operator-valued connections where the group action is allowed to be nonlinear and to act indiscriminately both on the physical state variables and on the space-time labels. Again, all relevant field equations and geometric structures are worked out, but this time specific application is made to gauging the Poincaré group.

Readers who are familiar with the exterior calculus may wish to proceed rapidly to the applications. To make this as simple and painless as possible, the following plan of abbreviated review is offered. Chapters Six through Ten listed below are followed by sequences of chapters. The boldface number refers to the chapter to be reviewed while the numbers to its right refer to the sections. It will often be necessary to read only the main theorems of the sections cited in order to proceed.

Chapter 6.	**2:**	1, 2, 3, 4, 6, 7,
	3:	1, 2, 4, 5, 6, 7,
	4:	1, 2, 5, 6, 7.
Chapter 7.	**2:**	1, 2, 3, 4, 6, 7,
	3:	1, 2, 4, 5, 6, 7,
	4:	1, 2, 5, 6, 7,
	6:	1, 2, 3, 4.
Chapter 8.	**2:**	1, 2, 3, 4, 6,
	3:	1, 2, 4, 5, 7,
	4:	1, 2, 4,
	5:	1, 2, 3, 4, 5.
Chapter 9.	**2:**	1, 2, 3, 4, 8,
	3:	1, 2, 4, 5,
	4:	1, 2, 4, 7,
	5:	1, 2, 3, 4, 5, 18, 19,
	7:	1, 2, 3.
Chapter 10.	**2:**	1, 2, 3, 4, 6, 7,
	3:	1, 2, 4, 5, 7,
	4:	1, 2, 5, 7,
	5:	1, 2, 3, 4, 5, 11, 15,
	6:	1, 2, 3, 4,
	7:	1, 2, 4,
	9:	1, 2, 3, 7.

This book is the outgrowth of a course on applied exterior calculus given at Lehigh University over the last eight years and owes much of its final form and substance to the many students who have helped remarkably. I am also indebted to the publishing house of Sijthoff and Noordhoff for their kind permission to use certain materials from the appendix of a previous book (*Isovector Methods for Equations of Balance*); in particular, Chapter Five on antiexact differential forms and linear homotopy operators. The labor in preparation of the many drafts of the manuscript was significantly lightened by the able secretarial assistance of Mary Connell and Lisa Ziegler. Finally, I wish to express my heartfelt thanks to Demetrios Lagoudas for his unstinting assistance in proofreading.

D. G. B. EDELEN

Bethlehem, Pennsylvania
September 1984

CONTENTS

PART THREE. APPLICATIONS TO PHYSICS

Applied Exterior Calculus

PART ONE

VECTORS AND FORMS

CHAPTER ONE

MATHEMATICAL PRELIMINARIES

1-1. NUMBER SPACE, POINTS, AND COORDINATES

The simplicity and intrinsic beauty of any mathematical discipline rests on the fact that it starts with the familiar or self-evident and builds an edifice of often unexpected logical implications. Whether the logical implications are viewed as an end in themselves, or as a structured stage for the play of physical reality, due care must be taken in the articulation of the underlying precepts. For the purposes of these discussions, the starting point is taken to be n-dimensional number space, E_n, together with mappings between E_n and the real line $\mathbb{R}$. Since E_n is the n-fold Cartesian product of the real line, $\mathbb{R}$, it may be given a system of Cartesian coordinates, in which case we say that E_n is referred to a Cartesian coordinate cover. Great care must be exercised at exactly this point, for the reader must not assume that the referral of E_n to a Cartesian coordinate cover carries along with it the ability to measure distances between points in E_n or to measure angles between intersecting lines in E_n. The fundamental concept here is that of a continuous swarm of points and that we have labeled the points of this swarm by a Cartesian coordinate cover as a matter of expediency and convenience. Once the Cartesian coordinate cover is in place, we can use it to introduce a topology and thereby discuss notions of nearness, continuity, and convergence. In particular, we shall say that a set in E_n is open if it is open in the Euclidean topology introduced on E_n by referral to a given Cartesian coordinate cover. For the time being, once E_n is referred to a Cartesian coordinate cover, this coordinate cover will remain fixed. What happens when the coordinate cover is changed will be taken up later.

Let U be an open set of E_n and let P be a generic point in U. The Cartesian coordinate cover of E_n can be used to assign any point P in U its coordinates $x^1(P), x^2(P), \ldots, x^n(P)$. This is conveniently written $P:(x^1, \ldots, x^n)$, or simply $P:(x^i)$ where it is understood that i runs from 1 through n. If f denotes a function that is defined on E_n and takes its values on the real line $\mathbb{R}$, we write

$$f(x^i) \quad \text{for } f(x^1, x^2, \ldots, x^n)$$

and sometimes even $f(x)$ when it is clear that x stands for $(x^i) = (x^1(P), x^2(P), \ldots, x^n(P))$. The function $f(x^i)$ then serves to define a map F from E_n into $\mathbb{R}$. We denote this situation by

$$F: E_n \to \mathbb{R} \,|\, t = f(x^i),$$

where the equation following the vertical slash gives a realization of the map F.

Suppose that we are given n real valued functions $\phi^1(t), \phi^2(t), \ldots, \phi^n(t)$ of the real variable t, where each of these functions is defined on some common open set $J \subset \mathbb{R}$. If we set $x^1 = \phi^1(t)$, $x^2 = \phi^2(t), \ldots, x^n = \phi^n(t)$, we obtain a curve in E_n; namely, a map Φ from the set J contained in the real line, $\mathbb{R}$, with coordinate t into the n-dimensional number space E_n. In this instance, we write

$$\Phi: J \subset \mathbb{R} \to E_n | x^i = \phi^i(t), \qquad i = 1, \ldots, n,$$

where the equations following the vertical slash give a realization of the map Φ. Simplification obtains if we write

$$\Phi: \mathbb{R} \to E_n | x^i = \phi^i(t),$$

where it is understood that the set $J \subset \mathbb{R}$, on which all n of the functions $\{\phi^i(t)\}$ are defined, is implied and that i runs from 1 through n.

There will be many instances in which we will have to write quantities such as

$$\sum_{i=1}^{n} g^i(x^k) W_i(x^k).$$

This can be significantly simplified by adopting the Einstein summation convention: *if an upstairs and a downstairs index are the same letter, then summation over the range of the index is implied*. Thus,

$$g^i(x^k) W_i(x^k) \equiv \sum_{i=1}^{n} g^i(x^k) W_i(x^k)$$

and

$$P^{ij}_{kr} H^k_j \equiv \sum_{j=1}^{n} \sum_{k=1}^{n} P^{ij}_{kr} H^k_j = U^i_r.$$

1-2. COMPOSITION OF MAPPINGS

The notion of a mapping will occur over and over again in these pages so the reader had better make his peace with this way of thinking from the start. We have already encountered two kinds of mappings in the previous section: mappings from the real line into E_n (curves), and mappings from E_n into the real line (scalar valued functions). These two kinds of mappings combine to yield a composition mapping, a notion that will also be fundamental throughout these discussions.

Let Φ be a mapping from the real line, $\mathbb{R}$, into E_n,

$$\Phi : \mathbb{R} \to E_n | x^i = \phi^i(t), \tag{1-2.1}$$

where $\mathbb{R}$ or a subset of $\mathbb{R}$ is the common domain of definition of all n functions $\{\phi^i(t)\}$ that define the mapping, and E_n contains the range of the mapping. Script R will be used to designate *range* and script D to designate *domain*. Thus $\mathscr{D}(\Phi)$ denotes the subset of $\mathbb{R}$ on which each of the n functions $\{\phi^i(t)\}$ is defined, and $\mathscr{R}(\Phi)$ denotes the (1-dimensional) subset of E_n whose points satisfy $x^i = \phi^i(t)$ for t in $\mathscr{D}(\Phi)$. We assume here and throughout these discussions that the n functions $\{\phi^i(t)\}$ are very smooth functions of the variable t for all t in $\mathscr{D}(\Phi)$. This will usually be taken to mean that each of the n functions $\{\phi^i(t)\}$ possesses continuous derivatives with respect to t of all orders for all t in $\mathscr{D}(\Phi)$.

A graphic realization of the mapping Φ is shown in Fig. 1. The reader is strongly advised to fix the graphic situation firmly in mind. This will provide a

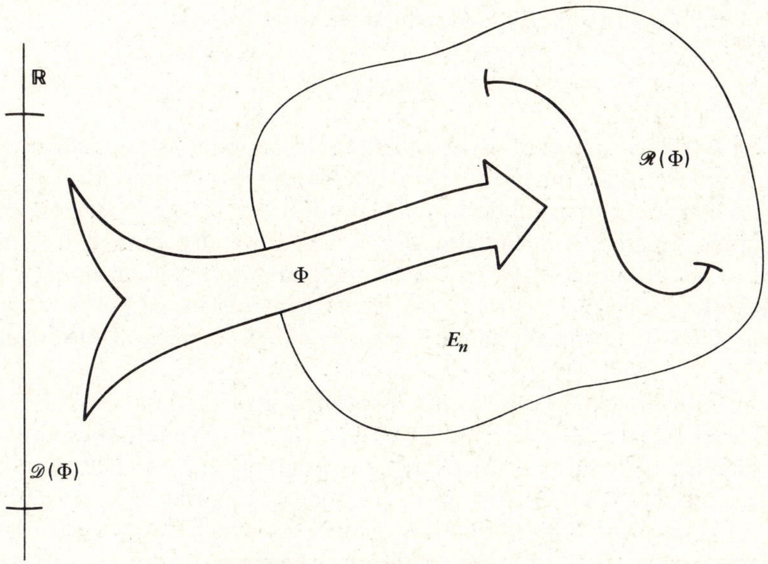

FIGURE 1. The map $\Phi : \mathbb{R} \to E_n$

ready mnemonic at this stage and later provide simple and direct means of visualizing almost immediate methods of proof for many important results.

Formula (1-2.1) involves two distinct constructs: the map Φ from $\mathbb{R}$ into E_n, and the realization or representation $x^i = \phi^i(t)$. Geometrically, the map Φ gives a curve in E_n and this curve knows nothing about how we have chosen to parametrize the real line $\mathbb{R}$ or how we have chosen the coordinates in E_n. In this sense, a curve in E_n may be thought of as a primitive geometric concept; it simply is. On the other hand, in order to give a quantitative description of the curve in E_n that is represented by Φ, it is necessary that we assign a coordinate cover to the domain $\mathbb{R}$ of Φ and a coordinate system on E_n. Now, there are many different ways of assigning coordinate covers to $\mathbb{R}$ and to E_n, and hence there are many ways in which the map Φ can be described. Thus when we write $x^i = \phi^i(t)$ in (1-2.1) what it means is that these n equations describe the map Φ for a given coordinate cover (t) of $\mathbb{R}$ and for the given coordinate cover (x^i) of E_n. In fact, the n functions $\{\phi^i(t)\}$ that enter into the equations $x^i = \phi^i(t)$ may be thought of as providing a *representation* of the map Φ relative to the given coordinate covers of $\mathbb{R}$ and E_n. Accordingly, if there is a change in the coordinate covers of either $\mathbb{R}$ or E_n, the functions $\{\phi^i(t)\}$ will change. A change in coordinate covers thus changes the representation of the map Φ but does not change the map Φ itself. This is because Φ designates the geometric construct called a curve in E_n and this geometric construct exists independent of the convenience of a quantitative description by means of coordinate covers. Thus although we have agreed to keep the coordinate cover of E_n fixed for the time being, it is still necessary to distinguish between the map Φ and its representation in terms of specific coordinate covers and specific functions.

Let F be a map from E_n to $\mathbb{R}$, which we write as

$$F\colon E_n \to \mathbb{R} \,|\, \tau = f(x^i). \tag{1-2.2}$$

Here, $\mathscr{D}(F)$ is the subset of E_n on which the function $f(x^i)$ is defined and $\mathbb{R}$ contains $\mathscr{R}(F)$. The function $f(x^i)$ is assumed to be smooth; that is, it possesses continuous partial derivatives of all orders for all $P\colon(x^i)$ in $\mathscr{D}(F)$. The graphic situation is depicted in Fig. 2, where we also show $f^{-1}(\tau_1)$ that is defined to be the subset of $\mathscr{D}(F)$ that is mapped onto the point in $\mathscr{R}(F)$ with coordinate τ_1. Clearly, $f^{-1}(\tau_1)$ is the locus of points in $\mathscr{D}(F)$ on which the function f has the constant value τ_1; that is, *a level surface* of the function f in $\mathscr{D}(F)$.

The primitive geometric construct associated with the map F is the collection of level surfaces of this map in $\mathscr{D}(F) \subset E_n$. It is therefore essential that we distinguish between the map F and the function $f(x^i)$ that is used in the *representation* $\tau = f(x^i)$ of the map F. The map F, in terms of its level surfaces in $\mathscr{D}(F)$, knows nothing about the manner in which we have chosen the coordinate cover of E_n or of $\mathbb{R}$, while the representation $\tau = f(x^i)$ gives a convenient quantification of F that is directly associated with a specific choice

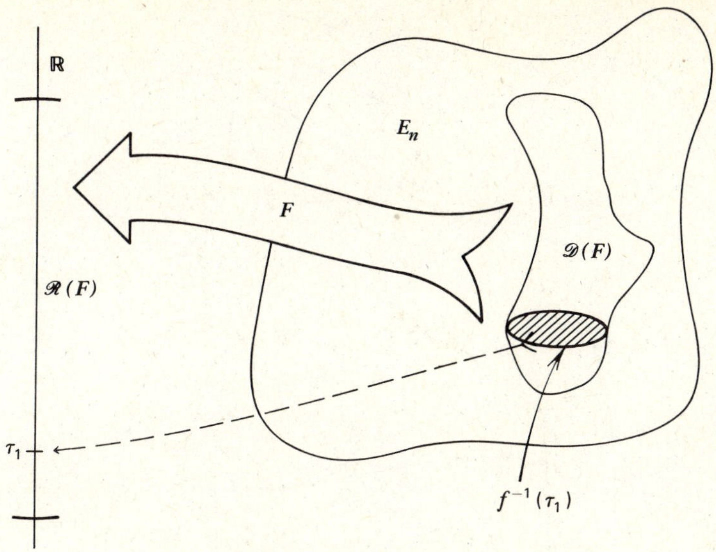

FIGURE 2. The map $F: E_n \to \mathbb{R}$

of coordinate covers for both E_n and for $\mathbb{R}$. If we change the coordinate covers, the representation of F will change, but the map F is still the same. A similar distinction has already been made between a map $\Phi: \mathbb{R} \to E_n$ and its representation, but there is already the basic distinction between an E_n-valued function and its components $\{\phi^i(x^k)\}$. Here, there is an increased possibility of confusion, for we have one map F and a single function f in any given representation of the map F, but different coordinate covers will yield different functions f for the same map F.

We now simply observe that the range of Φ and the domain of F are both contained in E_n. We can thus *compose* the maps Φ and F on $\mathscr{R}(\phi) \cap \mathscr{D}(F)$ if this intersection is not empty. When the composition is affected, we have

(1-2.3) $$F \circ \Phi = f(\phi^1(t), \ldots, \phi^n(t)) = f(\phi^i(t)),$$

namely, a function with values in $\mathbb{R}$. The composition of the maps Φ and F thus yields the *composition map*

(1-2.4) $$T = F \circ \Phi : \mathbb{R} \to \mathbb{R} \,|\, \tau = f(\phi^i(t)).$$

Clearly, the domain of T consists of all those points in $\mathscr{D}(\Phi)$ whose images under the map Φ are contained in $\mathscr{R}(\Phi) \cap \mathscr{D}(F)$. The graphic situation is shown in Fig. 3. For example, if $\Phi: \mathbb{R} \to E_3$ is defined by

$$x^1 = t + 6, \qquad x^2 = 6t - t^4, \qquad x^3 = 3t - e^{2t}$$

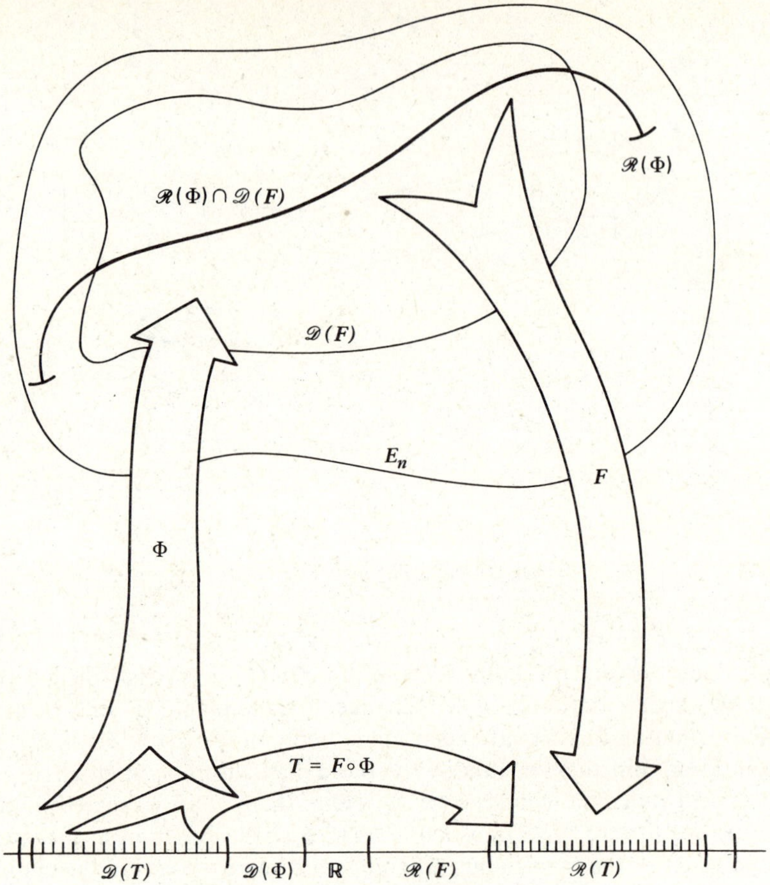

FIGURE 3. The composition map $T = F \circ \Phi$

and $F\colon E_3 \to \mathbb{R}$ is defined by

$$f(x^i) = (x^1)^3 + x^2 + 4(x^2)^2 - x^3 e^{x^1}$$

then the composition map $T = F \circ \Phi$ is given by

$$\tau = (t+6)^2 + 6t - t^4 + 4(6t - t^4)^2 - (3t - e^{2t})e^{t+6}.$$

The composition given above is somewhat special in view of the fact that both the domain of Φ and the range of F are contained in the same space, namely $\mathbb{R}$. This restriction is easily overcome. Let Φ be a map from a space A to a space B,

(1-2.5) $$\Phi\colon A \to B \mid y^\alpha = \phi^\alpha(x^i),$$

and let Ψ be a map from the space B to a space C,

$$\Psi : B \to C | z^w = \psi^w(y^\beta). \tag{1-2.6}$$

Here (x^i) is a coordinate cover of A, (y^β) is a coordinate cover of B, and (z^w) is a coordinate cover for C. If $\mathscr{R}(\Phi) \cap \mathscr{D}(\Psi)$ is not empty, then the *composition map* $T = \Psi \circ \Phi$ is a map from A to C,

$$T = \Psi \circ \Phi : A \to C | z^w = \psi^w(\phi^\beta(x^i)), \tag{1-2.7}$$

with $\mathscr{D}(T) = \Phi^{-1}(\mathscr{R}(\Phi) \cap \mathscr{D}(\Psi))$ contained in A and $\mathscr{R}(T) = \Psi(\mathscr{R}(\Phi) \cap \mathscr{D}(\Psi))$ contained in C. Here we have used the notation $\Phi^{-1}(\rho)$ to designate the totality of points in $\mathscr{D}(\Phi)$ that is mapped into the set ρ. This notation will be used throughout these discussions. Care must always be exercised to check that $\mathscr{R}(\Phi) \cap \mathscr{D}(\Psi)$ is not empty in order that $T = \Psi \circ \Phi$ be defined. The reader must also realize that $\mathscr{D}(T)$ is usually only a subset of $\mathscr{D}(\Phi)$ and that $\mathscr{R}(T)$ is usually only a subset of $\mathscr{R}(\Psi)$.

1-3. SOME USEFUL THEOREMS

Certain existence theorems will be required in subsequent discussions. These theorems will only be stated here, the reader being referred to the appendix for the proofs. In all instances, these existence theorems obtain, in one form or another, as direct applications of an important fixed point theorem. This theorem is stated below, even though it involves the notion of a complete metric space, for the simple reason that it has become a model for constructive proof of existence.

Let T be a map of a complete metric space, S, into itself that satisfies

$$\rho(T\phi, T\psi) < k\rho(\phi, \psi), \qquad k < 1, \tag{1-3.1}$$

where ρ is the distance function on S. Then there is a unique fixed point of T; that is $T\phi_0 = \phi_0$ for one and only one $\phi_0 \in S$. Further, this fixed point is given by

$$\phi_0 = \lim_{n \to \infty} T^n \phi \tag{1-3.2}$$

for any $\phi \in S$, where $T^n\phi \overset{\text{def}}{=} TT^{n-1}\phi$ so that T^n is the nth iteration of T.

The implicit function theorem tells us when we can solve a system of k equations in $k + n$ variables for k of the variables in terms of the remaining n variables.

Implicit Function Theorem

Let $F^1, \ldots, F^k$ be k functions of the variables $y^1, \ldots, y^k$, $x^1, \ldots, x^n$ that are defined in some neighborhood, N, of the origin of $E_k \times E_n$ and are continuous *together with their partial derivatives of all orders up to and including order r on N. Let*

$$F^i(0,\ldots,0\colon 0,\ldots,0) = 0, \qquad i = 1,\ldots,k, \tag{1-3.3}$$

and suppose that

$$\det(\partial F^i/\partial y^j)(0,\ldots,0;0,\ldots,0) \neq 0. \tag{1-3.4}$$

Then in a sufficiently small neighborhood of the origin in E_n, there exist functions $\phi^1(x^m), \phi^2(x^m),\ldots,\phi^k(x^m)$ of the n variables $x^1,\ldots,x^n$ that are uniquely determined by the equations

$$F^i(\phi^1(x^m),\phi^2(x^m),\ldots,\phi^k(x^m);x^1,x^2,\ldots,x^n) = 0, \tag{1-3.5}$$

$$\phi^i(0,\ldots,0) = 0, \qquad i = 1,\ldots,k. \tag{1-3.6}$$

Further, the functions $\{\phi^i(x^m)\}$ that serve to define the y's as functions of the x's on N through the relations

$$y^i = \phi^i(x^1,\ldots,x^n) = \phi^i(x^m), \qquad i = 1,\ldots,k \tag{1-3.7}$$

are continuous together with their partial derivatives of all orders up to and including r on N.

This theorem is very powerful. All we need to establish is that the k functions F^i of $k + n$ variables are such that

1. $F^i = 0$, $i = 1,\ldots,k$ are satisfied for at least one point $(y_0^l; x_0^m)$ of $E_k \times E_n$.
2. The functions F^i are of class C^r, and
3. That $\det(\partial F^i/\partial y^j)(y_0^l; x_0^m) \neq 0$.

If we then set $\bar{x}^m = x^m - x_0^m$, $\bar{y}^i = y^i - y_0^i$, all of the hypotheses of the theorem are satisfied. We may then conclude that there exist k functions $\{\phi^i(\bar{x}^m)\}$ on some neighborhood of the origin in E_n such that equations $F^i(\bar{y}^j; \bar{x}^m) = 0$ define the $\bar{y}$'s as functions of the $\bar{x}$'s through the relations

$$\bar{y}^i = \phi^i(\bar{x}^m); \quad \text{i.e., } y^i = y_0^i + \phi^i(x^m - x_0^m).$$

Note that the determinental condition (1-3.4) only has to be verified at the single point (y_0^i, x_0^m) of $E_k \times E_n$ in order to apply the theorem. It is then easily shown that

$$\det(\partial F^i/\partial y^j)(\phi^k(x^m); x^m) \neq 0$$

holds at all points of the neighborhood N over which the functions $\{\phi^i\}$ are defined.

Although we now know that such functions $\{\phi^i(x)\}$ exist on N, actually finding these functions explicitly is another matter altogether. This state of affairs is the rule rather than the exception, for existence theorems very rarely provide a specific algorithm for the actual solution of a problem. All they tell us is that the search for a solution is not in vain.

A special case of the implicit function theorem is the inverse mapping theorem that will figure heavily in our studies.

Inverse Mapping Theorem

Let

$$\Phi: E_n \to {}`E_n | y^i = F^i(x^j), \qquad i = 1, \dots, n \tag{1-3.8}$$

be a mapping for which the functions $F^i(x^j)$ *are defined and of class* C^r *in some neighborhood* N *of* (x_0^i) *in* E_n *and*

$$\det(\partial F^i / \partial x^j)(x_0^k) \neq 0. \tag{1-3.9}$$

Then there exists a neighborhood $`N$ *of* $y_0^i = F^i(x_0^j)$ *in* $`E_n$ *and a uniquely determined collection of n functions* $\{G^i(y^j)\}$ *of class* C^r *on* $`N$ *such that* $G^i(y_0^j) = x_0^i$ *and*

$$F^i(G^j(y^k)) = y^i, \qquad i = 1, \dots, n \tag{1-3.10}$$

for all (y^i) *in* $`N$*. Thus, if we restrict* Φ *according to*

$$\Phi_N : N \subset E_n \overset{\text{onto}}{\to} {}`N \subset {}`E_n | y^i = F^i(x^j), \qquad i = 1, \dots, n \tag{1-3.11}$$

then Φ_N *has an inverse mapping* Φ_N^{-1} *that is defined by*

$$\Phi_N^{-1} : {}`N \subset {}`E_n \overset{\text{onto}}{\to} N \subset E_n | x^i = G^i(y^j), \qquad i = 1, \dots, n. \tag{1-3.12}$$

The graphic situation is shown in Fig. 4. It is important to note that

$$\Phi_N^{-1} \circ \Phi_N : N \to N$$

is the identity map on N and that

$$\Phi_N \circ \Phi_N^{-1} : {}`N \to {}`N$$

is the identity map on $`N$. Again we have existence of the inverse mapping, but not necessarily any explicit realization. There are, however, important situations in which the explicit solution can be extracted. If we have a *linear*

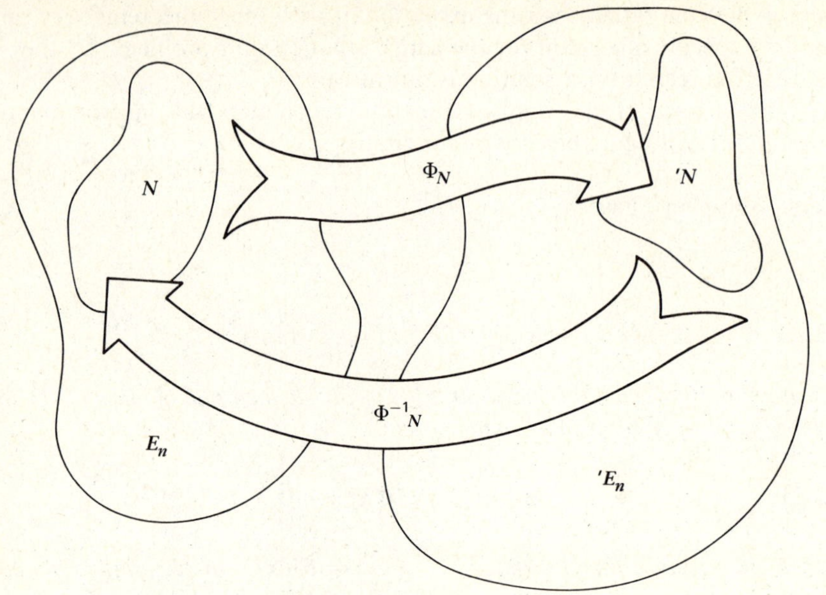

FIGURE 4. Inverse mapping

mapping

(1-3.13) $$\Phi: E_n \to {}^{\backprime}E_n | y^i = A^i_j x^j, \qquad \det\left(A^i_j\right) \neq 0$$

then

(1-3.14) $$\Phi^{-1}: {}^{\backprime}E_n \to E_n | x^i = A^{-1i}_{\ \ j} y^j$$

where $((A^{-1i}_{\ \ j}))$ is the inverse of the nonsingular matrix $((A^i_j))$. In this case, $N = E_n$ and ${}^{\backprime}N = {}^{\backprime}E_n$. We should also note that if $x^i = G^i(y^j)$ defines the inverse of the map $y^i = F^i(x^j)$, then the matrices $\{(\partial G^i/\partial y^j)\}$ and $\{(\partial F^i/\partial x^j)\}$ are inverses of each other, either considered as functions of the x's through use of the relations $y^i = A^i_j x^j$, or as functions of the y's through use of the relations $x^i = A^{-1i}_{\ \ j} y^j$.

1-4. SYSTEMS OF AUTONOMOUS FIRST-ORDER DIFFERENTIAL EQUATIONS

An essential part of these discussions will center around systems of autonomous first-order ordinary differential equations; namely, equations of the form

$$dx^i(t)/dt = f^i\left(x^1(t), \ldots, x^n(t)\right), \qquad i = 1, \ldots, n.$$

The name autonomous derives from the fact that the right-hand sides of these equations do not contain the independent variable t explicitly; that is, the values of the derivatives are uniquely determined by the values of the dependent variables.

Existence Theorem for Autonomous Ordinary Differential Equations

Let $f^1(x^1,\dots,x^n),\dots,f^n(x^1,\dots,x^n)$ be n functions of class C^r on some neighborhood, N, of the origin of E_n with $r > 1$. Then a neighborhood U of the origin of E_n and a neighborhood I of 0 in $\mathbb{R}$ can be found such that there exist unique functions $\phi^1(t; x_0^i),\dots,\phi^n(t; x_0^i)$, for any (x_0^i) in U and for all t in I such that

$$d\phi^i/dt = f^i(\phi^1,\dots,\phi^n) \tag{1-4.1}$$

and

$$\phi^i(0; x_0^j) = x_0^i \tag{1-4.2}$$

Further, the functions $\{\phi^i\}$ are of class C^{r+1} for all t in I and of class C^r in the initial data $\{x_0^i\}$ for all $\{x_0^i\}$ in U.

In many applications, the given functions $\{f^i\}$ will be of class C^∞, in which case the functions $\{\phi^i\}$ will be of class C^∞ in t and of class C^∞ in the initial data. In this case there is a representation of the functions $\{\phi^i\}$ that proves to be of significant use. Let us first observe that evaluation of (1-4.1) at $t = 0$ together with (1-4.2) yield

$$\left.\frac{d\phi^i}{dt}\right|_{t=0} = f^i(x_0^j).$$

If we differentiate (1-4.1) with respect to t and then use (1-4.1) again, we obtain

$$\frac{d^2\phi^i}{dt^2} = \frac{\partial f^i}{\partial \phi^j}\frac{d\phi^j}{dt} = \frac{\partial f^i}{\partial \phi^j} f^j,$$

and similarly

$$\frac{d^3\phi^i}{dt^3} = \frac{\partial^2 f^i}{\partial \phi^j \partial \phi^k} f^j f^k + \frac{\partial f^i}{\partial \phi^j}\frac{\partial f^j}{\partial \phi^k} f^k.$$

Thus an evaluation at $t = 0$ and use of (1-4.2) shows that

$$\left.\frac{d^2\phi^i}{dt^2}\right|_{t=0} = \frac{\partial f^i(x_0^m)}{\partial x_0^j} f^j(x_0^m),$$

$$\left.\frac{d^3\phi^i}{dt^3}\right|_{t=0} = \frac{\partial^2 f^i(x_0^m)}{\partial x_0^j \partial x_0^k} f^j(x_0^m) f^k(x_0^m) + \frac{\partial f^i(x_0^m)}{\partial x_0^j}\frac{\partial f^j(x_0^m)}{\partial x_0^k} f^k(x_0^m).$$

Now, Taylor's theorem tells us that in a sufficiently small neighborhood of $t = 0$, we may write

$$\phi^i(t; x_0^m) = x_0^i + t\left(\frac{d\phi^i}{dt}\right)\Bigg|_{t=0} + \frac{t^2}{2!}\left(\frac{d^2\phi^i}{dt^2}\right)\Bigg|_{t=0} + \cdots .$$

We now use the above expressions to evaluate the various derivatives at $t = 0$. The results can then be written in the suggestive form

$$\phi^i(t; x_0^m) = \left[\exp\left(tf^j(z^m)\frac{\partial}{\partial z^j}\right)z^i\right]_{z^m = x_0^m},$$

where $\exp(\xi)$ stands for the corresponding power series $\Sigma_{k=0}^{\infty}(1/k!)\xi^k$. It would thus appear that we may represent the solution (1-4.1) subject to the initial data (1-4.2) in the useful form

$$\phi^i(t; x_0^k) = \exp\left(tf^j(x_0^m)\frac{\partial}{\partial x_0^j}\right)\langle x_0^i\rangle \tag{1-4.3}$$

for t in some sufficiently small neighborhood of $t = 0$. The essential new aspect of these considerations is that the argument of the exponential is not a number or even a function; rather, it is an operator (with $z^m = x_0^m$)

$$tF = tf^j(z^m)\frac{\partial}{\partial z^j} = tf^1(z^m)\frac{\partial}{\partial z^1} + \cdots + tf^n(z^m)\frac{\partial}{\partial z^n}.$$

Thus $\exp(tF)$ is not defined if it does not operate on a function of the z's. It is for this reason that we must write $\exp(tF)\langle z^i\rangle$ in (1-4.3), for otherwise there would be just nonsense. This notation will be used throughout these discussions; the operator will be written to the left and the quantity acted upon by the operator will be enclosed in pointed brackets, $\langle \cdot \rangle$.

Let $g(x^i)$ be a smooth function that is defined on E_n and define $\{\bar{x}^i(t)\}$ by

$$\bar{x}^i(t) = \exp\left(tf^j(x_0^m)\frac{\partial}{\partial x_0^j}\right)\langle x_0^i\rangle, \qquad 1 \le i \le n$$

for t in a neighborhood J of $t = 0$ for which the power series in t on the right-hand side converges. It is then well-known that the derivative of the functions defined by the power series in t can be differentiated with respect to t for all t in J and the resulting derivative can be expressly evaluated by differentiation of the power series term by term. Now, $g(x^i)$ can be evaluated at (x_0^i) and at $(\bar{x}^i(t))$, for any t in J to give $g(x_0^i)$ and $g(\bar{x}^i(t))$. Noting that the function g is smooth, we can expand $g(\bar{x}^i(t))$ in a power series in t about $t = 0$. When this is done, an analysis similar to that leading to (1-4.3) shows that

$$g(\bar{x}^i(t)) = \exp\left(t f^j(x_0^m) \frac{\partial}{\partial x_0^j} \right) \langle g(x_0^i) \rangle$$

for all t in J. Thus knowledge of the n functions $\{f^j(x_0^m)\}$ allows us to evaluate $g(\bar{x}^i(t))$ in terms of $g(x_0^i)$ and its partial derivatives for all t in J.

We now have all of the information that is required in order to show that the n functions $\{\phi^i(t; x_0^k)\}$ given by (1-4.3) satisfy the autonomous system of first-order differential equations (1-4.1) subject to the initial data (1-4.2) for all t in a neighborhood J of $t = 0$. An evaluation of both sides of (1-4.3) at $t = 0$ gives $\phi^i(0; x_0^k) = x_0^i$, and hence the functions $\{\phi^i(t; x_0^k)\}$ satisfy the initial data (1-4.2). A differentiation of both sides of (1-4.3) with respect to t gives

$$\begin{aligned}\frac{d}{dt}\phi^i(t; x_0^m) &= \exp\left(t f^j(x_0^m) \frac{\partial}{\partial x_0^j} \right) \left\langle f^k(x_0^m) \frac{\partial}{\partial x_0^k} x_0^i \right\rangle \\ &= \exp\left(t f^j(x_0^m) \frac{\partial}{\partial x_0^j} \right) \langle f^i(x_0^m) \rangle\end{aligned}$$

for all t in J. If we identify $f^i(x_0^m)$ with $g(x_0^m)$ in the previous paragraph, we see that

$$\exp\left(t f^j(x_0^m) \frac{\partial}{\partial x_0^j} \right) \langle f^i(x_0^m) \rangle = f^i(\bar{x}^m(t)) = f^i(\phi^m(t; x_0^k))$$

for each value of i and all t in J. This shows that

$$\frac{d}{dt}\phi^i(t; x^m) = f^i(\phi^j(t; x^m)), \qquad i = 1, \ldots, n$$

for all t in J, and we are done.

Written in slightly different notation, we have the following important result:

If the n functions $\{f^i(x^m)\}$ are of class C^∞ in some neighborhood N of the origin of E_n, then, for t in a sufficiently small neighborhood of 0, *the system of autonomous ordinary differential equations*

$$\text{(1-4.4)} \qquad \frac{d\bar{x}^i(t)}{dt} = f^i(\bar{x}^m(t)), \qquad i = 1, \ldots, n$$

has a unique solution that satisfies the initial data

$$\text{(1-4.5)} \qquad \bar{x}^i(0) = x^i.$$

This solution may be represented by

$$\text{(1-4.6)} \quad \bar{x}^i(t) = \hat{x}^i(t; x^m) = \exp\left(t f^j(x^m) \frac{\partial}{\partial x^j} \right) \langle x^i \rangle, \qquad i = 1, \dots, n$$

and the functions $\hat{x}^i(t; x^m)$ are of class C^∞ in the arguments (x^m) in a neighborhood U of the origin of E_n.

1-5. VECTOR SPACES, DUAL SPACES, AND ALGEBRAS

The reader is assumed to be familiar with vector spaces of finite dimension and with the concepts of linear algebra. However, just to be certain that all mean the same thing by the well-worn words, the definitions will be given.

Let $\mathscr{V}$ be a collection of elements $U, V, W, \dots$ (written $U, V, W, \dots \in \mathscr{V}$) and let $a, b, c, \dots$ be elements of $\mathbb{R}$. Suppose that we are given an operation $+$ on $\mathscr{V} \times \mathscr{V}$ to $\mathscr{V}$,

$$+ : \mathscr{V} \times \mathscr{V} \to \mathscr{V} | W = U + V,$$

and an operation $\cdot$ on $\mathbb{R} \times \mathscr{V}$ to $\mathscr{V}$,

$$\cdot : \mathbb{R} \times \mathscr{V} \to \mathscr{V} | W = a \cdot U.$$

Then $\mathscr{V}$ is a *vector space* over the real number field $\mathbb{R}$ if and only if $\mathscr{V}$ and the operations $+$ and $\cdot$ satisfy the following conditions:

V1. $U + (V + W) = (U + V) + W$ for all $U, V, W \in \mathscr{V}$.

V2. $U + V = V + U$ for all $U, V \in \mathscr{V}$.

V3. There exists an element 0 of $\mathscr{V}$ such that

$$V + 0 = V \quad \text{for all } V \in \mathscr{V}.$$

V4. For each $U \in \mathscr{V}$ there exists one and only one $V \in \mathscr{V}$ such that $U + V = 0$.

V5. $(ab) \cdot U = a \cdot (b \cdot U)$ for all $a, b \in \mathbb{R}$ and all $U \in \mathscr{V}$.

V6. $(a + b) \cdot U = a \cdot U + b \cdot U$ for all $a, b \in \mathbb{R}$ and all $U \in \mathscr{V}$.

V7. $a \cdot (U + V) = a \cdot U + a \cdot V$ for all $a \in \mathbb{R}$ and all $U, V \in \mathscr{V}$.

V8. $1 \cdot U = U$ for all $U \in \mathscr{V}$.

Capital italic letters will be used exclusively for vectors throughout these discussions while capital script letters will designate vector spaces over the real number field.

A set of elements $\{U_1, U_2, \dots, U_k\}$ of a vector space $\mathscr{V}$ is said to be *linearly independent* if and only if the only values of the constants $c_1, c_2, \dots, c_k$ that satisfy

$$c_1 \cdot U_1 + c_2 \cdot U_2 + \cdots + c_k \cdot U_k = 0$$

are $c_1 = c_2 = \cdots = c_k = 0$. If the above equation can be satisfied without all of the c's being zero then the set $\{U_1, U_2, \ldots, U_k\}$ is said to be *linearly dependent*. The largest number of vectors in $\mathscr{V}$ that comprises a linearly independent set is the *dimension* of the vector space $\mathscr{V}$, which is written $\dim(\mathscr{V})$. If $\mathscr{V}$ is n-dimensional, then any system of n linearly independent elements $V_1, V_2, \ldots, V_n$ is a *basis* for $\mathscr{V}$, and any $U \in \mathscr{V}$ can be expressed as

$$U = k_1 \cdot V_1 + k_2 \cdot V_2 + \cdots + k_n \cdot V_n$$

where the constants $k_1, \ldots, k_n \in \mathbb{R}$ are uniquely determined by U and $V_1, \ldots, V_n$.

A *subspace* $\mathscr{S}$ of a vector space $\mathscr{V}$ is a subset of $\mathscr{V}$ that is closed under the operations $+$ and $\cdot$ (you cannot get outside of $\mathscr{S}$ by adding any two elements of $\mathscr{S}$ together or by multiplying any element of $\mathscr{S}$ by any element of $\mathbb{R}$). A *basis* for a subspace $\mathscr{S}$ is any maximal collection of linearly independent elements of $\mathscr{S}$, in which case we say that $\mathscr{S}$ is spanned by its basis. If $\{U_1, \ldots, U_k\}$ is a basis for $\mathscr{S}$, then we write

$$\mathscr{S} = \operatorname{span}(U_1, \ldots, U_k), \quad \dim(\mathscr{S}) = k,$$

where $\operatorname{span}(U_1, \ldots, U_k)$ is the collection of all linear combinations of the vectors $U_1, \ldots, U_k$. Conversely, if $\{W_1, W_2, \ldots, W_m\}$ is a linearly independent collection of elements of $\mathscr{V}$, then $\mathscr{W} = \operatorname{span}(W_1, \ldots, W_m)$ is a subspace of $\mathscr{V}$ of dimension m.

Notice that the definition of a vector space does not require it to be finite dimensional. A particularly important example of a nonfinite dimensional vector space is the collection $\Lambda^0(E_n)$ of *all* C^∞ mappings of E_n into $\mathbb{R}$ with $+$ and $\cdot$ defined by

$$(f + g)(x^k) = f(x^k) + g(x^k), \tag{1-5.1}$$

$$(a \cdot f)(x^k) = af(x^k). \tag{1-5.2}$$

Clearly, the sum of two C^∞ functions is a C^∞ function, and any numerical multiple of a C^∞ function is a C^∞ function. Thus $\Lambda^0(E_n)$ is closed under the operations $+$ and $\cdot$. It is then a simple matter to verify that axioms **V1** through **V8** are satisfied.

Let U^* be a mapping from a vector space $\mathscr{V}$ into $\mathbb{R}$,

$$U^* : \mathscr{V} \to \mathbb{R} \mid t = \langle U^*, V \rangle, \qquad V \in \mathscr{V}. \tag{1-5.3}$$

Such mappings are called *functionals* in order to distinguish them from ordinary functions. The notation $\langle U^*, V\rangle$ will be used to denote the evaluation of the functional U^* when it acts on an element V of $\mathscr{V}$.

A functional is said to be *linear* if and only if

$$\langle U^*, a \cdot V + b \cdot W\rangle = a\langle U^*, V\rangle + b\langle U^*, W\rangle \tag{1-5.4}$$

holds for all $V, W \in \mathscr{V}$ and all $a, b \in \mathbb{R}$. The collection of all linear functionals on a vector space $\mathscr{V}$ is referred to as the *dual space* of $\mathscr{V}$ and is usually designated by $\mathscr{V}^*$. Because the range of any linear functional is the real line, it is immediate that the definitions

$$\langle U^* + V^*, W\rangle = \langle U^*, W\rangle + \langle V^*, W\rangle, \tag{1-5.5}$$

$$\langle b \cdot U^*, W\rangle = b\langle U^*, W\rangle, \tag{1-5.6}$$

make $\mathscr{V}^*$ into a vector space over $\mathbb{R}$. Thus *the dual of any vector space is a vector space*. Further, if $\mathscr{V}$ is n-dimensional then $\mathscr{V}^*$ is also n-dimensional. The reason that dual spaces are important, beside their obvious computational efficiency, is that they almost invariably have nicer properties than the vector space with which one starts. This is particularly true whenever the original vector space is not finite dimensional.

A vector space over $\mathbb{R}$ is also the point of departure in the construction of an *algebra* over the real number field. This is simply a vector space $\mathscr{V}$ over the real number field that is equipped with a binary operation $(\cdot \mid \cdot)$ on $\mathscr{V} \times \mathscr{V}$ to $\mathscr{V}$ such that

$$(a \cdot U + b \cdot V \mid W) = a \cdot (U \mid W) + b \cdot (V \mid W), \tag{1-5.7}$$

$$(U \mid a \cdot V + b \cdot W) = a \cdot (U \mid V) + b \cdot (U \mid W) \tag{1-5.8}$$

for all $U, V, W \in \mathscr{V}$ and all $a, b \in \mathbb{R}$; that is, $(\cdot \mid \cdot)$ is linear both in the first factor and in the second factor. For example, the vector space $\Lambda^0(E_n)$ of all C^∞ functions on E_n becomes an algebra under the definition

$$(f \mid g)(x^i) = f(x^i)g(x^i). \tag{1-5.9}$$

Since the product of two C^∞ functions is a C^∞ function, (1-5.9) is on $\Lambda^0(E_n) \times \Lambda^0(E_n)$ to $\Lambda^0(E_n)$ and satisfies (1-5.7) and (1-5.8). An algebra is said to be *commutative* if and only if

$$(U \mid V) = (V \mid U) \tag{1-5.10}$$

for all $U, V \in \mathscr{V}$. Otherwise it is said to be *noncommutative*. An algebra is *associative* if and only if

$$(U \mid (V \mid W)) = ((U \mid V) \mid W) \tag{1-5.11}$$

for all $U, V, W \in \mathscr{V}$. The most familiar situation is that of an algebra that is both associative and commutative, such as $\Lambda^0(E_n)$ with the definition (1-5.9). Later we will see that an essential aspect of our studies will center around algebras that are both noncommutative and nonassociative.

The standard definitions of addition and multiplication of real numbers makes $\mathbb{R}$ into an associative, commutative algebra. Thus a vector space over $\mathbb{R}$ is a vector space over the associative, commutative algebra $\mathbb{R}$. This view is the foundation of a generalization of a vector space; simply drop the requirement that the underlying algebra be commutative. Let A be an associative algebra. A *module* $\mathscr{M}$ *over* A is a collection of elements $U, V, W, \ldots$ together with the two operations

$$+ : \mathscr{M} \times \mathscr{M} \to \mathscr{M} | W = U + V,$$

$$\cdot : A \times \mathscr{M} \to \mathscr{M} | W = \alpha \cdot U$$

that satisfy axioms **V1** through **V4** of a vector space and

M1. $(\alpha\beta) \cdot U = \alpha \cdot (\beta \cdot U)$ for all $\alpha, \beta \in A$ and all $U \in \mathscr{M}$.
M2. $(\alpha + \beta) \cdot U = \alpha \cdot U + \beta \cdot U$ for all $\alpha, \beta \in A$ and all $U \in \mathscr{M}$.
M3. $\alpha \cdot (U + V) = \alpha \cdot U + \alpha \cdot V$ for all $\alpha \in A$ and all $U, V \in \mathscr{M}$.
M4. $e \cdot U = U$ for all $U \in \mathscr{M}$ if A has a multiplicative identity e.

A familiar example of a module is the collection of all column matrices **u** with n entries and A is taken to be the ring of nonsingular n-by-n matrices **A**. Here addition is defined in the standard way and the operation $\alpha \cdot U$ is matrix multiplication **Au**.

The obvious similarities between the definitions of a vector space over $\mathbb{R}$ and a module over an associative algebra A are such that we will often speak of a vector space over an associative algebra A rather than a module over A. The reader is warned, however, that a vector space over any associative algebra other than $\mathbb{R}$ is technically a module. This broad use of the language often simplifies matters, but can hide essential differences in specific contexts. Explicit note will be made at those places in the discussion where the module structure leads to essential differences.

Let $\mathscr{V}$ be an algebra over the real number field, then $\mathscr{V}$ is a vector space over the real number field. Suppose that $\mathscr{U}$ is a subspace of $\mathscr{V}$, then $\mathscr{U}$ is a *subalgebra* of $\mathscr{V}$ if and only if $\mathscr{U}$ is closed under the binary operation $(\cdot | \cdot)$ of the algebra $\mathscr{V}$, in which case we write $\mathscr{U} \subset \mathscr{V}$. Thus $\mathscr{U}$ is a subalgebra of $\mathscr{V}$ if and only if $(P|Q)$ belongs to $\mathscr{U}$ for every $P, Q \in \mathscr{U}$.

Let $\mathscr{V}$ be an algebra and let $\mathscr{U}$ be a vector subspace of $\mathscr{V}$. The subspace $\mathscr{U}$ is said to be a *left ideal* of $\mathscr{V}$ if and only if $(W|V)$ belongs to $\mathscr{U}$ for every $W \in \mathscr{U}$ and for every $V \in \mathscr{V}$. The subspace $\mathscr{U}$ is said to be a *right ideal* if and only if $(V|W)$ belongs to $\mathscr{U}$ for every $W \in \mathscr{U}$ and for every $V \in \mathscr{V}$. If $\mathscr{U}$ is both a left ideal and a right ideal of $\mathscr{V}$ then $\mathscr{U}$ is a *two-sided ideal* or simply an *ideal* of $\mathscr{V}$. Clearly, if the multiplication law, $(\cdot | \cdot)$ of an algebra is either commutative or anticommutative (i.e., $(U|V) = -(V|U)$) then all ideals are two-sided ideals. Finally, if $\mathscr{U}$ is an ideal of $\mathscr{V}$ and $\mathscr{W}$ is both an ideal of $\mathscr{V}$ and a subalgebra of $\mathscr{U}$, then $\mathscr{W}$ is said to be a *subideal* of $\mathscr{U}$. In this event, we write $\mathscr{W} \subset \mathscr{U}$.

It is clear that we can go back and reformulate the notions of algebras, subalgebras, left ideals, right ideals, and two-sided ideals in terms of modules over an associative algebra A; simply replace "vector space over $\mathbb{R}$" by "module over A" throughout. As it turns out, this will often happen in a natural fashion.

PROBLEMS

1.1. Use (1-4.6) directly to prove that it satisfies the system (1-4.4) and the initial data (1-4.5).

1.2. Write out the first four terms of the series represented by (1-4.6) and compare them with what you obtain by successive differentiation of (1-4.4) followed by evaluation at $t = 0$.

1.3. Solve the following systems of first-order autonomous differential equations explicitly and compare these solutions with those obtained from (1-4.6):

(a) $d\bar{x}/dt = 4\bar{x} + 3$,

(b) $d\bar{x}^1/dt = 5\bar{x}^1 + 4, \quad d\bar{x}^2/dt = \bar{x}^1\bar{x}^2$,

(c) $d\bar{x}^1/dt = \bar{x}^1\bar{x}^2, \quad d\bar{x}^2/dt = (\bar{x}^2)^2$,

(d) $d\bar{x}^1/dt = 3(\bar{x}^1)^3, \quad d\bar{x}^2/dt = -5\bar{x}^1 + \bar{x}^2\bar{x}^3, \quad d\bar{x}^3/dt = 7$.

1.4. Verify that the collection of all C^∞ functions with the operations $+$ and $\cdot$ defined by (1-5.1) and (1-5.2) forms a vector space.

1.5. Verify that the dual space of a vector space with $+$ and $\cdot$ defined by (1-5.5) and (1-5.6) forms a vector space.

1.6. Verify that the collection of all C^∞ functions with $(\ |\ )$ defined by (1-5.9) is an associative, commutative algebra.

1.7. Does the ordinary three-dimensional vector space with the standard definition of the "cross product" form an algebra? Give the reasons for your answer.

1.8. Try to generalize the notion of a "cross product" to four and higher dimensional vector spaces so that they form algebras.

1.9. If $\langle U^*, \cdot\rangle$ is a fixed element of the dual, $\mathscr{V}^*$, of a vector space $\mathscr{V}$ and $f(t)$ is a function from $\mathbb{R}$ to $\mathbb{R}$, then $f(\langle U^*, \cdot\rangle)$ is a functional on $\mathscr{V}$. Show that $f(t) = bt + c$ is a linear functional if and only if $c = 0$ (a linear functional is both linear and homogeneous).

1.10. If $U_1, \ldots, U_n$ is a basis for $\mathscr{V}$, then show that the n scalars $c_1, \ldots, c_n$ in $V = c_1 \cdot U_1 + c_2 \cdot U_2 + c_n \cdot U_n$ are unique linear functionals of $V \in \mathscr{V}$.

CHAPTER TWO

VECTOR FIELDS

2-1. THE TANGENT TO A CURVE AT A POINT

Much of the material in this chapter hinges on the process of differentiation; hence due care must be exercised to make sure that a derivative under discussion actually exists. It is therefore convenient to adopt a convention that will save us many repetitions: any function that appears is assumed to be as many times differentiable as required. We will actually go one step further in the interests of simplicity, and assume that any function that occurs in these discussions is of class C^{∞} unless explicitly noted to the contrary. This assumption will not be unduly restrictive in most cases. The analysis of particular cases with significantly reduced continuity classes will usually be left to the reader, for the simple reason that the answers will usually depend heavily on the particular context in which the reduced continuity class arises.

Let J be an open interval of the real line, $\mathbb{R}$, that contains the point $t = 0$. The map

$$\Gamma: J \to E_n | \bar{x}^i = \gamma^i(t) \tag{2-1.1}$$

of J into E_n defines a smooth (C^{∞}) curve in E_n if all of the n functions $\{\gamma^i(t)\}$ are smooth functions on the common domain J. The representation $\bar{x}^i = \gamma^i(t)$ provides the information needed to compute "tangent vectors" to the curve: simply apply the operation d/dt to each of the n functions $\{\gamma^i(t)\}$ so as to obtain the n quantities $\{d\gamma^i(t)/dt\}$. Although this is indeed a simple operation, it can lead to a great deal of confusion, for $\{d\gamma^i(t)/dt\}$, in its most elementary interpretation, defines a one-parameter family of directed line segments, one of which starts at the point in E_n with coordinates $\bar{x}^i = \gamma^i(\bar{t})$

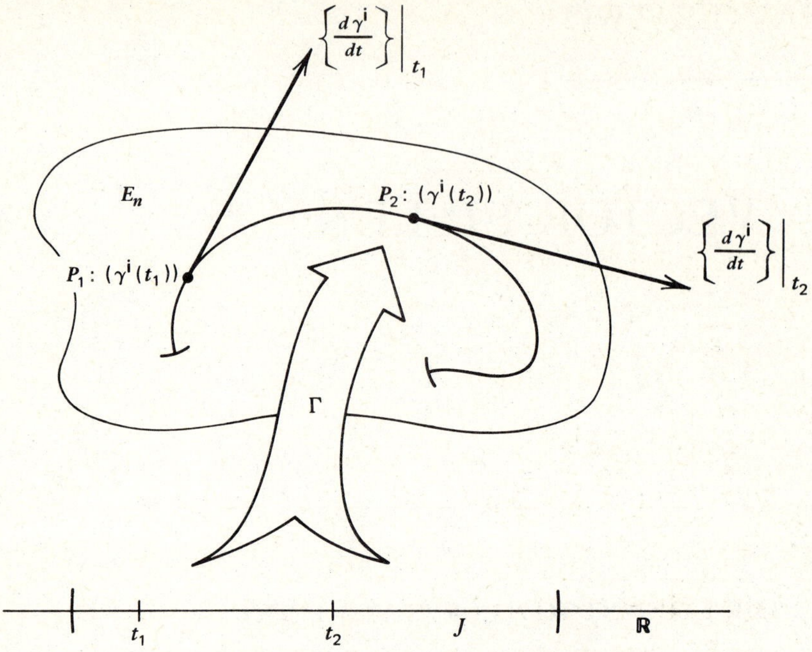

FIGURE 5. The curve Γ and its tangents

for given $\bar{t} \in J$. Thus if $\gamma^i(t_1) \neq \gamma^i(t_2)$ for any $t_1 \neq t_2$ and any value of the index i, then the quantities $\{d\gamma^i(t_1)/dt\}$ and $\{d\gamma^i(t_2)/dt\}$ have their points of origination at the distinct points $P_1 : (\gamma^i(t_1))$ and $P_2 : (\gamma^i(t_2))$ respectively. The geometric situation is shown in Fig. 5. It is thus clear that $\{d\gamma^i(t)/dt\}$ cannot be assigned the operations of a single vector space for all t in J, for vectors can be added only if they have a common point of origination. On the other hand, the n quantities $\{d\gamma^i(t_1)/dt\}$, $t_1 \in J$, are distinct from the n numbers that define a point in E_n for we can envision many different curves that pass through any given point in E_n. It should thus be clear that we have to start over from scratch if we are to have any success in associating a vector space structure with the intuitive notion of tangents to curves.

Let P denote a generic point of E_n and let the coordinates of P be (x_0^i) with respect to a given Cartesian coordinate cover of E_n, $P : (x_0^i)$. Again, let J be an open interval of the real line $\mathbb{R}$ that contains the point $t = 0$; but this time, let the map Γ be defined in terms of the representation

$$\text{(2-1.2)} \qquad \Gamma : J \to E_n | \bar{x}^i = x_0^i + \gamma^i(t), \qquad \gamma^i(0) = 0.$$

We then have, on applying d/dt to the functions $\bar{x}^i = x_0^i + \gamma^i(t)$,

$$d\bar{x}^i/dt = d\gamma^i(t)/dt,$$

and hence

$$\left.\frac{d\bar{x}^i}{dt}\right|_{t=0} = v^i, \qquad i = 1,\ldots,n, \tag{2-1.3}$$

where the n numbers $\{v^i\}$ are given by

$$v^i = \left.\frac{d\gamma^i(t)}{dt}\right|_{t=0}. \tag{2-1.4}$$

In view of the occurrence of the derivative with respect to t at $t = 0$, we have to look not only at the image $P:(x_0^i)$ of $t = 0$ in E_n for the given representation (2-1.2) of Γ, but also at the image of a neighborhood of $t = 0$ in E_n. Noting that the functions $\gamma^i(t)$ are smooth and that $\gamma^i(0) = 0$, we may expand the functions $\gamma^i(t)$ in a Taylor series in t about $t = 0$,

$$\begin{aligned}\gamma^i(t) &= \left(\left.\frac{d\gamma^i}{dt}\right|_{t=0}\right)t + \frac{1}{2}\left(\left.\frac{d^2\gamma^i}{dt^2}\right|_{t=0}\right)t^2 + \cdots \\ &= v^i t + \frac{1}{2}\left(\left.\frac{d^2\gamma^i}{dt^2}\right|_{t=0}\right)t^2 + \cdots,\end{aligned} \tag{2-1.5}$$

where the latter equality obtains by use of (2-1.4). In order to restrict our attention to a sufficiently small neighborhood of $t = 0$, we consider the collection of smooth functions

$$o(t) = \left\{f(t)\,|\,0 \subset \mathscr{D}(f(t)),\ \lim_{t\to 0}\left(\frac{f(t)}{t}\right) = 0\right\}. \tag{2-1.6}$$

It is then clear that (2-1.5) may be written as

$$\gamma^i(t) = v^i t + o(t). \tag{2-1.7}$$

There is more here, however, than first meets the eye. Suppose that

$$\bar{\Gamma}: J \to E_n\,|\,\bar{x}^i = x_0^i + \bar{\gamma}^i(t), \qquad \bar{\gamma}^i(0) = 0 \tag{2-1.8}$$

is another curve that passes through the point $P:(x_0^i)$. In this event, we may likewise write

$$\bar{\gamma}^i(t) = \bar{v}^i t + o(t) \tag{2-1.9}$$

where

$$\bar{v}^i = \left.\frac{d\bar{\gamma}^i(t)}{dt}\right|_{t=0} \tag{2-1.10}$$

Now, simply observe that (2-1.7) and (2-1.9) give

$$\bar{\gamma}^i(t) - \gamma^i(t) = (\bar{v}^i - v^i)t + o(t) \tag{2-1.11}$$

because (2-1.6) shows that the collection $o(t)$ is closed under addition and multiplication by numbers. On the other hand, Γ and $\bar{\Gamma}$ have the same tangent at $P:(x_0^i)$ whenever $v^i = \bar{v}^i$, in which case (2-1.11) gives

$$\bar{\gamma}^i(t) = \gamma^i(t) + o(t). \tag{2-1.12}$$

The relations (2-1.12) are clearly symmetric, reflexive and transitive, and hence they define an equivalence relation $\rho(P)$ on the collection of all representations of all curves passing through $P:(x_0^i)$. We may thus use the equivalence relation $\rho(P)$ to partition the collection of all representations of curves through $P:(x_0^i)$ into mutually disjoint equivalence classes. The following result is then clear. *If two curves passing through $P:(x_0^i)$ have representations that belong to the same equivalence class, then the two curves have the same tangent at the point $P:(x_0^i)$.*

The important thing to note here is that the point P in E_n is held fixed and that each choice of the functions $\{\gamma^i(t)\}$ with $\gamma^i(0) = 0$ in (2-1.2) gives a curve in E_n that passes through the point $P:(x_0^i)$ when the parameter t has the value zero. With this in mind we now consider the totality of all curves Γ in E_n that pass through the point P; that is, we allow the functions $\{\gamma^i(t)\}$ in (2-1.2) to range through all n-tuples of functions with domain J such that $\gamma^i(0) = 0$. Now, all such functions look like $\gamma^i(t) = v^i t + o(t)$, where the n-tuples of numbers $\{v^i\}$ range through the collection of all n-tuples of real numbers. An application of the operation d/dt to the functions $\bar{x}^i = x_0^i + \gamma^i(t) = x_0^i + v^i t + o(t)$, followed by an evaluation at $t = 0$, yields the n-tuple

$$v^i = \left.\frac{d\bar{x}^i}{dt}\right|_{t=0}. \tag{2-1.13}$$

This n-tuple $\{v^i\}$ characterizes the tangent to the curve $\bar{x}^i = x_0^i + \gamma^i(t) = x_0^i + v^i t + o(t)$ at the point $P:(x_0^i)$. Thus if we concentrate attention only on a sufficiently small neighborhood of the point $P:(x_0^i)$ for which we may neglect the $o(t)$ terms, any curve through $P:(x_0^i)$ looks like $\bar{x}^i = x_0^i + v^i t$, which is a representative element of the equivalence class that obtains under the equivalence relation $\rho(P)$. *In fact, the operation $d/dt|_{t=0}$ via* (2-1.2) *and* (2-1.13) *establishes a* 1-*to*-1 *correspondence between the equivalence classes of representations of curves through the point $P:(x_0^i)$ and the collection of all n-tuples of real numbers $\{v^i\}$.* The important thing here is that we can equip the collection of tangents to equivalence classes of curves through P with a natural vector space structure.

To this end, let

$$\bar{x}_1^i = x_0^i + \gamma^i(t)_1, \qquad \gamma^i(t)_1 = u^i t + o(t),$$

$$\bar{x}_2^i = x_0^i + \gamma^i(t)_2, \qquad \gamma^i(t)_2 = v^i t + o(t)$$

be two curves through $P:(x_0^i)$ that correspond to the n-tuples $\{u^i\}$ and $\{v^i\}$. It is then easily seen that the curve

$$\bar{x}_3^i = x_0^i + \gamma^i(t)_3, \qquad \gamma^i(t)_3 = (u^i + v^i)t + o(t)$$

passes through $P:(x_0^i)$ and corresponds with the n-tuple $\{u^i + v^i\}$. In fact, we have $\gamma^i(t)_3 = \gamma^i(t)_1 + \gamma^i(t)_2$. Thus the sum of two tangents to curves through P is a tangent to a curve through P. In like manner, for any $a \in \mathbb{R}$,

$$\bar{x}_4^i = x_0^i + \gamma^i(t)_4, \qquad \gamma^i(t)_4 = au^it + o(t)$$

is a curve in E_n that passes through $P:(x_0^i)$ and corresponds with the n-tuple $\{au^i\}$. This is also geometrically consistent since $\gamma^i(t)_4 = a\gamma^i(t)_1$. Thus any multiple of a tangent to a curve through P is a tangent to a curve through P. Accordingly, the collection of all tangents to all curves that pass through a point $P:(x_0^i)$ in E_n is naturally equipped with the operations of addition and multiplication by numbers. It is then an easy matter to verify that axioms **V1** through **V8** of a vector space are satisfied. We may thus identify the collection of all tangents to all curves through a point $P:(x_0^i)$ in E_n with an n-dimensional vector space $\mathscr{V}$. Further, and of greater importance, we may consider the vector space $\mathscr{V}$ to be *attached* to the space E_n at the point $P:(x_0^i)$. Under this interpretation we write $T_p(E_n)$ for the n-dimensional vector space $\mathscr{V}$ that is attached to E_n at the point P and refer to $T_P(E_n)$ as the *tangent space to* E_n *at* P.

The geometric notions are shown in Fig. 6. To show a clear distinction between the space itself and the tangent space that is attached at the point P,

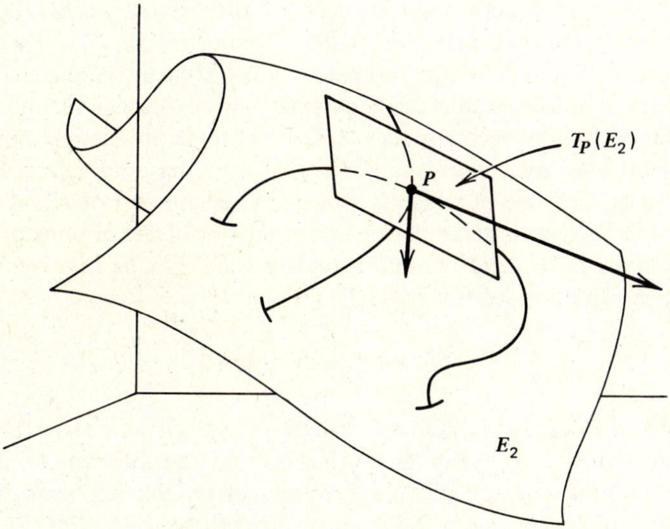

FIGURE 6. The space $T_P(E_2)$

we have taken the situation in which the space under consideration is a two-dimensional space E_2 that has been realized as a two-dimensional surface in three-dimensional Euclidean space.

2-2. THE TANGENT SPACE $T(E_n)$ AND VECTOR FIELDS ON E_n

An n-dimensional vector space $T_p(E_n)$ is attached to the space E_n at a generic point P of E_n in a natural way by allowing $d/dt|_{t=0}$ to act on all representations of all curves through the point $P:(x_0^i)$ such that $\bar{x}^i = x_0^i + \gamma^i(t)$, $\gamma^i(0) = 0$. On the other hand, the point P of E_n was generic in these considerations. Thus if we allow P to range through E_n or a smooth subset S of E_n, *we obtain a vector space attached to each point of* E_n *or to each point of* S. The union of these n-dimensional vector spaces, as P ranges over E_n or a subset S of E_n, is referred to as the *tangent space* of E_n or of S. The notation $T(E_n)$ or $T(S)$ is now standard for tangent spaces, so we write

$$T(E_n) = \bigcup_{P \in E_n} T_P(E_n) \tag{2-2.1}$$

and

$$T(S) = \bigcup_{P \in S} T_P(E_n) \qquad \text{for } S \subset E_n. \tag{2-2.2}$$

Great care must be exercised in the use of this notation, for otherwise it can lead to sheer nonsense. Clearly, the vector space operations of addition and multiplication by numbers occur in each of the vector spaces $T_P(E_n)$ separately. In particular, we may *not* "add" a vector in $T_P(E_n)$ to a vector in $T_Q(E_n)$ for P different from Q. The reader must keep in mind that $T(E_n)$ or $T(S)$ denotes a collection that contains many n-dimensional vector spaces; in fact, there are as many vector spaces in $T(S)$ as there are distinct points in S.

The reason why we have to consider $T(E_n)$ as the ensemble of n-dimensional vector spaces, one of which is attached to each point of E_n, is that such a collection is needed to make reasonable sense out of the notions of tangents to whole curves in E_n and of vector fields on E_n. Let us first return to the curve Γ in E_n that is specified by (2-1.1):

$$\Gamma : J \to E_n | \bar{x}^i = \gamma^i(t).$$

If we apply the operator d/dt to the functions $\{\gamma^i(t)\}$, we obtain the n-tuples of functions $\{d\gamma^i(t)/dt\}$. For any value of t in the interval J, say t_1, the n-tuple $d\gamma^i(t_1)/dt = v^i$ specifies the *selection* of an element from the vector space $T_P(E_n)$ with $P:(\gamma^i(t_1))$. Thus as t ranges through J, $d\gamma^i(t)/dt = v^i(t)$ specifies a selection of elements from $T(\Gamma)$ with one vector selected from

$T_{P:(\gamma^i(t))}(E_n)$ for each value of t. Clearly, this collection of vectors is only defined on the curve Γ itself since each vector comes from $T(\Gamma)$.

The tangent field to a curve $\Gamma|\bar{x}^i = \gamma^i(t)$ *is a map*

$$TΓ: J \to T(\Gamma) \tag{2-2.3}$$

that selects for each $t \in J$ *the element* $\{v^i(t) = d\gamma^i(t)/dt\}$, *from the vector space* $T_{P:(\gamma^i(t))}(E_n)$ *that belongs to*

$$T(\Gamma) = \bigcup_{t\in J} T_{P:(\gamma^i(t))}(E_n).$$

The tangent space $T(\Gamma)$ is simply where the tangent field to a curve Γ lives and that it is the map $T\Gamma$ of J into $T(\Gamma)$ that is the tangent field to the curve Γ. Care must also be taken to realize that the tangent field to the curve Γ depends in an intrinsic way on how the curve Γ is parameterized; that is, it depends upon the specific functions $\gamma^i(t)$ that occur in the representation $\bar{x}^i = \gamma^i(t)$. In order to see this, let $g(s)$ be a smooth function of the variable s such that $dg(s)/ds > 0$ and set $t = g(s)$. Since $dg(s)/ds > 0$, $t = g(s)$ can be solved for s in terms of t by the implicit function theorem so that we have $s = G(t)$. Finally let J_g be the image of J under the mapping defined by $s = G(t)$. Under these conditions, we may write the curve Γ in the equivalent form

$$\Gamma_g: J_g \to E_n|\bar{x}^i = \gamma^i(g(s)).$$

However, d/ds applied to $\bar{x}^i = \gamma^i(g(s))$ now gives the tangent field

$$T\Gamma_g: J_g \to T(\Gamma)$$

with $\bar{v}^i(s) = v^i(g(s))\,dg(s)/ds$. If you think for the moment of two different particles moving along the same curve Γ, then the "direction" in which either of the particles is moving at any point on the curve is fixed by the curve, but the speeds with which each particle moves along the curve can be different.

A straightforward generalization of the notion of a tangent field to a curve yields the important concept of a vector field.

Definition. A *vector field* V on a region S of E_n is a smooth (C^∞) map

$$V: S \to T(S)|v^i = v^i(x^j), \qquad i = 1, \ldots, n \tag{2-2.4}$$

such that if $P:(x^j) \in S$ then $\{v^i(x^j)\} \in T_P(E_n)$.

A vector field has nothing whatsoever to do with just one vector space. It is rather a selection of an element from each of a very large collection of vector spaces; indeed, there are as many vector spaces involved as there are points in the domain S of the vector field. In case the reader has any hesitation on this score, we hasten to point out that the vector space operations for vector fields also consist of as many operations as there are points in the domain S of the vector field. In fact, $T(S)$ inherits its vector space operations from the vector space operations of each $T_P(E_n)$ that makes up $T(S)$. Thus if

$$U: S \to T(S) | u^i = u^i(x^j),$$

$$V: S \to T(S) | v^i = v^i(x^j)$$

are two vector fields with domain S, then $W = U + V$ and $Z = a \cdot U$ are defined by

$$W: S \to T(S) | w^i = u^i(x^j) + v^i(x^j),$$

$$Z: S \to T(S) | z^i = au^i(x^j).$$

Addition and multiplication by numbers thus occur simultaneously in each of the n-dimensional vector spaces $T_{P:(x^j)}(E_n)$; that is, *addition and multiplication by numbers occur pointwise with respect to* E_n.

The definition of multiplication by numbers;

$$U: S \to T(S) | u^i = u^i(x^j),$$

$$W = a \cdot U: S \to T(S) | z^i = au^i(x^j), \qquad a \in \mathbb{R}$$

multiplies the vector $\{u^i(x^j)\}$ at $P:(x^j)$ by the number a and multiplies the vector $\{u^i(y^j)\}$ at $Q:(y^j)$ by the same number a. In this event, we refer to $T(S)$ as the *tangent space over* $\mathbb{R}$. On the other hand, if $a(x^i)$ is a function with E_n as domain, we can equally well multiply the vector $\{u^i(x^j)\}$ at $P:(x^j)$ by the number $a(x^j)$ and multiply the vector $\{u^i(y^j)\}$ at $Q:(y^j)$ by the number $a(y^j)$. This leads to the operation defined by

$$U: S \to T(S) | u^i = u^i(x^j),$$

$$W = f \cdot U: S \to T(S) | w^i = f(x^j) u^i(x^j).$$

Further, if we restrict $f(x^j)$ to be a member of the associative algebra $\Lambda^0(E_n)$ of C^∞ functions on E_n, multiplication by numbers extends directly to multiplication by elements of $\Lambda^0(E_n)$ through

$$\bullet: \Lambda^0(E_n) \times T(S) \to T(S) | w^i(x^j) = f(x^j) u^i(x^j)$$

for $U: S \to T(S) | u^i = u^i(x^j)$, $f(x^j) \in \Lambda^0(E_n)$. This situation will be referred to as *the tangent space over* $\Lambda^0(E_n)$. The tangent space over $\Lambda^0(E_n)$ is a much richer structure than the tangent space over $\mathbb{R}$. In fact, the tangent space over $\Lambda^0(E_n)$ is a *module* over $\Lambda^0(E_n)$, and this module will figure heavily in later discussions.

2-3. OPERATOR REPRESENTATIONS OF VECTOR FIELDS

Construction of $T_P(E_n)$ and of $T(E_n)$, although conceptually clear, gives vector fields as n-tuples of functions $\{v^i(x^j)\}$ that specify the selection of an appropriate element from each $T_P(E_n)$ as $P:(x^j)$ ranges over the domain of the vector field. This is cumbersome to work with in the extreme. We therefore proceed along different but equivalent lines that will allow us to work with only a single function at a time. This new approach, although more abstract, provides a formal structure that will prove to be very powerful in answering a number of important questions.

We return to the construction of the tangent to a curve through a point in E_n. Let

$$\Gamma : J \to E_n | \bar{x}^i = \gamma^i(t), \qquad \gamma^i(0) = x_0^i \tag{2-3.1}$$

define a curve through the point $P:(x_0^i)$. Further, let

$$F: E_n \to \mathbb{R} | r = f(x^1, \dots, x^n) = f(x^i), \tag{2-3.2}$$

where f is a smooth function, and let the range of Γ be contained in the domain of F. Under these conditions, we may compose the maps F and Γ to obtain the map

$$T = F \circ \Gamma : \mathbb{R} \to \mathbb{R} | \tau = \hat{F}(t) = f(\gamma^i(t)). \tag{2-3.3}$$

Now, the chain rule of the calculus shows that

$$\frac{d\hat{F}(t)}{dt} = \frac{d\gamma^i(t)}{dt} \frac{\partial f(x^j)}{\partial x^i}\bigg|_{x^i = \gamma^i(t)}. \tag{2-3.4}$$

The important thing to observe here is the occurrence of the coefficients $\{d\gamma^i(t)/dt\}$ on the right-hand side of (2-3.4). This suggests that vector fields can be defined in a manner similar to (2-3.4) provided we take advantage of the freedom in the choice of the function $f(x^i)$ that defines the map F. For example, if we make the choice $f(x^j) = x^k$, which is indeed a C^∞ function on E_n, then (2-3.4) yields

$$\frac{d\hat{F}(t)}{dt} = \frac{d\gamma^i(t)}{dt} \frac{\partial x^k}{dx^i}\bigg|_{x = \gamma(t)} = \frac{d\gamma^k(t)}{dt}.$$

We may thus recover the n-tuple of functions $\{d\gamma^i(t)/dt\}$ provided we take a sufficiently large collection of functions for the f's and evaluate the corresponding $d\hat{F}(t)/dt$'s (simply take the f's to be the n coordinate functions $x^1, x^2, \ldots, x^n$).

We now have to put these notions on a firm foundation. The essential ingredient in the above considerations is that of the functions f that represent the maps F from E_n to $\mathbb{R}$. Clearly, these should be as smooth as possible for otherwise spurious complications may be introduced by discontinuities in functions or some of their partial derivatives. The candidate for the choice of the collection from which the f's are to be picked has already been encountered; namely, the algebra $\Lambda^0(E_n)$ of all C^∞ functions on E_n. This algebra has the vector space operations

$$(f+g)(x^j) = f(x^j) + g(x^j), \qquad (b \cdot f)(x^j) = bf(x^j)$$

and the multiplication law

$$(f|g)(x^j) = f(x^j)g(x^j).$$

Further, since each element of $\Lambda^0(E_n)$ is a C^∞ function, the partial derivatives of any element of $\Lambda^0(E_n)$ is again a C^∞ function. The reader should note that choosing all C^{18} functions would not do since partial differentiation would give functions of class C^{17} and there would be difficulties.

If we start with any element $f(x^j)$ of $\Lambda^0(E_n)$, the action of an operator of the form

$$H = h^i(x^j)\frac{\partial}{\partial x^i}$$

is well defined when it acts on $f(x^j)$;

$$H\langle f\rangle(x^j) = h^i(x^j)\frac{\partial}{\partial x^i}f(x^j) = g(x^j).$$

If each of the n functions $h^i(x^j)$ belongs to $\Lambda^0(E_n)$, then $H\langle f\rangle = g$ gives back an element of $\Lambda^0(E_n)$. Further, it is clear from the calculus that

$$H\langle af + bg\rangle = aH\langle f\rangle + bH\langle g\rangle,$$

$$H\langle fg\rangle = H\langle f\rangle g + fH\langle g\rangle$$

(i.e., $H\langle(f|g)\rangle = (H\langle f\rangle|g) + (f|H\langle g\rangle)$). The first of these says that H is a linear operator on $\Lambda^0(E_n)$ and the second says that H satisfies the rule of Leibniz. Further, these two properties determine how the operator H combines with the vector space and multiplicative operations of the algebra $\Lambda^0(E_n)$.

Definition. An operation H on an algebra $\mathscr{A}$ is a *derivation* if and only if H maps $\mathscr{A}$ to $\mathscr{A}$ and satisfies the conditions

$$H\langle a \cdot f + b \cdot g\rangle = a \cdot H\langle f\rangle + b \cdot H\langle g\rangle, \tag{2-3.5}$$

$$H\langle (f|g)\rangle = (H\langle f\rangle|g) + (f|H\langle g\rangle) \tag{2-3.6}$$

for all $f, g \in \mathscr{A}$ and all $a, b \in \mathbb{R}$.

A glance at the form of the operator H,

$$H = h^i(x^j)\frac{\partial}{\partial x^i},$$

shows that writing equations that involve such operators will require many built up symbolic fractions. For this reason, and for elementary esthetics, we introduce the notation

$$\partial_i := \frac{\partial}{\partial x^i}. \tag{2-3.7}$$

Thus the operator H assumes the simpler form

$$H = h^i(x^j)\partial_i = h^1(x^j)\partial_1 + h^2(x^j)\partial_2 + \cdots + h^n(x^j)\partial_n.$$

With these facts at hand, we can give a new way of looking at vector fields on E_n.

Definition. Let $\{v^i(x^j)\}$ be a (C^∞) vector field on E_n. A derivation

$$V = v^i(x^j)\partial_i \tag{2-3.8}$$

on the algebra $\Lambda^0(E_n)$ of C^∞ functions on E_n is an *operator representation* (*representation*, for short) of the vector field $\{v^i(x^j)\}$.

This definition turns out to be both practical and powerful. The first thing to be noted is what happens when an operator representation of a vector field acts on the simplest possible functions other than constants; namely, the functions $x^1, x^2, \ldots, x^n$. It follows directly from (2-3.8) that

$$V\langle x^k\rangle = v^i(x^j)\partial_i x^k = v^k(x^j)$$

since $\partial_i x^k = \delta_i^k$ (where $\delta_i^k = \{1 \text{ for } i = k, 0 \text{ for } i \neq k\}$ are the Kronecker δ symbols). Clearly, any derivation $V = v^i(x^j)\partial_i$ reproduces its coefficients $\{v^i(x^j)\}$ in this way. This can be stated in the following equivalent fashion.

Lemma 2-3.1. *Let V be an operator representation of a vector field* $\{v^i(x^j)\}$ *on* E_n, *then V is uniquely determined by its action on the n functions* $x^1, x^2, \ldots, x^n$;

$$V = v^i(x^j)\,\partial_i = V\langle x^i \rangle(x^j)\,\partial_i, \tag{2-3.9}$$

where $V\langle x^i \rangle(x^j)$ *denotes the function that is obtained by allowing V to act on* x^i. *Thus* $v^i(x^j) = V\langle x^i \rangle(x^j)$, $i = 1, \ldots, n$, *and hence an operator representation of a vector field on* E_n *reproduces the vector field on* E_n.

Viewed from the other direction, Lemma 2-3.1 shows that any derivation on the algebra $\Lambda^0(E_n)$ determines a vector field on E_n. On the other hand, the collection of all vector fields on E_n is just the tangent space $T(E_n)$. It is therefore reasonable and consistent to identify a vector field with its operator representation and to identify $T(E_n)$ with the collection of all operator representations. Although it is necessary to distinguish between a vector field and its operator representation in certain contexts, the discussion presented here is at a level where the distinction is not essential. Bearing in mind that we have made this identification, using "vector field" in place of "operator representation of a vector field" leads to a marked economy. We therefore obtain the following equivalent definition.

Definition. The *tangent space*, $T(E_n)$ is the collection of all derivations on the algebra $\Lambda^0(E_n)$.

The reader should carefully note that vector fields as derivations on the algebra $\Lambda^0(E_n)$ do not have values. Evaluations can only be made by allowing the derivation to act on an element of $\Lambda^0(E_n)$. Thus $V = v^i(x^j)\,\partial_i$ is simply a collection of operations that are to be performed at some future date. On the other hand $V\langle f(x^j)\rangle = v^i(x^j)\,\partial_i f(x^j) = g(x^j)$ defines the new element $g(x^j)$ of $\Lambda^0(E_n)$ for any given $f \in \Lambda^0(E_n)$. Thus a vector field must act on a function f to give a new function g and this new function g can then be evaluated at some specific point of E_n to yield a value. In a more general sense of the term "value," we may view the action of a vector field on an element f of $\Lambda^0(E_n)$ to take its value g in $\Lambda^0(E_n)$. The two different interpretations of "value" will give the reader no difficulty in these discussions since the sense intended will always be clear from the context. In any event, it should now be abundantly clear that a vector field is something of an abstraction that does not have values of itself.

Let us go back and recall that $T(E_n)$ is the union over $P \in E_n$ of $T_P(E_n)$. Now, each of the vector spaces $T_P(E_n)$ being n-dimensional says that each $T_P(E_n)$ has various collections of systems of n linearly independent vectors and each of these collections constitutes a basis for $T_P(E_n)$. There is thus the obvious question as to whether we can choose a basis in each $T_P(E_n)$ so that they vary smoothly as P sweeps out E_n; that is, can we choose basis fields for all of $T(E_n)$ at once? It turns out that we can, as the next lemma shows.

Lemma 2-3.2. *The n vector fields* $\partial_1, \partial_2, \ldots, \partial_n$ *constitute a basis for* $T(E_n)$.

Proof. Each of the quantities ∂_i, $i = 1, \ldots, n$ is a derivation on $\Lambda^0(E_n)$, so each is a vector field on E_n. Lemma 2-3.1 then shows that any element of $T(E_n)$ is a linear combination of these n vector fields: $V = v^i(x^j)\partial_i$. It thus remains to show that these n vector fields are linearly independent. This means that we must show that

$$(c_1\partial_1 + c_2\partial_2 + \cdots + c_n\partial_n)\langle f \rangle = 0 \tag{2-3.10}$$

for *all* $f \in \Lambda^0(E_n)$ if and only if $c_1 = c_2 = \cdots = c_n = 0$ throughout E_n. Since (2-3.10) must hold for all $f \in \Lambda^0(E_n)$, it must likewise hold for $f = x^k$. However, $(c_1\partial_1 + \cdots + c_n\partial_n)\langle x^k \rangle = c_k$ and hence (2-3.10) is satisfied if and only if $c_k = 0$. Now, simply let k take each of the values between 1 and n to obtain the desired result that $c_1 = c_2 = \cdots = c_n = 0$ is the only choice of the c's for which (2-3.10) will be satisfied for all $f \in \Lambda^0(E_n)$. □

Definition. The basis fields $\{\partial_i, i = 1, \ldots, n\}$ of $T(E_n)$ are the *natural basis* of $T(E_n)$ with respect to the (x^i) coordinate cover of E_n.

We will see later on that the natural basis for $T(E_n)$ will change when we change the coordinate cover of E_n. The natural basis of $T(E_n)$ is thus seen to be intimately related to the coordinate cover of E_n. Since there are many coordinate covers for E_n, there will be many natural bases for $T(E_n)$. This should not be an unexpected circumstance as it is well known that a vector space has many bases. It is important to note, however, that a natural basis for $T(E_n)$ is fixed for a given coordinate cover. For the time being, the coordinate cover of E_n is also fixed.

The existence of a natural basis for $T(E_n)$ makes everything much simpler. Although we may still not add vectors in $T_P(E_n)$ to vectors in $T_Q(E_n)$ for P different from Q, verification of the vector space properties is now significantly simplified. In fact we do it once and for all for each P in E_n simultaneously. Simply observe if U and V belong to $T(E_n)$,

$$U = u^i(x^j)\partial_i, \qquad V = v^i(x^j)\partial_i,$$

then we define the vector space operations $+$ and $\cdot$ as would be done with respect to any basis of a finite dimensional vector space,

$$U + V = \left(u^i(x^j) + v^i(x^j)\right)\partial_i, \tag{2-3.11}$$

$$b \cdot U = \left(bu^i(x^j)\right)\partial_i. \tag{2-3.12}$$

It is then a trivial matter to verify that postulates **V1** through **V8** of a vector space are satisfied; indeed, the calculations are identical to those for an ordinary n-dimensional vector space of ordered n-tuples of real numbers. We note in passing that $u^i(x^j) + v^i(x^j)$ and $bu^i(x^j)$ on the right-hand sides of (2-3.11) and (2-3.12) are well defined at each point of E_n from the vector space properties of $\Lambda^0(E_n)$. Thus, $T(E_n)$ is closed under the vector space operations at each $P:(x^j)$ that are inherited from each $T_{P:(x^j)}(E_n)$, as indeed it must be. On the other hand, a $V \in T(E_n)$ only has a value at a point in E_n when it acts on a specific element of $\Lambda^0(E_n)$, in which case it has a single, unique value, *not* the n values of its coefficients: $V\langle f\rangle = v^i(x^j)\partial_i f(x^j) = g(x^j)$.

This view does not capture the full scope available by using the operator representation, for an arbitrary vector field $V = v^i(x^j)\partial_i$ is actually a linear combination of the basis fields $\{\partial_i,\ i = 1,\ldots,n\}$ with coefficients $\{v^i(x^j), i = 1,\ldots,n\}$ that are elements of $\Lambda^0(E_n)$. Now, simply note that we may multiply vectors at $P:(x^j)$ and $Q:(y^j)$ by different numbers, namely, $a(x^j)$ and $a(y^j)$. Thus, for $U = u^i(x^j)\partial_i$, (2-3.12) can be replaced by

$$f \cdot U = f(x^j)u^i(x^j)\partial_i. \tag{2-3.13}$$

On the other hand, (2-3.13) may be viewed as a map from $\Lambda^0(E_n) \times T(E_n)$ to $T(E_n)$ for $f(x^j) \in \Lambda^0(E_n)$. It is then a simple matter to verify that (2-3.13) satisfies **M1** through **M4** of a module over $\Lambda^0(E_n)$. Equipped in this way, *$T(E_n)$ becomes a module over* $\Lambda^0(E_n)$ and every element of $T(E_n)$ is generated from the basis fields $\{\partial_i,\ i = 1,\ldots,n\}$ by forming linear combinations with coefficients from $\Lambda^0(E_n)$; that is, $U = u^i(x^j)\partial_i$, $\{u^i(x^j) \in \Lambda^0(E_n),\ i = 1,\ldots,n\}$, $V = v^i(x^j)\partial_i$, $\{v^i(x^j) \in \Lambda^0(E_n),\ i = 1,\ldots,n\}$.

Lest the reader forget, $T(E_n)$ is the union over $P \in E_n$ of the vector spaces $T_P(E_n)$; there are as many vector spaces in $T(E_n)$ as there are points in E_n. Thus when we speak of $\{\partial_i,\ i = 1,\ldots,n\}$ as basis fields for $T(E_n)$, what we mean is that $\{\partial_i,\ i = 1,\ldots,n\}$ is a basis for each vector space $T_P(E_n)$ for $P \in E_n$. The wonderful and very useful thing here is that the operator representation allows us to get away with a single symbolic operator basis $\{\partial_i, i = 1,\ldots,n\}$ that can serve in each $T_P(E_n)$ by allowing it to operate on elements of $\Lambda^0(E_n)$ and then restrict the values that result upon evaluation at the point P. Thus for instance, $\partial_i\langle x^j\rangle = \delta_i^j$ is true at any point P in E_n, and hence the linear independence of the base elements $\{\partial_i\}$ at any $P \in E_n$ is established once and for all, no matter where in E_n P may happen to be.

2-4. ORBITS

We started these considerations with the notion of tangents to a curve and have wound up with the rather abstract view of a vector field as a derivation on an algebra. The full circle can be closed if we can show that a derivation on $\Lambda^0(E_n)$ leads to a unique system of curves in a natural manner. Above and beyond the question of theoretical nicety, such a demonstration will provide the basis for solving a number of important problems not the least of which is that of obtaining solutions to linear and nonlinear first-order partial differential equations.

A curve in E_n has been shown to be given in terms of a map

$$\Gamma: J \to E_n | \bar{x}^i = \bar{x}^i(t), \tag{2-4.1}$$

while a vector field on E_n is specified by a derivation

$$V = v^i(x^j)\partial_i.$$

If we think for the moment of the functions $\bar{x}^i(t)$ in terms of their values that are taken for a given value of t, then $\bar{P}:(\bar{x}^i)$ is a point in E_n. We now go one step further and ignore the t dependence altogether. The $(\bar{x}^i)$ may then be considered as new names for a point in E_n. Under this latter interpretation, we write the corresponding vector field as

$$\bar{V} = v^i(\bar{x}^j)\bar{\partial}_i$$

where the notation $\bar{\partial}_i := \partial/\partial\bar{x}^i$ has been used. Thus $\bar{V}$ is assigned values by allowing it to operate on functions $f(\bar{x}^i)$, $\bar{V}\langle f\rangle = v^i(\bar{x}^j)\bar{\partial}_i f$. In particular, we have

$$\bar{V}\langle\bar{x}^i\rangle = v^i(\bar{x}^j).$$

Suppose that we could find a family of curves in E_n such that one and only one curve of the family passes through any given point $\bar{P}:(\bar{x}^i)$ of E_n. Further, suppose that the curve that passes through $\bar{P}:(\bar{x}^i)$ has a tangent vector at $\bar{P}:(\bar{x}^i)$ that has values $\{v^i(\bar{x}^j)\}$. If this is the case, we have

$$\frac{d\bar{x}^i(t)}{dt} = v^i(\bar{x}^j(t))$$

from the definition of the tangent to a curve. The system of autonomous differential equations just written down shows that a solution will pass through the point $\bar{P}:(\bar{x}^i(t))$ at time t. Accordingly, the solution will pass through some point in E_n at $t = 0$. If it is kept clearly in mind that the curve in question is generated by $\{\bar{x}^i(t)\}$ for given $\{v^i(\bar{x}^j(t))\}$, we see that we can turn the problem around and specify the point through which the curve will pass when $t = 0$; that is, the problem of describing the curve can be made into an initial value problem. Since $P:(x^i)$ can be considered as a generic point in E_n, it is clearly advantageous to take the initial data as $\bar{x}^i(0) = x^i$. These considerations motivate the following definition.

Definition. The *orbits* of a vector field

$$V = V\langle x^i\rangle\partial_i = v^i(x^j)\partial_i \tag{2-4.2}$$

are all solutions of the autonomous system of differential equations

(2-4.3) $$\frac{d\bar{x}^i(t)}{dt} = \bar{V}\langle \bar{x}^i \rangle = v^i(\bar{x}^j)$$

subject to the initial data

(2-4.4) $$\bar{x}^i(0) = x^i.$$

Theorem 2-4.1. *The orbits of a vector field*

(2-4.5) $$V = V\langle x^i \rangle \partial_i = v^i(x^j)\partial_i$$

have the form

(2-4.6) $$\bar{x}^i(t) = \exp(tV)\langle x^i \rangle = \exp\big(tv^j(x^k)\partial_j\big)\langle x^i \rangle$$

for all t in some open interval J that contains $t = 0$. Thus a vector field determines and is determined by its orbits.

Proof. The representation (2-4.6) follows directly from application of the result (1-4.6) to the autonomous system of differential equations (2-4.3) subject to the initial data (2-4.4) for all t in some open interval J that contains $t = 0$. Thus, a vector field determines its orbits. Conversely, if we write (2-4.6) in the form

$$\bar{x}^i(t) = \exp(tV)\langle x^i \rangle,$$

then

$$\frac{d\bar{x}^i(t)}{dt} = V\langle \exp(tV)\langle x^i \rangle\rangle = \bar{V}\langle \bar{x}^i \rangle.$$

An evaluation for $t = 0$ then gives $\bar{x}^i(0) = x^i$ and we obtain

$$\left.\frac{d\bar{x}^i(t)}{dt}\right|_{t=0} = V\langle x^i \rangle = v^i(x^j).$$

Since the initial data $\bar{x}^i(0) = x^i$ is generic (i.e., (x^i) can have any values we please), the functions $\{v^i(x^j)\}$ are therefore determined for any $P:(x^i)$ in E_n. The orbits of a vector field V thus determine V. □

The proof of the last theorem has given us some important new information, for it tells us how to compute the vector field from its orbits. In order to do this, we simply differentiate with respect to t and then evaluate at $t = 0$. When the answer is written in terms of the generic initial data $\bar{x}^i(0) = x^i$, we have

$$v^i(x^j) = \frac{d}{dt}\bar{x}^i(t)\bigg|_{t=0}.$$

The vector field V is then reconstructed by setting $V = v^i(x^j)\partial_i$.

There is a very simple way of visualizing the orbits of a field and, in fact, the vector field itself. Suppose that we have a fluid that fills up E_n and at time $t = 0$ we paint the molecules of this fluid red. If the fluid undergoes a steady flow with velocity vector field V, then $\bar{x}^i(t) = \exp(tV)\langle x^i \rangle$ defines the coordinates of the fluid molecule at time t that was at the point with coordinates $\{x^i\}$ at time $t = 0$. The orbits of the velocity vector field thus become the trajectories of the red fluid molecules that undergo the steady flow with the given V.

A few examples may help to clarify what is going on. If we start with the vector field

$$V = -2y\frac{\partial}{\partial x} + xz\frac{\partial}{\partial z}$$

in E_3 with coordinates (x, y, z), we have to solve the system of differential equations

$$d\bar{x}/dt = -2\bar{y}, \qquad d\bar{y}/dt = 0, \qquad d\bar{z}/dt = \bar{x}\bar{z}$$

subject to the initial data

$$\bar{x}(0) = x, \qquad \bar{y}(0) = y, \qquad \bar{z}(0) = z.$$

This is quite an easy problem, so we simply write down the solution:

$$\bar{x}(t) = x - 2yt, \qquad \bar{y}(t) = y, \qquad \bar{z}(t) = z\exp(xt - yt^2).$$

A differentiation with respect to t followed by an evaluation at $t = 0$ then gives

$$d\bar{x}(0)/dt = -2y, \qquad d\bar{y}(0)/dt = 0, \qquad d\bar{z}(0)/dt = xz,$$

from which we recover the components $\{v^i(x^j)\}$ of our vector field; that is, $v^1(x^j) = -2y$, $v^2(x^j) = 0$, $v^3(x^j) = xz$ with $x^1 = x$, $x^2 = y$, $x^3 = z$.

In the second example, let the vector field be given by

$$V = y(1 - 2x)\frac{\partial}{\partial x} + (3z^2 - 5)\frac{\partial}{\partial y}$$

in E_4 with coordinates (x, y, z, r). The orbits are determined by solving

$$\frac{d\bar{x}}{dt} = \bar{y}(1 - 2\bar{x}), \qquad \frac{d\bar{y}}{dt} = 3\bar{z}^2 - 5, \qquad \frac{d\bar{z}}{dt} = 0, \qquad \frac{d\bar{r}}{dt} = 0$$

subject to the initial data

$$\bar{x}(0) = x, \qquad \bar{y}(0) = y, \qquad \bar{z}(0) = z, \qquad \bar{r}(0) = r.$$

The solution of this initial value problem is given by

$$\bar{x}(t) = (x - \tfrac{1}{2})\exp(-2yt - (3z^2 - 5)t^2) + \tfrac{1}{2},$$

$$\bar{y}(t) = (3z^2 - 5)t + y, \qquad \bar{z}(t) = z, \qquad \bar{r}(t) = r.$$

Again differentiation with respect to t and evaluation at $t = 0$ reproduces the given vector field V. It should also be noted that this solution can be obtained by summing the operator series representations that are given by

$$\bar{x}(t) = \exp(ty(1 - 2x)\partial/\partial x + t(3z^2 - 5)\partial/\partial y)\langle x\rangle,$$

$$\bar{y}(t) = \exp(ty(1 - 2x)\partial/\partial x + t(3z^2 - 5)\partial/\partial y)\langle y\rangle,$$

$$\bar{z}(t) = \exp(ty(1 - 2x)\partial/\partial x + t(3z^2 - 5)\partial/\partial y)\langle z\rangle,$$

$$\bar{r}(t) = \exp(ty(1 - 2x)\partial/\partial x + t(3z^2 - 5)\partial/\partial y)\langle r\rangle.$$

Explicit solutions rather than the operator series representations are preferable whenever the explicit solutions can be obtained. Obtaining explicit solutions is not always an easy matter, however, and so it is often the case that the operator series representations are all that can be given.

2-5. LINEAR AND QUASILINEAR FIRST-ORDER PARTIAL DIFFERENTIAL EQUATIONS

A linear (homogeneous) first-order partial differential equation for the determination of a function $f(x^j)$ on E_n is a partial differential equation of the form

$$v^i(x^j)\frac{\partial f}{\partial x^i} = 0, \tag{2-5.1}$$

where the n coefficient functions $\{v^i(x^j)\}$ are given functions of the n variables (x^j) relative to a given coordinate cover of E_n. If we use the given functions $\{v^i(x^j)\}$ to construct the vector field

$$V = v^i(x^j)\,\partial_i, \tag{2-5.2}$$

then we can rewrite (2-5.1) in the equivalent form

$$V\langle f\rangle = 0. \tag{2-5.3}$$

Thus the problem of finding all solutions to the partial differential equation (2-5.1) is equivalent to the problem of finding all elements $f(x^j)$ of $\Lambda^0(E_n)$ that are mapped onto the zero function 0 by the action of the vector field V. This does not solve the problem, it gives only a restatement of it. However, this restatement in terms of a vector field makes available all of the information we already know concerning vector fields.

Of particular importance is the system of orbital equations associated with the vector field V that is defined by the partial differential equation (2-5.1):

$$d\bar{x}^i/dt = \bar{V}\langle\bar{x}^i\rangle = v^i(\bar{x}^j), \tag{2-5.4}$$

$$\bar{x}^i(0) = x^i. \tag{2-5.5}$$

Suppose that $g(x^j) = c$ is a given hypersurface in E_n. This surface is taken into itself by the orbits of the vector field V if and only if $g(\bar{x}^i(t)) = g(x^i) = c$ holds for all t in some open interval J of $t = 0$. When this expression is differentiated with respect to t and we make use of (2-5.4), we have

$$0 = \frac{dc}{dt} = \frac{dg(\bar{x}^j(t))}{dt} = \frac{\partial g}{\partial\bar{x}^i}\frac{d\bar{x}^i(t)}{dt} = v^i(\bar{x}^j)\frac{\partial g}{\partial\bar{x}^i};$$

that is, $\bar{V}\langle g\rangle = 0$. Thus we come full circle to our original partial differential equation (2-5.1). This relationship is codified in the following way.

Definition. The vector field

$$V = v^i(x^j)\,\partial_i \tag{2-5.6}$$

it said to be a *characteristic* vector field of the linear partial differential equation

$$v^i(x^j)\,\partial_i f = V\langle f\rangle = 0 \tag{2-5.7}$$

and the orbits of the vector field V are the *characteristic curves* (characteristics) of the partial differential equation $V\langle f\rangle = 0$.

We have seen that the characteristics (orbits) of a vector field V can be written in the form

$$\bar{x}^i(t) = \exp(tV)\langle x^i\rangle. \tag{2-5.8}$$

Suppose that we can solve one of these equations for t as a function of the x's and the $\bar{x}$'s that occur in that equation. We could then write $t = T(\bar{x}^i; x^j)$ and use this to eliminate the variable t from the remaining $n - 1$ $\bar{x}$'s so that we would wind up with $n - 1$ functions of the x's and the $\bar{x}$'s that do not involve the independent variable t explicitly. This in turn shows that we can find $n - 1$ functionally independent functions $g_1, g_2, \ldots, g_{n-1}$ so that the $n - 1$ relations that obtain among the x's and the $\bar{x}$'s can be written as

$$g_a(\bar{x}^j) = g_a(x^j), \qquad a = 1, \ldots, n - 1. \tag{2-5.9}$$

(If this were not the case, it would easily follow that putting $\bar{x}^i = \bar{x}^i(t)$ would give an explicit dependence on the independent variable t.) However, we also have

$$dg_a(x^j)/dt = 0,$$

because the x's are the initial data and hence independent of t, and hence a differentiation of both sides of (2-5.9) yields

$$0 = \frac{dg_a(x^j)}{dt} = \frac{\partial g_a(\bar{x}^j(t))}{\partial \bar{x}^i(t)} \frac{d\bar{x}^i}{dt} = \bar{V}\langle g_a\rangle|_{\bar{x}^i = \bar{x}^i(t)}. \tag{2-5.10}$$

Each of the $n - 1$ functions $g_a(\bar{x}^j)$ that is defined in this way is such that it satisfies $\bar{V}\langle g_a\rangle = 0$ and each function g_a defines a hypersurface in E_n by (2-5.9) that is mapped into itself by the orbits of the vector field V. The functions g_a are referred to as *primitive integrals* of the partial differential equation (2-5.1). Finally, if we set

$$f = \Psi(g_1(x^j), g_2(x^j), \ldots, g_{n-1}(x^j)), \tag{2-5.11}$$

then f satisfies $V\langle f\rangle = 0$ provided function Ψ is of class C^1 in its $n - 1$ arguments; simply observe that

$$V\langle f\rangle = V\langle \Psi\rangle = \sum_{a=1}^{n-1} \frac{\partial \Psi}{\partial g_a} V\langle g_a\rangle = 0$$

because $V\langle g_a\rangle = 0$ for $a = 1, \ldots, n - 1$. This method of solving $V\langle f\rangle = 0$ is known as the *method of characteristics*.

What now remains to be shown is that every C^1 solution of the first order partial differential equation $V\langle f\rangle = 0$ can be written in the form (2-5.11) for some C^1 function Ψ of the $n-1$ primitive integrals $\{g_a(x^j),\ a=1,\dots,n-1\}$. We have seen that each of the $n-1$ primitive integrals g_a satisfies $V\langle g_a\rangle = 0$; that is,

$$v^i(x^j)\frac{\partial g_a}{\partial x^i} = 0, \qquad a = 1,\dots,n-1. \tag{2-5.12}$$

Further, if f is a solution of $V\langle f\rangle = 0$, then we also have

$$v^i(x^j)\frac{\partial f}{\partial x^i} = 0. \tag{2-5.13}$$

The system (2-5.12) and (2-5.13) together constitute a system of n simultaneous equations that are satisfied at every point in E_n. On the other hand, we may view (2-5.12) and (2-5.13) as a system of n simultaneous equations for the determination of n quantities $\{v^i(x^j)\}$ since f and the $n-1$ g_a's are presumed known. Now, the functions $\{v^i(x^j)\}$ cannot vanish identically on E_n for the problem would then dissolve (the zero vector as an operator maps all functions into zero). This can be the case, however, only if the functional determinant

$$\frac{\partial(f, g_1, g_2,\dots, g_{n-1})}{\partial(x^1, x^2,\dots, x^n)}$$

vanishes identically (the system (2-5.12), (2-5.13) has rank less than n). This latter condition is both necessary and sufficient that the functions $f, g_1, g_2,\dots, g_{n-1}$ be functionally dependent; that is, $f = \Psi(g_1,\dots,g_{n-1})$ for some Ψ. Finally, the fact that any C^1 solution of $V\langle f\rangle = 0$ can be expressed as a C^1 function of only $n-1$ functionally independent primitive integrals makes reasonable sense since $V\langle f\rangle = 0$ constitutes only one restriction on the function f of the n variables (x^i).

For example, suppose that we are given the partial differential equation

$$xz\frac{\partial f}{\partial z} - 2y\frac{\partial f}{\partial x} = 0.$$

The characteristic vector field of this equation is obviously

$$V = -2y\frac{\partial}{\partial x} + xz\frac{\partial}{\partial z}$$

and the equations for the characteristics are

$$d\bar{x}(t)/dt = -2\bar{y}(t), \qquad d\bar{y}(t)/dt = 0,$$

$$d\bar{z}(t)/dt = \bar{x}(t)\bar{z}(t),$$

subject to the initial data $\bar{x}(0) = x$, $\bar{y}(0) = y$, $\bar{z}(0) = z$. We have already solved this problem in a previous example:

$$\bar{x}(t) = x - 2yt, \qquad \bar{y}(t) = y, \qquad \bar{z}(t) = z\exp(xt - yt^2).$$

Since the first of these relations yields

$$t = \frac{x - \bar{x}}{2y},$$

we obtain

$$\bar{z} = z\exp\left(\frac{x - \bar{x}}{2y}x - y\left(\frac{x - \bar{x}}{2y}\right)^2\right) = z\exp\left(\frac{x^2 - \bar{x}^2}{4y}\right)$$

$$= z\exp\left(\frac{x^2}{4y} - \frac{\bar{x}^2}{4\bar{y}}\right) = z\exp\left(\frac{x^2}{4y}\right)\exp\left(-\frac{\bar{x}^2}{4\bar{y}}\right)$$

and $y = \bar{y}$. Thus

$$y = \bar{y}, \qquad \bar{z}\exp\left(\frac{\bar{x}^2}{4\bar{y}}\right) = z\exp\left(\frac{x^2}{4y}\right)$$

and the two primitive integrals g_1 and g_2 are

$$g_1 = y, \qquad g_2 = z\exp\left(\frac{x^2}{4y}\right).$$

The general solution of $V\langle f\rangle = 0$ is accordingly given by

$$f = \Psi\left(y, z\exp\left(\frac{x^2}{4y}\right)\right).$$

We now turn to somewhat more complicated kinds of first-order partial differential equations. Let $V = v^i(x^j)\partial_i$ be a given vector field on E_n and let $m(x^j; y)$ be a given function of the $n + 1$ arguments $(x^1, x^2, \ldots, x^n; y)$. A partial differential equation of the form

$$V\langle f\rangle = m(x^j; f) \tag{2-5.14}$$

is said to be *quasilinear*. Thus even if the function $m(x^j; f)$ is a constant function, $m(x^j; f) = k$, the partial differential equation

$$V\langle f\rangle = k$$

is quasilinear; anything other than zero on the right-hand side of (2-5.14) gives a quasilinear partial differential equation.

The method whereby quasilinear partial differential equations are solved is to replace such equations by equivalent linear ones that can be solved by the method of characteristics. This is accomplished by changing the underlying space. We started with the n-dimensional space E_n referred to the coordinate cover (x^i) while the function $m(x^j; f)$ that occurs on the right-hand side of (2-5.14) is defined over the space $E_n \times \mathbb{R}$. This suggests that it would be useful to consider the problem defined over the new space $E_{n+1} = E_n \times \mathbb{R}$ with the coordinate cover $(x^i; y)$. This new space is a natural one to use since it may be viewed as the *graph space* of the solutions of (2-5.14). To see this we suppose that $f(x^j)$ is a solution of (2-5.14) and construct the map

$$M : E_n \to E_{n+1} | x^i = x^i, \qquad i = 1, \ldots, n, \qquad y = f(x^i).$$

This map represents an n-dimensional hypersurface in E_{n+1} that is the graph of the solution $f(x^j)$.

Let $F(x^j; y)$ be a function defined on E_{n+1} and consider the equation

$$F(x^j; y) = C = \text{constant}. \tag{2-5.15}$$

The implicit function theorem tells us that we can solve (2-5.15) for y in terms of the x's so as to obtain

$$y = f(x^j) \tag{2-5.16}$$

for any value of the constant C provided the condition

$$\frac{\partial}{\partial y} F(x^j; y) \neq 0 \tag{2-5.17}$$

is satisfied. When this evaluation of y is put back into (2-5.15), we have the identical satisfaction of

$$F(x^j; f(x^j)) = C \tag{2-5.18}$$

for all points in E_n. Accordingly, (2-5.18) gives the identity

$$\frac{dF(x^j; f(x^j))}{dx^i} = \partial_i F(x^j; y)|_{y=f} + \left.\frac{\partial F}{\partial y}\right|_{y=f} \partial_i f = 0. \tag{2-5.19}$$

We now multiply the ith of these equations by $v^i(x^j)$ and sum i from 1 through n. The result is

$$v^i \partial_i F|_{y=f} + \left.\frac{\partial F}{\partial y}\right|_{y=f} v^i \partial_i f = 0. \tag{2-5.20}$$

Thus if $f(x^j)$ satisfies the given quasilinear partial differential equation (2-5.14), then the function $F(x^j; f(x^j))$ satisfies

$$v^i \partial_i F|_{y=f} + m(x^j, y) \left.\frac{\partial F}{\partial y}\right|_{y=f} = 0. \tag{2-5.21}$$

We now shift our considerations back to the space E_{n+1}. The relation (2-5.21) can be written in the equivalent form

$$\left.\left\{\left(V + m\frac{\partial}{\partial y}\right)\langle F(x^j; y)\rangle\right\}\right|_{y=f} = 0. \tag{2-5.22}$$

If we now drop the restriction $y = f(x^j)$, (2-5.22) becomes the linear first-order partial differential equation

$$\left(V + m\frac{\partial}{\partial y}\right)\langle F(x^j; y)\rangle = 0 \tag{2-5.23}$$

for the determination of the function $F(x^j; y)$. This equation can be solved by the method of characteristics and hence we can find all C^1 solutions $F(x^j; y)$. We now show that each function $F(x^j; y)$ that satisfies (2-5.23) and the condition $\partial F/\partial y \neq 0$ defines a function $y = f(x^j)$ implicitly through the equation

$$F(x^j; f(x^j)) = C \tag{2-5.24}$$

that satisfies the given quasilinear equation $V\langle f\rangle = m$. Because $F(x^j; y)$ satisfies (2-5.23), restriction to the hypersurface $y = f(x^j)$ gives

$$\left(V + m\frac{\partial}{\partial y}\right)\langle F\rangle|_{y=f} = V\langle F\rangle|_{y=f} + \left.\left(m\frac{\partial F}{\partial y}\right)\right|_{y=f} = 0. \tag{2-5.25}$$

On the other hand, (2-5.24) gives

$$V\langle F(x^j; f(x^j))\rangle = V\langle F\rangle|_{y=f} + \left.\frac{\partial F}{\partial y}\right|_{y=f} V\langle f\rangle = 0. \tag{2-5.26}$$

An elimination of the common term $V\langle F\rangle|_{y=f}$ between (2-5.25) and (2-5.26) then yields

$$\left.\frac{\partial F}{\partial y}\right|_{y=f}\left(V\langle f\rangle - m(x^j; f(x^j))\right) = 0 \tag{2-5.27}$$

and we are done.

The general solution of the quasilinear partial differential equation

$$V\langle f\rangle = m(x^j; f) \tag{2-5.28}$$

is obtained in the implicit form

$$F(x^j; f) = C \tag{2-5.29}$$

by finding the general solution of the linear partial differential equation

$$\left(V + m(x^j; y)\frac{\partial}{\partial y}\right)F(x^j; y) = 0 \tag{2-5.30}$$

on the $(n+1)$*-dimensional space* E_{n+1} *such that*

$$\frac{\partial}{\partial y}F(x^j; y) \neq 0. \tag{2-5.31}$$

This new formulation has replaced E_n by the new space E_{n+1} with the coordinate cover $(x^1, \ldots, x^n; y)$ and the operator V by the operator $V + m(x^j; y)\partial/\partial y$, which is an element of $T(E_{n+1})$. When this new operator is written out, we have

$$v^i(x^j)\,\partial_i + m(x^j; y)\frac{\partial}{\partial y},$$

so that it is immaterial that the functions $v^i(x^j)$ are defined on E_n only. The restriction that $V \in T(E_n)$ is clearly unnecessary. We therefore consider a general $\hat{V} \in T(E_{n+1})$,

$$\hat{V} = v^i(x^j; y)\,\partial_i + m(x^j; y)\frac{\partial}{\partial y}. \tag{2-5.32}$$

The same argument that was given above shows that any solution of

$$\hat{V}\langle F\rangle = 0 \tag{2-5.33}$$

on E_{n+1} such that $\partial F/\partial y \neq 0$ serves to define a solution of

$$v^i(x^j; f)\,\partial_i f = m(x^j; f) \tag{2-5.34}$$

implicitly through the relation

$$F(x^j; f(x^j)) = C. \tag{2-5.35}$$

The method of characteristics is thus applicable to the *general quasilinear* partial differential equation

$$v^i(x^j; f)\,\partial_i f = m(x^j; f).$$

A simple but illustrative example is provided by

$$\left(2x\frac{\partial}{\partial x} - 3y\frac{\partial}{\partial y}\right)\langle f\rangle = (f)^2.$$

In this case, we have to solve the linear equation

$$\left(2x\frac{\partial}{\partial x} - 3y\frac{\partial}{\partial y} + f^2\frac{\partial}{\partial f}\right)\langle F\rangle = 0$$

whose characteristic vector field is

$$\hat{V} = 2x\frac{\partial}{\partial x} - 3y\frac{\partial}{\partial y} + f^2\frac{\partial}{\partial f}.$$

The orbital equations of this vector field are

$$d\bar{x}/dt = 2\bar{x}, \qquad d\bar{y}/dt = -3\bar{y}, \qquad d\bar{f}/dt = (\bar{f})^2$$

subject to the initial data

$$\bar{x}(0) = x, \qquad \bar{y}(0) = y, \qquad \bar{f}(0) = f.$$

Since the general solutions of these equations are

$$\bar{x}(t) = xe^{2t}, \qquad \bar{y}(t) = ye^{-3t}, \qquad -(\bar{f}(t))^{-1} = t - (f)^{-1},$$

using the third to eliminate t from the first two yields

$$\bar{x}/x = e^{-2/\bar{f}}e^{2/f}, \qquad \bar{y}/y = e^{3/\bar{f}}e^{-3/f}.$$

The primitive integrals are thus given by

$$g_1 = \bar{x}e^{2/\bar{f}} = xe^{2/f}, \qquad g_2 = \bar{y}e^{-3/\bar{f}} = ye^{-3/f},$$

and hence the general solution is

$$F(x, y; f) = \Psi(xe^{2/f}, ye^{-3/f}).$$

The general solution of the original equation is thus given in implicit form by

$$\Psi(xe^{2/f}, ye^{-3/f}) = C$$

when Ψ ranges over all C^1 functions of its arguments such that $\partial\Psi/\partial f \neq 0$.

For the quasilinear equation

$$\frac{2x}{f}\frac{\partial f}{\partial x} + (1 - f^2)\frac{\partial f}{\partial y} = 1,$$

the corresponding linear equation is

$$\frac{2x}{f}\frac{\partial F}{\partial x} + (1 - f^2)\frac{\partial F}{\partial y} + \frac{\partial F}{\partial f} = 0.$$

The characteristic vector field is therefore

$$\hat{V} = \frac{2x}{f}\frac{\partial}{\partial x} + (1 - f^2)\frac{\partial}{\partial y} + \frac{\partial}{\partial f}$$

whose orbital equations are

$$d\bar{x}/dt = 2\bar{x}/\bar{f}, \qquad d\bar{y}/dt = (1 - \bar{f}^2), \qquad d\bar{f}/dt = 1$$

subject to the initial data

$$\bar{x}(0) = x, \qquad \bar{y}(0) = y, \qquad \bar{f}(0) = f.$$

Since the general solutions of these equations are $\bar{f} = f + t$,

$$\bar{x} = x\left(\frac{f + t}{f}\right)^2, \qquad \bar{y} = y + t + \frac{1}{3}f^3 - \frac{1}{3}(f + t)^3,$$

use of the first to eliminate the variable t yields

$$\bar{x}/\bar{f} = x/f, \qquad \bar{y} - \bar{f} + \tfrac{1}{3}\bar{f}^3 = y - f + \tfrac{1}{3}f^3.$$

The primitive integrals are thus given by

$$g_1 = x/f, \qquad g_2 = y - f + \tfrac{1}{3}f^3,$$

and hence the general solution is

$$F(x, y; f) = \Psi\left(x/f, y - f + \tfrac{1}{3}f^3\right).$$

The general solution of the original equation is thus given in implicit form by

$$\Psi\left(x/f, y - f + \tfrac{1}{3}f^3\right) = C$$

where Ψ ranges over all C^1 functions of its arguments such that $\partial\Psi/\partial f \neq 0$.

The next level of difficulty is where we are given a system of k quasilinear partial differential equations for the determination of k functions $\{f^\alpha, \alpha = 1,\ldots,k\}$ of n variables (x^i):

$$v^i(x^j; f^\beta)\partial_i f^\alpha(x^j) = m^\alpha(x^j; f^\beta), \qquad \alpha = 1,\ldots,k. \tag{2-5.36}$$

Here, Greek indices have the range 1 through k and we make particular note of the fact that there is only the *single* operator $v^i\partial_i$ that occurs on the left-hand side. Analogy with the previous discussion suggests that we consider a new space E_{n+k} with the coordinate cover

$$(x^1, x^2,\ldots, x^n : y^1, y^2,\ldots, y^k) = (x^i; y^\alpha)$$

and introduce the operator

$$\hat{V} = v^i(x^j; y^\beta)\partial_i + m^\alpha(x^j; y^\beta)\partial_\alpha \tag{2-5.37}$$

where we have used the notation $\partial_\alpha := \partial/\partial y^\alpha$. Clearly, we have $\hat{V} \in T(E_{n+k})$ and hence the problem

$$\hat{V}\langle F(x^j; y^\beta)\rangle = 0$$

can be solved by the method of characteristics. Let $\{F^1(x^j; y^\beta),\ldots, F^k(x^j; y^\beta)\} = \{F^\gamma(x^j; y^\beta)\}$ be k functions that satisfy

$$\hat{V}\langle F^\gamma(x^j; y^\beta)\rangle = 0, \qquad \gamma = 1,\ldots,k \tag{2-5.38}$$

and the condition

$$\frac{\partial(F^1, F^2,\ldots, F^k)}{\partial(y^1, y^2,\ldots, y^k)} \neq 0. \tag{2-5.39}$$

Under these conditions, the implicit function theorem shows that we may solve the k equations

$$F^\gamma(x^j; y^\beta) = C^\gamma, \qquad \gamma = 1,\ldots,k \tag{2-5.40}$$

for the y's in terms of the x's so as to obtain

$$y^\alpha = f^\alpha(x^j), \tag{2-5.41}$$

in which case

$$F^{\gamma}(x^j; f^{\beta}(x^j)) = C^{\gamma}, \qquad \gamma = 1,\dots,k \tag{2-5.42}$$

will be identically satisfied on E_n. If we apply the operator $v^i(x^j; f^{\beta})\partial_i$ to (2-5.42), we thus obtain

$$v^i(x^j; f^{\beta})\partial_i F^{\gamma}|_{y=f} + \left.\frac{\partial F^{\gamma}}{\partial y^{\alpha}}\right|_{y=f} v^i(x^j; f^{\beta})\partial_i f^{\alpha} = 0. \tag{2-5.43}$$

On the other hand, $F^{\gamma}(x^j; y^{\beta})$ satisfies (2-5.38), and hence use of (2-5.37) and restriction to $y^{\alpha} = f^{\alpha}(x^j)$ gives

$$v^i(x^j; f^{\beta})\partial_i F^{\gamma}|_{y=f} + \left.\frac{\partial F^{\gamma}}{\partial y^{\alpha}}\right|_{y=f} m^{\alpha}(x^j; f^{\beta}) = 0. \tag{2-5.44}$$

An elimination of the common terms between (2-5.43) and (2-5.44) gives the system of k relations

$$\left.\frac{\partial F^{\gamma}}{\partial y^{\alpha}}\right|_{y=f} \{v^i(x^j; f^{\beta})\partial_i f^{\alpha} - m^{\alpha}(x^j; f^{\beta})\} = 0. \tag{2-5.45}$$

Accordingly, since the matrix of coefficients in (2-5.45) is nonsingular by (2-5.39), the functions $\{f^{\alpha}(x^j)\}$ satisfy

$$v^i(x^j; f^{\beta})\partial_i f^{\alpha} = m^{\alpha}(x^j; f^{\beta});$$

that is, (2-5.36).

The general solution of the system of quasilinear partial differential equations

$$v^i(x^j; f^{\beta})\partial_i f^{\alpha} = m^{\alpha}(x^j; f^{\beta}), \qquad \alpha = 1,\dots,k$$

is obtained in the implicit form

$$F^{\gamma}(x^j; f^{\beta}) = C^{\gamma}, \qquad \gamma = 1,\dots,k$$

by finding k solutions $\{F^{\gamma}(x^j; y^{\beta})\}$ *of the linear partial differential equation*

$$v^i(x^j; y^{\beta})\partial_i F^{\gamma} + m^{\alpha}(x^j; y^{\beta})\partial_{\alpha} F^{\gamma} = 0$$

that satisfy

$$\frac{\partial(F^1, F^2,\dots, F^k)}{\partial(y^1, y^2,\dots, y^k)} \neq 0.$$

For example, suppose that we have to find two functions $f(u, v)$ and $g(u, v)$ that satisfy the first-order equations

$$(u\partial_u - u^2\partial_v)\langle f\rangle = ug^2, \qquad (u\partial_u - u^2\partial_v)\langle g\rangle = 3g.$$

In this case, (2-5.37) gives the vector field

$$\hat{V} = u\partial_u - u^2\partial_v + ug^2\partial_f + 3g\partial_g$$

on the four-dimensional space E_4 with coordinate cover (u, v, f, g). Since the orbital equations are $d\bar{u}/dt = \bar{u}$, $d\bar{v}/dt = -\bar{u}^2$, $d\bar{f}/dt = \bar{u}\bar{g}^2$, $d\bar{g}/dt = 3\bar{g}$, subject to the initial data $\bar{u}(0) = u$, $\bar{v}(0) = v$, $\bar{f}(0) = f$, $\bar{g}(0) = g$, we have $\bar{u} = ue^t$, $\bar{v} = v + u^2(1 - e^{2t})/2$, $\bar{f} = f + ug^2(e^{7t} - 1)/7$, $\bar{g} = ge^{3t}$. An elimination of the variable t then gives the three primitive integrals $v + u^2/2$, $f - ug^2/7$, gu^{-3}. The general solution of $\hat{V}\langle F\rangle = 0$ is thus given by

$$F = \Psi\left(v + \tfrac{1}{2}u^2, f - \tfrac{1}{7}ug^2, gu^{-3}\right).$$

A direct application of the theory given above shows that the general solution of the original problem is given in implicit form by

$$\Psi_1\left(v + \tfrac{1}{2}u^2, f - \tfrac{1}{7}ug^2, gu^{-3}\right) = c_1,$$

$$\Psi_2\left(v + \tfrac{1}{2}u^2, f - \tfrac{1}{7}ug^2, gu^{-3}\right) = c_2,$$

for any C^1 choice of the functions Ψ_1, Ψ_2 such that $\partial(\Psi_1, \Psi_2)/\partial(f, g) \neq 0$. In this particular instance, we can actually go considerably further. Let us choose Ψ_1 to be independent of its third argument and Ψ_2 to be independent of its second argument:

$$\Psi_1\left(v + \tfrac{1}{2}u^2, f - \tfrac{1}{7}ug^2\right) = c_1, \qquad \Psi_2\left(v + \tfrac{1}{2}u^2, gu^{-3}\right) = c_2.$$

When the second of these is solved for g, we obtain

$$g = u^3 M_1\left(v + \tfrac{1}{2}u^2\right)$$

for some function M_1. If this is substituted into the first and f is solved for, we have

$$f = M_2\left(v + \tfrac{1}{2}u^2\right) + \tfrac{1}{7}u^7 M_1\left(v + \tfrac{1}{2}u^2\right)$$

for some function M_2. We have thus obtained explicit solutions that depend on two arbitrary functions M_1, M_2 of the single argument $v + \frac{1}{2}u^2$.

2-6. THE NATURAL LIE ALGEBRA OF $T(E_n)$

The tangent space $T(E_n)$ has a natural vector space structure that is inherited through its construction as the union over $P \in E_n$ of the vector spaces $T_P(E_n)$. The question thus arises as to whether a binary operation $(\cdot \mid \cdot)$ can be defined over $T(E_n) \times T(E_n)$ for which $T(E_n)$ becomes an algebra. The easiest way of answering this question is to use the realization of $T(E_n)$ as the collection of all derivations

(2-6.1) $$V = v^i(x^j)\,\partial_i$$

on the algebra $\Lambda^0(E_n)$.

We begin by noting that any derivation $V \in T(E_n)$ on the algebra $\Lambda^0(E_n)$ means that for any $f(x^j) \in \Lambda^0(E_n)$,

(2-6.2) $$V\langle f\rangle(x^j) = g(x^j)$$

gives a new element $g(x^j)$ of $\Lambda^0(E_n)$. Thus if

(2-6.3) $$U = u^i(x^j)\,\partial_i$$

is any other element of $T(E_n)$, we may apply U to $g(x^j)$. When this is done, we have

(2-6.4) $$U\langle g\rangle = U\langle V\langle f\rangle\rangle = u^i\partial_i\big(v^j\partial_j f\big)$$
$$= u^i\big(\partial_i v^j\big)\partial_j f + u^i v^j \partial_i \partial_j f,$$

where $\partial_i\partial_j f = \partial^2 f/\partial x^i \partial x^j$ so that

(2-6.5) $$\partial_i\partial_j f = \partial_j \partial_i f.$$

The first term on the right-hand side of (2-6.4), $u^i(\partial_i v^j)\partial_j f$, is again a derivation on $\Lambda^0(E_n)$; simply put

$$w^j(x^k) = u^i(x^k)\,\partial_i v^j(x^k)$$

to obtain $u^i(\partial_i v^j)\partial_j f = w^j \partial_j f = W\langle f\rangle$. The second term, $u^i v^j \partial_i\partial_j f$, is not a derivation; simply observe that

$$\partial_i\partial_j(pq) = \partial_i p\,\partial_j q + p\,\partial_i\partial_j q + \partial_i q\,\partial_j p + q\,\partial_i\partial_j p$$
$$\neq p\,\partial_i\partial_j q + q\,\partial_i\partial_j p.$$

It thus follows that $U\langle V\langle f\rangle\rangle$ does not define a binary operation on $T(E_n) \times T(E_n)$ to $T(E_n)$. The work thus far is not in vain, however, if we can manage to eliminate the disruptive terms $u^i v^j \partial_i \partial_j f$.

Let us apply U and V in the opposite order. This gives, in direct analogy with (2-6.4),

$$V\langle U\langle f\rangle\rangle = v^i(\partial_i u^j)\partial_j f + v^i u^j \partial_i \partial_j f. \tag{2-6.6}$$

We now subtract (2-6.6) from (2-6.4) to obtain

$$\begin{aligned} U\langle V\langle f\rangle\rangle - V\langle U\langle f\rangle\rangle &= (u^i \partial_i v^j - v^i \partial_i u^j)\partial_j f \\ &\quad + (u^i v^j - u^j v^i)\partial_i \partial_j f. \end{aligned} \tag{2-6.7}$$

If we set

$$S^{ij} = u^i v^j - u^j v^i, \qquad H_{ij} = \partial_i \partial_j f, \tag{2-6.8}$$

then

$$S^{ij} = -S^{ji}, \qquad H_{ij} = H_{ji} \tag{2-6.9}$$

where the latter follows from (2-6.5). We thus have

$$(u^i v^j - u^j v^i)\partial_i \partial_j f = S^{ij} H_{ij}. \tag{2-6.10}$$

The following lemma shows that this double sum vanishes identically.

Lemma 2-6.1. *If $S^{ij} = -S^{ji}$ and $H_{ij} = H_{ji}$, then the double sum $M = S^{ij} H_{ij}$ vanishes identically.*

Proof. We simply note that the symmetry of H_{ij} and the antisymmetry of S_{ij} in the indices (i, j) yields

$$M = S^{ij} H_{ij} = -S^{ji} H_{ij} = -S^{ji} H_{ji} = -M$$

so that $M = 0$. □

When this result is put back into (2-6.7), we see that

$$U\langle V\langle f\rangle\rangle - V\langle U\langle f\rangle\rangle = (u^i \partial_i v^j - v^i \partial_i u^j)\partial_j f. \tag{2-6.11}$$

It is now only necessary to put

$$W = (u^i \partial_i v^j - v^i \partial_i u^j)\partial_j \tag{2-6.12}$$

in order to see that

$$U\langle V\langle f\rangle\rangle - V\langle U\langle f\rangle\rangle = W\langle f\rangle \tag{2-6.13}$$

defines a binary operation on $T(E_n) \times T(E_n)$ to $T(E_n)$. The first step in defining a multiplication on $T(E_n)$ with the required properties is thus accomplished.

Lemma 2-6.2. *The binary operation* $[\cdot, \cdot]$, *defined by*

$$[U,V]\langle f\rangle = U\langle V\langle f\rangle\rangle - V\langle U\langle f\rangle\rangle \tag{2-6.14}$$

for all $f \in \Lambda^0(E_n)$, *has the representation*

$$[U,V] = \left(u^j\partial_j v^i - v^j\partial_j u^i\right)\partial_i \tag{2-6.15}$$

for every U, V *in* $T(E_n)$. *Thus* $[\cdot,\cdot]$ *defines a binary map of* $T(E_n) \times T(E_n)$ *to* $T(E_n)$.

All that now remains is to check that the binary operation $[\cdot,\cdot]$ satisfies the conditions (1-5.7) and (1-5.8) in order to conclude that $T(E_n)$ equipped with the multiplication $[\cdot,\cdot]$ forms an algebra over the real number field. This task is simplified by noting that the following result is a direct consequence of (2-6.14).

Lemma 2-6.3. *The binary operation* $[\cdot,\cdot]$ *satisfies the identity*

$$[U,V] = -[V,U] \tag{2-6.16}$$

for all U, $V \in T(E_n)$.

The final result goes as follows.

Lemma 2-6.4. *The binary operation* $[\cdot,\cdot]$ *satisfies the identities*

$$[U, a \cdot V + b \cdot W] = a \cdot [U,V] + b \cdot [U,W], \tag{2-6.17}$$

$$[a \cdot U + b \cdot V, W] = a \cdot [U,W] + b \cdot [V,W] \tag{2-6.18}$$

for all U, V, $W \in T(E_n)$ *and all* a, $b \in \mathbb{R}$.

Proof. Use of Lemma 2-6.3 shows that (2-6.18) is a consequence of (2-6.17). Thus we need only establish (2-6.17). This, however, is straightforward, for the definition (2-6.14) gives (note that $a \in \mathbb{R}$ implies $U\langle a \cdot f\rangle = a \cdot U\langle f\rangle$)

$$\begin{aligned}
[U, a \cdot V + b \cdot W]\langle f\rangle &= U\langle (a \cdot V + b \cdot W)\langle f\rangle\rangle \\
&\quad -(a \cdot V + b \cdot W)\langle U\langle f\rangle\rangle \\
&= U\langle a \cdot V\langle f\rangle + b \cdot W\langle f\rangle\rangle \\
&\quad -a \cdot V\langle U\langle f\rangle\rangle - b \cdot W\langle U\langle f\rangle\rangle \\
&= a \cdot U\langle V\langle f\rangle\rangle + b \cdot U\langle W\langle f\rangle\rangle \\
&\quad -a \cdot V\langle U\langle f\rangle\rangle - b \cdot W\langle U\langle f\rangle\rangle \\
&= a \cdot (U\langle V\langle f\rangle\rangle - V\langle U\langle f\rangle\rangle) \\
&\quad +b \cdot (U\langle W\langle f\rangle - W\langle U\langle f\rangle\rangle) \\
&= a \cdot [U, V]\langle f\rangle + b \cdot [U, W]\langle f\rangle
\end{aligned}$$

for all $f \in \Lambda^0(E_n)$. □

The multiplication that is defined on $T(E_n)$ by (2-6.14) occurs so often that it has been given a special name.

Definition. The multiplication $[\cdot,\cdot\,]$ that is defined on $T(E_n)$ by

(2-6.19) $$[U, V]\langle f\rangle = U\langle V\langle f\rangle\rangle - V\langle U\langle f\rangle\rangle$$

for all $f \in \Lambda^0(E_n)$ is called the *Lie product*.

It is now customary to use the notation $[\cdot,\cdot\,]$ for the Lie product, and this is why we have deviated from the notation established in Chapter One where $(\cdot\,|\,\cdot)$ was used. We can now state our principal result in the following form.

Theorem 2-6.1. *The tangent space, $T(E_n)$, equipped with the Lie product $[\cdot, \cdot]$, is an algebra over the real number field.*

The algebra induced on $T(E_n)$ by the Lie product possesses certain properties that makes it quite different from the algebras that you may have already encountered. We already know that $T(E_n)$ is an *anticommutative* algebra, for Lemma 2-6.3 has shown that $[U, V] = -[V, U]$. This is not too unusual, for the classic vector product of two vectors in three-dimensional Euclidean space exhibits this property: $\vec{a} \times \vec{b} = -\vec{b} \times \vec{a}$. The telling result is that $T(E_n)$ is a nonassociative algebra, as the following lemma demonstrates.

Lemma 2-6.5. *The Lie product satisfies the Jacobi identity*

$$[U,[V,W]]+[V,[W,U]]+[W,[U,V]]=0. \tag{2-6.20}$$

The proof of this result is a straightforward but lengthy calculation based on the definition (2-6.15). After each of the three terms is evaluated by this formula, it is found that the sum given by (2-6.20) leads to a collection of terms all of which cancel in pairs. The details of this calculation are left to the reader as an exercise.

The lack of associativity now is established easily. If we use the anticommutativity of the Lie product in the third term in (2-6.20), we have

$$[U,[V,W]]+[V,[W,U]]-[[U,V],W]=0,$$

and hence

$$[U,[V,W]]-[[U,V],W]=-[V,[W,U]]. \tag{2-6.21}$$

On the other hand, the Lie product is associative only if

$$[U,[V,W]]-[[U,V],W]\equiv 0,$$

which is definitely not the case here. Particular care must thus be exercised in the use of the algebra $T(E_n)$ for it is something of a revelation just how many familiar results cease to hold as soon as the multiplication becomes nonassociative.

Anticommutative algebras with the Jacobi identity occur so frequently that they have been given a special name that derives from their historical origins in the work of S. Lie on continuous groups.

Definition. An algebra whose binary operation $(\cdot\mid\cdot)$ satisfies

$$(U|V)=-(V|U), \tag{2-6.22}$$

$$(U|(V|W))+(V|(W|U))+(W|(U|V))=0 \tag{2-6.23}$$

is a *Lie algebra*.

The basic Theorem 2-6.1 can thus be restated in the following equivalent form.

Theorem 2-6.2. *The tangent space, $T(E_n)$, equipped with the Lie product is a Lie algebra.*

The following result is recorded for future reference.

Lemma 2-6.6. *If $p(x^j)$ and $q(x^j)$ belong to $\Lambda^0(E_n)$, then*

(2-6.24) $$[pU, qV] = pq[U, V] + (pU\langle q\rangle)V - (qV\langle p\rangle)U.$$

Proof. The result follows directly from (2-6.14):

$$\begin{aligned}[pU, qV]\langle f\rangle &= pU\langle qV\langle f\rangle\rangle - qV\langle pU\langle f\rangle\rangle \\ &= pU\langle q\rangle V\langle f\rangle + pqU\langle V\langle f\rangle\rangle - qV\langle p\rangle U\langle f\rangle - qpV\langle U\langle f\rangle\rangle \\ &= pq\{U\langle V\langle f\rangle\rangle - V\langle U\langle f\rangle\rangle\} + pU\langle q\rangle V\langle f\rangle - qV\langle p\rangle U\langle f\rangle \\ &= pq[U, V]\langle f\rangle + \{pU\langle q\rangle V - qV\langle p\rangle U\}\langle f\rangle.\end{aligned}$$ □

2-7. LIE SUBALGEBRA OF $T(E_n)$ AND SYSTEMS OF LINEAR FIRST-ORDER PARTIAL DIFFERENTIAL EQUATIONS

The concept of a Lie subalgebra of a Lie algebra is possibly even more useful than the corresponding concept of a subspace of a vector space.

Definition. A *Lie subalgebra* of a Lie algebra $\mathscr{A}$ is a vector subspace of $\mathscr{A}$, considered as a vector space over $\Lambda^0(E_n)$ (as a $\Lambda^0(E_n)$ module), that is closed under the multiplication operation of the algebra $\mathscr{A}$.

Suppose that $U_1, \ldots, U_r$ belong to $T(E_n)$ and let $\mathscr{S}$ denote the vector subspace of $T(E_n)$ that is spanned by the elements $U_1, \ldots, U_r$. For simplicity, let us write

$$\{U_1, \ldots, U_r\} = \{U_a\}$$

where it is understood that the index a runs from 1 through r. The definition of a Lie subalgebra shows that the subspace $\mathscr{S}$ forms a Lie subalgebra only if $\mathscr{S}$ is closed under $[\cdot, \cdot]$; that is, the Lie product of any two elements of $\mathscr{S}$ belongs to $\mathscr{S}$. Since $\mathscr{S}$ is spanned by $\{U_a\}$, this can be the case if and only if $[U_a, U_b]$ can be expressed as a linear combination of the U_a's.

Lemma 2-7.1. *If $\mathscr{S}$ is a subspace of $T(E_n)$ that is spanned by the r elements $\{U_1, \ldots, U_r\} = \{U_a, a = 1, \ldots, r\}$, then $\mathscr{S}$ is a Lie subalgebra of $T(E_n)$ if and only if there exist r^3 elements $\{C^e_{ab}(x^j)\}$ of $\Lambda^0(E_n)$ such that*

(2-7.1) $$[U_a, U_b] = C^e_{ab}U_e,$$

(2-7.2) $$C^e_{ab} = -C^e_{ba},$$

and

(2-7.3)

$$0 = \left(U_a\langle C^h_{bc}\rangle + U_b\langle C^h_{ca}\rangle + U_c\langle C^h_{ab}\rangle + C^e_{bc}C^h_{ae} + C^e_{ca}C^h_{be} + C^e_{ab}C^h_{ce}\right)U_h\langle f\rangle$$

for all $f \in \Lambda^0(E_n)$.

Proof. We have already seen that $\mathscr{S}$ can be a subalgebra of $T(E_n)$ only if $[U_a, U_b]$ can be expressed as a linear combination of the U's, and this is exactly what (2-7.1) says. Since $[U_a, U_b] = -[U_b, U_a]$, we must also have (2-7.2). Similarly, the Jacobi identity, (2-6.20) and (2-6.24), give the conditions

$$\begin{aligned} 0 &= [U_a, [U_b, U_c]] + [U_b, [U_c, U_a]] + [U_c, [U_a, U_b]] \\ &= [U_a, C^e_{bc}U_e] + [U_b, C^e_{ca}U_e] + [U_c, C^e_{ab}U_e] \\ &= U_a\langle C^e_{bc}\rangle U_e + U_b\langle C^e_{ca}\rangle U_e + U_c\langle C^e_{ab}\rangle U_e \\ &\quad + C^e_{bc}[U_a, U_e] + C^e_{ca}[U_b, U_e] + C^e_{ab}[U_c, U_e]; \end{aligned}$$

that is, with e set equal to h in the first three terms,

$$0 = \left(U_a\langle C^h_{bc}\rangle + U_b\langle C^h_{ca}\rangle + U_c\langle C^h_{ab}\rangle + C^e_{bc}C^h_{ae} + C^e_{ca}C^h_{be} + C^e_{ab}C^h_{ce}\right)U_h\langle f\rangle$$

must hold for all $f \in \Lambda^0(E_n)$. □

This lemma shows that a Lie subalgebra is a much more constrained structure than a subspace; simply look at the additional requirements (2-7.1) through (2-7.3). For example, the subspace of $T(E_3)$ that is spanned by the elements

$$U_1 = x\partial_y - y\partial_x, \qquad U_2 = y\partial_z - z\partial_y$$

gives

$$[U_1, U_2] = x\partial_z - z\partial_x \stackrel{\text{def}}{=} U_3.$$

However, $xU_2 - yU_3 + zU_1 \equiv 0$ and the coefficient functions in this expression all vanish only when all of the U's vanish, namely at $x = y = z = 0$. Hence the subspace spanned by U_1 and U_2 forms a Lie subalgebra of $T(E_3)$. On the other hand, for the subspace spanned by

$$V_1 = x\partial_y - y\partial_x + \partial_z, \qquad V_2 = \partial_y$$

we have

$$[V_1, V_2] = \partial_x \overset{\text{def}}{=} V_3.$$

Since $\alpha V_1 + \beta V_2 + \gamma V_3 = 0$ only when $\alpha = \beta = \gamma = 0$, the subspace spanned by V_1 and V_2 does *not* form a Lie subalgebra of $T(E_3)$. We must thus consider the enlarged set V_1, V_2, V_3. However, $\partial_z = V_1 + yV_3 - xV_2$ and hence the subspace of $T(E_3)$ that is spanned by V_1, V_2, V_3 is also spanned by $\partial_x, \partial_y, \partial_z$; the completion of the subspace spanned by V_1, V_2 to a Lie subalgebra gives the original Lie algebra $T(E_3)$.

It should be specifically noted that *so long as the* $\{C^e_{ab}\}$ *in* (2-7.1) *are not restricted to be constants, we can form linear combinations of the elements of* $T(E_n)$ *whose coefficients are themselves elements of* $\Lambda^0(E_n)$; that is, the vector subspace may be formed by terms of the form

$$\alpha_1(x^j)U_1 + \alpha_2(x^j)U_2 + \cdots;$$

it is a submodule of the $\Lambda^0(E_n)$ module $T(E_n)$. The formation of a Lie subalgebra is invariant for replacements of constants by functions in forming linear combinations. This follows from (2-6.24), for

$$[\alpha U_a, \beta U_b] = \alpha\beta[U_a, U_b] + (\alpha U_a\langle\beta\rangle)U_b - (\beta U_b\langle\alpha\rangle)U_a$$

shows that if the subspace spanned by $\{U_1, \ldots, U_r\}$ forms a Lie subalgebra, then the space spanned by $\{\alpha^a_1(x^j)U_a, \ldots, \alpha^a_r(x^j)U_a\}$ also forms a Lie subalgebra.

Suppose that we start with a subspace $\mathscr{S}_1$ of $T(E_n)$ that is spanned by the elements $U_1, \ldots, U_r$ that does not form a Lie subalgebra. In this event, there will be certain Lie products of the elements $U_1, \ldots, U_r$ that can not be expressed as linear combinations of the set $U_1, \ldots, U_r$. If there are s such Lie products, each of them is a new element of $T(E_n)$. Designating these new elements by $U_{r+1}, \ldots, U_{r+s}$, we form the new set of elements $U_1, \ldots, U_r$, $U_{r+1}, \ldots, U_{r+s}$. Designate the subspace of $T(E_n)$ that is spanned by these $r + s$ elements as $\mathscr{S}_2$. Now either $\mathscr{S}_2$ forms a Lie subalgebra or it does not. If it does, then we are through. If it does not, we adjoin all new elements of $T(E_n)$ that are formed by taking Lie products to obtain a new set of spanning elements for the new subspace $\mathscr{S}_3$. Clearly, this process can be continued indefinitely. When it is performed, we say that the original subspace $\mathscr{S}_1$ has been *completed* to a Lie subalgebra $\mathscr{S}$.

The notion of the completion of a subspace of $T(E_n)$ to a Lie subalgebra is useful in many contexts. One of the simplest is that associated with the solvability of a *system* of first-order linear partial differential equations. Suppose that we are given r elements $\{U_a\}$ of $T(E_n)$, and we wish to find all functions $f(x^j)$ on E_n that satisfy the system of r first-order linear partial differential equations

(2-7.4) $$U_a\langle f\rangle = 0, \qquad a = 1,\dots,r.$$

Each of these first-order equations can be solved, at least in principle, by the method of characteristics that was given in Section 2-5. What we now have to do is to determine whether there exist solutions of any one of them that are also solutions of the remaining $r-1$ equations of the system (2-7.4); that is, whether the general solution

$$f = \Psi(g_{1,a}(x^j),\dots,g_{n-1,a}(x^j))$$

of the ath equation can be forced to satisfy the remaining $r-1$ equations by an appropriate choice of the function Ψ.

Suppose that f is a solution of $U_a\langle f\rangle = 0$ and also a solution of $U_b\langle f\rangle = 0$. It then follows that f will satisfy $U_a\langle U_b\langle f\rangle\rangle = 0$ and $U_b\langle U_a\langle f\rangle\rangle = 0$. These conditions contain second derivatives of the function f. The dependence on second derivatives is easily eliminated, however, for we need only form

(2-7.5) $$[U_a, U_b]\langle f\rangle = U_a\langle U_b\langle f\rangle\rangle - U_b\langle U_a\langle f\rangle\rangle = 0.$$

Thus *if f is to satisfy the system $U_a\langle f\rangle = 0$, $a = 1,\dots,r$, then it must also satisfy the linear first-order partial differential equations*

(2-7.6) $$[U_a, U_b]\langle f\rangle = 0, \qquad 1 \le a < b \le r.$$

If the subspace $\mathscr{S}_1$ of $T(E_n)$ that is spanned by $\{U_a\}$ is a Lie subalgebra of $T(E_n)$, then

(2-7.7) $$[U_a, U_b]\langle f\rangle = C^e_{ab}U_e\langle f\rangle,$$

in which case *satisfaction of the conditions* (2-7.6) *is implied by satisfaction of the original equations* $U_a\langle f\rangle = 0, a = 1,\dots,r$. On the other hand, if the subspace $\mathscr{S}_1$ does not form a Lie subalgebra of $T(E_n)$, then we have to adjoin new equations to our original system that come from the conditions (2-7.6). This yields a subspace $\mathscr{S}_2$ of $T(E_n)$ that is spanned by $\{U_1,\dots,U_r,U_{r+1},\dots,U_{r+s}\}$ if there are s new equations. The process just described now has to be repeated starting with the subspace $\mathscr{S}_2$; that is, we have to complete the subspace $\mathscr{S}_1$ to a Lie subalgebra of $T(E_n)$.

Theorem 2-7.1. *Let $\mathscr{S}_1$ be a subspace of $T(E_n)$ that is spanned by a given system of r elements $U_1,\dots,U_r$ and let $\mathscr{S}$ be the completion of $\mathscr{S}_1$ to a Lie subalgebra of $T(E_n)$. Necessary conditions for the existence of a solution to the* *system of first-order linear partial differential equations*

(2-7.8) $$U_a\langle f\rangle = 0, \qquad a = 1,\dots,r$$

is that f satisfy

(2-7.9) $$U\langle f\rangle = 0$$

for every U in the subspace $\mathscr{S}$. *Thus a solution of* (2-7.8) *exists only if there is at least one function on* E_n *that is annihilated by the Lie subalgebra that is the completion of the subspace* $\mathscr{S}_1$.

Theorem 2-7.1 shows that there is quite a difference between a single first-order partial differential equation and a system of two or more such equations. It is also clear that everything depends on the structure of the Lie subalgebra $\mathscr{S}$ of $T(E_n)$ that is obtained by completion of the subspace $\mathscr{S}_1$. Suppose that $\mathscr{S}$ has a basis that is comprised of k elements

$$V_1, V_2, \dots, V_k$$

with $k \geq r$ necessarily. We then have the problem of solving the system of k equations $V_b\langle f\rangle = 0, \quad b = 1,\dots,k$. On the other hand, for any P in E_n, $\dim(T_P(E_n)) = n$, so that $k \leq n$. We thus have two cases, $k < n$ and $k = n$. For the case $k = n$, we have n linearly independent elements $\{V_1,\dots,V_n\} = \{V_i\}$ with

$$V_i = v_i^j(x^k)\,\partial_j, \qquad \det\big(v_i^j(x^k)\big) \neq 0$$

on some open set N in E_n. Thus for all P in N we can solve

$$V_i\langle f\rangle = v_i^j\partial_j f = 0$$

for $\partial_j f$ and obtain $\partial_j f = 0$, $j = 1,\dots,n$. In this event, the only solution is the trivial solution $f =$ constant.

Corollary 2-7.1. *Let* $\mathscr{S}_1$ *be a subspace of* $T(E_n)$ *that is associated with a given system of first order linear partial differential equations, as in Theorem* 2-7.1, *and let* $\mathscr{S}$ *be the completion of* $\mathscr{S}_1$ *to a Lie subalgebra. If* $\mathscr{S}$ *has an n-dimensional basis on some neighborhood N of* E_n, *then the only solution of the given system of partial differential equations on N is the trivial solution* $f(x^j) =$ *constant.*

The result just obtained makes clear that it is the case $k < n$ that is of interest. In this case, we have the subspace $\mathscr{S}$ of dimension $k < n$ that is the completion of the subspace $\mathscr{S}_1$ to a Lie algebra. We thus have the $k < n$

vector fields $V_1, \ldots, V_k$ that are linearly independent over $\Lambda^0(E_n)$ and are such that $\mathscr{S} = \text{span}(V_1, \ldots, V_k)$,

$$[V_a, V_b] = C^e_{ab} V_e, \qquad a, b, e = 1, \ldots, k.$$

The problem of solving the original system of first-order partial differential equations (2-7.4) now becomes the problem of finding all functions f of class C^1 on E_n such that $V\langle f \rangle = 0$ for all $V \in \mathscr{S}$.

The k vectors $V_1, \ldots, V_k$ are a basis for the subspace $\mathscr{S}$, but are not necessarily the best or most convenient basis. Another basis can be obtained by the relations

$$\overline{V}_\alpha = K^a_\alpha V_a, \qquad \alpha = 1, \ldots, k,$$

with $K^a_\alpha \in \Lambda^0(E_n)$ such that $\det(K^a_\alpha) \neq 0$. Let the entries of the inverse matrix of $((K^a_\alpha))$ be denoted by K^β_a:

$$K^a_\alpha K^\beta_a = \delta^\beta_\alpha, \qquad K^a_\alpha K^\alpha_b = \delta^a_b, \qquad V_a = K^\alpha_a \overline{V}_\alpha.$$

We then have

$$\begin{aligned}
[\overline{V}_\alpha, \overline{V}_\beta] &= K^a_\alpha K^b_\beta [V_a, V_b] + \left(K^a_\alpha V_a\langle K^b_\beta\rangle\right) V_b - \left(K^b_\beta V_b\langle K^a_\alpha\rangle\right) V_a \\
&= \left\{ K^a_\alpha K^b_\beta K^\gamma_e C^e_{ab} + K^a_\alpha K^\gamma_b V_a\langle K^b_\beta\rangle - K^b_\beta K^\gamma_a V_b\langle K^a_\alpha\rangle \right\} \overline{V}_\gamma \\
&= \overline{C}^\gamma_{\alpha\beta} \overline{V}_\gamma,
\end{aligned}$$

and hence the formation of a Lie algebra is uneffected by such a change of basis although the quantities $\overline{C}^\gamma_{\alpha\beta}$ will generally be quite different from the quantities C^e_{ab} with which we start. Now, it is well known that we can arrive at a new basis with a maximal number of zero components by the Gauss reduction process applied to the matrix whose rows are the components of the vectors $V_1, V_2, \ldots, V_k$. After a possible reordering of the independent variables, we can thus obtain a basis for $\mathscr{S}$ of the form

$$\overline{V}_\alpha = \partial_\alpha + \sum_{m=k+1}^{n} v^m_\alpha \partial_m, \qquad \alpha = 1, \ldots, k.$$

A basis with this structure is said to be in *Jacobi normal form* and the system of partial differential equations $\overline{V}_\alpha\langle f \rangle = 0$ is also said to be in *Jacobi normal form*.

Gauss reduction of Jacobi normal form carries with it a very surprising consequence. Direct computation of the Lie product of any two $\overline{V}$'s gives

$$[\overline{V}_\alpha, \overline{V}_\beta] = \sum_{m=k+1}^{n} \left\{ \overline{V}_\alpha\langle v^m_\beta\rangle - \overline{V}_\beta\langle v^m_\alpha\rangle \right\} \partial_m$$

and the operator on the right-hand side does not contain any of the operators

$\partial_1, \partial_2, \ldots, \partial_k$. We know, however, that $[\overline{V}_\alpha, \overline{V}_\beta] = \overline{C}^\gamma_{\alpha\beta}\overline{V}_\gamma$ and the right-hand sides of these equations will contain the operators $\partial_1, \partial_2, \ldots, \partial_k$ unless $\overline{C}^\gamma_{\alpha\beta} = 0$. Accordingly, *a basis $\overline{V}_1, \ldots, \overline{V}_k$ of a Lie algebra over $\Lambda^0(E_n)$ in Jacobi normal form satisfies*

$$[\overline{V}_\alpha, \overline{V}_\beta] = 0;$$

that is,

$$\overline{V}_\alpha\langle v^m_\beta\rangle = \overline{V}_\beta\langle v^m_\alpha\rangle.$$

The result just obtained leads to a direct process of integration of a system of partial differential equations in Jacobi normal form:

$$\overline{V}_\alpha\langle f\rangle = 0, \qquad \alpha = 1, \ldots, k.$$

We start with the first of these equations,

$$\overline{V}_1\langle f\rangle = 0.$$

Since this is a linear first-order equation, it can be solved by the method of characteristics. This gives the representation

$$f = \Psi(g_1(x^i), g_2(x^i), \ldots, g_{n-1}(x^i)),$$

where the g's are a system of $n-1$ independent primitive integrals of $\overline{V}_1\langle f\rangle = 0$. Starting with this representation, we put it into the second equation, $\overline{V}_2\langle f\rangle = 0$. This gives

$$\overline{V}_2\langle g_1\rangle \frac{\partial\Psi}{\partial g_1} + \overline{V}_2\langle g_2\rangle \frac{\partial\Psi}{\partial g_2} + \cdots + \overline{V}_2\langle g_{n-1}\rangle \frac{\partial\Psi}{\partial g_{n-1}} = 0,$$

which will be a linear partial differential equation for the determination of Ψ as a function of the $n-1$ variables $g_1, \ldots, g_{n-1}$ provided we can show that the coefficients $\{\overline{V}_2\langle g_s\rangle\}$ are functions of the $n-1$ variables $g_1, \ldots, g_{n-1}$ *only*. However, because $\{\overline{V}_\alpha\}$ are in Jacobi normal form,

$$\overline{V}_1\langle \overline{V}_2\langle g_s\rangle\rangle = \overline{V}_2\langle \overline{V}_1\langle g_s\rangle\rangle = \overline{V}_2\langle 0\rangle = 0,$$

and hence each $\overline{V}_2\langle g_s\rangle$ is a solution of the equation $\mathrm{V}_1\langle u\rangle = 0$. Thus, since every solution of $\overline{V}_1\langle u\rangle = 0$ is expressible as a function of the primitive integrals $g_1, \ldots, g_{n-1}$, we see that

$$\overline{V}_2\langle g_s\rangle = h_s(g_1, \ldots, g_{n-1}), \qquad s = 1, \ldots, n-1.$$

The linear partial differential equation for $\Psi(g_1, \ldots, g_{n-1})$ can be solved by the method of characteristics. This will give

$$\Psi = \Phi(m_1, \ldots, m_{n-2})$$

where $\{m_1, \ldots, m_{n-2}\}$ is a system of $n - 2$ primitive integrals, each of which is a function of the variables $g_1, \ldots, g_{n-1}$. Continuing in this fashion with the next equation, and so forth, we obtain the following basic result.

Theorem 2-7.2. *Let $\mathscr{S}$ be the Lie algebra that is generated by the completion of a system of linear first-order partial differential equations on E_n and let $\mathscr{S}$ have a k-dimensional basis on a neighborhood N of E_n with $k < n$. The given system of partial differential equations has exactly $n - k$ simultaneous primitive integrals $h_1(x^i), \ldots, h_{n-k}(x^i)$ on N and the general solution on N is given by $\Phi(h_1(x^i), \ldots, h_{n-k}(x^i))$.*

We have seen that the subspace of $T(E_3)$ spanned by

$$U_1 = x\partial_y - y\partial_x, \qquad U_2 = y\partial_z - z\partial_y$$

forms a Lie subalgebra of $T(E_3)$. The method of characteristics applied to $U_1\langle f\rangle = 0$ yields $f = \Psi(x^2 + y^2, z)$, while $U_2\langle f\rangle = 0$ yields $f = \phi(x, y^2 + z^2)$. Thus, the common integrals of $U_1\langle f\rangle = U_2\langle f\rangle = 0$ are given by $f = \rho(x^2 + y^2 + z^2)$, namely an arbitrary C^1 function of the $3 - 2 = 1$ argument $x^2 + y^2 + z^2$. On the other hand, the completion of the subspace spanned by

$$V_1 = x\partial_y - y\partial_x + \partial_z, \qquad V_2 = \partial_y$$

to a Lie subalgebra gives all of $T(E_3)$. Thus f satisfies $V_1\langle f\rangle = V_2\langle f\rangle = 0$ only if $\partial_1 f = \partial_2 f = \partial_3 f = 0$, in which case the only solutions are the trivial ones $f =$ constant.

Another occurrence of a Lie subalgebra, which arises in almost any physical discipline, is in the theory of Lie groups. In fact, the whole theory actually started there. *A Lie subalgebra of $T(E_n)$ which is spanned by a finite number of elements $V_1, \ldots, V_r$ such that*

$$[V_a, V_b] = C^e_{ab} V_e, \qquad \partial_i C^e_{ab} = 0$$

generates a Lie group that can be realized as a group of point transformations acting on E_n (a group of transformations that moves the points of E_n along the orbits of all linear combinations of the vector fields $V_1, \ldots, V_r$ with constant coefficients). In this event, the C's are referred to as the *structure constants* of the group.

The careful reader will have perceived that we started with vector spaces over the real number field, $\mathbb{R}$, and have progressed to a module (vector spaces over the associative algebra $\Lambda^0(E_n)$ of C^∞ functions on E_n). A case in point is

$T(E_n)$, for it is closed under multiplication of its elements by elements of $\Lambda^0(E_n)$ rather than just by constants (elements of $\mathbb{R}$); if $V \in T(E_n)$ then $f(x^j)V \in T(E_n)$ for any $f(x^j) \in \Lambda^0(E_n)$. The meaning of linear dependence in a vector space over $\mathbb{R}$ and in a vector space over $\Lambda^0(E_n)$ are quite different, however.

Definition. A collection $\{V_a, a = 1, \ldots, r\}$ of elements of $T(E_n)$ is said to be *linearly dependent over* $\mathbb{R}$ (linearly dependent with constant coefficients) if and only if

$$c_1V_1 + c_2V_2 + \cdots + c_rV_r = 0$$

and

$$\partial_i c_a = 0, \qquad a = 1, \ldots, r$$

can be satisfied simultaneously at every point of E_n without all of the c's being zero. A collection of elements of $T(E_n)$ is said to be *linearly independent over* $\mathbb{R}$ if it is not linearly dependent over $\mathbb{R}$.

Definition. A collection $V_1, \ldots, V_r$ of elements of $T(E_n)$ is said to be *linearly dependent over* $\Lambda^0(E_n)$ if and only if

$$f_1(x^j)V_1 + f_2(x^j)V_2 + \cdots + f_r(x^j)V_r = 0$$

can be satisfied simultaneously at every point of E_n by a collection of r elements $\{f_a(x^j)\}$ of $\Lambda^0(E_n)$ not all of which vanish identically over E_n. A collection of elements of $T(E_n)$ is said to be *linearly independent over* $\Lambda^0(E_n)$ if and only if it is not linearly dependent over $\Lambda^0(E_n)$.

We first note that there can be at most n elements in a linearly independent system over $\Lambda^0(E_n)$, because each $T_P(E_n)$ for each $P \in E_n$ is n-dimensional. This is also seen from the fact that any element of $T(E_n)$ is a linear combination over $\Lambda^0(E_n)$ of the natural basis $\{\partial_i\}$. On the other hand, there can be significantly more that n elements of $T(E_n)$ that are linearly independent over $\mathbb{R}$. For example,

$$\partial_x, \partial_y, \partial_z, x\partial_y - y\partial_x, y\partial_z - z\partial_y, z\partial_x - x\partial_z$$

is a system of six elements of $T(E_3)$ that is linearly independent over $\mathbb{R}$. Simply note that $x\partial_y - y\partial_x$ cannot be written as a linear combination of $\partial_x, \partial_y, \partial_z$ with constant coefficients for all $(x, y, z) \in E_3$.

2-8. BEHAVIOR UNDER MAPPINGS

Our considerations have been confined thus far to a domain space E_n with a fixed (Cartesian) coordinate cover. The question thus naturally arises as to what happens when we change the coordinate cover. As with most things the obvious question is not necessarily the easiest to answer, so we pose a less obvious but easier question to answer first: what happens when we perform mappings from E_n to E_m?

As before, we assume that E_n is covered by the (x^i) coordinate system. Let E_m be covered by the (y^A) coordinate system with capital Latin indices ranging over the integers from 1 to m. If Φ is a smooth map from E_n to E_m that is represented by the m functions $\{\phi^A(x^j)\}$, then we write

$$\Phi: E_n \to E_m | y^A = \phi^A(x^j). \tag{2-8.1}$$

When we recall that $T(E_n)$ is the collection of all derivations on the algebra $\Lambda^0(E_n)$, it is clear that the behavior of elements of $T(E_n)$ is to a great extent controlled by the relationship between the two algebras $\Lambda^0(E_n)$ and $\Lambda^0(E_m)$ that is induced by the mapping Φ. We therefore take up this matter first.

We know that any element $f(x^j)$ of $\Lambda^0(E_n)$ can be viewed as generating a mapping F from E_n to $\mathbb{R}$,

$$F: E_n \to \mathbb{R} | \tau = f(x^j), \tag{2-8.2}$$

and likewise, any element $g(y^A)$ of $\Lambda^0(E_m)$ can be viewed as generating a mapping G from E_m to $\mathbb{R}$,

$$G: E_m \to \mathbb{R} | \tau = g(y^A). \tag{2-8.3}$$

Combining these three mappings generates the graphic situation shown in Fig. 7. It is clear from this figure, or from the fact that $\mathscr{R}(\Phi) = \mathscr{D}(G) = E_m$, that we may compose the mappings G and Φ. This, however, yields the map

$$H = G \circ \Phi: E_n \to \mathbb{R} | \tau = g(\phi^A(x^j)). \tag{2-8.4}$$

When this is done for every g in $\Lambda^0(E_m)$, a map from $\Lambda^0(E_n)$ to $\Lambda^0(E_m)$ is obtained, that is commonly denoted by Φ^*;

$$\Phi^*: \Lambda^0(E_m) \to \Lambda^0(E_n) | H = G \circ \Phi;\ h(x^j) = g(\phi^A(x^j)). \tag{2-8.5}$$

Thus, *any map Φ from E_n to E_m induces a map Φ^* of any element of* $\Lambda^0(E_m)$ onto an element of $\Lambda^0(E_n)$; that is, Φ induces the map Φ^* that is in the *opposite* direction to that of Φ. This gives rise to the graphic situation that is shown in Fig. 8. The important thing to be noted here is that there is a unique

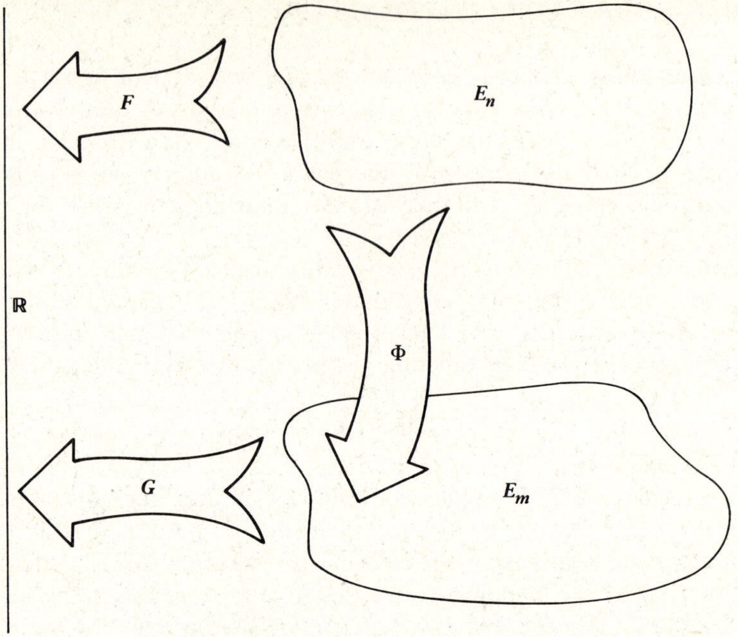

FIGURE 7. The maps Φ, F and G

element $h(x^j) = g(\phi^A(x^j))$ of $\Lambda^0(E_n)$, for any given element $g(y^A)$ of $\Lambda^0(E_m)$, that is induced by the given map Φ.

We are now in a position to determine how a vector field behaves under the map Φ of E_n to E_m that is given by (2-8.1). Let V be an element of $T(E_n)$, so that we have

$$V = v^i(x^j)\,\partial_i \tag{2-8.6}$$

for some given n-tuple of functions $\{v^i(x^j)\}$. Since V acts on all elements of $\Lambda^0(E_n)$, it certainly acts on those elements $h(x^j) = g(\phi^A(x^j))$ that are induced by Φ^*, and we have

$$V\langle h\rangle = v^i(x^j)\,\partial_i g\big(\phi^A(x^j)\big) = v^i\,(x^j)\,\partial_i\phi^A\partial_A g|_{y=\phi(x)}. \tag{2-8.7}$$

Now, set

$$v^A(x^j) = v^i(x^j)\,\partial\phi^A(x^j)/\partial x^i, \tag{2-8.8}$$

in which case (2-8.7) becomes

$$V\langle h\rangle = v^A(x^j)\,\partial_A g\,(y^B)|_{y=\phi(x)}. \tag{2-8.9}$$

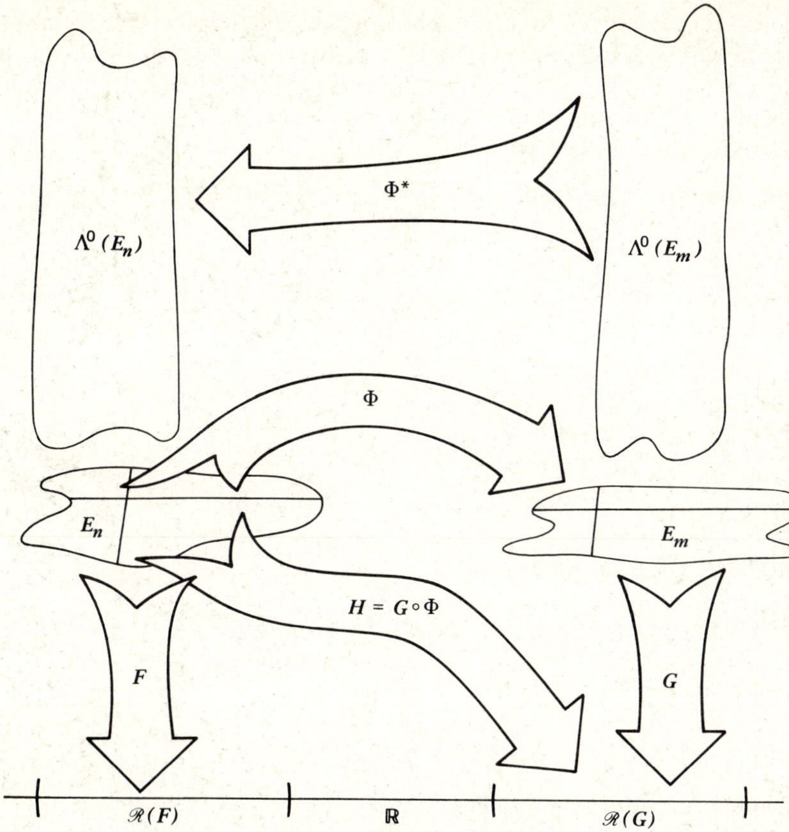

FIGURE 8. The induced map Φ^*

On the other hand, a generic element of $T(E_m)$ looks like

(2-8.10) $$`V\langle g\rangle = `v^A(y^B)\,\partial_A g(y^B)$$

for any $g \in \Lambda^0(E_m)$. A comparison of (2-8.9) and (2-8.10) shows that we will have

(2-8.11) $$V\langle h\rangle = `V\langle g\rangle, \qquad h = g\circ\Phi$$

if and only if we may achieve satisfaction of the relations

(2-8.12) $$`v^A(y^B) = v^A(x^j) = v^i(x^j)\,\partial\phi^A(x^j)/\partial x^i$$

as a consequence of the relations $y^A = \phi^A(x^j)$ that define the map Φ. In other words we have to be able to solve the equations

(2-8.13) $$y^A - \phi^A(x^j) = 0$$

for the x's as functions of the y's. The implicit function theorem (Section 1.3) tells us that this is only possible if $m = n$, in which case $E_m = {}^{`}E_n$ with ${}^{`}E_n$ denoting a replica of E_n. Further, the inverse mapping theorem (Section 1.3) also requires satisfaction of the condition $\det(\partial_i \phi^A(x^j)) \neq 0$ in order to be able to determine the inverse mapping

$$\Phi^{-1} : {}^{`}E_n = E_m \to E_n | x^i = \phi^{-1i}(y^B). \tag{2-8.14}$$

Theorem 2-8.1. *If Φ is an invertible mapping of E_n to ${}^{`}E_n$, then Φ induces the map*

$$\Phi_* : T(E_n) \to T({}^{`}E_n) \tag{2-8.15}$$

such that

$$V\langle h \rangle = {}^{`}V\langle g \rangle, \qquad h = g \circ \Phi \tag{2-8.16}$$

for all $g \in \Lambda^0({}^{`}E_n)$, in which case

$$V = v^i(x^j)\,\partial_i, \qquad {}^{`}V = {}^{`}v^A(y^B)\,\partial_A \tag{2-8.17}$$

are related by

$${}^{`}v^A(y^B) = \{v^i \partial_i \phi^A\} \circ \Phi^{-1}. \tag{2-8.18}$$

The geometric situation is shown in Fig. 9.

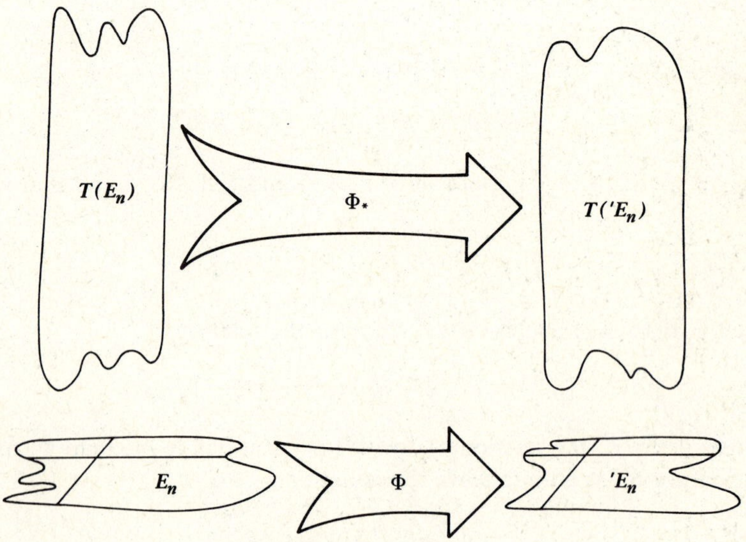

FIGURE 9. The map Φ_*

The "fly in the ointment" in these considerations is obviously the need to express the coefficient functions $v^A(x^j)$ *as functions of the* y's so that $v^A(x^j)\partial_A$ becomes a derivation $`v^A(y^B)\partial_A$ on $\Lambda^0(E_m)$. In other words, implementation of the requirement that we map whole vector fields to vector fields. If we back off of this requirement, then everything becomes much simpler. A case in the extreme is the tangent field to a specific curve in E_n. In this instance, we have the mappings

(2-8.19) $$\Gamma: J \to E_n | x^i = \gamma^i(t),$$

(2-8.20) $$\Phi: E_n \to E_m | y^A = \phi^A(x^i),$$

and the induced composite map (see Fig. 10)

(2-8.21) $$`\Gamma = \Phi \circ \Gamma: J \to E_m | y^A = \phi^A(\gamma^i(t)).$$

It is then a simple matter to see that

(2-8.22) $$T\Gamma: J \to T_{\Gamma(J)}(E_n) | v^i(t) = d\gamma^i(t)/dt,$$

(2-8.23) $$`(T\Gamma): J \to T_{\Gamma(J)}(E_m) | `v^A(t) = \frac{d}{dt}\phi^A(\gamma^i(t)),$$

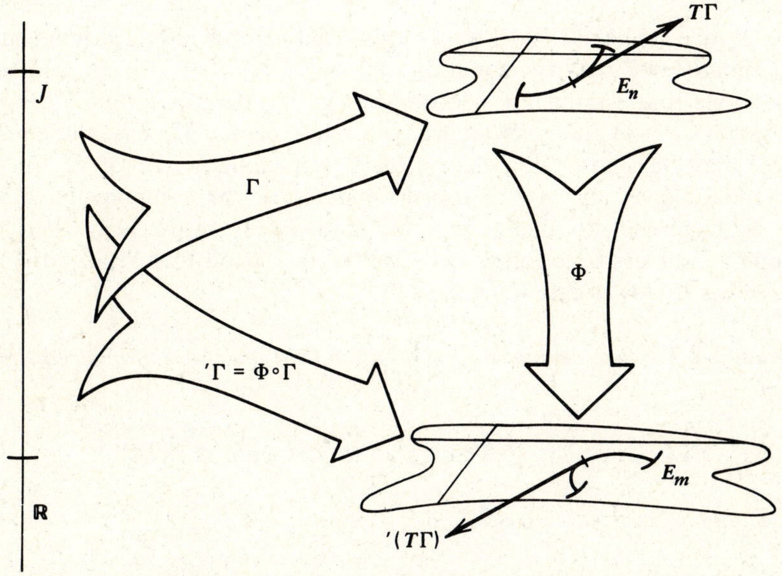

FIGURE 10. The map $T\Phi$ of the tangent field to a curve

and hence

$$(2\text{-}8.24)\quad {}^{`}v^A(t) = \left.\frac{\partial\phi^A(x^j)}{\partial x^k}\right|_{x=\gamma} d\gamma^k(t)/dt = \left.\frac{\partial\phi^A(x^j)}{\partial x^k}\right|_{x=\gamma(t)} v^k(t).$$

Thus, the map Φ induces a map of $T\Gamma$ onto ${}^{`}(T\Gamma)$ that is given by (2-8.24). It follows that a map Φ from any E_n to any E_m induces a map of any one orbit of a given vector field on E_n onto the tangent field to the image in E_m of the given orbit. What we cannot do, for $n \neq m$ is to map all orbits of a vector field on E_n onto curves in E_m in such a fashion that a vector field is induced on E_m. From the practical point of view, what happens is that the images in E_m of all orbits of a vector field in E_n result in a system of curves that have a large number of intersections or do not fill E_m. If there is even one intersection of the images of two different orbits, say at $\bar{y}_0^A$ in E_m, then there are two different tangents to two different curves through $\bar{y}_0^A$ and it becomes impossible to describe such curves by an autonomous system of equations $d\bar{y}^A(t) = v^A(\bar{y}^B(t))$, $\bar{y}^A(0) = \bar{y}_0^A$. Thus our results are consistent.

Let us now restrict attention to the case in which Φ is invertible. We then have the maps

$$(2\text{-}8.25)\qquad \Phi: E_n \to {}^{`}E_n | y^A = \phi^A(x^j), \qquad A = 1,\ldots,n,$$

$$\det(\partial_i\phi^A) \neq 0,$$

$$(2\text{-}8.26)\qquad \Phi_*: T(E_n) \to T({}^{`}E_n) | {}^{`}v^A(y^B) = \{v^i\partial_i\phi^A\} \circ \Phi^{-1},$$

where ${}^{`}E_n$ is a replica of E_n. If we were not told beforehand to the contrary, we could equally well read the equations $y^A = \phi^A(x^j)$, $A = 1,\ldots,n$ as a system of equations that assigns new coordinates (y^A) to the point P of E_n that had coordinates (x^i) in the given coordinate cover of E_n. That is, the equations (2-8.25) may be viewed as a change of the coordinate cover of E_n, in which case (2-8.26) gives the law of transformation for the components of vector fields with respect to the bases $\{\partial_i\}$ and $\{\partial_A\}$, respectively. Perhaps it becomes a little easier to follow if we set ${}^{`}x^A = y^A$ and then replace the upper case indices by lower case indices.

Theorem 2-8.2. *The components of a vector field undergo the induced transformation*

$$(2\text{-}8.27)\quad (\Phi C)_*: T(E_n) \to T(E_n) | {}^{`}v^i({}^{`}x^k) = \{v^j\partial_j\phi^i\} \circ (\Phi^{-1}C)$$

for a change of coordinate cover

$$(2\text{-}8.28)\qquad \Phi C: {}^{`}x^k = \phi^k(x^j), \qquad k = 1,\ldots,n,$$

where

$$V = v^i \partial_i, \qquad `V = `v^i `\partial_i, \qquad `\partial_i \equiv \frac{\partial}{\partial `x^i},$$

and we have

(2-8.29) $$V\langle h \rangle = `V\langle `h \rangle, \qquad h = `h \circ \Phi = (\Phi C)^* `h.$$

A restriction of enormous severity would be imposed if we were able to work only with problems involving invertible mappings of E_n to $`E_n$. It is for this reason that every effort is made to shift as many considerations as possible to the *dual space* of $T(E_n)$, for this space possesses very nice properties under mappings from E_n to E_m with $n \neq m$.

There is another situation in which a map $\Phi: E_n \to E_m$ has a well defined induced action on $T(E_n)$, and this comes about through the notion of a coordinate transformation. Let

(2-8.30) $$\Phi: E_n \to E_m | y^A = \phi^A(x^j)$$

be a given map with $m > n$. In this case, the range of Φ in E_m can not sweep out all of E_m. What it does sweep out is an n-dimensional surface in E_m if

(2-8.31) $$\operatorname{rank}\big(\partial_i \phi^A(x^j)\big) = n$$

for all points $P:(x^j)$ in some neighborhood (possibly all) of E_n. Simply observe that $y^A = \phi^A(x^j)$, for $x^1 = c^1$, $x^2 = c^2, \ldots,$ $x^{k-1} = c^{k-1}$, $x^{k+1} = c^{k+1}, \ldots,$ $x^n = c^n$ defines a curve in E_m whose tangent vector field has components $\{\partial_k \phi^A\}$. Thus each of the n vectors $\{\partial_1 \phi^A \partial_A, \ldots, \partial_n \phi^A \partial_A\}$ is tangent to the range of Φ at the image of $P:(x^j)$ in E_m. The condition (2-8.31) then tells us that these n vectors are linearly independent and hence $\mathscr{R}(\Phi)$ is an n-dimensional surface in E_m. The important thing to realize here is that the map Φ may be considered as inducing $(x^1, \ldots, x^n)$ as new coordinates on $\mathscr{R}(\Phi)$. If $V = v^i(x^j)\partial_i \in T(E_n)$ then $V\langle h \rangle = v^i(x^j)\partial_i \phi^A(x^j)\partial_A g$ for $h = g \circ \Phi = \Phi^* g$. However, $\{\partial_j \phi^A \partial_A, j = 1, \ldots, n\}$ are n linearly independent vector fields tangent to $\mathscr{R}(\Phi)$ and hence constitute a basis for $T(\mathscr{R}(\Phi))$. We may thus define the action of Φ_* as this induced map from $T(E_n)$ to $T(\mathscr{R}(\Phi))$.

Theorem 2-8.3. *If* $\Phi: E_n \to E_m | y^A = \phi^A(x^j)$ *is a given map such that*

(2-8.32) $$m > n,$$

(2-8.33) $$\operatorname{rank}\big(\partial_i \phi^A(x^j)\big) = n,$$

then the induced map

$$(2\text{-}8.34)\qquad \Phi_*: T(E_n) \to T(\mathscr{R}(\Phi))\,|\,V\langle h\rangle = `V\langle g\rangle, \qquad h = \Phi^* g$$

is well defined with respect to the induced basis

$$(2\text{-}8.35)\qquad `\partial_i \stackrel{\text{def}}{=} \partial_i\phi^A\partial_A$$

of $T(\mathscr{R}(\Phi))$.

In all instances where Φ_* is well defined, it induces a vector field on the range of Φ; that is, it pulls a vector field from the domain of Φ up to the range of Φ. This direct graphic property is reinforced with the following definition.

Definition. If Φ is a map for which Φ_* is well defined, then $\Phi_* V$ is referred to as the vector field *pulled forward by* Φ, and Φ_* is referred to as the *pull forward* map.

All that now remains is to codify how Φ_* combines with the Lie algebra operations of $T(E_n)$.

Theorem 2-8.4. *The pull forward map has the following properties whenever the maps* Φ *and* Ψ *are such that* Φ_*, Ψ_* *and* $(\Psi \circ \Phi)_*$ *are well defined*:

PF1. $\Phi_*(f \cdot U + g \cdot V) = \Phi^{-1*}f \cdot \Phi_* U + \Phi^{-1*}g \cdot \Phi_* V$,

PF2. $\Phi_*[U, V] = [\Phi_* U, \Phi_* V]$,

PF3. $(\Psi \circ \Phi)_* = \Psi_* \circ \Phi_*$.

Proof. We have already seen that Φ induces the map Φ^* of $\Lambda^0(\mathscr{R}(\Phi))$ to $\Lambda^0(\mathscr{D}(\Phi))$. Hence Φ^{-1*} maps $\Lambda^0(\mathscr{D}(\Phi))$ into $\Lambda^0(\mathscr{R}(\phi))$. Property **PF1** is thus a direct consequence that Φ_* maps $T(\mathscr{D}(\Phi))$ into $T(\mathscr{R}(\Phi))$ and that $T(\mathscr{R}(\Phi))$ is a vector space over $\Lambda^0(\mathscr{R}(\Phi))$. By definition, $[U, V]\langle h\rangle = U\langle V\langle h\rangle\rangle - V\langle U\langle h\rangle\rangle$. When this is combined with $U\langle h\rangle = `U\langle g\rangle$, $V\langle h\rangle = `V\langle g\rangle$, $h = \Phi^* g$, we obtain

$$\begin{aligned}[`U, `V]\langle g\rangle &= `U\langle `V\langle g\rangle\rangle - `V\langle `U\langle g\rangle\rangle \\ &= U\langle V\langle h\rangle\rangle - V\langle U\langle h\rangle\rangle = [U, V]\langle h\rangle.\end{aligned}$$

This establishes property **PF2**. Property **PF3** is a direct consequence of Fig. 11. □

The following direct proof of **PF2** may prove informative. We have

$$`[U, V] = \left(u^k\partial_k v^j - v^k\partial_k u^j\right)\partial_j\phi^B\partial_B,$$

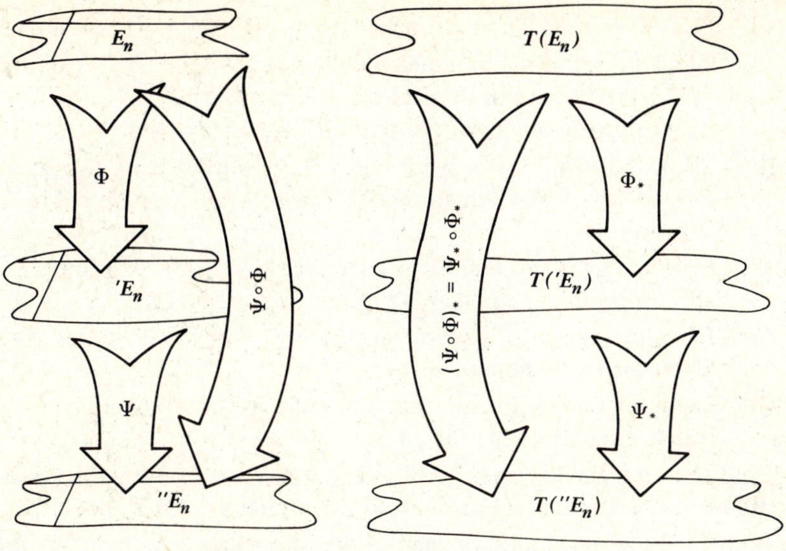

FIGURE 11. Pull forward by composition maps

and (2-6.15) and (2-8.18). However,

$$u^k\left(\partial_k v^j\right)\partial_j\phi^B = u^k\partial_k\left(v^j\partial_j\phi^B\right) - u^k v^j\partial_k\partial_j\phi^B$$

$$= u^k\partial_k\phi^A\partial_A\left(v^j\partial_j\phi^B\right) - u^k v^j\partial_k\partial_j\phi^B$$

and hence

$$`[U,V] = \left\{u^k\partial_k\phi^A\partial_A\left(v^j\partial_j\phi^B\right) - v^k\partial_k\phi^A\partial_A\left(u^j\partial_j\phi^B\right)\right\}\partial_B$$

$$-\left\{u^k v^j\partial_k\partial_j\phi^B - v^k u^j\partial_j\partial_k\phi^B\right\}\partial_B$$

$$= \left\{u^k\partial_k\phi^A\partial_A\left(v^j\partial_j\phi^B\right) - v^k\partial_k\phi^A\partial_A\left(u^j\partial_j\phi^B\right)\right\}\partial_B$$

$$= \left\{`u^A\partial_A `v^B - `v^A\partial_A `u^B\right\}\partial_B$$

$$= [`U, `V].$$

PROBLEMS

2.1. Let $T(p)$ be the one parameter family of maps

$$T(p): E_2 \to E_2 | \bar{x}^1(p) = e^p x^1, \qquad \bar{x}^2(p) = e^{-p}x^2 + b(1-e^p)x^1.$$

(a) Find $T^{-1}(p): E_2 \to E_2$ (find the inverse map for all values of p).

(b) Find $v^i(x^j; p)$ such that $d\bar{x}^i/dp = v^i(x^j; p)$.

(c) Use (b) to find $\bar{v}^i(\bar{x}^j(p); p)$ such that $d\bar{x}^i/dp = \bar{v}^i(\bar{x}^j(p); p)$.

(d) What values of the parameter b give $\bar{v}^i(\bar{x}^j(p); p) = V^i(\bar{x}^j(p))$ (i.e. $\bar{v}(\bar{x}(p); p)$ does not depend on p explicitly).
From now on, choose b so that $\bar{v}^i(\bar{x}^j; p) = V^i(\bar{x}^j)$.

(e) Show that $\bar{x}^i(p) = \exp(pV^k(x^j)\partial_k)\langle x^i\rangle$ by solving the given system of first order equations $d\bar{x}^i/dp = V^i(\bar{x}^j)$ and by direct substitution.

(f) Find the image of the curve $\Gamma: [0, 3\pi/2] \to E_2 | x^1 = e^t \ln(2t)$, $x^2 = e^{2t}\cos(2t)$ under the map $T(p)$ for each value of p.

(g) Find the image of the surface $(x^1)^2 + (x^2)^2 = 9$ under the map $T(p)$ for each value of p.

(h) Find all curves in E_2 that are mapped into themselves by the transformation $T(p)$ for all p.

2.2. Find the vector fields whose orbital equations are given by the systems of differential equations (a) through (d) of problem 1.3.

2.3. Find the general solution of each of the linear first order partial differential equation $\mathscr{V}\langle f\rangle = 0$, where $\mathscr{V}$ is each of the vector fields determined in problem 2.2.

2.4. Compute the Lie product and then find the Lie subalgebra that is the closure of the subspace of $T(E_3)$ that is spanned by

(a) $\partial_x,\ (z^2 - y)\partial_z$;

(b) $\partial_x,\ x\partial_y - y\partial_x$;

(c) $\partial_x - 3\partial_y,\ a\partial_y - y\partial_x$;

(d) $\partial_x,\ x^2\partial_y - y\partial_x$;

(e) $\partial_y,\ y^3\partial_x + \partial_z$;

(f) $\sin(x)\partial_y + \cos(y)\partial_x,\ \partial_x - \partial_z$;

(g) $\partial_x + 2\partial_y - 7\partial_z,\ \partial_x - e^{-x}\partial_y$;

(h) $\dfrac{1}{1+x^2}\partial_y,\ \dfrac{1}{1+y^2}\partial_x$;

(i) $x\partial_y - y\partial_x,\ y\partial_z - z\partial_y$;

(j) $x^2\partial_y - y^2\partial_x,\ y^2\partial_z - z^2\partial_y$;

(k) $x^3\partial_y - y^3\partial_x,\ y^3\partial_z - z^3\partial_y$;

(l) $x^4\partial_y - y^4\partial_x,\ y^4\partial_z - z^4\partial_y$.

Answers

The Lie products are

(h) $2\{x(1 + y^2)\partial_y - y(1 + x^2)\partial_x\}/(1 + x^2)^2(1 + y^2)^2$,

(i) $x\partial_z - z\partial_x$,

(j) $2y(x^2\partial_z - z^2\partial_x)$,

(k) $3y^2(x^3\partial_z - z^3\partial_x)$,

(l) $4y^3(x^4\partial_z - z^4\partial_x)$.

2.5. Determine the mapping Φ_* of $T(E_3)$ for each of the maps Φ given below, and then determine $\Phi_* V$ explicitly for $V = (x - 2y^2)\partial_x + (1 - 4e^{3z})\partial_y + (z^2 - 11xy)\partial_z$:

(a) $y^A = A_i^A x^i$, $\det(A_i^A) \neq 0$, $\partial_j A_i^A = 1$;

(b) $y^1 = x^1\cos(x^2)$, $y^2 = x^1\sin(x^2)$, $y^3 = 3 - 2e^{x^3}$;

(c) $y^1 = x^2\sin(x^1)$, $y^2 = x^2\cos(x^1)$, $y^3 = -6x^3$;

(d) $y^1 = 2x^2 - 5$, $y^2 = 3e^{x^1}$, $y^3 = -2e^{-x^3+x^2}$;

(e) $y^1 = x^2\cos(x^3)\cos(x^1)$, $y^2 = x^2\cos(x^3)\sin(x^1)$, $y^3 = x^2\sin(x^3)$.

2.6. Find the orbits of the vector field V and all C^1 solutions of $V\langle f\rangle = 0$, where V is given by

(a) $x\partial_x + y\partial_y + z\partial_z$;

(b) $x\partial_x + y\partial_y + 2z\partial_z$;

(c) $x\partial_x - 3y\partial_y + z\partial_z$;

(d) $(1 + x^2)\partial_x + xy\partial_y + xz\partial_z$;

(e) $x\partial_y - y\partial_x$;

(f) $x\partial_y - 2y\partial_z$;

(g) $x\partial_y + y\partial_x$;

(h) $(1 - 4x^2)\partial_x + x\partial_y - (xy + 5)\partial_z$;

(i) $\partial_x - (3y - 2e^{-x})\partial_y + y\partial_z$;

(j) $x(1 + 3x)^2\partial_x + x\partial_y$;

(k) $x\partial_x + x^2\partial_y + x^3\partial_z$;

(l) $\partial_x + z^2\partial_y$;

(m) $x\partial_x + (y + z)\partial_z$.

2.7. Find all C^1 solutions of the quasilinear equation $V\langle f\rangle = m$ for

(a) $V = x\partial_x - (1 + 3x)\partial_y + 4z^2\partial_z$, $m = 1 - 3f$;

(b) $V = z^2\partial_x + 3\partial_y + 2y\partial_z$, $m = xzf$;

(c) $V = xy^2\partial_x - 3y\partial_y + (1 + y^2)\partial_z$, $m = 3y(1 + 2f^2)$;

(d) $V = f^2\partial_x - f\partial_y + z\partial_z$, $m = 1 - 3z$;

(e) $V = y^3\partial_x + (3 - 2y)\partial_y + f^2\partial_z$, $m = x - yf$;

(f) $V = 3x\partial_x + 2e^{-z}f^2\partial_y - x^2\partial_z$, $m = 3x^3 - 2xf$.

2.8. Let $V_a, 1 \le a \le r$, be $r < n$ given elements of $T(E_n)$ such that

$$[V_a, V_b] = C_{ab}^c V_c,$$

and let $R_a(x^j, y), 1 \le a \le r$, be r given elements of $\Lambda^0(E_n \times \mathbb{R})$. Show that the system of quasilinear partial differential equations

$$V_a\langle f\rangle = R_a(x^j, f), \qquad 1 \le a \le r$$

has a solution only if the functions $\{R_a(x^j, f)\}$ satisfy the system of integrability conditions

$$(V_a + R_a \partial_f)\langle R_b \rangle - (V_b + R_b \partial_f)\langle R_a \rangle = C_{ab}^c R_c.$$

If these conditions are satisfied, show that the given system can be solved by the method of characteristics applied to the linear system

$$(V_a + R_a \partial_f)\langle F \rangle = 0, \qquad 1 \le a \le r.$$

CHAPTER THREE

EXTERIOR FORMS

3-1. THE DUAL SPACE $T^*(E_n)$

As before, E_n is an n-dimensional number space with a given Cartesian coordinate cover (x^i). Let $T(E_n)$ be the tangent space of E_n that was studied in the preceding chapter. $T(E_n)$ is thus a Lie algebra of derivations on $\Lambda^0(E_n)$ with the globally defined basis fields $\{\partial_i, i = 1, \ldots, n\}$. Since $T(E_n)$ is also the union over $P \in E_n$ of the vector spaces $T_P(E_n)$, the concept of a dual space introduced in Section 1-5 can be extended to that of the dual space of $T(E_n)$ as the union over $P \in E_n$ of the vector spaces $T_P^*(E_n)$ dual to $T_P(E_n)$.

Definition. The *dual space*, $T^*(E_n)$, is the union over $P \in E_n$ of the dual vector spaces $T_P^*(E_n)$:

$$T^*(E_n) = \bigcup_{P \in E_n} T_P^*(E_n). \tag{3-1.1}$$

This definition shows that $T^*(E_n)$ inherits a definite structure from the vector space structures of each of its constituent $T_P^*(E_n)$'s. Thus if $\omega \in T^*(E_n)$ and $V \in T(E_n)$, the notation introduced in Section 1-5 yields

$$\omega : T(E_n) \to \mathbb{R} \,|\, r = \langle \omega, V \rangle, \tag{3-1.2}$$

$$\langle \omega, f \cdot U + g \cdot V \rangle = f\langle \omega, U \rangle + g\langle \omega, V \rangle, \tag{3-1.3}$$

$$\langle f \cdot \omega + g \cdot \rho, V \rangle = f\langle \omega, V \rangle + g\langle \rho, V \rangle \tag{3-1.4}$$

for all $\omega, \rho \in T^*(E_n)$, all $U, V \in T(E_n)$, and all $f, g \in \Lambda^0(E_n)$. Because the vector space operations of $T^*(E_n)$ are inherited from those of each $T_P^*(E_n)$, addition and multiplication by numbers occur pointwise with respect to E_n. Further, an "element of $T^*(E_n)$" means a selection of an element from $T_P^*(E_n)$ for each $P \in E_n$, in exactly the same manner that $V \in T(E_n)$ means a selection of an element $\{v^i\}$ from $T_P(E_n)$ for each P in E_n. This brings us to another point that can be easily missed in use of the notation introduced by (3-1.1) through (3-1.4). Both $\omega \in T^*(E_n)$ and $V \in T(E_n)$ are fields defined on E_n. In order to take specific note of this fact, let us write $\omega(x^j)$ and $V(x^j)$ for the evaluations of these fields at $P:(x^j)$. When we write $\langle \omega, V \rangle$, it will have the value $\langle \omega(x^j), V(x^j) \rangle = h(x^j)$ at $P:(x^j)$, where $h(x^j)$ may be considered to be an element of $\Lambda^0(E_n)$. Thus $\omega \in T^*(E_n)$, viewed as the field of selections of elements from each $T_P^*(E_n)$, defines a map from $T(E_n)$ into $\Lambda^0(E_n)$:

$$\langle \cdot, \cdot \rangle : T^*(E_n) \times T(E_n) \to \Lambda^0(E_n) | \langle \omega(x^j), V(x^j) \rangle = h(x^j).$$

With these caveats in mind, we will henceforth simply use the term "vector space" for $T^*(E_n)$.

Lemma 3-1.1. *The dual space $T^*(E_n)$ is a vector space over $\Lambda^0(E_n)$.*

We now need to clothe this abstract result in a realizable and computationally useful garb. This is most easily accomplished by the introduction of a "natural basis."

Definition. A collection of n fields $\{\theta^i,\ i = 1, \ldots, n\}$ is a *natural basis* for $T^*(E_n)$ (with respect to the (x^i) coordinate cover) if and only if

$$\langle \theta^i, \partial_j \rangle = \delta^i_j. \tag{3-1.5}$$

The fact that we can define something does not necessarily mean that we can use it. In the present case, the definition of a natural basis gives us the equations (3-1.5) to work with, but we have no guarantee of either the existence or the uniqueness of such a collection of elements of $T^*(E_n)$.

Lemma 3-1.2. *If a solution of the system* (3-1.5) *exists, then it is unique.*

Proof. Suppose that there are two natural bases, θ^i and ρ^i; that is, $\langle \theta^i, \partial_j \rangle = \delta^i_j$, $\langle \rho^i, \partial_j \rangle = \delta^i_j$. Subtracting the corresponding equations of these two systems gives us

$$\langle \theta^i, \partial_j \rangle - \langle \rho^i, \partial_j \rangle = \langle \theta^i - \rho^i, \partial_j \rangle = 0.$$

Thus, for every n-tuple of C^∞ functions $\{v^i(x^k)\}$, we have

$$v^j\langle\theta^i - \rho^i, \partial_j\rangle = \langle\theta^i - \rho^i, V\rangle = 0$$

for every $V = v^j(x^k)\partial_j \in T(E_n)$. It then follows from the uniqueness of the zero element 0^* of each $T_P^*(E_n)$ that $\theta^i - \rho^i = 0^*$ and uniqueness is established. □

We must now turn to the question of existence. The simplest way to establish existence is to exhibit a solution explicitly. Now, for any $f \in \Lambda^0(E_n)$ and any $V = v^i\partial_i \in T(E_n)$, we know that $V\langle f\rangle = v^i\partial f/\partial x^i$ takes its values in $\mathbb{R}$. Thus $V\langle f\rangle$ may be viewed as a linear functional on $T(E_n)$ since it has the required linearity properties: $V\langle f+g\rangle = V\langle f\rangle + V\langle g\rangle$, $(U+V)\langle f\rangle = U\langle f\rangle + V\langle f\rangle$.

The symbol "df," defined by

$$\langle df, V\rangle = V\langle f\rangle \tag{3-1.6}$$

for all $V \in T(E_n)$, is an element of $T^*(E_n)$ for each $f \in \Lambda^0(E_n)$.

The result we have just obtained sometimes gives a good bit of trouble the first time around, so the following comments may prove useful. The key to the situation is that $V = v^i(x^j)\partial_i$ realized any element V of $T(E_n)$ as an *operator*, and hence V does not possess values per se. On the other hand, when V acts on an element $f(x^j)$ of $\Lambda^0(E_n)$, we obtain a new element $V\langle f\rangle = v^i\partial_i f = g$ of $\Lambda^0(E_n)$. Finally, observe that the values of $g(x^i)$ belong to $\mathbb{R}$, and hence it is consistent to view the full quantity $V\langle f\rangle$ as a mapping form $T(E_n)$ to $\mathbb{R}$; namely, a functional on $T(E_n)$. This, however, is exactly what (3-1.6) says if we call the element of $T^*(E_n)$ that is thus defined the symbol df.

From yet another point of view, we note that the quantity $V\langle f\rangle$ has two ingredients, a vector field V and a function f. The work of the previous chapter kept V fixed and allowed f to range over the collection $\Lambda^0(E_n)$. This served to show that V was a derivation on the algebra $\Lambda^0(E_n)$. The present considerations keep f fixed and allow V to range over $T(E_n)$. This serves to define the associated symbol df and to show that df is an element of $T^*(E_n)$. The concept of the duality of $T(E_n)$ and $T^*(E_n)$ is therefore used directly in this process and is, in point of fact, the underpinning of all that we do in this chapter.

Lemma 3-1.3. *The unique natural dual basis of $T^*(E_n)$ is given by the n elements $\{dx^i,\ i = 1,\ldots,n\}$. Any element ω of $T^*(E_n)$ can be written uniquely as*

$$\omega = \omega_i(x^j)\,dx^i, \tag{3-1.7}$$

where the coefficients $\{\omega_i(x^j)\}$ *are determined by*

$$\omega_i = \langle \omega, \partial_i \rangle, \qquad i = 1, \ldots, n \tag{3-1.8}$$

and thus belong to $\Lambda^0(E_n)$.

Proof. The quantities dx^i are defined by

$$\langle dx^i, V \rangle = V\langle x^i \rangle$$

and hence $\langle dx^i, \partial_j \rangle = \partial_j x^i = \delta^i_j$. This establishes the existence of the n quantities dx^i that satisfy (3-1.5). Incidentally, $\langle dx^i, \partial_j \rangle = \delta^i_j$ also shows that $\{dx^i\}$ is a linearly independent system (i.e., $\langle c_i\, dx^i, V \rangle = 0$ for all $V \in T(E_n)$ if and only if $c_i = 0, i = 1, \ldots, n$). Since $T^*(E_n)$ is an n-dimensional vector space over $\Lambda^0(E_n)$, any $\omega \in T^*(E_n)$ can be written in the form (3-1.7). Hence $\langle \omega, \partial_j \rangle = \langle \omega_i\, dx^i, \partial_j \rangle = \omega_i \langle dx^i, \partial_j \rangle = \omega_i \delta^i_j = \omega_j$ and (3-1.8) is established. □

All that is now needed in order to make these ideas complete is a concrete realization of the abstract symbol df. If we apply df to the basis $\{\partial_i\}$, (3-1.6) gives

$$\langle df, \partial_i \rangle = \partial_i f. \tag{3-1.9}$$

On the other hand, $df \in T^*(E_n)$ and Lemma 3-1.3 shows that

$$df = \gamma_i(x^k)\, dx^i \tag{3-1.10}$$

with

$$\gamma_i(x^k) = \langle df, \partial_i \rangle = \partial_i f(x^k) \tag{3-1.11}$$

by (3-1.8) and (3-1.9). Thus (3-1.10) yields

$$df = \partial_i f(x^k)\, dx^i. \tag{3-1.12}$$

This, however, is the familiar total differential of the function f of the n variables $x^1, x^2, \ldots, x^n$.

Lemma 3-1.4. *The abstract element* df *of* $T^*(E_n)$ *that is defined by* (3-1.6) *for any* $f \in \Lambda^0(E_n)$ *has the realization*

$$df = \partial_i f(x^k)\, dx^i \tag{3-1.13}$$

as the total differential of the function $f(x^k)$.

There is still some degree of ambiguity, for (3-1.13) defines df in terms of the "differentials" $dx^1, dx^2, \ldots, dx^n$. Thus either we interpret the quantities $\{dx^i\}$ as abstract basis elements—the notion of the differential of a coordinate function is a primitive in the logical sense—or else dx^i is interpreted as the differential of the coordinate function x^i and (3-1.13) evaluates something in terms of itself. The reader should keep this carefully in mind, for it will not be resolved until the next chapter when we introduce the operation of exterior differentiation.

There is another interpretation that can be attached to (3-1.13) that is important if for no other reason than to show consistency with what has gone before. Suppose we divide both sides of (3-1.13) by dt so as to obtain

$$df(x^k)/dt = \partial_i f(x^k)\, dx^i/dt.$$

Read in this way, (3-1.13) may be viewed as providing an approximation to the change in the value of the function $f(x^k)$ that obtains from evaluation at $P:(x^j)$ and at $P:(x^j + v^j t + o(t))$ with $v^j = dx^j/dt|_{t=0}$. We are thus back to the familiar ground of the differentiation of the composition of a map from $\mathbb{R}$ to E_n with a map from E_n to $\mathbb{R}$. There is no possibility of inconsistency, as should be evident from the various possible interpretations of the symbols on the two sides of (3-1.6).

This same view, when applied to an arbitrary element

(3-1.14) $$\omega = \omega_i(x^j)\, dx^i$$

of $T^*(E_n)$ shows something quite different. If we again divide both sides by dt, the result is

(3-1.15) $$\omega/dt = \omega_i(x^j)\, dx^i/dt.$$

The right-hand side of this expression is well defined whenever the x's are functions of the variable t, but the left-hand side, ω/dt, makes no sense (is not d/dt of any single function) unless we are sufficiently charitable to interpret ω as an infinitesimal of the same order as dt whenever the x's are functions of the variable t. This should not be too surprising, for it is hoped that everyone has heard of an inexact differential. It is simply an expression of the form

(3-1.16) $$\omega = \omega_i(x^j)\, dx^i$$

that is not the derivative with respect to t of any function $f(x^j)$ when the x's become functions of the variable t. Quantities such as (3-1.16) are referred to as *differential forms* or Pfaffian forms. In fact it is this terminology, namely differential forms, that is the most widely used today. Accordingly, the symbol $\Lambda^1(E_n)$ will be used instead of $T^*(E_n)$. The following definition is given here in order to codify this situation.

Definition. $\Lambda^1(E_n)$, the space of *differential forms* of degree one on E_n, coincides with the dual space $T^*(E_n)$. The elements of $\Lambda^1(E_n)$ have the representation

$$\omega = \omega_i(x^j)\,dx^i \tag{3-1.17}$$

in terms of the natural basis $\{dx^i\}$, and are referred to as *differential forms of degree one*. $\Lambda^0(E_n)$ is referred to as the *space of forms of degree zero*.

There is one additional and very useful result that obtains as a direct consequence of Lemmas 3-1.1 through 3-1.4.

Lemma 3-1.5. *Let $V = v^i(x^k)\partial_i$ be a given element of $T(E_n)$ and let $\omega = \omega_i(x^k)\,dx^i$ be a given element of $\Lambda^1(E_n)$. Then ω applied to V has the evaluation*

$$\langle \omega, V\rangle = \omega_i(x^k)v^i(x^k). \tag{3-1.18}$$

Proof. $\langle \omega, V\rangle = \langle \omega_i\,dx^i, v^j\partial_j\rangle = \omega_i v^j\langle dx^i, \partial_j\rangle = \omega_i v^j\delta^i_j = \omega_i v^i$. □

The simplest physical example of an exterior form arises in mechanics. Let E_3 designate three-dimensional Euclidean space with the Cartesian coordinate cover (x^1, x^2, x^3). By a field of force in E_3 we mean a collection of three functions $F_1(x^i)$, $F_2(x^i)$, $F_3(x^i)$ such that a particle at $P:(x^i)$ will experience the force that is specified by evaluating the three functions $\{F_i(x^j)\}$. With this information, let us define the differential form

$$W = F_i(x^j)\,dx^i. \tag{3-1.19}$$

If we view $\{dx^i\}$ as the components of an infinitesimal displacement of the particle along its trajectory in E_3 through the point $P:(x^j)$, then the differential form W that is defined by (3-1.19) is the work done in this infinitesimal displacement. Further, if $V = v^i(t)\partial_i$ is the velocity vector field of the trajectory $x^i = \phi^i(t)$ of the particle, then $dx^i = v^i\,dt$ at the point $P:(\phi^i(t))$ and W becomes $W = F_i(\phi^j(t))v^i\,dt$. This same result can also be obtained by allowing W to act on V

$$\langle W, V\rangle = F_i(\phi^j(t))v^i(t)$$

and this expression is simply the instantaneous rate at which work is done on the particle by the force field. In particular, we have

$$W = F_i(\phi^j(t))v^i\,dt = \langle W, V\rangle\,dt$$

in this case. From yet another point of view, work is a scalar while "velocity" is a vector field. Accordingly, since work is known physically to be linear in the forces, "force" has to be a linear functional of the mechanical state variables, velocity; that is $F \in \Lambda^1(E_3)$. A little reflection on this example will quickly convince you that each of the various interpretations of elements $\Lambda^1(E_3)$ that we have obtained above is of use, depending on the nature of the problem and the kind of questions being asked.

3-2. THE EXTERIOR OR "VECK" PRODUCT

We know that the vector space $T(E_n)$ equipped with the Lie product becomes an algebra. It is therefore natural to ask whether $\Lambda^1(E_n)$ can be made into an algebra. The answer is unequivocally no, so we must proceed along other lines.

The vector space structure of $\Lambda^1(E_n)$ is delineated by the fact that $\Lambda^1(E_n)$ has the natural basis $\{dx^i\}$ as a vector space over $\Lambda^0(E_n)$; every element of $\Lambda^1(E_n)$ has the form $\omega_i(x^j)\,dx^i$ with $\omega_i(x^j) \in \Lambda^0(E_n)$. A rule of combination can thus be defined by defining it for the elements of the basis $\{dx^i\}$ and requiring that it behave properly with respect to the vector space operations.

Definition. The *exterior* or *veck* product $\wedge$ is defined on $\Lambda^1(E_n) \times \Lambda^1(E_n)$ by the requirements

(3-2.1) $$dx^i \wedge dx^j = -dx^j \wedge dx^i;$$

(3-2.2) $$dx^i \wedge f(x^k)\,dx^j = f(x^k)\,dx^i \wedge dx^j$$

for all $f(x^k) \in \Lambda^0(E_n)$;

(3-2.3) $$\alpha \wedge (\beta + \gamma) = \alpha \wedge \beta + \alpha \wedge \gamma, \qquad \alpha \wedge \beta = -\beta \wedge \alpha,$$

for all $\alpha, \beta, \gamma \in \Lambda^1(E_n)$.

Let us first show that this law of combination is actually defined for any pair of elements $\alpha = \alpha_i\,dx^i$ and $\beta = \beta_i\,dx^i$ of $\Lambda^1(E_n)$. Now, $\alpha \wedge \beta$ is defined by the relation

$$\begin{aligned}\alpha \wedge \beta &= (\alpha_i\,dx^i) \wedge (\beta_j\,dx^j)\\ &= (\alpha_1\,dx^1 + \cdots + \alpha_n\,dx^n) \wedge (\beta_1\,dx^1 + \cdots + \beta_n\,dx^n).\end{aligned}$$

Thus use of (3-2.3) followed by (3-2.2) yields

$$\begin{aligned}\alpha \wedge \beta &= (\alpha_1\,dx^1 + \cdots + \alpha_n\,dx^n) \wedge \beta_1\,dx^1 + \cdots\\ &\quad + (\alpha_1\,dx^1 + \cdots + \alpha_n\,dx^n) \wedge \beta_n\,dx^n\\ &= \beta_1(\alpha_1\,dx^1 + \cdots + \alpha_n\,dx^n) \wedge dx^1 + \cdots\\ &\quad + \beta_n(\alpha_1\,dx^1 + \cdots + \alpha_n\,dx^n) \wedge dx^n.\end{aligned}$$

However, (3-2.3) imply that

$$(\alpha + \beta) \wedge \gamma = -\gamma \wedge (\alpha + \beta) = -\gamma \wedge \alpha - \gamma \wedge \beta = \alpha \wedge \gamma + \beta \wedge \gamma,$$

and hence

(3-2.4)

$$\begin{aligned} \alpha \wedge \beta &= \beta_1 \alpha_1 \, dx^1 \wedge dx^1 + \beta_1 \alpha_2 \, dx^2 \wedge dx^1 + \cdots \\ &\quad + \beta_n \alpha_1 \, dx^1 \wedge dx^n + \beta_n \alpha_2 \, dx^2 \wedge dx^n + \cdots + \beta_n \alpha_n \, dx^n \wedge dx^n \\ &= \alpha_i \beta_j \, dx^i \wedge dx^j. \end{aligned}$$

However, $dx^i \wedge dx^j = -dx^j \wedge dx^i$ by (3-2.1), and hence (3-2.4) also gives

(3-2.5) $$\alpha \wedge \beta = -\alpha_i \beta_j \, dx^j \wedge dx^i = -\alpha_j \beta_i \, dx^i \wedge dx^j.$$

The following result is then obtained by adding (3-2.4) and (3-2.5).

Lemma 3-2.1. *The exterior product of* $\alpha = \alpha_i \, dx^i$ *and* $\beta = \beta_j \, dx^j$ *has the evaluation*

(3-2.6) $$2\alpha \wedge \beta = \left(\alpha_i \beta_j - \alpha_j \beta_i\right) dx^i \wedge dx^j.$$

We certainly can, at least in principle, form the veck product of all pairs of elements of $\Lambda^1(E_n)$. If we form all linear combinations of the quantities thus obtained with coefficients from $\Lambda^0(E_n)$, we obtain a vector space over $\Lambda^0(E_n)$ that is usually denoted by $\Lambda^2(E_n)$. Since $dx^i \wedge dx^j = -dx^j \wedge dx^i$ it follows that this vector space has the basis $\{dx^i \wedge dx^j, \, i < j\}$ so it is a vector space of dimension $n(n-1)/2$.

Definition. The $n(n-1)/2$ dimensional vector space $\Lambda^2(E_n)$ over $\Lambda^0(E_n)$ with the basis

(3-2.7) $$\{dx^i \wedge dx^j, \, i < j\}$$

is the vector space of exterior 2-forms over E_n. The elements of $\Lambda^2(E_n)$ are referred to as 2-*forms*, or *exterior forms of degree* 2.

Definition. The elements of $\Lambda^0(E_n)$ are referred to as *exterior forms of degree* 0. The elements of $\Lambda^1(E_n)$ are referred to as *exterior forms of degree* 1.

With these definitions at hand, it is permissible to say that the veck product, $\wedge$, defines a map of $\Lambda^1(E_n) \times \Lambda^1(E_n)$ to $\Lambda^2(E_n)$.

What we have done once, we can certainly do again. The vector space $\Lambda^3(E_n)$ of *forms of degree* 3 is thus generated by forming all linear combinations of all veck products of the basis elements dx^i taken three at a time. Since there are now three products involved, it is clearly necessary to say something about how such multiple multiplications associate. The simplest thing is to require that the multiplications be associative,

(3-2.8) $$\alpha \wedge (\beta \wedge \gamma) = (\alpha \wedge \beta) \wedge \gamma.$$

From now on we simply continue this process and obtain the sequence of vector spaces $\Lambda^4(E_n), \Lambda^5(E_n), \ldots$. We note, however, that (3-2.1) implies $dx^i \wedge dx^i = 0$. Thus when we come to $\Lambda^n(E_n)$, we can only form the single independent element

$$dx^1 \wedge dx^2 \wedge dx^3 \wedge \cdots \wedge dx^{n-1} \wedge dx^n.$$

If we veck this with any dx^i, the result is

$$dx^i \wedge dx^1 \wedge dx^2 \wedge \cdots \wedge dx^n = 0$$

since dx^i occurs twice in any such product. The process thus terminates with $\Lambda^n(E_n)$.

Definition. The space $\Lambda^k(E_n)$, $0 < k \leq n$, of *exterior forms of degree k* is the vector space of dimension $\binom{n}{k} = n!/k!(n-k)!$ over $\Lambda^0(E_n)$ with the natural basis

(3-2.9) $$\{dx^{i_1} \wedge dx^{i_2} \wedge \cdots \wedge dx^{i_k}, \quad i_1 < i_2 < \cdots < i_k\}.$$

If $\alpha \in \Lambda^k(E_n)$ then we write

(3-2.10) $$\deg(\alpha) = k.$$

The value of $\binom{n}{k}$ for the dimension of Λ^k comes directly from (3-2.9), namely n distinct things taken k at a time. Table 1 summarizes these results.

The sequence of spaces $\Lambda^0(E_n), \ldots, \Lambda^n(E_n)$ allows us to collect together the properties of the veck product in a more elegant formulation.

Lemma 3-2.2. *The operation* $\wedge$ *of exterior multiplication generates a map*

(3-2.11) $$\wedge : \Lambda^r(x^j) \times \Lambda^s(x^j) \to \Lambda^{r+s}(x^j)$$

and exhibits the following properties:

P1. $\alpha \wedge (\beta + \gamma) = \alpha \wedge \beta + \alpha \wedge \gamma$.
P2. $\alpha \wedge \beta = (-1)^{\deg(\alpha)\deg(\beta)} \beta \wedge \alpha$.
P3. $\alpha \wedge (\beta \wedge \gamma) = (\alpha \wedge \beta) \wedge \gamma$.

TABLE 1. BASES AND DIMENSIONS

Space	Basis	Dimension
$\Lambda^0(E_n)$	1	$1 = \binom{n}{0}$
$\Lambda^1(E_n)$	$\{dx^k\}$	$n = \binom{n}{1}$
$\Lambda^2(E_n)$	$\{dx^k \wedge dx^m,\ k < m\}$	$\frac{n(n-1)}{2} = \binom{n}{2}$
$\Lambda^3(E_n)$	$\{dx^k \wedge dx^m \wedge dx^r,\ k < m < r\}$	$\frac{n(n-1)(n-2)}{3!} = \binom{n}{3}$
$\vdots$	$\vdots$	$\vdots$
$\Lambda^n(E_n)$	$dx^1 \wedge dx^2 \wedge \cdots \wedge dx^n$	$1 = \binom{n}{n}$

Proof. The only thing that has not already been verified is that **P2** implies (3-2.2). Since $f(x^k)$ belongs to $\Lambda^0(E_n)$, $\deg(f) = 0$, and **P2** implies $dx^i \wedge f(x^k) = (-1)^{\deg(f)\deg(dx^i)} \cdot f(x^k) \wedge dx^i = f(x^k)\, dx^i$. Thus *in exterior multiplications by elements of* $\Lambda^0(E_n)$ *we may omit the symbol* $\wedge$. The associative property P3 then gives $dx^i \wedge f \wedge dx^j = f \wedge dx^i \wedge dx^j = f\, dx^i \wedge dx^j$. □

The sequence of vector spaces $\Lambda^0(E_n)$, $\Lambda^1(E_n), \ldots, \Lambda^n(E_n)$ thus possess the following properties:

1. Each $\Lambda^k(E_n)$ is a vector space over $\Lambda^0(E_n)$ with the operations of addition and multiplication by elements of $\Lambda^0(E_n)$, but elements from $\Lambda^k(E_n)$ and $\Lambda^m(E_n)$ with $k \neq m$ cannot be added.
2. Any element of $\Lambda^k(E_n)$ can be vecked with any element from $\Lambda^m(E_n)$, but the answer belongs to $\Lambda^{k+m}(E_n)$.

These results are similar in structure to what obtains with the algebra of homogeneous polynomials. There we can add two homogeneous polynomials of the same degree together to obtain a homogeneous polynomial of the same degree, and we can also multiply homogeneous polynomials of degrees k and m together to obtain a homogeneous polynomial of degree $k + m$. Now, the theory of homogeneous polynomials is significantly simplified by introducing an all-encompassing space $\mathscr{P}$ of all homogeneous polynomials that is *graded* by the degrees of the polynomials that constitute its subspaces. We do the same thing here.

Definition. The *graded exterior algebra of differential forms* is the direct sum

$$\Lambda(E_n) = \Lambda^0(E_n) \oplus \Lambda^1(E_n) \oplus \Lambda^2(E_n) \oplus \cdots \oplus \Lambda^n(E_n) \tag{3-2.12}$$

with the vector space operations of each $\Lambda^k(E_n)$ together with the exterior product $\wedge$ as a map from $\Lambda(E_n) \times \Lambda(E_n)$ to $\Lambda(E_n)$.

The definition of each $\Lambda^k(E_n)$ shows that the natural basis for $\Lambda(E_n)$ consists of the direct sum of the natural bases for each of its constituent graded subspaces. The basis is thus given by

$$1 \oplus \{dx^i\} \oplus \{dx^i \wedge dx^j,\ i < j\} \oplus \cdots \oplus dx^1 \wedge dx^2 \wedge \cdots \wedge dx^n,$$

and thus $\dim(\Lambda) = \Sigma_{k=0}^{n}\binom{n}{k}$. However, $(1+t)^n = \Sigma_{k=0}^{n}\binom{n}{k}t^k$ so that $2^n = \Sigma_{k=0}^{n}\binom{n}{k}$ and hence

$$\dim(\Lambda(E_n)) = 2^n. \tag{3-2.13}$$

There is one word of caution that must be sounded when it comes to writing an element of $\Lambda^k(E_n)$, for there are actually two conventions in current use. We know that $\Lambda^k(E_n)$ has the basis

$$\{dx^{i_1} \wedge dx^{i_2} \wedge \cdots \wedge dx^{i_k}, \qquad i_1 < i_2 < \cdots < i_k\}$$

and hence any $\omega \in \Lambda^k(E_n)$ is uniquely written as

$$\omega = \sum_{i_1 < i_2 < \cdots < i_k} \omega_{i_1 i_2 \cdots i_k}(x^m)\, dx^{i_1} \wedge dx^{i_2} \wedge \cdots \wedge dx^{i_k} \tag{3-2.14}$$

in terms of the $\binom{n}{k}$ functions

$$\{\omega_{i_1 i_2 \cdots i_k}(x^m), \qquad i_1 < i_2 < \cdots < i_k\}. \tag{3-2.15}$$

Now, we can extend the $\binom{n}{k}$ functions (3-2.15) to n^k functions $\{\omega_{i_1 i_2 \cdots i_n}(x^m)\}$ for all values of the indices $i_1, \ldots, i_k$ by requiring the new functions to be antisymmetric in every pair of indices. If we then sum over all k indices from 1 through n we will repeat each term of (3-2.15) $k!$ times. Thus we must write

$$\omega = \frac{1}{k!}\omega_{i_1 i_2 \cdots i_k}(x^m)\, dx^{i_1} \wedge dx^{i_2} \wedge \cdots \wedge dx^{i_k} \tag{3-2.16}$$

with the operable summation convention in order that the result agree with (3-2.14). Please note the $1/k!$ that arises in this alternative way of writing the answer, and the fact that *the new n^k coefficient function $\{\omega_{i_1 i_2 \cdots i_k}\}$ are now antisymmetric in every pair of indices.* This situation has already been in evidence in the process of exterior multiplication, as the factor of two on the left-hand side of (3-2.6) clearly shows. The first time through, the situation appears unduly complex, for there are actually two different representations involved: representations in terms of the natural basis, and representations that use the convenience of the summation convention. A little practice with the manipulations, however, easily sets the mind at rest.

3-3. ALGEBRAIC RESULTS

Exterior forms of degree higher than one were actually invented in order to obtain simple characterizations of subspaces of $\Lambda^1(E_n)$. This is accomplished by use of the notion of simple elements of $\Lambda^k(E_n)$.

Definition. An element ω of $\Lambda^k(E_n)$ is said to be *simple* if and only if there exist k elements $\{\eta^1, \eta^2, \ldots, \eta^k\} = \{\eta^a\}$ of $\Lambda^1(E_n)$ such that

$$\omega = \eta^1 \wedge \eta^2 \wedge \cdots \wedge \eta^k. \tag{3-3.1}$$

We have already encountered simple elements of $\Lambda^k(E_n)$; namely, the basis elements of $\Lambda^k(E_n)$. In fact, the representation

$$\omega = \sum_{i_1 < i_2 < \cdots < i_k} \omega_{i_1 i_2 \cdots i_k}(x^m)\, dx^{i_1} \wedge dx^{i_2} \wedge \cdots \wedge dx^{i_k} \tag{3-3.2}$$

shows that *every element of $\Lambda^k(E_n)$ consists of linear combination of simple elements of $\Lambda^k(E_n)$ with coefficients from $\Lambda^0(E_n)$.*

Lemma 3-3.1. *The collection of all simple elements of $\Lambda^k(E_n)$ is closed under multiplication by elements of $\Lambda^0(E_n)$. The collection of all simple elements of $\Lambda(E_n)$ is closed under exterior multiplication. Every $\omega \in \Lambda^n(E_n)$ is simple.*

Proof. Any simple element of $\Lambda^k(E_n)$ has the form (3-3.1), and hence

$$f\omega = f\eta^1 \wedge \eta^2 \wedge \cdots \wedge \eta^k.$$

However, $f\eta^1 = \gamma$ belongs to $\Lambda^1(E_n)$ so that

$$f\omega = \gamma \wedge \eta^2 \wedge \cdots \wedge \eta^k$$

and hence $f\omega$ is also simple. If $\omega \in \Lambda^k(E_n)$ and $\beta \in \Lambda^m(E_n)$ are simple, then

$$\omega \wedge \beta = \eta^1 \wedge \cdots \wedge \eta^k \wedge \rho^1 \wedge \cdots \wedge \rho^m$$

with $\beta = \rho^1 \wedge \cdots \wedge \rho^m$. This shows that the exterior product of any two simple elements of $\Lambda(E_n)$ is also a simple element of $\Lambda(E_n)$. Finally, if $\omega \in \Lambda^n(E_n)$, then

$$\omega = f(x^m)\, dx^1 \wedge \cdots \wedge dx^n$$

since $\Lambda^n(E_n)$ is one-dimensional, and hence every element of $\Lambda^n(E_n)$ is simple. □

Lemma 3-3.2. *If $\omega \in \Lambda^k(E_n)$ with k an odd integer, then $\omega \wedge \omega = 0$.*

Proof. If k is an odd integer, then $\deg(\omega) = 2r + 1$. Property **P2** gives

$$\omega \wedge \omega = (-1)^{(2r+1)^2} \omega \wedge \omega = -\omega \wedge \omega$$

and the result is established. □

Note. If $\omega \in \Lambda^k(E_n)$ with k an even integer, then $\omega \wedge \omega$ need not vanish. Simply observe that

$$(dx \wedge dy + dz \wedge dt) \wedge (dx \wedge dy + dz \wedge dt) = 2\, dx \wedge dy \wedge dz \wedge dt \neq 0.$$

Lemma 3-3.3. *Let $\omega^1, \ldots, \omega^k$ be k given elements of $\Lambda^1(E_n)$ and construct the simple k-form*

$$\Omega = \omega^1 \wedge \omega^2 \wedge \cdots \wedge \omega^k. \tag{3-3.3}$$

A necessary and sufficient condition that the k given 1-forms be linearly dependent is

$$\Omega = 0. \tag{3-3.4}$$

A necessary and sufficient condition that the k given 1-forms be linearly independent is

$$\Omega \neq 0. \tag{3-3.5}$$

Proof. If $\omega^1, \ldots, \omega^k$ are linearly dependent, then one of them, say the first, can be expressed as a linear combination of the remaining ones:

$$\omega^1 = f_2 \omega^2 + \cdots + f_k \omega^k. \tag{3-3.6}$$

Now simply veck (3-3.6) with $\omega^2 \wedge \omega^3 \wedge \cdots \wedge \omega^k$ to obtain

$$\Omega = f_2 \omega^2 \wedge \omega^2 \wedge \omega^3 \wedge \cdots \wedge \omega^k + \cdots + f_k \omega^k \wedge \omega^2 \wedge \omega^3 \wedge \cdots \wedge \omega^k.$$

Each of the $k - 1$ terms on the right-hand side contains two identical 1-form factors, so that each term vanishes by Lemma 3-3.2. Conversely, consider the equation

$$g_1 \omega^1 + g_2 \omega^2 + \cdots + g_k \omega^k = 0. \tag{3-3.7}$$

If we veck this equation with the simple element $\omega^2 \wedge \omega^3 \wedge \cdots \wedge \omega^k$, the only term on the left-hand side that does not contain two identical 1-form factors is $g_1 \omega^1 \wedge \omega^2 \wedge \cdots \wedge \omega^k$, and hence

$$g_1 \Omega = 0$$

if (3-3.7) is to be satisfied. In like manner, vecking (3-3.7) with each of the remaining $k-1$ possible simple elements of $\Lambda^{k-1}(E_n)$ that can be formed from $\omega^1, \ldots, \omega^k$ gives the requirements

$$g_m \Omega = 0, \qquad m = 1, \ldots, k \tag{3-3.8}$$

in order that (3-3.7) be satisfied. Thus if $\Omega \neq 0$, then (3-3.7) can be satisfied only by $g_1 = g_2 = \cdots = 0$ and the result is established. □

Lemma 3-3.4. *If $\eta^1, \ldots, \eta^k$ are k elements of $\Lambda^1(E_n)$ and another collection of k elements of $\Lambda^1(E_n)$ is defined by*

$$\omega^a = K^a_b \eta^b, \tag{3-3.9}$$

then

$$\omega^1 \wedge \cdots \wedge \omega^k = \det(K^a_b)\eta^1 \wedge \cdots \wedge \eta^k. \tag{3-3.10}$$

Proof. The simplest way of establishing this result is by induction on the number of 1-forms. For one 1-form we obviously have

$$\omega^1 = K^1_1 \eta^1 = \det(K^1_1)\eta^1,$$

and for two,

$$\begin{aligned} \omega^1 \wedge \omega^2 = K^1_a K^2_b \eta^a \wedge \eta^b &= \left(K^1_1 K^2_2 - K^1_2 K^2_1\right)\eta^1 \wedge \eta^2 \\ &= \det(K^a_b)\eta^1 \wedge \eta^2. \end{aligned}$$

Let us assume that the result is true for $r-1$ 1-forms:

$$\omega^1 \wedge \cdots \wedge \omega^{r-1} = \det(K^p_q)\eta^1 \wedge \cdots \wedge \eta^{r-1} \tag{3-3.11}$$

where (K^p_q) is an $(r-1)$-by-$(r-1)$ matrix. In this event, we have

(3-3.12)

$$\begin{aligned} \omega^1 \wedge \cdots \wedge \omega^{r-1} \wedge \omega^r &= K^1_{a_1} K^2_{a_2} \cdots K^{r-1}_{a_{r-1}} K^r_{a_r} \eta^{a_1} \wedge \cdots \wedge \eta^{a_{r-1}} \wedge \eta^{a_r} \\ &= K^1_{a_1} K^2_{a_2} \cdots K^{r-1}_{a_{r-1}} K^r_1 \eta^{a_1} \wedge \cdots \wedge \eta^{a_{r-1}} \wedge \eta^1 \\ &\quad + K^1_{a_1} K^2_{a_2} \cdots K^{r-1}_{a_{r-1}} K^r_2 \eta^{a_1} \wedge \cdots \wedge \eta^{a_{r-1}} \wedge \eta^2 + \cdots \\ &\quad + K^1_{a_1} K^2_{a_2} \cdots K^{r-1}_{a_{r-1}} K^r_r \eta^{a_1} \wedge \cdots \wedge \eta^{a_{r-1}} \wedge \eta^r. \end{aligned}$$

Now, observe that none of the indices $a_1, \ldots, a_{r-1}$ in the first term can be

equal to 1 for otherwise η^1 would appear twice as a factor and the answer would be zero. Likewise, none of $a_1, \ldots, a_{r-1}$ in the second term can be $2, \ldots,$ and none of $a_1, \ldots, a_{r-1}$ in the last term can be equal to r. Thus if we write (3-3.12) in the equivalent form

$$\omega^1 \wedge \cdots \wedge \omega^r = K_1^r \big(K_{a_1}^1 K_{a_2}^2 \cdots K_{a_{r-1}}^{r-1} \eta^{a_1} \wedge \cdots \wedge \eta^{a_{r-1}} \big) \wedge \eta^1 + \cdots$$

$$+ K_r^r \big(K_{a_1}^1 K_{a_2}^2 \cdots K_{a_{r-1}}^{r-1} \eta^{a_1} \wedge \cdots \wedge \eta^{a_{r-1}} \big) \wedge \eta^r$$

then reordering the factors η^p and using the induction hypothesis gives

$$= K_1^r C_1^r \eta^1 \wedge \eta^2 \wedge \cdots \wedge \eta^r + \cdots + K_r^r C_r^r \eta^1 \wedge \eta^2 \wedge \cdots \eta^r$$

$$= \det(K_b^a) \eta^1 \wedge \cdots \wedge \eta^r$$

where C_b^r is the cofactor of K_b^r. □

This last result is quite useful in the study of determinants. For example, suppose that $\eta^b = L_c^b \rho^c$ for another collection $\rho^1, \ldots, \rho^k$ of 1-forms. We then have

$$\omega^a = K_b^a L_c^b \rho^c$$

from (3-3.9). Using Lemma 3-3.4 then gives

$$\omega^1 \wedge \cdots \wedge \omega^r = \det(K_b^a) \eta^1 \wedge \cdots \wedge \eta^r,$$

$$\eta^1 \wedge \cdots \wedge \eta^r = \det(L_b^a) \rho^1 \wedge \cdots \wedge \rho^r,$$

$$\omega^1 \wedge \cdots \wedge \omega^r = \det(K_b^a L_c^b) \rho^1 \wedge \cdots \wedge \rho^r$$

from which we may conclude that

$$\det(K_b^a L_c^b) = \det(K_b^a) \det(L_e^c).$$

All of the necessary results are now at hand in order to obtain the basic characterization of subspaces of $\Lambda^1(E_n)$.

Theorem 3-3.1. *There exists a nonzero simple r-form Ω for each r-dimensional subspace U_r of $\Lambda^1(E_n)$, that is determined by U_r up to a nonzero scalar factor, such that α belongs to U_r if and only if*

(3-3.13) $$\alpha \wedge \Omega = 0.$$

Let U_{r_1} and U_{r_2} be two subspaces of $\Lambda^1(E_n)$ with simple r_1- and r_2-forms Ω_1 and

Ω_2, *respectively. A necessary and sufficient condition that* U_{r_1} *be contained in* U_{r_2} *is that there exists an* $(r_2 - r_1)$*-form* γ *such that* $\Omega_2 = \Omega_1 \wedge \gamma$. *A necessary and sufficient condition that* U_{r_1} *and* U_{r_2} *have only the zero* 1*-form in common is* $\Omega_1 \wedge \Omega_2 \neq 0$. *If this condition is satisfied, then* $\Omega_1 \wedge \Omega_2$ *is a simple* $(r_1 + r_2)$*-form of the subspace* $U_{r_1} \cup U_{r_2}$.

Proof. If $\omega^1, \ldots, \omega^{r_1}$ is a basis for U_{r_1}, then set $\Omega_1 = \omega^1 \wedge \cdots \wedge \omega^{r_1}$. If $\eta^1, \ldots, \eta^{r_1}$ is any other basis for U_{r_1}, then $\eta^a = K^a_b \omega^b$ with $\det(K^a_b) \neq 0$. Lemma 3-3.4 shows that $\eta^1 \wedge \cdots \wedge \eta^{r_1} = \det(K^a_b)\Omega_1$, and hence Ω_1 is determined by U_{r_1} up to a scalar factor. An $\alpha \in \Lambda^1(E_n)$ belongs to U_{r_1} if and only if α is a linear combination of the basis elements of U_{r_1}, in which case $\alpha, \omega^1, \ldots, \omega^{r_1}$ are linearly dependent. Lemma 3-3.3 shows that a necessary and sufficient condition for this to hold is that $0 = \alpha \wedge \omega^1 \wedge \cdots \wedge \omega^{r_1} = \alpha \wedge \Omega_1$. A necessary and sufficient condition that U_{r_1} be contained in U_{r_2} is that a basis $\omega^1, \ldots, \omega^{r_1}$ of U_{r_1} can be extended to a basis $\omega^1, \ldots, \omega^{r_1}, \omega^{r_1+1}, \ldots, \omega^{r_2}$ of U_{r_2}. In this event, $\Omega_1 = \omega^1 \wedge \cdots \wedge \omega^{r_1}$, $\Omega_2 = \omega^1 \wedge \cdots \wedge \omega^{r_1} \wedge \omega^{r_1+1} \wedge \cdots \wedge \omega^{r_2} = \Omega_1 \wedge \omega^{r_1+1} \wedge \cdots \wedge \omega^{r_2} = \Omega_1 \wedge \gamma$ with $\gamma = \omega^{r_1+1} \wedge \cdots \wedge \omega^{r_2} \in \Lambda^{r_2 - r_1}(E_n)$. A necessary and sufficient condition that $U_{r_1} \cap U_{r_2}$ be the zero element of $\Lambda^1(E_n)$ is that $\omega^1, \ldots, \omega^{r_1}, \beta^1, \ldots, \beta^{r_2}$ be independent, where $\omega^1, \ldots, \omega^{r_1}$ is a basis for U_{r_1} and $\beta^1, \ldots, \beta^{r_2}$ is a basis for U_{r_2}. This is the case if and only if $\Omega_1 \wedge \Omega_2 = \omega^1 \wedge \cdots \wedge \omega^{r_1} \wedge \beta^1 \wedge \cdots \wedge \beta^{r_2} \neq 0$, by Lemma 3-3.3. If this condition is satisfied, then $\Omega_1 \wedge \Omega_2$ is a simple characteristic $(r_1 + r_2)$-form of $U_{r_1} \cup U_{r_2}$. □

The formation of forms of degree higher than one is indeed a useful construct. This is particularly true when we realize that most algebraic problems revolve around characterizations of subspaces of solutions, etc. In this regard, the following Lemma of E. Cartan proves to be quite useful.

Lemma 3-3.5. *Let* $\omega^1, \ldots, \omega^r$ *be r linearly independent* 1*-forms and suppose that the r* 1*-forms* $\gamma_1, \ldots, \gamma_r$ *satisfy*

(3-3.14) $$\omega^1 \wedge \gamma_1 + \omega^2 \wedge \gamma_2 + \cdots + \omega^r \wedge \gamma_r = 0.$$

Then each of the 1*-forms* $\gamma_1, \ldots, \gamma_r$ *belongs to the subspace spanned by* $\omega^1, \ldots, \omega^r$. *Thus there exists a matrix* $((A_{ab}))$ *such that*

(3-3.15) $$\gamma_a = A_{ab}\omega^b$$

and, in addition, we have

(3-3.16) $$A_{ab} = A_{ba}, \qquad a, b = 1, \ldots, r.$$

Proof. Since $\omega^1, \ldots, \omega^r$ are linearly independent,

(3-3.17) $$\Omega = \omega^1 \wedge \omega^2 \wedge \cdots \wedge \omega^r \neq 0.$$

Thus if we veck (3-3.14) with $\omega^2 \wedge \omega^3 \wedge \cdots \wedge \omega^r$, we obtain

$$\gamma^1 \wedge \Omega = 0. \tag{3-3.18}$$

Indeed, proceeding in like fashion we see that

$$\gamma^a \wedge \Omega = 0, \qquad a = 1, \dots, r \tag{3-3.19}$$

and hence each of the γ's is linearly dependent on the ω's. This establishes (3-3.15). A substitution of (3-3.15) back into (3-3.14) yields the additional condition (3-3.16). □

3-4. INNER MULTIPLICATION

The graded exterior algebra $\Lambda(E_n)$ can be considered as generated from $\Lambda^0(E_n)$ and $\Lambda^1(E_n)$ by the exterior product, $\wedge$, together with the vector space properties that are thereby implied. On the other hand, the exterior product is only an ascending operation,

$$\wedge : \Lambda^k(E_n) \times \Lambda^m(E_n) \to \Lambda^{k+m}(E_n),$$

and hence we cannot come back down the ladder once we have reached the topmost collection $\Lambda^n(E_n)$. Serious difficulties would thus arise if we did not define an operation that lowers the degrees of forms.

Definition. *Inner multiplication*, or the *pull down* is a map

$$\lrcorner : T(E_n) \times \Lambda^k(E_n) \to \Lambda^{k-1}(E_n) \tag{3-4.1}$$

with the following properties:

I1. $V \lrcorner f = 0$ for all $V \in T(E_n)$ and all $f \in \Lambda^0(E_n)$.
I2. $V \lrcorner \omega = \langle \omega, V \rangle$ for all $V \in T(E_n)$ and all $\omega \in \Lambda^1(E_n)$.
I3. $V \lrcorner (\alpha + \beta) = V \lrcorner \alpha + V \lrcorner \beta$ for all $\alpha, \beta \in \Lambda^k(E_n)$, $k = 1, \dots, n$ and all $V \in T(E_n)$.
I4. $V \lrcorner (\alpha \wedge \beta) = (V \lrcorner \alpha) \wedge \beta + (-1)^{\deg(\alpha)} \alpha \wedge (V \lrcorner \beta)$ for all $\alpha, \beta \in \Lambda(E_n)$ and all $V \in T(E_n)$.

We have taken a somewhat different tack here in that $\lrcorner$ is defined in terms of its axioms **I1** through **I4**, rather than define the operation directly and then establish its properties. It is incumbent upon us to show that **I1** through **I4** result in $\lrcorner$ being well defined.

Lemma 3-4.1. *The operation* $\lrcorner$ *that satisfies axioms* **I1** *through* **I4** *is well defined on* $T(E_n) \times \Lambda(E_n)$ *and yields a map of* $T(E_n) \times \Lambda^k(E_n)$ *to* $\Lambda^{k-1}(E_n)$. *It also satisfies*

I5. $(f \cdot U + g \cdot V) \lrcorner \omega = f \cdot (U \lrcorner \omega) + g \cdot (V \lrcorner \omega)$.

Proof. Axiom **I1** defines $\lrcorner$ for all forms of degree zero and axiom **I2** defines $\lrcorner$ for all forms of degree one. In particular,

$$V = v^i \partial_i, \qquad \omega = w_i \, dx^i, \tag{3-4.2}$$

then

$$V \lrcorner \omega = \langle \omega, V \rangle = w_i v^i \in \Lambda^0(E_n). \tag{3-4.3}$$

Since $\langle \omega, V \rangle$ is a linear functional, **I5** is clearly satisfied for forms of degree zero and one. On the other hand, any form of degree k is the sum of simple forms of degree k, and hence it is sufficient to show that $\lrcorner$ is well defined for a simple k-form by axiom **I3**. However, axioms **I2** and **I4** yield

(3-4.4)

$$\begin{aligned}
&V \lrcorner (\omega^1 \wedge \omega^2 \wedge \cdots \wedge \omega^k)\\
&= (V \lrcorner \omega^1) \wedge \omega^2 \wedge \cdots \wedge \omega^k - \omega^1 \wedge (V \lrcorner (\omega^2 \wedge \omega^3 \wedge \cdots \wedge \omega^k))\\
&= \langle \omega^1, V \rangle \omega^2 \wedge \cdots \wedge \omega^k\\
&\quad - \omega^1 \wedge \{(V \lrcorner \omega^2) \wedge \omega^3 \wedge \cdots \wedge \omega^k - \omega^2 \wedge (V \lrcorner (\omega^3 \wedge \cdots \wedge \omega^k))\}\\
&= \langle \omega^1, V \rangle \omega^2 \wedge \cdots \wedge \omega^k - \langle \omega^2, V \rangle \omega^1 \wedge \omega^3 \wedge \cdots \wedge \omega^k\\
&\quad + \langle \omega^3, V \rangle \omega^1 \wedge \omega^2 \wedge \omega^4 \wedge \cdots \wedge \omega^k\\
&\quad + \cdots + (-1)^{k-1} \langle \omega^k, V \rangle \omega^1 \wedge \cdots \wedge \omega^{k-1}
\end{aligned}$$

and this belongs to $\Lambda^{k-1}(E_n)$ since each term belongs to $\Lambda^{k-1}(E_n)$. It is also clear from (3-4.4) that property **I5** is likewise satisfied since $\langle \omega^k, V \rangle$ is a linear functional for each value of k. □

Lemma 3-4.2. *We have*

I6. $U \lrcorner (V \lrcorner \omega) = -V \lrcorner (U \lrcorner \omega)$

for all $U, V \in T(E_n)$ *and all* $\omega \in \Lambda(E_n)$, *and hence*

$$V \lrcorner (V \lrcorner \omega) = 0. \tag{3-4.5}$$

Proof. Since any element of $\Lambda^k(E_n)$ is the sum of simple elements, it is sufficient to establish property **I6** for simple elements of $\Lambda^k(E_n)$. In this event, simply allow $U \lrcorner$ to act on both sides of (3-4.4), reverse the roles of U and V and then add the results. The answer is zero which is the same thing as **I6**, and this implies (3-4.5). □

It does prove useful to go through a direct proof of Lemma 3-4.2 for basis elements of $\Lambda^2(E_n)$ in order to improve our ability to work with the pull down operation:

$$V \lrcorner (dx^i \wedge dx^j) = (V \lrcorner dx^i) \wedge dx^j - dx^i \wedge (V \lrcorner dx^j)$$

$$= \langle dx^i, V \rangle \, dx^j - \langle dx^j, V \rangle \, dx^i = v^i dx^j - v^j dx^i;$$

$$U \lrcorner (V \lrcorner (dx^i \wedge dx^i)) = v^i U \lrcorner dx^j - v^j U \lrcorner dx^i = v^i u^j - v^j u^i;$$

$$V \lrcorner (U \lrcorner (dx^i \wedge dx^j)) = u^i v^j - u^j v^i = -U \lrcorner (V \lrcorner (dx^i \wedge dx^j)).$$

We have already seen that a linear first order partial differential equation can be written in the form

$$V\langle f \rangle = 0 \tag{3-4.6}$$

for a given $V \in T(E_n)$. If we further recall that df is an element of $\Lambda^1(E_n)$ that is defined by

$$\langle df, V \rangle = V\langle f \rangle, \tag{3-4.7}$$

then the definition of $\lrcorner$ shows that

$$V \lrcorner df = \langle df, V \rangle = V\langle f \rangle. \tag{3-4.8}$$

The partial differential equation (3-4.6) can thus be written in the equivalent form

$$V \lrcorner df = 0, \tag{3-4.9}$$

and the vector field V is the characteristic vector field of this partial differential equation. The equivalence of the formulations (3-4.6) and (3-4.9) clearly provide a means of viewing partial differential equations in terms of exterior forms. In this vein the following definition proves to be useful.

Definition. A vector field $V \in T(E_n)$ is said to be a *characteristic* vector field of the given exterior form ω if and only if

$$V \lrcorner \omega = 0. \tag{3-4.10}$$

Lemma 3-4.3. *If a given exterior form ω has one nonzero characteristic vector field, then the collection of all characteristic vector fields of ω forms a subspace, $\omega T(E_n)$, of $T(E_n)$.*

Proof. If $V \neq 0$ is a characteristic vector field of ω then $f \cdot (V \lrcorner \omega) = (f \cdot V) \lrcorner \omega = 0$ and hence fV is a characteristic vector field of ω. The exterior form ω thus has at least a one-dimensional subspace of characteristic vector fields. If U and V are characteristic vector fields, then

$$(fU + gV) \lrcorner \omega = fU \lrcorner \omega + gV \lrcorner \omega = 0$$

and hence the collection of all characteristic vector fields of ω is closed under the vector space operations of $T(E_n)$; that is $\omega T(E_n)$ is a subspace of $T(E_n)$. □

Lemma 3-4.4. *If*

$$\omega = w_i(x^j)\, dx^i \tag{3-4.11}$$

is a 1*-form, then each of the* $n - 1$ *vector fields*

$$V_k = w_1(x^j)\,\partial_k - w_k(x^j)\,\partial_1, \qquad k = 2, \ldots, n \tag{3-4.12}$$

is a characteristic vector field of ω. *Thus if* $w_1(x^j) \neq 0$ *on some neighborhood* N *in* E_n, *then* ω *has an* $(n - 1)$*-dimensional characteristic subspace on* N.

Proof. If we use (3-4.12) to form $V_k \lrcorner \omega$, we obtain

$$V_k \lrcorner \omega = w_1 \partial_k \lrcorner \omega - w_k \partial_1 \lrcorner \omega = w_1 w_k - w_k w_1 = 0.$$

Since it is clear by inspection of (3-4.12) that these $n - 1$ vectors are linearly independent whenever $w_1(x^j) \neq 0$, the result follows. □

The important thing to be realized here is that the $n - 1$ characteristic vector fields need not mesh on the neighborhood N so that they are tangent to a family of surfaces of dimension $n - 1$. We will come back to this question later when we have more definite information concerning the possible shapes that a 1-form may take.

Some examples may prove helpful at this point, if

$$\omega = \tfrac{1}{2} a_{ij}\, dx^i \wedge dx^j, \qquad a_{ij} = -a_{ji}$$

then

$$\begin{aligned}
V \lrcorner \omega &= \tfrac{1}{2} a_{ij} V \lrcorner (dx^i \wedge dx^j) = \tfrac{1}{2} a_{ij} \{ (V \lrcorner dx^i) \wedge dx^j - dx^i \wedge (V \lrcorner dx^j) \} \\
&= \tfrac{1}{2} a_{ij} \{ v^i\, dx^j - v^j\, dx^i \} = \tfrac{1}{2} a_{ij} v^i\, dx^j - \tfrac{1}{2} a_{ij} v^j\, dx^i \\
&= \tfrac{1}{2} a_{ji} v^j\, dx^i - \tfrac{1}{2} a_{ij} v^j\, dx^i \\
&= \tfrac{1}{2} (a_{ji} - a_{ij}) v^j\, dx^i = v^j a_{ji}\, dx^i;
\end{aligned}$$

and if

$$\omega = \frac{1}{3!} a_{ijk}\, dx^i \wedge dx^j \wedge dx^k, \qquad a_{ijk} = -a_{jik},$$

$$a_{ijk} = -a_{ikj}, \qquad a_{ijk} = -a_{kji}$$

then

$$V \lrcorner \omega = \frac{1}{3!} a_{ijk}\left(v^i\, dx^j \wedge dx^k - v^j\, dx^i \wedge dx^k + v^k\, dx^i \wedge dx^j\right)$$

$$= \frac{1}{3!}\left(a_{ijk} v^i\, dx^j \wedge dx^k - a_{jik} v^i\, dx^j \wedge dx^k + a_{jki} v^i\, dx^j \wedge dx^k\right)$$

$$= \frac{1}{3!} v^i \left(a_{ijk} - a_{jik} + a_{jki}\right) dx^j \wedge dx^k$$

$$= \frac{1}{3!} v^i \left(a_{ijk} + a_{ijk} + a_{ijk}\right) dx^j \wedge dx^k$$

$$= \frac{1}{2} v^i a_{ijk}\, dx^j \wedge dx^k.$$

Thus, in general, if

(3-4.13) $$\omega = \frac{1}{k!} a_{i_1 i_2 \cdots i_k}\, dx^{i_1} \wedge dx^{i_2} \wedge \cdots \wedge dx^{i_k}$$

with $\{a_{i_1 i_2 \cdots i_k}\}$ antisymmetric in all pairs of indices, then

(3-4.14) $$V \lrcorner \omega = \frac{1}{(k-1)!} v^j a_{j i_2 \cdots i_k}\, dx^{i_2} \wedge dx^{i_3} \wedge \cdots \wedge dx^{i_k}.$$

An immediate use of the pull down operator $\lrcorner$ is that it provides a means of testing whether a given exterior form is simple or not.

Lemma 3-4.5. *Let ω be a given nonzero element of $\Lambda^k(E_n)$ and let $(\omega | V_1, V_2, \ldots, V_{k-1})$ be the 1-form that is defined by*

(3-4.15) $$(\omega | V_1, V_2, \ldots, V_{k-1}) = V_{k-1} \lrcorner V_{k-2} \lrcorner \cdots \lrcorner V_2 \lrcorner V_1 \lrcorner \omega.$$

The k-form ω is simple if and only if

(3-4.16) $$\omega \wedge (\omega | V_1, V_2, \ldots, V_{k-1}) = 0$$

for all $(k-1)$-tuples of elements $V_1, \ldots, V_{k-1}$ of $T(E_n)$.

Proof. The proof will be given for the case $k = 2$ since it is representative of the general case. If $\omega \in \Lambda^2(E_n)$ is simple, then

$$\omega = \omega_1 \wedge \omega_2 \tag{3-4.17}$$

with ω_1 and ω_2 linearly independent 1-forms. Pick a basis $\{U^1, U^2, \dots, U^n\}$ of $T(E_n)$ such that

$$U^1 \lrcorner \omega_1 = 1, \qquad U^1 \lrcorner \omega_2 = 0, \qquad U^2 \lrcorner \omega_1 = 0, \qquad U^2 \lrcorner \omega_2 = 1; \tag{3-4.18}$$

$$U^b \lrcorner \omega_1 = 0, \qquad U^b \lrcorner \omega_2 = 0, \qquad b = 3, \dots, n,$$

then any $V \in T(E_n)$ may be expressed as

$$V = p_1 U^1 + p_2 U^2 + p_3 U^3 + \cdots + p_n U^n.$$

Now, any V belonging to the subspace of $T(E_n)$ that is spanned by $\{U_3, U_4, \dots, U_n\}$ has the property that

$$V \lrcorner \omega = V \lrcorner (\omega_1 \wedge \omega_2) = (V \lrcorner \omega_1)\omega_2 - (V \lrcorner \omega_2)\omega_1 = 0.$$

Hence we need consider only those V's that belong to the subspace $\mathscr{S}_2$ that is spanned by $\{U_1, U_2\}$:

$$V = r_1 U^1 + r_2 U^2. \tag{3-4.19}$$

But (3-4.15) and (3-4.18) give

$$(\omega | U^1) = U^1 \lrcorner \omega = U^1 \lrcorner (\omega_1 \wedge \omega_2) = \omega_2,$$

$$(\omega | U^2) = U^2 \lrcorner \omega = U^2 \lrcorner (\omega_1 \wedge \omega_2) = -\omega_1,$$

and hence

$$(\omega | V) = r_1 \omega_2 - r_2 \omega_1 \tag{3-4.20}$$

for V in $\mathscr{S}_2$. A combination of (3-4.17) with (3-4.20) now gives

$$\omega \wedge (\omega | V) = (\omega_1 \wedge \omega_2) \wedge (r_1 \omega_1 - r_2 \omega_2) = 0$$

for all r_1, r_2 and hence for all V in $\mathscr{S}_2$. Thus if ω is simple then (3-4.16) is satisfied. On the other hand, a general $\omega \in \Lambda^2(E_n)$ is given by

$$\omega = \tfrac{1}{2} w_{ij}\, dx^i \wedge dx^j, \qquad w_{ij} = -w_{ji}, \tag{3-4.21}$$

so that any $V = v^m \partial_m$ yields

$$(\omega|V) = v^m w_{mk}\, dx^k \tag{3-4.22}$$

by (3-4.14). Thus

$$\omega \wedge (\omega|V) = \tfrac{1}{2} w_{ij}\, dx^i \wedge dx^j \wedge \left(v^m w_{mk}\, dx^k\right) \tag{3-4.23}$$

has to vanish for all possible values of the n quantities $\{v^m\}$ if (3-4.16) is to be satisfied for all $v \in T(E_n)$:

$$w_{mk} w_{ij}\, dx^k \wedge dx^i \wedge dx^j = 0, \qquad m = 1, \ldots, n. \tag{3-4.24}$$

This is a complicated system of $n\binom{n}{3}$ *quadratic equations* for the determination of the $\{w_{ij}\}$ due to the antisymmetry of $dx^k \wedge dx^i \wedge dx^j$ in each pair of indices. We will not actually obtain the solution to this system. Suffice it to say that the general solution is given by

$$w_{ij} = \tfrac{1}{2}(\omega_{1i}\omega_{2j} - \omega_{1j}\omega_{2i}), \tag{3-4.25}$$

in which case $\omega = \omega_1 \wedge \omega_2$ with $\omega_1 = \omega_{1i}\, dx^i$, $\omega_2 = \omega_{2i}\, dx^i$. □

Note. For a form $\omega = (1/r!) w_{i_1 \cdots i_r}\, dx^{i_1} \wedge \cdots \wedge dx^{i_r}$ of degree r, (3-4.16) yields the *quadratic* conditions

$$w_{i_1 \cdots i_{r-1} j_1} w_{j_2 \cdots j_{r+1}}\, dx^{j_1} \cdots dx^{j_{r+1}} = 0. \tag{3-4.26}$$

The condition that an r-form be simple is always a nonlinear requirement on its coefficients. This nonlinearity is what makes a number of problems very difficult.

We have seen that $\Lambda(E_n)$, with the vector space operations on each of its subspaces $\Lambda^k(E_n)$ over $\Lambda^0(E_n)$ and the exterior or veck product $\wedge$ on $\Lambda^k(E_n) \times \Lambda^m(E_n)$ to $\Lambda^{k+m}(E_n)$, forms a graded algebra, where the grading is by the degree operation deg($\cdot$). Axiom **I3**, $V \lrcorner(\alpha + \beta) = V \lrcorner \alpha + V \lrcorner \beta$, shows how the pulldown operation behaves with respect to the vector space operation $+$. What really makes $V \lrcorner$ work, however, is axiom **I4**,

$$V \lrcorner(\alpha \wedge \beta) = (V \lrcorner \alpha) \wedge \beta + (-1)^{\deg(\alpha)} \alpha \wedge (V \lrcorner \beta), \tag{3-4.27}$$

since it tells us how $V \lrcorner$ behaves with respect to exterior multiplication. An inspection of (3-4.27) shows that V would act as a derivation if the factor $(-1)^{\deg(\alpha)}$ were not present (i.e., $V \lrcorner$ would act on $\alpha \wedge \beta$ by the rule of Leibniz). The factor $(-1)^{\deg(\alpha)}$ is present, however, and we must learn to live with it. In point of fact, the factor $(-1)^{\deg(\alpha)}$ is so important that it has given rise to a whole class of new operations for graded algebras.

Definition. Let $\mathscr{A}$ be a graded algebra with grading gr$(\cdot)$ and multiplication $(\cdot|\cdot)$. An operation ρ: is an *antiderivation* on $\mathscr{A}$ if and only if

$$\rho:(\alpha+\beta)=\rho:\alpha+\rho:\beta, \tag{3-4.28}$$

$$\rho:(\alpha|\beta)=(\rho:\alpha|\beta)+(-1)^{\mathrm{gr}(\alpha)}(\alpha|\rho:\beta). \tag{3-4.29}$$

It is now only necessary to note that gr$(\cdot)=$ deg$(\cdot)$ and $(\alpha|\beta)=\alpha\wedge\beta$ for $\mathscr{A}=\Lambda(E_n)$ in order to establish the following result.

Lemma 3-4.6. *Inner multiplication $V\lrcorner$, acts as an antiderivation on the exterior algebra $\Lambda(E_n)$.*

This way of looking at the pulldown operation will prove to be very useful later on. It will turn out that there is another important antiderivation on the exterior algebra and that we will be able to prove things about successive applications of different antiderivations in a simple and direct way.

3-5. TOP DOWN GENERATION OF BASES

The whole graded exterior algebra $\Lambda(E_n)$ has been generated from $\Lambda^0(E_n)$ and $\Lambda^1(E_n)$ by constructing sums of exterior products. In fact, the basis for $\Lambda^k(E_n)$ is constructed by forming the maximal set of linearly independent k-fold exterior products of the basis elements $\{dx^i\}$ of $\Lambda^1(E_n)$. Now that we have the pull down operator $\lrcorner$ that lowers the degree of an exterior form by one, it might be expected that we could generate bases by starting at the top with $\Lambda^n(E_n)$. This is indeed the case as we now show.

Definition. The natural basis of $\Lambda^n(E_n)$ is referred to as the *volume element* of E_n and is denoted by μ:

$$\mu=dx^1\wedge dx^2\wedge\cdots dx^n. \tag{3-5.1}$$

Lemma 3-5.1. *The collection of* $(n-1)$*-forms*

$$\mu_i=\partial_i\lrcorner\mu,\qquad i=1,\dots,n \tag{3-5.2}$$

has the following properties:

1. *The set* $\{\mu_i,\ i=1,\dots,n\}$ *forms a basis for* $\Lambda^{n-1}(E_n)$ *so that any* $\alpha\in\Lambda^{n-1}(E_n)$ *can be uniquely expressed in the form*

$$\alpha=\alpha^i\mu_i. \tag{3-5.3}$$

2. *We have*

(3-5.4) $$dx^i \wedge \mu_j = \delta^i_j \mu.$$

3. *If* $V = v^i \partial_i \in T(E_n)$, *then*

(3-5.5) $$V \lrcorner \mu = v^i \mu_i.$$

4. *If we write* μ *in the equivalent form*

(3-5.6) $$\mu = \frac{1}{n!} e_{i_1 i_2 \cdots i_n} dx^{i_1} \wedge dx^{i_2} \wedge \cdots \wedge dx^{i_n},$$

then

(3-5.7) $$\mu_j = \frac{1}{(n-1)!} e_{j i_2 \cdots i_n} dx^{i_2} \wedge dx^{i_3} \wedge \cdots \wedge dx^{i_n}.$$

5. *If we write* $\alpha \in \Lambda^{n-1}(E_n)$ *in the equivalent form*

(3-5.8) $$\alpha = \frac{1}{(n-1)!} \alpha_{i_2 i_3 \cdots i_n} dx^{i_2} \wedge dx^{i_3} \wedge \cdots \wedge dx^{i_n},$$

then equality of (3-5.3) *and* (3-5.7) *imply*

(3-5.9) $$\alpha_{i_2 i_3 \cdots i_n} = \alpha^j e_{j i_2 i_3 \cdots i_n}.$$

Proof. We first establish (3-5.4). Since μ is a basis for $\Lambda^n(E_n)$, $dx^i \wedge \mu \in \Lambda^{n+1}(E_n)$ and hence $dx^i \wedge \mu = 0$. It thus follows from property **I4** that

$$0 = \partial_j \lrcorner (dx^i \wedge \mu) = (\partial_j \lrcorner dx^i)\mu - dx^i \wedge (\partial_j \lrcorner \mu)$$

$$= \delta^i_j \mu - dx^i \wedge \mu_j.$$

If the equation $0 = c^j \mu_j$ is to be satisfied, we must also satisfy

$$0 = dx^i \wedge (c^j \mu_j) = c^j dx^i \wedge \mu_j = c^j \delta^i_j \mu = c^j \mu$$

for all $j = 1, \ldots, n$. However, $\mu \neq 0$ and hence $0 = c^j \mu_j$ is satisfied if and only if $c^1 = c^2 = \cdots = c^n = 0$; that is, $\{\mu_i, i = 1, \ldots, n\}$ are n linearly independent elements of $\Lambda^{n-1}(E_n)$. Since $\Lambda^{n-1}(E_n)$ is n-dimensional, $\{\mu_i, i = 1, \ldots, n\}$ is a basis for $\Lambda^{n-1}(E_n)$ and (3-5.3) holds. If we veck both sides of (3-5.3) with dx^j, the result is $dx^j \wedge \alpha = \alpha^i dx^j \wedge \mu_i = \alpha^i \delta^j_i \mu = \alpha^j \mu$ by (3-5.4) and hence

(3-5.10) $$\alpha^j \mu = dx^j \wedge \alpha.$$

Thus since $dx^j \wedge \alpha \in \Lambda^n(E_n)$ and $\Lambda^n(E_n)$ is one-dimensional with basis μ,

(3-5.10) serves to uniquely determine $\{\alpha^i\}$. The result (3-5.5) follows directly from the definition of μ_i, for $V \lrcorner \mu = v^i \partial_i \lrcorner \mu = v^i \mu_i$. The remaining results are straightforward calculations that are left to the reader. □

The almost trivial result given by (3-5.5) is the basis for the solution of a number of important problems when (3-5.5) is interpreted properly.

Lemma 3-5.2. *Resolution of $\Lambda^{n-1}(E_n)$ on the basis $\{\mu_i\}$ induces a 1-to-1 correspondence between $T(E_n)$ and $\Lambda^{n-1}(E_n)$. In particular, any $V = v^i \partial_i \in T(E_n)$ induces the $(n-1)$-form $\alpha = v^i \mu_i = V \lrcorner \mu$ and any $\alpha = \alpha^i \mu_i \in \Lambda^{n-1}(E_n)$ induces the vector field $V = \alpha^i \partial_i$.*

This whole process can now be repeated starting with the $(n-1)$-forms $\{\mu_i\}$. The proof of the following results follows exactly the lines given for the proof of Lemma 3-5.1.

Lemma 3-5.3. *The collection of $(n-2)$-forms*

$$\mu_{ji} = \partial_j \lrcorner \mu_i \tag{3-5.11}$$

has the following properties:

1. *We have*

$$\mu_{ji} = -\mu_{ij}. \tag{3-5.12}$$

2. *The set $\{\mu_{ij} | i < j\}$ forms a basis for $\Lambda^{n-2}(E_n)$ and hence any $\alpha \in \Lambda^{n-2}(E_n)$ can be uniquely expressed in the form*

$$\alpha = \tfrac{1}{2}\alpha^{ij}\mu_{ij}, \qquad \alpha^{ij} = -\alpha^{ij}. \tag{3-5.13}$$

3. *We have*

$$dx^i \wedge \mu_{jk} = \delta^i_j \mu_k - \delta^i_k \mu_j. \tag{3-5.14}$$

4. *If $V = v^i \partial_i$ is any element of $T(E_n)$ and $\beta = \beta^i \mu_i$ is any element of $\Lambda^{n-1}(E_n)$, then*

$$V \lrcorner \beta = v^i \beta^j \mu_{ij}. \tag{3-5.15}$$

A direct proof of (3-5.14) can be obtained as follows: $0 = dx^i \wedge \mu_k - \delta^i_k \mu$ implies $0 = \partial_j \lrcorner (dx^i \wedge \mu_k - \delta^i_k \mu) = \delta^i_j \mu_k - dx^i \wedge \mu_{jk} - \delta^i_k \mu_j$. □

Lemma 3-5.4. *The collection of $(n-3)$-forms*

$$\mu_{kji} = \partial_k \lrcorner \mu_{ji} \tag{3-5.16}$$

has the following properties:

1. *We have*

$$\mu_{kij} = -\mu_{ikj}, \qquad \mu_{kij} = -\mu_{kji}. \tag{3-5.17}$$

2. *The set* $\{\mu_{kij} | k < i < j\}$ *forms a basis for* $\Lambda^{n-3}(E_n)$ *and hence any* $\alpha \in \Lambda^{n-3}(E_n)$ *can be uniquely expressed in the form*

$$\alpha = \sum_{i<j<k} \alpha^{ijk} \mu_{ijk}. \tag{3-5.18}$$

3. *We have*

$$dx^i \wedge \mu_{mjk} = \delta^i_m \mu_{jk} + \delta^i_j \mu_{km} + \delta^i_k \mu_{mj}. \tag{3-5.19}$$

Again, the proof of (3-5.19) follows directly from (3-5.14): $0 = dx^i \wedge \mu_{jk} - \delta^i_j \mu_k + \delta^i_k \mu_j$ implies

$$\begin{aligned} 0 &= \partial_m \lrcorner \left(dx^i \wedge \mu_{jk} - \delta^i_j \mu_k + \delta^i_k \mu_j \right) \\ &= \delta^i_m \mu_{jk} - dx^i \wedge \mu_{mjk} - \delta^i_j \mu_{mk} + \delta^i_k \mu_{mj} \\ &= -dx^i \wedge \mu_{mjk} + \delta^i_j \mu_{km} + \delta^i_m \mu_{jk} + \delta^i_k \mu_{mj}. \end{aligned}$$

As an example, consider E_3 with the coordinate cover (x, y, z). We then have

$$\mu = dx \wedge dy \wedge dz$$

and hence

$$\mu_1 = dy \wedge dz, \qquad \mu_2 = -dx \wedge dz, \qquad \mu_3 = dx \wedge dy,$$

$$\mu_{12} = -dz, \qquad \mu_{13} = dy, \qquad \mu_{23} = -dx,$$

$$\mu_{123} = -1.$$

Thus if $\alpha = 3x^4y^2 dx \wedge dy + (z^2 - xy)\, dx \wedge dz + xe^{3y} dy \wedge dz$, then α has the representation

$$\alpha = 3x^4y^2\mu_3 - (z^2 - xy)\mu_2 + xe^{3y}\mu_1,$$

while $\beta = z\,dx + 7x^2e^{-z}\,dy + 4(x^2 - y^2)\,dz$ admits the representation

$$\beta = -z\mu_{23} + 7x^2e^{-z}\mu_{13} - 4(x^2 - y^2)\mu_{12}.$$

For E_4 with the coordinate cover (x, y, z, t), we have $\mu = dx \wedge dy \wedge dz \wedge dt$ and hence

$$\mu_1 = dy \wedge dz \wedge dt, \qquad \mu_2 = -dx \wedge dz \wedge dt, \qquad \mu_3 = dx \wedge dy \wedge dt,$$

$$\mu_4 = -dx \wedge dy \wedge dz.$$

3-6. IDEALS OF $\Lambda(E_n)$

There is a construct that uses the full scope of $\Lambda(E_n)$ as a graded algebra and which will prove to be of considerable importance in later studies; namely, an ideal. A brief discussion of ideals has been given in Chapter one. There we distinguished between right ideals and left ideals. However, since the exterior product on $\Lambda(E_n)$ has the property

$$\alpha \wedge \beta = (-1)^{\deg(\alpha)\deg(\beta)} \beta \wedge \alpha,$$

a left ideal is always a right ideal and conversely. In this circumstance we need only speak of an ideal, where it is understood that it is both a right and a left ideal; there are only two-sided ideals of the algebra $\Lambda(E_n)$. We therefore recall the definition of an ideal in this simpler context.

Definition. A (homogeneous) *ideal*, I, of the exterior algebra $\Lambda(E_n)$ is a set of elements of $\Lambda(E_n)$ such that:

1. If $\alpha \in I$ and $\beta \in I$, then $\alpha + \beta \in I$ if the sum is defined.
2. If $\alpha \in I$, then $\gamma \wedge \alpha \in I$ for every $\gamma \in \Lambda(E_n)$.

This definition, simply put, says that an ideal of $\Lambda(E_n)$ is a graded vector subspace of $\Lambda(E_n)$ such that exterior multiplication of any element of $\Lambda(E_n)$ by an element of the ideal gives back an element of the ideal. Thus in particular, if $\alpha \in I$ then $f \cdot \alpha \in I$ for every $f \in \Lambda^0(E_n)$. An ideal of $\Lambda(E_n)$ is therefore a subset of $\Lambda(E_n)$ that is closed under the vector space operation and is also closed under exterior multiplication by any element of $\Lambda(E_n)$. You cannot get outside of an ideal of $\Lambda(E_n)$ by exterior multiplication by anything in $\Lambda(E_n)$.

Suppose we are given k elements $\alpha_1, \alpha_2, \ldots, \alpha_k$ of $\Lambda(E_n)$. We can then construct all elements of the form

$$\beta = \gamma_1 \wedge \alpha_1 + \gamma_2 \wedge \alpha_2 + \cdots + \gamma_k \wedge \alpha_k \tag{3-6.1}$$

provided $\deg(\gamma_a) + \deg(\alpha_a) = k$ so that the additions are defined. This condition is the "homogeneity" part; namely, homogeneous with respect to degree.

Let $I\{\alpha_1,\dots,\alpha_k\}$ denote the collection of all such elements. If ρ^1 and ρ^2 are two elements of $I\{\alpha_1,\dots,\alpha_k\}$ of the same degree,

$$\rho^1 = \gamma_1 \wedge \alpha_1 + \cdots + \gamma_k \wedge \alpha_k, \qquad \deg(\gamma_a) + \deg(\alpha_a) = k,$$

$$\rho^2 = \sigma_1 \wedge \alpha_1 + \cdots + \sigma_k \wedge \alpha_k, \qquad \deg(\sigma_a) + \deg(\alpha_a) = k,$$

then

$$\rho^1 + \rho^2 = (\gamma_1 + \sigma_1) \wedge \alpha_1 + \cdots + (\gamma_k + \sigma_k) \wedge \alpha_k.$$

Accordingly the set $I\{\alpha_1,\dots,\alpha_k\}$ is closed under addition. Further if σ is an arbitrary element of $\Lambda(E_n)$, and $\beta \in I\{\alpha_1,\dots,\alpha_k\}$, we have

$$\begin{aligned}\sigma \wedge \beta &= \sigma \wedge (\gamma_1 \wedge \alpha_1 + \cdots + \gamma_k \wedge \alpha_k)\\ &= (\sigma \wedge \gamma_1) \wedge \alpha_1 + \cdots + (\sigma \wedge \gamma_k) \wedge \alpha_k,\end{aligned}$$

and hence the exterior produce of any element of $I\{\alpha_1,\dots,\alpha_k\}$ with any element of $\Lambda(E_n)$ belongs to $I\{\alpha_1,\dots,\alpha_k\}$. The definition of an ideal shows that $I\{\alpha_1,\dots,\alpha_k\}$ *is an ideal of* $\Lambda(E_n)$. Ideals that are formed in this way will occur over and over again, so it is useful to have a convenient manner of referring to them.

Definition. An ideal I is said to be *generated* by the elements $\alpha_1, \alpha_2, \dots, \alpha_k$ if and only if every element of I is the sum of terms, each of which contains at least one of the elements $\alpha_1,\dots,\alpha_k$ as a factor. If I is generated by $\alpha_1,\dots,\alpha_k$, then we write $I\{\alpha_1,\dots,\alpha_k\}$.

If we are in E_4 with the coordinate cover (x, y, z, t) and we are given

$$\alpha_1 = 2\,dx - 3y\,dz, \qquad \alpha_2 = x\,dy - z\,dt,$$

$$\alpha_3 = x^2 t\,dx \wedge dt - t\,dy \wedge dz,$$

then any 1-form in $I\{\alpha_1, \alpha_2, \alpha_3\}$ looks like

$$f(2\,dx - 3y\,dz) + g(x\,dy - z\,dt)$$

for some $f, g \in \Lambda^0(E_n)$, and any 2-form in $I\{\alpha_1, \alpha_2, \alpha_3\}$ looks like

$$\beta \wedge (2\,dx - 3y\,dz) + \gamma \wedge (x\,dy - z\,dt) + f(x^2 t\,dx \wedge dt - t\,dy \wedge dz)$$

for some $\beta, \gamma \in \Lambda^1(E_n)$, $f \in \Lambda^0(E_n)$.

One of the primary uses of ideals is that it allows us to talk about forms that are determined up to sums of products of given forms in a simple and direct manner. Suppose that β is determined by the relation

$$\beta = \rho + \gamma_1 \wedge \alpha_1 + \gamma_2 \wedge \alpha_2 + \cdots + \gamma_k \wedge \alpha_k, \tag{3-6.2}$$

where ρ does not belong to $I\{\alpha_1, \ldots, \alpha_k\}$ and the forms $\gamma_1, \gamma_2, \ldots, \gamma_k$ are not known. It is still meaningful to say that β is determined by ρ to within an element of $I\{\alpha_1, \ldots, \alpha_k\}$. This statement can be made more precise if we rewrite (3-6.2) in the equivalent form

$$\beta - \rho \in I\{\alpha_1, \ldots, \alpha_k\}. \tag{3-6.3}$$

In order to recover an actual equation, it is useful to write

$$\beta - \rho \equiv 0 \bmod I\{\alpha_1, \ldots, \alpha_k\} \tag{3-6.4}$$

(read, $\beta - \rho$ is congruent to zero modulo the ideal $I\{\alpha_1, \ldots, \alpha_k\}$), or

$$\beta \equiv \rho \bmod I\{\alpha_1, \ldots, \alpha_k\}. \tag{3-6.5}$$

The nice thing about this notation is that any such relation remains true under exterior multiplication just as zero multiplied by anything is still zero. Simply observe that $\gamma \wedge (\beta - \rho) \in I\{\alpha_1, \ldots, \alpha_k\}$ if $\beta - \rho \in I\{\alpha_1, \ldots, \alpha_k\}$ since $I\{\alpha, \ldots, \alpha_k\}$ is generated by $\alpha_1, \ldots, \alpha_k$ and is closed under arbitrary exterior multiplication. Thus

$$\beta - \rho \equiv 0 \bmod I \Rightarrow \gamma \wedge (\beta - \rho) \equiv 0 \bmod I \tag{3-6.6}$$

for all $\gamma \in \Lambda(E_n)$, and significant computational simplifications result.

Although the concept of an ideal is an algebraic one, the principle use of ideals of $\Lambda(E_n)$ arises in conjunction with the calculus of exterior forms. A full realization of the utility of ideals must thus wait until we have mastered the material of the next chapter. Suffice it to say that we will use the notion of an ideal over and over again in the applications.

There is an example that can be given at this point that is in some respects typical of the use of ideals. Suppose that $\omega^1, \ldots, \omega^k$ are k given elements of $\Lambda^1(E_n)$ and we wish to determine all vector fields $V \in T(E_n)$ that are simultaneous characteristic vector fields of all of the given 1-forms;

$$V \lrcorner \omega^a = 0, \qquad a = 1, \ldots, k. \tag{3-6.7}$$

In this case, we say that V is a *characteristic vector field of the exterior system* $\{\omega^a,\ a = 1, \ldots, k\}$.

Lemma 3-6.1. *A vector field V is a characteristic vector field of an exterior system of 1-forms $\{\omega^a,\ a = 1, \ldots, k\}$ if and only if $V \lrcorner \rho$ belongs to the ideal $I\{\omega^1, \ldots, \omega^k\}$ for every $\rho \in I\{\omega^1, \ldots, \omega^k\}$.*

Proof. If $\rho \in I\{\omega^1, \dots, \omega^k\}$, then

$$\rho = \beta_a \wedge \omega^a, \tag{3-6.8}$$

and hence

$$\begin{aligned} V \lrcorner \rho &= (V \lrcorner \beta_a) \wedge \omega^a + (-1)^{\deg(\beta_a)} \beta_a \wedge (V \lrcorner \omega^a) \\ &\equiv (-1)^{\deg(\beta_a)} (V \lrcorner \omega^a) \beta_a \bmod I. \end{aligned} \tag{3-6.9}$$

Thus if V is a characteristic vector field of the given exterior system, then $V \lrcorner \rho$ belongs to $I\{\omega^1, \dots, \omega^k\}$ for every $\rho \in I\{\omega^1, \dots, \omega^k\}$. Conversely, if $V \lrcorner \rho \equiv 0 \bmod I\{\omega^1, \dots, \omega^k\}$ for all $\rho \in I\{\omega^1, \dots, \omega^k\}$ then it must be true for all $\rho = K_a \omega^a$, $K_a \in \Lambda^0(E_n)$; that is,

$$K_a V \lrcorner \omega^a \equiv 0 \bmod I\{\omega^1, \dots, \omega^k\}. \tag{3-6.10}$$

However, I is generated by $\omega^1, \dots, \omega^k$ so that the only 0-form in the ideal is the zero function:

$$K_a V \lrcorner \omega^a = 0. \tag{3-6.11}$$

Since this must hold for all possible choices of K_a, we obtain

$$V \lrcorner \omega^a = 0 \qquad a = 1, \dots, k. \tag{3-6.12}$$

□

Lemma 3-6.2. *The set of all characteristic vector fields of an exterior system of 1-forms $\{\omega^a,\ a = 1, \dots, k\}$ forms a vector subspace $\mathscr{S}$ of $T(E_n)$ over $\Lambda^0(E_n)$. If*

$$\Omega = \omega^1 \wedge \omega^2 \wedge \cdots \wedge \omega^k \neq 0 \tag{3-6.13}$$

so that the k given 1-forms are linearly independent, then $\mathscr{S}$ has dimension $n - k$.

Proof. If U and V are characteristic vector fields of ω^a, $a = 1, \dots, k$, then $(f \cdot U + g \cdot V) \lrcorner \rho = f \cdot (U \lrcorner \rho) + g \cdot (V \lrcorner \rho) \equiv 0 + 0 \bmod I\{\omega^1, \dots, \omega^k\}$ for all $\rho \in I\{\omega^1, \dots, \omega^k\}$. Hence characteristic vector fields form a vector subspace of $T(E_n)$ over $\Lambda^0(E_n)$. Under satisfaction of (3-6.13), the k given ω's are linearly independent. Complete them to a basis, $\omega^1, \omega^2, \dots, \omega^k;\ \omega^{k+1}, \dots, \omega^n$, for $\Lambda^1(E_n)$. The basis for $T(E_n)$ that is dual to this one is given by the system of n vector fields V_i such that

$$V_i \lrcorner \omega^j = \delta_i^j. \tag{3-6.14}$$

Thus

$$V_i \lrcorner \omega^j = 0, \qquad i = k+1, \dots, n, \qquad j = 1, \dots, k. \tag{3-6.15}$$

The $n - k$ vector fields $V_{k+1}, \dots, V_n$ are linearly independent by construction and are also characteristic vector fields of the given exterior system. On the other hand, $V_1 \lrcorner \omega^1 = V_2 \lrcorner \omega^2 = \cdots = V_k \lrcorner \omega^k = 1$ and hence $V_1, \ldots, V_k$ are not characteristic vectors of the given exterior system. Accordingly, $V_{k+1}, \ldots, V_n$ span the subspace $\mathscr{S}$ and hence $\dim(\mathscr{S}) = n - k$. □

The method of proof of this lemma requires further comment since it relies upon a method that is invaluable in actual calculations. The basic idea was that $\omega^1, \ldots, \omega^k$ are linearly independent and hence they can be completed to form a basis

$$\omega^1, \ldots, \omega^k;\ \omega^{k+1}, \ldots, \omega^n$$

for all of $\Lambda^1(E_n)$. The dual basis $V_1, \ldots, V_k;\ V_{k+1}, \ldots, V_n$ for $T(E_n)$ is then defined directly from duality by

$$V_i \lrcorner \omega^j = \delta_i^j.$$

These latter equations show that $V_{k+1}, \ldots, V_n$ span the characteristic vector subspace of the differential system $\{\omega^1, \ldots, \omega^k\}$. Incidentally, we also see that $V_1, \ldots, V_k$ span the characteristic vector subspace of the differential system $\{\omega^{k+1}, \ldots, \omega^n\}$. Thus the direct sum decomposition

$$\Lambda^1(E_n) = \operatorname{span}(\omega^1, \ldots, \omega^k) \oplus \operatorname{span}(\omega^{k+1}, \ldots, \omega^n)$$

of $\Lambda^1(E_n)$ leads to a corresponding direct sum decomposition

$$T(E_n) = \operatorname{span}(V_1, \ldots, V_k) \oplus \operatorname{span}(V_{k+1}, \ldots, V_n)$$

of $T(E_n)$ with the characteristic duality properties

$$\operatorname{span}(V_{k+1}, \ldots, V_n) \lrcorner \operatorname{span}(\omega^1, \ldots, \omega^k) = 0,$$

$$\operatorname{span}(V_1, \ldots, V_k) \lrcorner \operatorname{span}(\omega^{k+1}, \ldots, \omega^n) = 0.$$

It is essential to realize that things are only this simple when the ideal is generated by forms of degree one. In order to see this, suppose that I were generated by

$$\alpha_1 = dx - y\,dz, \qquad \alpha_2 = t\,dx \wedge dz - x\,dy \wedge dt$$

and we are in E_4 with the coordinate cover (x, y, z, t). If $V \in T(E_4)$, then $V = v^1\partial_x + v^2\partial_y + v^3\partial_z + v^4\partial_t$ and

$$V \lrcorner \alpha_1 = v^1 - yv^3,$$

$$V \lrcorner \alpha_2 = t(v^1\,dz - v^3\,dx) - x(v^2\,dt - v^4\,dy).$$

Because $V \lrcorner \alpha_1$ is an element of $\Lambda^0(E_4)$ and $I\{\alpha_1, \alpha_2\}$ does not contain elements $\Lambda^0(E_n)$, V will be a characteristic vector field of $I\{\alpha_1, \alpha_2\}$ only if $V \lrcorner \alpha_1 = 0$; that is, $v^1 = yv^3$. On the other hand, $V \lrcorner \alpha_2$ is an element of $\Lambda^1(E_4)$, and hence $V \lrcorner \alpha_2$ can belong to $I\{\alpha_1, \alpha_2\}$ provided we can find an $f \in \Lambda^0(E_4)$ such that $V \lrcorner \alpha_2 = f \cdot \alpha_1$ since $\deg(\alpha_1) = 1$. We thus have the condition

$$t(v^1\,dz - v^3\,dx) - x(v^2\,dt - v^4\,dy) = f(\,dx - y\,dz).$$

When this equation is resolved on the basis $\{dx, dy, dz, dt\}$ and $v^1 = yv^3$ is used, we obtain the conditions

$$-tv^3 = f, \qquad xv^4 = 0, \qquad tyv^3 = -fy, \qquad -xv^2 = 0,$$

respectively. The first serves for the determination of f which then renders the third an identity. Since the second and fourth give $v^2 = v^4 = 0$, the collection of all characteristic vector fields of $I\{\alpha_1, \alpha_2\}$ is of the form

$$V = v^3 \cdot (y\partial_x + \partial_z)$$

for any $v^3 \in \Lambda^0(E_4)$. The set of characteristic vector fields of the ideal $I\{\alpha_1, \alpha_2\}$ is thus the one-dimensional vector subspace of $T(E_4)$ that is spanned by $y\partial_x + \partial_z$. Notice that the dimension here is one rather than two as Lemma 3-6.2 might suggest.

3-7. BEHAVIOR UNDER MAPPINGS

Our discussions of exterior forms have taken place in E_n with the fixed coordinate cover, namely the (x^i) coordinate cover. This coordinate cover has been all pervasive, for the basis elements of $\Lambda^k(E_n)$ and for all of $\Lambda(E_n)$ have been written in terms of exterior products of the differentials of these coordinate functions, $\{dx^i\}$. It is essential that we study how elements of the graded exterior algebra behave under changes of the coordinate cover and under mappings in general, for otherwise our results would be restricted to that one special case of E_n with the (x^i) coordinate system. We shall follow the same procedure as given in Section 2-8 and for the same reasons; that is, we first study the general mapping properties of exterior forms.

Let E_m be an m-dimensional number space with coordinate functions (y^A) and let Φ be a given map from E_n to E_m,

(3-7.1) $$\Phi: E_n \to E_m | y^A = \phi^A(x^i).$$

It was shown in Section 2-8 that Φ induces a map Φ^* of $\Lambda^0(E_m)$ to $\Lambda^0(E_n)$ by composition

(3-7.2) $$\Phi^*: \Lambda^0(E_m) \to \Lambda^0(E_n) | h(x^i) = g \circ \Phi = g(\phi^A(x^i));$$

that is, Φ^* maps in the direction *opposite* to that of Φ. Now, $\Lambda^0(E_n)$ is the first entry of the sequence $\Lambda^0(E_n), \Lambda^1(E_n), \ldots, \Lambda^n(E_n)$ that makes up $\Lambda(E_n)$, and any form of degree k is the sum of terms, each of which is an element of $\Lambda^0(E_n)$ multiplied by a k-fold exterior product of the elements $dx^1, \ldots, dx^n$. It thus follows that the behavior of $\Lambda(E_n)$ under the map Φ will be determined once we obtain the mapping properties of the elements $dx^1, dx^2, \ldots, dx^n$.

The starting point for our considerations is obviously E_m since the domain of Φ^* is $\Lambda^0(E_m)$. Because each of the coordinate functions $\{y^A\}$ is an element of $\Lambda^0(E_m)$, we can combine (3-7.1) and (3-7.2) by writing

(3-7.3) $$\Phi^* y^A = \phi^A(x^j), \qquad A = 1, \ldots, m$$

since $\phi^A(x^j)$ is an element of $\Lambda^0(E_n)$. On the other hand, the base elements $\{dy^A\}$ of $\Lambda^1(E_m)$ are simply the differentials of the functions $\{y^A\}$, while the differentials of $\Phi^* y^A$ are

(3-7.4) $$d\Phi^* y^A = d\phi^A(x^j) = \frac{\partial \phi^A(x^j)}{\partial x^i}\, dx^i$$

because $\Phi^* y^A \in \Lambda^0(E_n)$. However, the right-hand side of (3-7.4) is a linear combination of the base elements $\{dx^i\}$ of $\Lambda^1(E_n)$, and hence $d\Phi^* y^A$ belongs to $\Lambda^1(E_n)$. In other words, we just apply the chain rule of the calculus. These considerations do, nevertheless, establish the consistency of the following definition of the action of Φ^* on the basis $\{dy^A\}$:

(3-7.5) $$\Phi^*\, dy^A \stackrel{\text{def}}{=} d\Phi^* y^A = \left(\partial_i \phi^A(x^j)\right) dx^i.$$

An arbitrary 1-form in E_m has the structure

(3-7.6) $$`\beta = b_A(y^B)\, dy^A$$

where $\{b_A(y^B)\}$ is an m-tuple of elements of $\Lambda^0(E_m)$. Since we know how such functions behave under Φ^* and how the dy^A behave under Φ^*, we would know how $`\beta$ behaves under Φ^* if we require Φ^* of products to be equal to the products of Φ^* acting on each factor separately:

(3-7.7) $$\Phi^* `\beta = \Phi^* b_A \Phi^*\, dy^A.$$

A combination of (3-7.7) with (3-7.4) now gives

(3-7.8) $$\Phi^* `\beta = b_A\left(\phi^B(x^j)\right) \frac{\partial \phi^A(x^j)}{\partial x^i}\, dx^i$$

so that $\Phi^* `\beta$ certainly belongs to $\Lambda^1(E_n)$. Thus if we denote the image of $`\beta$ in

$\Lambda^1(E_n)$ by $\beta = b_i(x^j)\,dx^i$, then (3-7.8) shows that

$$`\beta = b_A(y^B)\,dy^A, \qquad \beta = b_i(x^j)\,dx^i, \qquad \Phi^*`\beta = \beta, \tag{3-7.9}$$

$$b_i(x^j) = b_A(\phi^B(x^j))\frac{\partial\phi^A(x^j)}{\partial x^i}. \tag{3-7.10}$$

It now remains only to check that these definitions are consistent with the fact that elements of $\Lambda^1(E_m)$ are linear functionals on $T(E_m)$ and that elements of $\Lambda^1(E_n)$ are linear functionals on $T(E_n)$. Let

$$V = v^i\partial_i \tag{3-7.11}$$

be an arbitrary element of $T(E_n)$ and suppose that Φ is such that Φ_* is well defined on $T(E_n)$ to $T(E_m)$. In this case we have

$$\Phi_*V = v^i(\partial\phi^A/\partial x^i)\,\partial_A = `v^A\partial_A. \tag{3-7.12}$$

If $`\beta$ and $\Phi^*`\beta$ are as given in (3-7.9), then

$$\langle `\beta, \Phi_*V\rangle = b_A`v^A = b_A\frac{\partial\phi^A}{\partial x^i}v^i,$$

$$\langle \Phi^*`\beta, V\rangle = b_iv^i = b_A\frac{\partial\phi^A}{\partial x^i}v^i$$

and hence

$$\langle `\beta, \Phi_*V\rangle = \langle \Phi^*`\beta, V\rangle; \tag{3-7.13}$$

that is, *the value of the linear functional $\langle\beta, V\rangle$ is invariant under the mappings induced by Φ.*

The result also goes the other way. Suppose that (3-7.13) holds for all $V \in T(E_n)$ and that the action of Φ_* on $T(E_n)$ is well defined. If we write

$$`\beta = b_A(y^B)\,dx^A, \qquad \Phi^*`\beta = b_i(x^j)\,dx^i, \tag{3-7.14}$$

then (3-7.13) gives

$$\begin{aligned} 0 = \langle `\beta, \Phi_*V\rangle - \langle \Phi^*`\beta, V\rangle &= b_A(y^B)`v^A - b_i(x^j)v^i \\ &= b_A(\phi^B(x^j))\frac{\partial\phi^A(x^j)}{\partial x^i}v^i - b_i(x^j)v^i \end{aligned} \tag{3-7.15}$$

for all n-tuples $\{v^1,\ldots,v^n\}$. This is the case, however, if and only if $b_i = b_A\partial\phi^A/\partial x^i$, namely (3-7.10) holds. It should now be clear that the action of Φ^*

can be extended to all of $\Lambda(E_m)$ by the requirement $\Phi^*(\alpha \wedge \beta) = \Phi^*\alpha \wedge \Phi^*\beta$.

Theorem 3-7.1. *A smooth map*

$$\Phi: E_n \to E_m | y^A = \phi^A(x^j) \tag{3-7.16}$$

induces the smooth pull back map

$$\Phi^*: \Lambda(E_m) \to \Lambda(E_n) \tag{3-7.17}$$

by the requirements

$$\Phi^*\, dy^A = d\Phi^* y^A, \qquad A = 1, \ldots, m, \tag{3-7.18}$$

$$\Phi^*(\alpha \wedge \beta) = \Phi^*\alpha \wedge \Phi^*\beta, \qquad \Phi^*(\alpha + \beta) = \Phi^*\alpha + \Phi^*\beta. \tag{3-7.19}$$

Proof. All that remains to be shown is that (3-7.18) and (3-7.19) define Φ^* on all of $\Lambda(E_m)$. Now, any element of Λ is the sum of simple elements and hence $\Phi^*(\alpha + \beta) = \Phi^*\alpha + \Phi^*\beta$ shows that we need only consider simple elements. On the other hand, simple elements are exterior products of 1-forms and hence $\Phi^*(\alpha \wedge \beta) = \Phi^*\alpha \wedge \Phi^*\beta$ shows that we need only establish the result for 1-forms. This has already been done for arbitrary 1-forms ${}'\beta = b_A(y^B)\, dy^A$, and so the result is established. □

We make particular note of the fact that Φ^* *is well defined for any smooth map* Φ. There is no restriction that $m = n$ or that Φ be invertible for the construction of the map Φ^*. This is in strong contrast to the situation with vector fields, for there Φ_* was only well defined for certain kinds of maps. Exterior forms thus have very nice behavior for all smooth mappings of E_n to E_m no matter what the values of n and m are. This pleasant circumstance is one of the reasons why we will use exterior forms rather than vector fields whenever possible.

A detailed examination of the mapping properties of each of the vector spaces $\Lambda^1(E_n), \Lambda^2(E_n), \ldots, \Lambda^n(E_n)$ is most helpful in understanding the full import of the map Φ^*. For 1-forms, let us write ${}'\alpha = {}'\alpha_A(y^B)\, dy^A$ and $\alpha = \Phi^*\,{}'\alpha = \alpha_i(x^j)\, dx^i$, then the previous results give

$$\alpha_i(x^j) = \frac{\partial \phi^A(x^j)}{\partial x^i}\,{}'\alpha_A(\phi^B(x^j)). \tag{3-7.20}$$

For 2-forms, we write

$${}'\beta = \tfrac{1}{2}\,{}'\beta_{AB}(y^C)\, dy^A \wedge dy^B, \qquad \beta = \Phi^*\,{}'\beta = \tfrac{1}{2}\beta_{ij}(x^k)\, dx^i \wedge dx^j. \tag{3-7.21}$$

The relations between the coefficients are obtained directly from how elements of $\Lambda^0(E_m)$ and the dy^A's behave under Φ^*:

$$\Phi^*\,`\beta = \tfrac{1}{2}\Phi^*(`\beta_{AB}\,dy^A \wedge dy^B) = \tfrac{1}{2}\Phi^*`\beta_{AB}\Phi^*\,dy^A \wedge \Phi^*\,dy^B$$

$$= \frac{1}{2}`\beta_{AB}(\phi^C(x^j))\frac{\partial\phi^A}{\partial x^i}\,dx^i \wedge \frac{\partial\phi^B}{\partial x^j}\,dx^j$$

$$= \frac{1}{2}`\beta_{AB}(\phi)\frac{\partial\phi^A}{\partial x^i}\frac{\partial\phi^B}{\partial x^j}\,dx^i \wedge dx^j = \frac{1}{2}\beta_{ij}(x^k)\,dx^i \wedge dx^j,$$

so that

$$\beta_{ij}(x^k) = \frac{\partial\phi^A(x^k)}{\partial x^i}\frac{\partial\phi^B(x^k)}{\partial x^j}`\beta_{AB}(\phi^C(x^k)). \tag{3-7.22}$$

In the case of 3-forms, we have

$$\beta_{ijk} = \frac{\partial\phi^A}{\partial x^i}\frac{\partial\phi^B}{\partial x^j}\frac{\partial\phi^C}{\partial x^k}`\beta_{ABC}(\phi^D(x)), \tag{3-7.23}$$

and so forth. The graphic situation is shown in Fig. 12. An inspection of this figure establishes the following result.

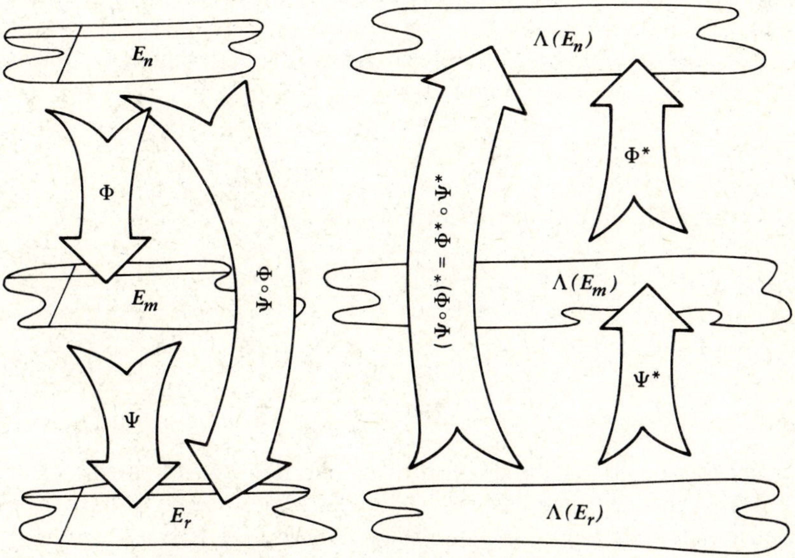

FIGURE 12. The induced pull back maps of Λ

Theorem 3-7.2. *If $\Phi: E_n \to E_m$ and $\Psi: E_m \to E_r$ are smooth and composable, then their composition induces the map $(\Psi \circ \Phi)^*$ with*

$$(\Psi \circ \Phi)^* = \Phi^* \circ \Psi^*. \tag{3-7.24}$$

Let us look in more detail at two specific cases. In the first,

$$\Phi: \mathbb{R} \to E_3 | y^A = \phi^A(t)$$

so that Φ^* is a map from $\Lambda(E_3)$ to $\Lambda(\mathbb{R}) = \Lambda^0(\mathbb{R}) \oplus \Lambda^1(\mathbb{R})$. Suppose that

$$`\beta = f(x, y, z)\, dx + g(x, y, z)\, dy + h(x, y, z)\, dz, \qquad \in \Lambda^1(E_3)$$

then

$$\Phi^* `\beta = \{ f(\phi^1, \phi^2, \phi^3)\, d\phi^1/dt + g(\phi)\, d\phi^2/dt + h(\phi)\, d\phi^3/dt \}\, dt.$$

Further if $`\gamma \in \Lambda^2(E_3)$, then $\Phi^* `\gamma = 0$ because every 2-form on the one-dimensional space $\mathbb{R}$ vanishes identically. In general *if $\Phi: E_n \to E_m$, $n < m$ then $\Phi^* \omega = 0$ for every $\omega \in \Lambda^{n+r}(E_m)$ with $r > 0$*. However,

$$\Phi: E_n \to E_m | y^A = \phi^A(x^1, \ldots, x^n), \qquad A = 1, \ldots, m$$

with $n < m$ defines an n-dimensional surface in E_m, and hence, $\Phi^* \alpha = 0$ if $\deg(\alpha) \leq n$, may be viewed as what it means to "solve" an exterior equation $\alpha = 0$ when α does not vanish identically on E_n.

Definition. Let $\alpha \in \Lambda^r(E_m)$ be a given r-form on E_m that does not vanish identically on E_m. An exterior equation

$$\alpha = 0 \tag{3-7.25}$$

is said to be *solved* in parametric form by a map

$$\Phi: E_k \to E_m | y^A = \phi^A(x^j) \tag{3-7.26}$$

if and only if

$$\Phi^* \alpha = 0 \tag{3-7.27}$$

is satisfied identically on the domain of Φ. We note that $\Phi^* \alpha \in \Lambda^r(E_k)$ and hence $\Phi^* \alpha$ vanishes identically if $k < r$; in which case we have a *trivial* parametric solution: *nontrivial parametric solutions are possible only if $k \geq r$*. We will have much more to say about this important topic later.

An example will probably make the situation as clear as it is going to be at this point. Let α be given by

$$\alpha = x\,dy - 3y\,dx$$

and let $\Phi: \mathbb{R} \to E_2 | x = \phi(t),\ y = \psi(t)$, where $\phi(t)$ and $\psi(t)$ are as yet unknown. We then have

$$\Phi^*\alpha = \left\{ \phi(t)\frac{d\psi}{dt} - 3\psi(t)\frac{d\phi}{dt} \right\} dt,$$

and hence $\Phi^*\alpha = 0$ throughout $\mathbb{R}$ if and only if

$$\phi\frac{d\psi}{dt} = 3\psi(t)\frac{d\phi}{dt}.$$

Thus $\alpha = 0$ is solved by every map $\Phi: \mathbb{R} \to E_2 | x = \phi(t),\ y = \psi(t)$ (i.e., every curve) for which the functions $\phi(t)$ and $\psi(t)$ stand in the relation $\phi\, d\psi/dt = 3\psi\, d\phi/dt$. There are uncountably many such curves, for simply assign the function $\phi(t)$ and then use the differential equation to compute $\psi(t)$. The exterior equation $x\,dy - 3y\,dx = 0$ has innumerably many parametric solutions of dimension one.

Invertible maps

$$\Phi: E_n \to {}^{`}E_n | y^A = \phi^A(x^j), \qquad \det(\partial\phi^A/\partial x^i) \neq 0 \tag{3-7.28}$$

are of particular importance, for (y^A) may be interpreted as new coordinates of a point that had coordinates (x^i). We have already seen that

$$\Phi^*\,dy^A = \frac{\partial\phi^A(x^j)}{\partial x^i}\,dx^i, \tag{3-7.29}$$

and hence

$$\Phi^*\,{}^{`}\mu = \Phi^*(dy^1 \wedge \cdots \wedge dy^n) = \Phi^*\,dy^1 \wedge \cdots \wedge \Phi^*\,dy^n. \tag{3-7.30}$$

The actual calculation is significantly simplified by use of Lemma 3-3.4 since (3-7.29) gives a direct relation between the 1-forms $d\Phi^*y^A$ and the 1-forms dx^i. When (3-3.10) is used, the right-hand side of (3-7.30) becomes $\det(\partial\phi^A/\partial x^i)\mu$, which can also be written as $\det(\partial y^A/\partial x^i)\mu$:

$$\Phi^*\,{}^{`}\mu = \det(\partial y^A/\partial x^i)\mu. \tag{3-7.31}$$

A mnemonic convenience obtains if we write

$$\mu(y) = dy^1 \wedge \cdots \wedge dy^n, \qquad \mu(x) = dx^1 \wedge \cdots \wedge dx^n,$$

for (3-7.31) then becomes

$$\Phi^*\mu(y) = \det(\partial y^A/\partial x^i)\mu(x). \tag{3-7.32}$$

If $\alpha(y)$ is an n-form on $`E_n$, then

$$\alpha(y) = f(y^A)\mu(y). \tag{3-7.33}$$

Thus if we write

$$\alpha(x) = g(x^j)\mu(x) = \Phi^*\alpha(y), \tag{3-7.34}$$

then (3-7.32) shows that the coefficient functions f and g stand in the relation

$$g(x^j) = \det(\partial y^B/\partial x^k)f(y^A(x^j)). \tag{3-7.35}$$

The case of invertible maps allows us to obtain the mapping properties of the "top down" basis $\{\mu_i\}$ of Λ^{n-1}. In this regard, it is useful to introduce the notation

$$\mu_i(x) = \partial_i \lrcorner \mu(x), \qquad \mu_A(y) = \partial_A \lrcorner \mu(y). \tag{3-7.36}$$

For $V \in T(E_n)$, Φ induces the map Φ_* so that $\Phi_* V \in T(`E_n)$. This shows that the action of Φ^* on the pull down of a form α by V is given by

$$\Phi^*(\Phi_* V \lrcorner `\alpha) = V \lrcorner \Phi^* `\alpha. \tag{3-7.37}$$

Now,

$$\Phi_* V = v^i \frac{\partial \phi^A}{\partial x^i} \partial_A, \tag{3-7.38}$$

so that

$$\Phi_* V \lrcorner \mu(y) = v^i \frac{\partial \phi^A}{\partial x^i} \partial_A \lrcorner \mu(y) = v^i \frac{\partial \phi^A}{\partial x^i} \mu_A(y), \tag{3-7.39}$$

$$V \lrcorner \Phi^*\mu(y) = v^i \partial_i \lrcorner \det\left(\frac{\partial \phi^A}{\partial x^j}\right)\mu(x) = \det\left(\frac{\partial \phi^A}{\partial x^i}\right) v^i \mu_i(x), \tag{3-7.40}$$

with

$$\Phi^*(\Phi_* V \lrcorner \mu(y)) = V \lrcorner \Phi^*\mu(y) \tag{3-7.41}$$

for all $V \in T(E_n)$, implies

(3-7.42) $$\frac{\partial \phi^A}{\partial x^i} \Phi^* \mu_A(y) = \det\left(\frac{\partial \phi^B}{\partial x^k}\right) \mu_i(x).$$

These results, in turn, allow us to obtain the relations between elements of $\Lambda^{n-1}(E_n)$ and $\Lambda^{n-1}(`E_n)$ when resolved on the bases $\mu_i(x)$ and $\mu_A(y)$, respectively. If $`\alpha(y)$ is an element of $\Lambda^{n-1}(`E_n)$, then

(3-7.43) $$`\alpha(y) = F^A(y^B)\mu_A(y).$$

Thus if we set

(3-7.44) $$\alpha(x) = F^i(x^j)\mu_i(x) = \Phi^* `\alpha(y),$$

then

$$F^i(x)\mu_i(x) = F^A(\phi^B(x))\Phi^*\mu_A(y)$$

while (3-7.42) shows that

$$F^i(x)\mu_i(x) = F^i(x)\frac{\partial \phi^A}{\partial x^i} \frac{1}{\det\left(\frac{\partial \phi^B}{\partial x^k}\right)} \Phi^*\mu_A(y).$$

We thus obtain the relations

(3-7.45) $$F^A(\phi^B(x)) = \frac{1}{\det\left(\frac{\partial \phi^B}{\partial x^k}\right)} \frac{\partial \phi^A}{\partial x^i} F^i(x).$$

An additional insight into the intrinsic structure of the basis $\{\mu_i(x)\}$ of $\Lambda^{n-1}(E_n)$ can now be obtained. A map

(3-7.46) $$\Phi: E_{n-1} \to E_n | x^i = \phi^i(u^\alpha)$$

defines an $(n-1)$-dimensional surface $\mathscr{S}$ in E_n if we require that $\operatorname{rank}(\partial \phi^i / \partial u^\alpha) = n - 1$ throughout the domain of Φ. In these circumstances, the coordinates $(u^\alpha, \alpha = 1, \ldots, n-1)$ of E_{n-1} are referred to as the surface coordinates of $\mathscr{S}$ in E_n. If we allow Φ_* to act on the element ∂_α, we obtain

(3-7.47) $$\Phi_* \partial_\alpha = \frac{\partial \phi^i}{\partial u^\alpha} \partial_i$$

on $\mathscr{S}$, which is a vector field on $\mathscr{S}$ that is tangent to $\mathscr{S}$ in E_n for each value of $\alpha = 1, \ldots, n-1$. However (3-7.37) shows that

$$\Phi^*(\Phi_* \partial_\alpha \lrcorner \mu(x)) = \partial_\alpha \lrcorner \Phi^* \mu(x) = 0, \tag{3-7.48}$$

since $\Phi^* \mu(x) \in \Lambda^n(E_{n-1})$ vanishes identically. A combination of (3-7.47) and (3-7.48) thus gives

$$\begin{aligned} 0 = \Phi^*(\Phi_* \partial_\alpha \lrcorner \mu(x)) &= \Phi^*\left(\frac{\partial \phi^i}{\partial u^\alpha} \partial_i \lrcorner \mu(x)\right) \\ &= \Phi^*\left(\frac{\partial \phi^i}{\partial u^\alpha} \mu_i\right). \end{aligned} \tag{3-7.49}$$

On the other hand, $\Phi^*(\mu_i \partial \phi^i / \partial u^\alpha)$ can be interpreted as having values on the surface $\mathscr{S}$ when $\mathscr{S}$ is given the surface coordinates (u^α). If we denote this restriction to $\mathscr{S}$ by $\mu_i|_{\mathscr{S}}$, then (3-7.49) give

$$\left.\frac{\partial \phi^i(u^\beta)}{\partial u^\alpha} \mu_i\right|_{\mathscr{S}} = 0, \qquad \alpha = 1, \ldots, n-1. \tag{3-7.50}$$

In view of the fact that $(\partial \phi^i / \partial u^\alpha) \partial_i$ is tangent to $\mathscr{S}$ for $\alpha = 1, \ldots, n-1$, (3-7.50) simply states that $\{\mu_i|_{\mathscr{S}}\}$ may be viewed as the *normal* element of the surface $\mathscr{S}$.

A terminology has grown up in the current literature that should be noted here.

Definition. A map

$$\Phi : E_m \to E_n | x^i = \phi^i(u^1, \ldots, u^m) \tag{3-7.51}$$

with $m < n$ is called a *section* of E_n. If

$$\operatorname{rank}(\partial \phi^i / \partial u^\alpha) = m, \tag{3-7.52}$$

on $\mathscr{D}(\Phi)$ then the section Φ of E_n is said to be *regular*.

The definition of a regular section $\Phi : E_m \to E_n$ shows that it defines a regular m-dimensional surface in E_n, in which case the coordinates (u^α) of E_m may be viewed as surface coordinates of the regular section. If a regular section $\Phi : E_m \to E_n$ is such that $\Phi^* \omega = 0$ then the section is said to *solve* the exterior equation $\omega = 0$. Thus a regular section of E_n that solves $\omega = 0$ gives a solution of $\omega = 0$ in parametric form. The important thing to note here is that the solution of $\omega = 0$ is now interpreted as a regular m-dimensional surface in E_n with the property that $\Phi^* \omega = 0$ is now realized as the evaluation of ω on the section Φ of E_n. The ability to realize solutions of exterior equations as regular sections of the original space (i.e., as surfaces in the original space) provides significant insights into problems, for it allows full use of one's geometric intuition.

PROBLEMS

The following exterior forms are given on E_4 with the coordinate cover (x, y, z, t);

$$\omega_1 = x\,dy + 3e^{xy}\,dx + 2y\,dz - dt,$$

$$\omega_2 = xz^2\,dx \wedge dy + te^x\,dx \wedge dt + 4\sin(3t)\,dy \wedge dt,$$

$$\omega_3 = (x^2 + 2xt)\,dx - 5t^3\,dz + (t - x^2 + y^2)\,dt,$$

$$\omega_4 = t\,dx \wedge dy \wedge dt - 8t^2x\,dy \wedge dz \wedge dt + dx \wedge dy \wedge dz,$$

$$\omega_5 = (x^5 - 5yt)\,dx \wedge dz + 3x\cos(2yz)\,dy \wedge dt.$$

3.1. Evaluate the following exterior products:

(a) $\omega_1 \wedge \omega_2$; (b) $\omega_1 \wedge \omega_3$; (c) $\omega_1 \wedge \omega_4$; (d) $\omega_1 \wedge \omega_5$;
(e) $\omega_2 \wedge \omega_3$; (f) $\omega_2 \wedge \omega_4$; (g) $\omega_2 \wedge \omega_5$;
(h) $\omega_1 \wedge \omega_2 \wedge \omega_3$; (i) $\omega_4 \wedge \omega_5$; (j) $\omega_3 \wedge \omega_2 \wedge \omega_1$.

Answers

(a) $-8y\sin(3t)\,dy \wedge dz \wedge dt - 2yte^x\,dx \wedge dz \wedge dt$
$+ (12e^{xy}\sin(3t) - xte^x - xz^2)\,dz \wedge dy \wedge dt$
$+ 2xyz^2\,dx \wedge dy \wedge dz$;

(b) $(2y^3 - 2x^2y + 2yt - 5t^3)\,dz \wedge dt + (t - x^2 + y^2)x\,dy \wedge dt$
$- 5xt^3\,dy \wedge dz - (x + 2t)x^2\,dx \wedge dy$
$+ (x^2 - 3x^2e^{xy} + 3y^2e^{xy} + 3te^{xy} + 2xt)\,dx \wedge dt$
$- (15t^3e^{xy} + 2x^2y + 4xyt)\,dx \wedge dz$;

(c) $(1 + 2yt - 24xt^2e^{xy})\,dx \wedge dy \wedge dz \wedge dt$;

(d) $-6xy\cos(2yz)\,dy \wedge dz \wedge dt + (5yt - x^5)\,dx \wedge dz \wedge dt$
$+ 9x\cos(2yz)e^{xy}\,dx \wedge dy \wedge dt + x(5yt - x^5)\,dx \wedge dy \wedge dz$;

(e) $20t^3\sin(3t)\,dy \wedge dz \wedge dt + 5t^4e^x\,dx \wedge dz \wedge dt$
$+ x(4x\sin(3t) + 8t\sin(3t) - x^2z^2 + y^2z^2 + tz^2)\,dx \wedge dy \wedge dt$
$- 5xz^2t^3\,dx \wedge dy \wedge dz$;

(f) 0;

(g) $4(5yt - x^5)\sin(3t)\, dx \wedge dy \wedge dz \wedge dt$.

3.2. Find all 1-forms that are contained in the subspace spanned by ω_1 and ω_3. Find all 1-forms that do not belong to the subspace spanned by ω_1 and ω_3.

3.3. Complete ω_1, ω_3 to a basis for $\Lambda^1(E_4)$ (i.e., find 1-forms γ and δ such that $\omega_1, \omega_2, \gamma, \delta$ are linearly independent). Express the 1-form $x^3\, dy + z^2\, dx - e^{-2x}\, dt + (1 + y^4)\, dz$ in terms of this basis.

3.4. Let $V = \partial_x + 2x^2\partial_y - (x^2 + t^2)\partial_z + xe^{3z}\partial_t$. Evaluate the following inner multiplications:

(a) $V \lrcorner \omega_1$; (b) $V \lrcorner \omega_2$; (c) $V \lrcorner \omega_3$;
(d) $V \lrcorner \omega_4$; (e) $V \lrcorner \omega_5$; (g) $V \lrcorner (\omega_2 \wedge \omega_5)$.

Answers

(a) $3e^{xy} - xe^{3z} + 2x^3 - 2x^2y - 2yt^2$;
(b) $(te^x + 8x^2\sin(3t))\, dt + x(z^2 - 4\sin(3t)e^{3z})\, dy$
$\quad - x(te^{x+3z} + 2x^2z^2)\, dx$;
(c) $5t^5 - x^3e^{3z} + xy^2e^{3z} + xte^{3z} + 5x^2t^3 + x^2 + 2xt$;
(d) $-16x^3t^3\, dz \wedge dt + t(1 - 8x^3t - 8xt^3)\, dy \wedge dt$
$\quad + (1 - 8x^2t^2e^{3z})\, dy \wedge dz - 2x^2t\, dx \wedge dt - 2x^3\, dx \wedge dz$
$\quad + (xte^{3z} - x^2 - t^2)\, dx \wedge dy$;
(e) $6x^3\cos(2yz)\, dt + (x^5 - 5yt)\, dz$
$\quad - 3x^2\cos(2yz)e^{3z}\, dy + (x^7 + x^5t^2 - 5x^2yt - 5yt^3)\, dx$;
(f) $4\sin(3t)(5yt - x^5)\, dy \wedge dz \wedge dt$
$\quad + 8x^2\sin(3t)(x^5 - 5yt)\, dx \wedge dz \wedge dt$
$\quad + 4\sin(3t)(x^7 + x^5t^2 - 5x^2yt - 5yt^3)\, dx \wedge dy \wedge dt$
$\quad + 4e^{3z}\sin(3t)x(x^5 - 5yt)\, dx \wedge dy \wedge dz$.

3.5. Find all vector fields V such that

(a) $V \lrcorner \omega_1 = 0$; (b) $V \lrcorner \omega_2 = 0$; (c) $V \lrcorner \omega_3 = 0$;
(d) $V \lrcorner \omega_4 = 0$; (e) $V \lrcorner \omega_5 = 0$; (f) $V \lrcorner \omega_1 = V \lrcorner \omega_3 = 0$.

3.6. Find the basis for $T(E_4)$ that is dual to the basis $\omega_1, \omega_3, \gamma, \delta$ of $\Lambda^1(E_4)$ that was found in Problem 3.3.

3.7. Find $\mu_i, \mu_{ij}, \mu_{ijk}$ for E_4 with $\mu = dx \wedge dy \wedge dz \wedge dt$. Find $\bar{\mu}_i, \bar{\mu}_{ij}, \bar{\mu}_{ijk}$ for E_4 with $\mu = dx \wedge dz \wedge dt \wedge dy$. Express $\omega_1, \omega_2, \omega_3, \omega_4, \omega_5$ in terms of these top down generated bases for $\Lambda(E_4)$.

3.8. Find all characteristic vector fields of the following ideals: (a) $I\{\omega_1, \omega_3\}$; (b) $I\{\omega_1, \omega_2\}$; (c) $I\{\omega_1, \omega_2, \omega_3\}$; (d) $I\{\omega_2, \omega_5\}$; (e) $I\{\omega_2, \omega_4\}$.

3.9. The map ϕ is defined by

$$\phi: (r, \theta, \rho) \to (x, y, z, t) \mid x = r\cos(\theta), \qquad y = r\sin(\theta),$$
$$z = \rho - 2, \qquad t = r\rho.$$

(a) Compute ϕ^* of $\omega_1, \omega_2, \omega_3, \omega_4, \omega_5$.
(b) Compute $\mathscr{R}(\phi)$, the range of ϕ in E_4.
(c) Compute ϕ^{-1} with domain $\mathscr{R}(\phi)$.
(d) Compute $\phi_* \partial_r$, $\phi_* \partial_\theta$, $\phi_* \partial_\rho$.
(e) With $\pi = dx \wedge dy \wedge dz$, compute $\phi^*(\phi_* \partial_\theta \lrcorner \pi)$ and $\phi^*(\phi_* \partial_r \lrcorner \pi)$.
(f) Compute $\phi^*\mu$, $\phi^*\mu_i$ with $\mu = dx \wedge dy \wedge dz \wedge dt$.
(g) Compute $\phi^* I\{\omega_1, \omega_3\}$, $\phi^* I\{\omega_1, \omega_2\}$, $\phi^* I\{\omega_2, \omega_5\}$.

3.10. Find all maps $\phi: E_k \to E_4$ with $1 \le k \le 4$ for which
(a) $\phi^*\omega_1 = 0$; (b) $\phi^4\omega_2 = 0$; (c) $\phi^*\omega_3 = 0$;
(d) $\phi^*\omega_4 = 0$; (e) $\phi^*\omega_5 = 0$.

3.11. If $\omega \in \Lambda^k(E_n)$, and $(\omega | V_1, \ldots, V_{k-1})$ is defined by (3-4.15), what is the dimension of the subspace of $\Lambda^1(E_n)$ that is generated by allowing $V_1, \ldots, V_{k-1}$ to sweep out $T(E_n)$ (i.e., consider the map $Z: \Lambda^k \times (T(E_n))^{(n-1)} \overset{(3\text{-}4.15)}{\to} \Lambda^1$)? How does this relate to the condition (3-4.16) for simplicity of ω?

CHAPTER FOUR

EXTERIOR DERIVATIVES, LIE DERIVATIVES, AND INTEGRATION

4-1. THE EXTERIOR DERIVATIVE

Elements of the graded exterior algebra $\Lambda(E_n)$ are fields of exterior forms that are defined over the base space E_n:

$$\omega(x^j) = w_{i_1 \cdots i_k}(x^j)\, dx^{i_1} \wedge dx^{i_2} \wedge \cdots \wedge dx^{i_k}.$$

Noting that the coefficient functions $\{w_{i_1 \cdots i_k}(x^j)\}$ are elements of $\Lambda^0(E_n)$, a calculus of exterior forms would seem to be a reasonable expectation. In one respect, such a calculus is already partially in evidence, for the natural basis fields $\{dx^i\}$ of $\Lambda^1(E_n)$ were obtained in Section 3-1 by introducing the notion of the differential of a function as a primitive concept. We know, however, that functions are elements of $\Lambda^0(E_n)$ and hence belong to $\Lambda(E_n)$. A direct road to a calculus of exterior forms will thus be provided if we can find an operation "d" for which the notion of a differential is not restricted solely to forms of degree zero.

We begin by noting that any element of $\Lambda^k(E_n)$ is the sum of terms of the form $f(x^j)\, dx^{i_1} \wedge dx^{i_2} \wedge \cdots \wedge dx^{i_k}$, and that we already know the meaning of $df(x^j)$. The operation d can thus be introduced by specifying how it combines with the operations + and $\wedge$ of the exterior algebra. This observation and a good deal of hindsight lead to the requirements

$$d(\alpha + \beta) = d\alpha + d\beta, \tag{4-1.1}$$

$$d(\alpha \wedge \beta) = (d\alpha) \wedge \beta + (-1)^{\deg(\alpha)} \alpha \wedge d\beta \tag{4-1.2}$$

for all $\alpha, \beta \in \Lambda(E_n)$. The requirement (4-1.1) demands that d distribute with respect to addition and hence *d is a linear operator*. On the other hand, (4-1.2) says that *d is an antiderivation* because of the factor $(-1)^{\deg(\alpha)}$. Consistency with what has gone before demands that

$$df = \frac{\partial f}{\partial x^j} dx^j \qquad \forall f \in \Lambda^0(E_n). \tag{4-1.3}$$

If d is allowed to act on both sides of (4-1.3) and (4-1.2) is used, we obtain

$$\begin{aligned} d\,df &= d\left(\frac{\partial f}{\partial x^j} dx^j\right) = d\left(\frac{\partial f}{\partial x^j}\right) \wedge dx^j + \frac{\partial f}{\partial x^j} d\,dx^j \\ &= \frac{\partial^2 f}{\partial x^k \partial x^j} dx^k \wedge dx^j + \frac{\partial f}{\partial x^j} d\,dx^j \\ &= \frac{1}{2}\left(\frac{\partial^2 f}{\partial x^k \partial x^j} - \frac{\partial^2 f}{\partial x^j \partial x^k}\right) dx^k \wedge dx^j + \frac{\partial f}{\partial x^j} d\,dx^j \\ &= \frac{\partial f}{\partial x^j} d\,dx^j \end{aligned} \tag{4-1.4}$$

because $(\partial f/\partial x^j) \in \Lambda^0(E_n)$ and hence $\partial^2 f/\partial x^i \partial x^j = \partial^2 f/\partial x^j \partial x^i$. Thus, the operator d will provide a simple and direct statement of the necessary symmetry $\partial^2 f/\partial x^i \partial x^j = \partial^2 f/\partial x^j \partial x^i$ if we require

$$d\,dx^i = 0, \qquad i = 1, \ldots, n. \tag{4-1.5}$$

Such a requirement is clearly consistent in view of (4-1.4), and it achieves maximal simplicity because there will never be "derivatives" in our calculus of any order but the first.

We must now show that an operator d with these properties is defined for all elements of $\Lambda(E_n)$ and that it is unique. Uniqueness is essential here, as a little reflection will clearly show. If $\beta \in \Lambda^k(E_n)$, then β is a sum of terms of the form $f(x^j)\,dx^{i_1} \wedge dx^{i_2} \wedge \cdots \wedge dx^{i_k}$. Thus if d satisfies (4-1.1), it is sufficient to check that d is well defined for monomials of the form

$$\gamma = f(x^m)\,dx^{i_1} \wedge dx^{i_2} \wedge \cdots \wedge dx^{i_k}. \tag{4-1.6}$$

If we apply d to both sides of (4-1.6) and use property (4-1.2), we have

$$d\gamma = df \wedge dx^{i_1} \wedge dx^{i_2} \wedge \cdots \wedge dx^{i_k} + f d\{dx^{i_1} \wedge dx^{i_2} \wedge \cdots \wedge dx^{i_k}\}. \tag{4-1.7}$$

Thus (4-1.2) and (4-1.5) imply

$$d\gamma = \frac{\partial f}{\partial x^j} dx^j \wedge dx^{i_1} \wedge \cdots \wedge dx^{i_k}, \tag{4-1.8}$$

from which we conclude that d is defined for every element of $\Lambda(E_n)$; in particular, $d: \Lambda^k(E_n) \to \Lambda^{k+1}(E_n)$. If we apply d to both sides of (4-1.8) and make use of (4-1.5), it follows that

$$d d\gamma = 0 \tag{4-1.9}$$

for all $\gamma \in \Lambda(E_n)$, and this in turn implies (4-1.5). Finally, uniqueness is secured through the final requirement that $df = (\partial f/\partial x^i)\, dx^i$ hold for all $f \in \Lambda^0(E_n)$.

Theorem 4-1.1. *There is one and only one operator d on $\Lambda(E_n)$ with the following properties*

D1. $d(\alpha + \beta) = d\alpha + d\beta$,
D2. $d(\alpha \wedge \beta) = (d\alpha) \wedge \beta + (-1)^{\deg(\alpha)}\alpha \wedge (d\beta)$,
D3. $df = (\partial f/\partial x^j)\, dx^j, \qquad f \in \Lambda^0(E_n)$,
D4. $d d\alpha = 0$.

If $\alpha \in \Lambda^k(E_n)$, then $d\alpha \in \Lambda^{k+1}(E_n)$ and hence d may be viewed as the map

$$d: \Lambda^k(E_n) \to \Lambda^{k+1}(E_n) \tag{4-1.10}$$

that satisfies properties **D1** *through* **D4**.

Definition. The operator d, given in Theorem 4-1.1 is called *exterior differentiation*.

The actual calculation of the exterior derivative of any given exterior form follows directly from properties **D1** through **D4**. For example, with a 1-form

$$\alpha = \alpha_i(x^k)\, dx^i, \tag{4-1.11}$$

we have

$$d\alpha = d\alpha_i \wedge dx^i = \frac{\partial \alpha_i}{\partial x^j} dx^j \wedge dx^i \tag{4-1.12}$$

$$= \frac{1}{2}(\partial_j \alpha_i - \partial_i \alpha_j) dx^j \wedge dx^i$$

because $dx^j \wedge dx^i = -dx^i \wedge dx^j$. Similarly, if

$$\beta = \tfrac{1}{2}\beta_{ij} dx^i \wedge dx^j, \qquad \beta_{ij} = -\beta_{ji},$$

then

$$d\beta = \frac{1}{2} d\beta_{ij} \wedge dx^i \wedge dx^j = \frac{1}{2}(\partial_k \beta_{ij}) dx^k \wedge dx^i \wedge dx^j \tag{4-1.13}$$

$$= \frac{1}{3!}(\partial_k \beta_{ij} + \partial_i \beta_{jk} + \partial_j \beta_{ki}) dx^k \wedge dx^i \wedge dx^j.$$

What about elements from the top of $\Lambda(E_n)$?

Theorem 4-1.2. *If $\omega \in \Lambda^n(E_n)$, then $d\omega = 0$.*

Proof. Since d maps $\Lambda^k(E_n)$ into $\Lambda^{k+1}(E_n)$, d maps $\Lambda^n(E_n)$ into $\Lambda^{n+1}(E_n)$ and $\Lambda^{n+1}(E_n)$ contains the single element 0. □

Elements of $\Lambda^{n-1}(E_n)$ are most easily handled by first establishing the following result.

Lemma 4-1.1. *The basis elements $\{\mu_i\}$ and $\{\mu_{ij}\}$ satisfy*

$$d\mu_i = 0, \qquad d\mu_{ij} = 0. \tag{4-1.14}$$

Proof. By definition (see Section 3-5)

$$\mu_i = \partial_i \lrcorner \mu = \partial_i \lrcorner (dx^1 \wedge dx^2 \wedge \cdots dx^n) \tag{4-1.15}$$

so that each μ_i is the product of $n - 1$ simple factors of the form dx^{i_j}. Since $d\,dx^{i_j} = 0$, properties **D2** and **D4** show that $d\mu_i = 0$. The same argument applies to the $(n-2)$-forms μ_{ij}. □

Theorem 4-1.3. *If $\alpha \in \Lambda^{n-1}(E_n)$, then resolution of α on the basis $\{\mu_i\}$ by*

$$\alpha = \alpha^i \mu_i \tag{4-1.16}$$

gives

$$d\alpha = \left(\partial_i \alpha^i\right)\mu. \tag{4-1.17}$$

If $\beta \in \Lambda^{n-2}$, *then resolution of* β *on the basis* $\{\mu_{ij}, i < j\}$ *by*

$$\beta = \tfrac{1}{2}\beta^{ij}\mu_{ij}, \qquad \beta^{ij} = -\beta^{ji} \tag{4-1.18}$$

gives

$$d\beta = \left(\partial_i \beta^{ij}\right)\mu_j. \tag{4-1.19}$$

Proof. Applying d to both sides of (4-1.16) and using properties **D1** through **D4** gives $d\alpha = d\alpha^i \wedge \mu_i + \alpha^i d\mu_i$. However $d\mu_i = 0$ by Lemma 4-1.1 and hence $d\alpha = (\partial_j \alpha^i)\, dx^j \wedge \mu_i = (\partial_j \alpha^i)\delta_i^j \mu = (\partial_i \alpha^i)\mu$ by (3-5.4). Similarly, (4-1.18) gives $d\beta = \frac{1}{2}\, d\beta^{ij} \wedge \mu_{ij} = \frac{1}{2}(\partial_k \beta^{ij})\, dx^k \wedge \mu_{ij} = \frac{1}{2}(\partial_k \beta^{ij})(\delta_i^k \mu_j - \delta_j^k \mu_i) = \frac{1}{2}\partial_i \beta^{ij}\mu_j - \frac{1}{2}\partial_j \beta^{ij}\mu_i$ by (3-5.13). Thus $\beta^{ij} = -\beta^{ji}$ yields $d\beta = \frac{1}{2}\partial_i \beta^{ij}\mu_j - \frac{1}{2}\partial_i \beta^{ji}\mu_j = \frac{1}{2}\partial_i(\beta^{ij} - \beta^{ji})\mu_j = \partial_i \beta^{ij}\mu_j$ and the result is established. □

The important thing to note in the context of Theorem 4-1.3 is the emergence of the "divergence" type quantities $\partial_i \alpha^i$ and $\partial_i \alpha^{ij}$ as the coefficients of μ and μ_j, respectively. There is no difficulty in obtaining "divergence" type quantities in the exterior calculus provided the right choice of basis is made; namely, $\{\mu_i\}$ for $\Lambda^{n-1}(E_n)$ and $\{\mu_{ij}, i < j\}$ for $\Lambda^{n-2}(E_n)$.

Let E_3 with coordinates $(X^A) = (X^1, X^2, X^3)$ be the three-dimensional Euclidean space of reference configurations of a deformable body and let ${}^{\backprime}E_3$ with coordinates $(x^i) = (x^1, x^2, x^3)$ be the three-dimensional Euclidean space of current configurations of the body. If we denote the components of the Piola-Kirchhoff stress tensor by $((\sigma_i^{\ A}))$, we can form the collection of three 2-forms $\{\sigma_i\} \in \Lambda^2(E_3)$ by

$$\sigma_i = \sigma_i^{\ A}\mu_A,$$

where $\mu_A = \partial_A \lrcorner \mu = \partial_A \lrcorner (dX^1 \wedge dX^2 \wedge dX^3)$ are the elements of oriented surface in E_3 and a basis for $\Lambda^2(E_3)$. The standard interpretation of the Piola-Kirchhoff stress tensor then shows that the "traction vector" $\mathcal{T}$ (element of $\Lambda^1({}^{\backprime}E_3)$) for a unit element of bounding surface of the deformable body is given by

$$\mathcal{T} = \sigma_i\, dx^i = \left(\sigma_i^{\ A}\mu_A\right) dx^i.$$

It is now an elementary calculation to see that

$$d\sigma_i = d\left(\sigma_i^{\ A}\mu_A\right) = \left(\partial_A \sigma_i^{\ A}\right)\mu$$

are the components of the resulting force that acts on an element of unit volume in the reference configuration as a consequence of the application of tractions to the boundary of the body. Since $d\,d\omega \equiv 0$, it follows that any $\{\sigma_i\}$ given by

$$\sigma_i = d\beta_i, \qquad \beta_i = \beta_{iA}\,dX^A \in \Lambda^1(E_3)$$

yields identically zero components of the resulting force per unit volume in the reference configuration; i.e., a null stress distribution. Thus

$$\sigma_i^A\mu_A = d\beta_i = \partial_B\beta_{iA}\,dX^B \wedge dX^A = \tfrac{1}{2}(\partial_B\beta_{iA} - \partial_A\beta_{iB})\,dX^B \wedge dx^A$$

defines the collection of all null Piola-Kirchhoff stress distributions.

We have seen that any smooth map $\Phi: E_n \to E_m$ induces a smooth map $\Phi^*: \Lambda(E_m) \to \Lambda(E_n)$ simply by composition and application of the chain rule, and this came about by setting

$$d\Phi^* y^A = \Phi^* dy^A.$$

It would thus appear that d and Φ^* should commute. Care must be exercised here, however, because a statement such as $d(\Phi^*\omega) = \Phi^*(d\omega)$ involves a possibly misleading abuse of notation. First of all since Φ^* acts on ω, the exterior form ω is defined on the range of Φ, namely E_m with the coordinate cover (y^A). Accordingly, the "d" in $d\omega$ is therefore the exterior derivative operator on $\Lambda(E_m)$. Second, it is clear that $\Phi^*\omega$ is an exterior form that is defined over the domain of Φ and hence $\Phi^*\omega$ belongs to $\Lambda(E_n)$. It thus follows that the "d" in $d(\Phi^*\omega)$ is the exterior derivative operator on $\Lambda(E_n)$. Finally, with these distinctions clearly in mind, we see that

$$\Phi^* dy^E = d\Phi^* y^E = d\phi^E(x^k) = \frac{\partial\phi^E}{\partial x^i}\,dx^i$$

because $\Phi^* y^E = \phi^E(x^k)$ is an element of $\Lambda^0(E_n)$.

Theorem 4-1.4. *If $\Phi: E_n \to E_m$ is a smooth map, then*

(4-1.20) $$d(\Phi^*\omega) = \Phi^*(d\omega)$$

for all $\omega \in \Lambda(E_m)$.

Proof. Since every element of $\Lambda(E_m)$ is a sum of simple elements of $\Lambda(E_m)$ and $\Phi^*(\alpha + \beta) = \Phi^*\alpha + \Phi^*\beta$, $\Phi^*(\alpha \wedge \beta) = (\Phi^*\alpha) \wedge (\Phi^*\beta)$, it suffices to establish (4-1.20) for a generic 1-form

$$\alpha = \alpha_A(y^B)\,dy^A, \qquad d\alpha = \partial_E\alpha_A(y^B)\,dy^E \wedge dy^A.$$

Thus if $y^A = \phi^A(x^k)$ defines the map Φ, we have

$$\Phi^*\alpha = \alpha_A\big(\phi^B(x^k)\big)\Phi^* \, dy^A,$$

$$\Phi^*(d\alpha) = \partial_E \alpha_A\big(\phi^B\big)\Phi^* \, dy^E \wedge \Phi^* \, dy^A,$$

$$d(\Phi^*\alpha) = d\alpha_A\big(\phi^B(x^k)\big) \wedge \Phi^* \, dy^A + \alpha_A\big(\phi^B\big) \, d\big(\Phi^* \, dy^A\big).$$

However, $d(\Phi^* \, dy^A) = d(d\Phi^* y^A) = 0$, and hence we have

$$d(\Phi^*\alpha) = d\alpha_A\big(\phi^B(x^k)\big) \wedge \Phi^* \, dy^A = \partial_E \alpha_A\big(\phi^B\big) \, d\phi^E \wedge \Phi^* \, dy^A$$

$$= \partial_E \alpha_A\big(\phi^B\big)\Phi^* \, dy^E \wedge \Phi^* \, dy^A.$$

□

The importance of this is that the order of application of mappings and of exterior differentiation is immaterial. In particular, the action of Φ^* is well defined for any smooth map Φ whether or not exterior differentiation is involved. For example, if $\alpha \in \Lambda^k(E_m)$ and $\beta \in \Lambda^{k+1}(E_m)$ are given, then $\gamma = d\alpha + \beta$ belongs to $\Lambda^{k+1}(E_m)$ and $\Phi^*\gamma = d\Phi^*\alpha + \Phi^*\beta \in \Lambda^{k+1}(E_n)$ is well defined.

The operation d of exterior differentiation leads to the definition of two subspaces of $\Lambda(E_n)$ in a natural way.

Definition. An element α of $\Lambda(E_n)$ is said to be *closed* if and only if α is in the kernel of d;

(4-1.21) $$d\alpha = 0.$$

Definition. An element α of $\Lambda(E_n)$ is said to be *exact* if and only if α is in the range of d; there exists a $\beta \in \Lambda(E_n)$ such that

(4-1.22) $$\alpha = d\beta.$$

Theorem 4-1.5. *The collection of all closed elements of $\Lambda(E_n)$ forms a subspace $\mathscr{C}(E_n)$ of $\Lambda(E_n)$ over $\mathbb{R}$ but not over $\Lambda^0(E_n)$. The collection of all exact elements of $\Lambda(E_n)$ forms a subspace $\mathscr{E}(E_n)$ of $\Lambda(E_n)$ over $\mathbb{R}$ but not over $\Lambda^0(E_n)$ and*

(4-1.23) $$\mathscr{E}(E_n) \subset \mathscr{C}(E_n).$$

Proof. If α and β are closed elements of $\Lambda^k(E_n)$, then $d\alpha = 0$, $d\beta = 0$ and hence

$$d\big(f(x)\alpha + g(x)\beta\big) = df \wedge \alpha + dg \wedge \beta.$$

Thus if $df = 0$, $dg = 0$, that is $f =$ constant, $g =$ constant, then $d(f\alpha + g\beta) = 0$ and $\mathscr{C}(E_n)$ is a subspace over $\mathbb{R}$. On the other hand, it is easily seen that for $df \neq 0$, $dg \neq 0$, then $d\alpha = 0$, $d\beta = 0$ do not imply $d(f\alpha + g\beta) = 0$ for all $f, g \in \Lambda^0(E_n)$. Thus, $\mathscr{C}(E_n)$ is not a vector subspace over $\Lambda^0(E_n)$. Similarly, if α and β belong to $\mathscr{E}(E_n)$, there exist elements ρ and η such that $\alpha = d\rho$, $\beta = d\eta$ and $f\alpha + g\beta = f\,d\rho + g\,d\eta = d(f\rho) + d(g\eta) - df \wedge \rho - dg \wedge \eta$. The same argument then shows that $\mathscr{E}(E_n)$ is a vector subspace over $\mathbb{R}$ but not over $\Lambda^0(E_n)$. Finally if $\beta \in \mathscr{E}(E_n)$, then $\beta = d\eta$ for some η and hence $d\beta = d\,d\eta = 0$; that is β is closed. □

The inclusion of $\mathscr{E}(E_n)$ in $\mathscr{C}(E_n)$ is the basis by which a large number of significant problems are solved. A typical such problem is the following. Suppose that we are given a 1-form

(4-1.24) $$F = F_i(x^j)\,dx^i$$

and we would like to find an element η of $\Lambda^0(E_n)$ such that

(4-1.25) $$F = d\eta = \partial_i\eta\,dx^i.$$

A comparison of (4-1.24) and (4-1.25) shows that this is the case if and only if the function η satisfies the simultaneous system of partial differential equations

(4-1.26) $$F_i(x^j) = \partial_i\eta(x^j);$$

that is the "force" with components $\{F_i\}$ admits a potential function η. From our new point of view, (4-1.25) says that such a function η can exist if and only if F is exact, in which case F must be closed; that is,

(4-1.27) $$0 = dF = \tfrac{1}{2}(\partial_i F_j - \partial_j F_i)\,dx^i \wedge dx^j$$

is a system of necessary conditions for the existence of the function η. Therefore, F_i must satisfy

(4-1.28) $$\partial_i F_j = \partial_j F_i.$$

This generalized $\vec{F} = \vec{\nabla}\eta$ only if $\vec{\nabla} \times \vec{F} = \vec{0}$ in three dimensions to any number of dimensions. There is quite a bit more here, however. Suppose that α is a given 2-form and we ask when can we find a 1-form β such that

(4-1.29) $$\alpha = d\beta.$$

Since (4-1.29) says that α is exact if there exists such a β, such α must likewise be closed in order that there exist such a β:

(4-1.30) $$d\alpha = 0.$$

Thus if $\alpha = \frac{1}{2}\alpha_{ij}\,dx^i \wedge dx^j$, then the coefficient functions α_{ij}, with $\alpha_{ij} = -\alpha_{ji}$ must satisfy

$$\partial_i \alpha_{jk} + \partial_j \alpha_{ki} + \partial_k \alpha_{ij} = 0 \tag{4-1.31}$$

(see (4-1.13)). Similarly, if $\alpha = \alpha^i \mu_i \in \Lambda^{n-1}(E_n)$ then $\alpha = d\beta$ only if $0 = d\alpha = (\partial_i \alpha^i)\mu$; that is $\partial_i \alpha^i = 0$.

Suppose that we satisfy these necessary conditions for the existence of solutions; $\alpha = d\beta$ only if $d\alpha = 0$. The question then arises as to whether we can actually find a β that makes the equation $\alpha = d\beta$ true for given α. This would indeed be the case if we could show that every closed form is an exact form; $d\alpha = 0$ implies a β such that $\alpha = d\beta$. A little reflection will easily show that such wishful thinking is just that, for such a result cannot be true in general. For instance, suppose that we are in two dimensions and $\vec{F}$ satisfies $\vec{\nabla} \times \vec{F} = \vec{0}$ on a region with a hole in it. We then know that $\vec{F} = \vec{\nabla}\eta$ cannot necessarily be satisfied by a single-valued function η because of the hole: the line integral of $\vec{F} \cdot d\vec{r}$ around the hole need not be zero! On the other hand, there is a partial converse of the inclusion $\mathscr{E}(E_n) \subset \mathscr{C}(E_n)$ that comes about if the region of definition is sufficiently nice. This is known as the Poincaré lemma, which we prove in the next chapter.

Lemma 4-1.2. (Poincaré Lemma) *If $\mathscr{S}$ is a region of E_n that can be shrunk to a point in a smooth way ($\mathscr{S}$ is star-shaped with respect to one of its points), then*

$$\mathscr{C}(\mathscr{S}) \subset \mathscr{E}(\mathscr{S}); \tag{4-1.32}$$

that is, if $d\alpha = 0$ on $\mathscr{S}$ then there exists a β on $\mathscr{S}$ such that $\alpha = d\beta$.

Significantly more information is actually available. In fact, the form β can be computed directly from the given form α by application of the homotopy operator H that is the subject of the next chapter.

We noted in Section 3-1 that dx^i could be given two different interpretations. The first was that of an element of the natural basis for $\Lambda^1(E_n)$. For this interpretation, let us write $(dx)^i$ since the definition of the dual basis gives $\langle (dx)^i, \partial_j \rangle = \delta^i_j$ with $\langle (dx)^i, \partial_j \rangle = \partial_j \langle x^i \rangle$. The second interpretation was that of the differential of the coordinate function x^i. To make a clear distinction, let us write $d(x^i)$ for this differential. The considerations up to this point have been based on the identification $(dx)^i = d(x^i)$ without a proof of consistency, as the careful reader has already realized! This situation is easily remedied, however, once the axioms **D1** through **D4** of the exterior derivative have been given. In order to see this, let us return to the original definition of the base elements $(dx)^i$ by means of $\langle (dx)^i, \partial_j \rangle = \delta^i_j$. The only axiom that is affected is **D3**, which must now be given in the following form

D3. $$d(f) = \partial f / \partial x^j (dx)^j.$$

Written in this way, **D3** defines the action of the operator d on elements of $\Lambda^0(E_n)$. Now, simply observe that x^i, for given i, is an element of $\Lambda^0(E_n)$. We may thus apply **D3** to obtain the desired result

$$d(x^i) = \partial x^i/\partial x^j(dx)^j = (dx)^i.$$

The two interpretations of dx^i are thus seen to be consistent. Their simultaneous employment in the previous chapter thus leads to no internal inconsistency, rather only an economy.

A few explicit examples may help at this point. Let E_4 be referred to the coordinate cover (x, y, z, t). We then have the following specific results:

$$f = x^5y^2e^{8x+t} - 5y^4\sin(3z^7 + 9),$$

$$df = x^4y^2(8x+5)e^{8x+t}\,dx + 2y\big(x^5e^{8x+t} - 10y^2\sin(3z^7+9)\big)\,dy$$

$$-105y^4z^6\cos(3z^7+9)\,dz + x^5y^2e^{8x+t}\,dt;$$

$$\alpha = x^7ye^t\,dx - t^3\,dy + (x - 7yz)\,dz + z^7\,dt,$$

$$d\alpha = 7z^6\,dz \wedge dt + 3t^2\,dy \wedge dt - 7z\,dy \wedge dz$$

$$-x^7ye^t\,dx \wedge dt + dx \wedge dz - x^7e^t\,dx \wedge dy;$$

$$\beta = zt\,dx \wedge dy + e^y\,dx \wedge dz + x^9\,dz \wedge dt,$$

$$d\beta = z\,dx \wedge dy \wedge dt + (t - e^y)\,dx \wedge dy \wedge dz + 9x^8\,dx \wedge dz \wedge dt;$$

$$\gamma = \exp(x + 2y + 3z + 4t)\,dx \wedge dy \wedge dz - x^2y^3\,dx \wedge dz \wedge dt,$$

$$d\gamma = \big(3x^2y^2 - 4\exp(x + 2y + 3z + 4t)\big)\,dx \wedge dy \wedge dz \wedge dt.$$

4-2. CLOSED IDEALS AND A CONFLUENCE OF IDEAS

A very useful class of ideals of the exterior algebra are those that remain unchanged under exterior differentiation.

Definition. An ideal I of $\Lambda(E_n)$ is said to be *closed* if and only if $d\rho \in I$ for every $\rho \in I$, in which case we write

$$dI \subset I. \tag{4-2.1}$$

Any finitely generated ideal $I\{\omega^1,\dots,\omega^k\}$ can be closed by formation of a new ideal $\bar{I}\{\omega^1,\dots,\omega^k; d\omega^1,\dots,d\omega^k\}$ called the *closure* of $I\{\omega^1,\dots,\omega^k\}$. If $\rho \in \bar{I}$, then

$$\rho = \gamma_a \wedge \omega^a + \Gamma_a \wedge d\omega^a$$

in which case

$$d\rho = d\gamma_a \wedge \omega^a + \left\{(-1)^{\deg(\gamma_a)}\gamma_a + d\Gamma_a\right\} \wedge d\omega^a$$

and hence $d\rho \in \bar{I}$.

The important question is as follows: when is a given ideal $I\{\omega^1,\dots,\omega^k\}$ a closed ideal as it stands? This question is answered by the following theorem for the case in which all of the generators of the ideal are of the same degree.

Theorem 4-2.1. *Let $I\{\omega^1,\dots,\omega^k\} \overset{\text{def}}{=} I\{\omega^a\}$ be an ideal of $\Lambda(E_n)$ such that each of its generators is of the same degree. Then $I\{\omega^a\}$ is a closed ideal if and only if there exist k^2 1-forms $\{\Gamma^a_b\}$ such that the generators satisfy*

$$d\omega^a = \Gamma^a_b \wedge \omega^b. \tag{4-2.2}$$

Proof. If $\rho \in I\{\omega^a\}$, then $\rho = \gamma_a \wedge \omega^a$ for some k-tuple of forms $\{\gamma_a\}$ with $\deg(\gamma_a) = b$ and hence

$$d\rho = d\gamma_a \wedge \omega^a + (-1)^b \gamma_a \wedge d\omega^a \equiv (-1)^b \gamma_a \wedge d\omega^a \bmod I\{\omega^k\};$$

$d\rho \in I\{\omega^a\}$ for every $\rho \in I\{\omega^a\}$ if and only if each of the forms $d\omega^a$ belongs to $I\{\omega^a\}$. Thus since $d\omega^a$ has degree one greater than ω^a, $d\omega^a \in I\{\omega^a\}$ for $a = 1,\dots,k$ if and only if there exist k^2 1-forms Γ^a_b such that the generators satisfy (4-2.2). □

We note in passing that the Γ^a_b in (4-2.2) cannot be entirely arbitrary, for $dd\omega^a$ must vanish:

$$0 = dd\omega^a = d\Gamma^a_b \wedge \omega^b - \Gamma^a_b \wedge d\omega^b = d\Gamma^a_b \wedge \omega^b - \Gamma^a_b \wedge \Gamma^b_c \wedge \omega^c,$$

that is,

$$(d\Gamma^a_b - \Gamma^a_c \wedge \Gamma^c_b) \wedge \omega^b = 0. \tag{4-2.3}$$

The question of existence and uniqueness of solutions of systems of exterior differential equations of the form (4-2.2) subject to the integrability conditions (4-2.3) will be studied in the next chapter. It suffices at this point simply to note that there need not exist a system of k^2 1-forms Γ^a_b for which (4-2.2) can be satisfied for given $\{\omega^a\}$; *not every ideal generated by forms of the same degree is closed.*

An alternative and sometimes more useful criterion for the closure of an ideal is available when the ideal is generated by forms of degree 1. Suppose that I is generated by 1-forms $\omega^1, \dots, \omega^r$. If these r 1-forms are not linearly independent, $\omega^1 \wedge \cdots \wedge \omega^r = 0$, we may replace the generators $\omega^1, \dots, \omega^r$ by a basis $\gamma^1, \dots, \gamma^k$ for the subspace of $\Lambda^1(E_n)$ that is spanned by $\omega^1, \dots, \omega^r$ and the ideal I will be unchanged. We may therefore assume that the generators of I are linearly independent over $\Lambda^0(E_n)$ without loss of generality.

Theorem 4-2.2. *Let $I\{\omega^a\}$ be an ideal of $\Lambda(E_n)$ whose generators $\omega^1, \dots, \omega^k$ are 1-forms such that*

$$\omega^1 \wedge \omega^2 \wedge \cdots \wedge \omega^k = \Omega(I) \neq 0 \tag{4-2.4}$$

and $k < n - 1$, then $dI\{\omega^a\} \subset I\{\omega^a\}$ if and only if

$$d\omega^a \wedge \Omega(I) = 0, \qquad a = 1, \dots, k. \tag{4-2.5}$$

Proof. If the generators satisfy (4-2.2), then they also satisfy (4-2.5). In order to establish the converse, complete $\omega^1, \dots, \omega^k$ to a basis $\omega^1, \dots, \omega^k, \omega^{k+1}, \dots, \omega^n$ of $\Lambda^1(E_n)$ and label the additional 1-forms with the index $\alpha, \beta, \dots$. Since $d\omega^a \in \Lambda^2(E_n)$, we may always write

$$d\omega^a = \xi^a_{ij}\omega^i \wedge \omega^j = \xi^a_{bc}\omega^b \wedge \omega^c + 2\xi^a_{\alpha c}\omega^\alpha \wedge \omega^c + \xi^a_{\alpha\beta}\omega^\alpha \wedge \omega^\beta,$$

and hence

$$\Omega(I) \wedge d\omega^a = \xi^a_{\alpha\beta}\Omega(I) \wedge \omega^\alpha \wedge \omega^\beta.$$

Since $k < n - 1$, $\deg(\Omega(I) \wedge \omega^\alpha \wedge \omega^\beta) \le n$ and hence $\Omega(I) \wedge \omega^\alpha \wedge \omega^\beta$ is a simple nonzero $(k+2)$-form because ω^α and ω^β are independent and not members of the subspace spanned by $\omega^1, \dots, \omega^k$. Thus (4-2.5) is satisfied only when $\xi^a_{\alpha\beta} = 0$, $\alpha, \beta = k+1, \dots, n$, $\quad a = 1, \dots, k$; that is,

$$d\omega^a = \left(\xi^a_{bc}\omega^b + 2\xi^a_{\alpha c}\omega^\alpha\right) \wedge \omega^c = \Gamma^a_c \wedge \omega^c$$

with $\Gamma^a_c = \xi^a_{bc}\omega^b + 2\xi^a_{\alpha c}\omega^\alpha$. □

The case $k \ge n - 1$ was excluded because $\Omega(I) \wedge d\omega^a$ would have degree greater than n and hence vanish identically. This case is taken care of by the following.

Theorem 4-2.3. *If I is generated by either $n - 1$ or n linearly independent 1-forms, then $dI \subset I$.*

Proof. For $k = n$, $\{\omega^a\}$ constitute a basis for $\Lambda^1(E_n)$ and hence $d\omega^a = \xi^a_{bc}\omega^b \wedge \omega^c$. The result then follows by Theorem 4-2.1. For the case $k = n - 1$, let γ be an additional 1-form such that $\omega^1, \ldots, \omega^{n-1}, \gamma$ is a basis for $\Lambda^1(E_n)$. We then have

$$d\omega^a = \xi^a_{bc}\omega^b \wedge \omega^c + \xi^a_c \gamma \wedge \omega^c$$

and the result again follows by Theorem 4-2.1. □

A simple example at this point will allow us to pull together the various ideas that have been introduced and will provide the springboard for the considerations of the next section. Consider the 1-form ω on E_n that is defined by

$$\omega = f(x^j)\, dg(x^j) = f\partial_i g\, dx^i, \qquad f \neq 0. \tag{4-2.6}$$

Exterior differentiation yields

$$d\omega = df \wedge dg \tag{4-2.7}$$

so that

$$\omega \wedge d\omega = 0. \tag{4-2.8}$$

Theorem 4-2.2 shows that the ideal, $I\{\omega\}$ generated by ω is closed: $dI \subset I$. Indeed since $f \neq 0$, we can rewrite (4-2.7) in the equivalent form

$$d\omega = \frac{1}{f}\, df \wedge f dg = \frac{1}{f}\, df \wedge \omega, \tag{4-2.9}$$

which is just Theorem 4-2.1.

Suppose that we need to solve the exterior equation

$$\omega = 0 \tag{4-2.10}$$

with ω given by (4-2.6). Since $f \neq 0$, it follows directly from (4-2.6) that $\omega = 0$ holds on every $(n-1)$-dimensional surface

$$g(x^j) = \text{constant}. \tag{4-2.11}$$

Accordingly, if the map

$$\Phi: E_r \to E_n | x^i = \phi^i(u^1, \ldots, u^r) \tag{4-2.12}$$

is such that

$$g(\phi^j(u^a)) = \text{constant}, \tag{4-2.13}$$

then

(4-2.14) $$\Phi^*\omega = f(\phi^j)\,dg(\phi^j) = f(\phi^j)\frac{\partial g(\phi^j)}{\partial \phi^i}\frac{\partial \phi^i}{\partial u^a}\,du^a = 0$$

and hence Φ gives the solutions of $\omega = 0$ in parametric form.

Starting with $\omega = \omega^1$, we can complete to a basis $\omega^1, \omega^2, \dots, \omega^n$ of $\Lambda^1(E_n)$. The dual basis is given by

(4-2.15) $$V_i \lrcorner \omega^j = \delta_i^j,$$

and hence we obtain

(4-2.16) $$V_\alpha \lrcorner \omega = 0, \qquad \alpha = 2, \dots, n;$$

that is, each of the $n - 1$ vector fields V_α, $\alpha = 2, \dots, n$ is a characteristic vector of ω. The characteristic subspace $\mathscr{S}(I\{\omega\})$ of the ideal $I\{\omega\}$ is thus of dimension $n - 1$. When (4-2.6) is substituted into (4-2.16) and we note that $f \neq 0$, we obtain

(4-2.17) $$V_\alpha \lrcorner dg = V_\alpha\langle g\rangle = 0, \qquad \alpha = 1, \dots, n.$$

The function g thus satisfies the system (4-2.17) of $n - 1$ simultaneous first order partial differential equations. Since $g(x^j)$ is not identically constant (i.e., ω is not the zero element of $\Lambda^1(E_n)$) the system (4-2.17) is complete. This implies that

(4-2.18) $$[V_\alpha, V_\beta] = C^\gamma_{\alpha\beta} V_\gamma,$$

for if one of these Lie products were not expressible in terms of the original set of $n - 1$ linearly independent vector fields $\{V_\alpha\}$, we would obtain n linearly independent vector fields that would annihilate g, and g would have to be identically constant. The characteristic subspace $\mathscr{S}(I\{\omega\})$ is thus a Lie subalgebra of $T(E_n)$. Conversely, if we are given a Lie subalgebra of $T(E_n)$ that is spanned by $n - 1$ linearly independent vector fields $\{V_\alpha, \alpha = 2, \dots, n\}$, we can reverse the process by (4-2.15) and obtain the 1-form $\omega^1 = \omega$ for which the subalgebra is the subalgebra of the characteristic subspace $\mathscr{S}(I\{\omega\})$. This form will then satisfy $\omega \wedge d\omega = 0$ and hence there exists a 1-form Γ such that

(4-2.19) $$d\omega = \Gamma \wedge \omega.$$

Exterior differentiation then yields

(4-2.20)

$$0 = dd\omega = d\Gamma \wedge \omega - \Gamma \wedge d\omega = d\Gamma \wedge \omega - \Gamma \wedge \Gamma \wedge \omega = d\Gamma \wedge \omega.$$

Since (4-2.19) also implies $d\omega = (\Gamma + \lambda\omega) \wedge \omega$, (4-2.20) is also equivalent to $0 = d(\Gamma + \lambda\omega) \wedge \omega$ and we can always choose λ so that $\Gamma + \lambda\omega$ is exact,

$$\Gamma = d\eta - \lambda\omega. \tag{4-2.21}$$

In this event, (4-2.19) becomes

$$d\omega = d\eta \wedge \omega \tag{4-2.22}$$

which is satisfied by $\omega = e^{\eta}\, dg$ for some g. The function g can then be obtained from the given ω by $dg = e^{-\eta}\omega$; that is $e^{\eta} = f$. The system of simultaneous complete first-order differential equations $V_\alpha\langle g\rangle = 0$ is thus solved by solving the equivalent problem $\omega = 0$.

The duality between $\omega = 0$, $dI\{\omega\} \subset I\{\omega\}$ and the complete system $V_\alpha\langle f\rangle = 0$, $\alpha = 2, \ldots, n$ is what allows us to actually solve the complete system $V_\alpha\langle f\rangle = 0$, $\alpha = 2, \ldots, n$ explicitly. The ability to obtain explicit solutions is very valuable and it is thus natural to ask whether there are similar dualities when the given data is more complicated; that is, when we have more than 1 form of degree one that generate the ideal I, or when we have a complete system of partial differential equations whose associated vector fields span a subspace of dimension less than $n - 1$. This question is answered in the affirmative by the celebrated Frobenius theorem in the next section.

4-3. THE FROBENIUS THEOREM

The reason why everything worked in the example given at the end of the last section was the shape that was assumed for the 1-form,

$$\omega = f(x^j)\, dg(x^j);$$

that is, ω was proportional to the exact 1-form $dg(x^j)$. This idea clearly generalizes to the case of more than a single 1-form,

$$\omega^a = K^a_b(x^j)\, dg^b(x^j), \qquad K^a_b(x^j) \in \Lambda^0(E_n).$$

Further, the ideal $I\{\omega^a\}$ that is generated by the 1-forms $\{\omega^a\}$ is the same as the ideal $I\{dg^a\}$ that is generated by the exact 1-forms $\{dg^a\}$ in this case. Thus if $\rho \in I\{\omega^a\}$, then $\rho \in I\{dg^a\}$ and hence $\rho = \eta_b \wedge dg^b$. It is then a trivial matter to see that $d\rho = d\eta_b \wedge dg^b \in I\{dg^a\}$ and hence the ideal $I\{dg^a\}$ is a closed ideal.

There are several technical details that must be taken care of before we can make full use of these ideas. We have already seen that any given system of 1-forms can be replaced by an equivalent system of 1-forms that spans the same subspace of $\Lambda^1(E_n)$ as was spanned by the given system. Linear independence of any given system of 1-forms may thus be assumed without loss of generality.

Definition. A collection of r linearly independent 1-forms, $\{\omega^a, a = 1, \dots, r\}$, is an *exterior system of dimension r*. The symbol D_r will be used to designate a given exterior system of dimension r and the ideal that is generated by $\{\omega^a\}$ will be denoted by $I\{D_r\}$.

Definition. An exterior system D_r with 1-forms $\{\omega^a\}$ is said to be *completely integrable* if and only if there exist r independent functions $\{g^a(x^j)\}$ such that each of the r 1-forms $\{\omega^a\}$ vanishes on each of the r-parameter family of $(n - r)$-dimensional surfaces

$$\{g^a(x^j) = c^a, \qquad a = 1, \dots, r\} \tag{4-3.1}$$

generated by allowing the r constants $\{c^a\}$ to range over all r-tuples of real numbers.

Theorem 4-3.1. *An exterior system D_r with 1-forms $\{\omega^a\}$ is completely integrable if and only if there exists a nonsingular r-by-r matrix of functions $((A^a_b(x^j)))$ and r independent functions $\{g^b(x^j)\}$ such that*

$$\omega^a = A^a_b \, dg^b. \tag{4-3.2}$$

Proof. If $\{\omega^a\}$ are given by (4-3.2), then they are completely integrable. Suppose therefore that $\{\omega^a\}$ is completely integrable so that r linearly independent functions $g^a(x^j)$ exist. We can then construct the ideal $I\{dg^a\}$ and this ideal is the largest closed ideal such that every one of its elements vanishes on the surfaces $\{g^a(x^j) = \text{constant}, a = 1, \dots, r\}$. However, complete integrability of $\{\omega^a\}$ says that each ω^a vanishes on the surfaces $\{g^a(x^j) = \text{constant}, a = 1, \dots, r\}$, and hence $I\{D_r\} \subset I\{dg^a\}$. Thus since both the ω's and the dg's are 1-forms, the ideal inclusion is satisfied only if there exists a matrix $((A^a_b))$ such that $\omega^a = A^a_b \, dg^b$. Since the ω's are linearly independent and the g's are independent, $\omega^1 \wedge \cdots \omega^r \neq 0$, $dg^1 \wedge \cdots \wedge dg^r \neq 0$ and hence $\omega^1 \wedge \cdots \wedge \omega^r = \det(A^a_b) \, dg^1 \wedge \cdots \wedge dg^r$ implies $\det(A^a_b) \neq 0$. □

Let us start with (4-3.2). Exterior differentiation and $\det(A^a_b) \neq 0$ then give

$$d\omega^a = dA^a_b \wedge dg^b = (dA^a_e) A^{-1e}_c \wedge A^c_b \, dg^b = (dA^a_e) A^{-1e}_c \wedge \omega^c. \tag{4-3.3}$$

Theorem 4-2.1 shows that *the ideal $I\{D_r\}$ is closed if D_r is completely integrable*. The converse of this statement, *D_r is completely integrable if the ideal $I\{D_r\}$ is closed*, is the famous Frobenius theorem. We defer the proof of this theorem until later since it involves arguments whose groundwork has not yet been prepared.

Theorem 4-3.2. (Frobenius) *An exterior system D_r that is defined by r 1-forms $\{\omega^a\}$ is completely integrable if and only if the ideal $I\{D_r\}$ is closed; that is $dI\{D_r\} \subset I\{D_r\}$ which is equivalent to*

$$d\omega^a = \Gamma^a_b \wedge \omega^b \tag{4-3.4}$$

or

$$\omega^1 \wedge \omega^2 \wedge \cdots \wedge \omega^r \wedge d\omega^a = 0, \qquad a = 1, \ldots, r. \tag{4-3.5}$$

We now turn to the question of equivalent systems of first-order partial differential equations. If $\{\omega^a\}$ define a differential system D_r, then the linear independence of the ω^a, $a = 1, \ldots, r$ allows us to complete this system to a basis $\omega^1, \ldots, \omega^r, \omega^{r+1}, \ldots, \omega^n$ of $\Lambda^1(E_n)$. The dual basis of $T(E_n)$ is then defined by

$$V_i \lrcorner \omega^j = \delta_i^j. \tag{4-3.6}$$

Thus, $\{V_\alpha, \alpha = r+1, \ldots, n\}$ satisfies

$$V_\alpha \lrcorner \omega^a = 0, \qquad a = 1, \ldots, r, \qquad \alpha = r+1, \ldots, n, \tag{4-3.7}$$

so that $\{V_\alpha\}$ spans the subspace of characteristic vector fields of D_r.

Lemma 4-3.1. *The characteristic subspace, $\mathscr{S}(D_r)$, of an exterior system D_r is a subspace of $T(E_n)$ of dimension $n - r$ and a basis for $\mathscr{S}(D_r)$ is given by $\{V_\alpha, \alpha = r+1, \ldots, n\}$ where $\{V_i, i = 1, \ldots, n\}$ is the dual basis that is defined by* (4-3.6).

Proof. By definition $\mathscr{S}(D_r)$ consists of all elements of $T(E_n)$ such that $V \lrcorner \rho \in I\{D_r\}$ for all $\rho \in I\{D_r\}$. The construct immediately above shows that $\{V_\alpha\}$ is a basis for $\mathscr{S}(D_r)$ and hence $\mathscr{S}(D_r)$ is of dimension $n - r$. □

We now need a technical result in order to connect complete integrability of D_r with the corresponding notion of complete integrability of the system $V_\alpha\langle f \rangle = 0$, $\alpha = r+1, \ldots, n$.

Lemma 4-3.2. *If η is any element of $\Lambda^1(E_n)$ and U, V are any elements of $T(E_n)$, then*

$$U\langle V \lrcorner \eta \rangle - V\langle U \lrcorner \eta \rangle = [U, V] \lrcorner \eta + V \lrcorner U \lrcorner d\eta. \tag{4-3.8}$$

Proof. Set $\eta = \eta_i dx^i$, $U = u^i \partial_i$, $V = v^i \partial_i$, then $U \lrcorner \eta = u^i \eta_i$, $V \lrcorner \eta = v^i \eta_i$ and hence

$$\begin{aligned} U\langle V \lrcorner \eta \rangle - V\langle U \lrcorner \eta \rangle &= \{U\langle v^i \rangle - V\langle u^i \rangle\} \eta_i + v^i U\langle \eta_i \rangle - u^i V\langle \eta_i \rangle \\ &= [U, V] \lrcorner \eta + v^i U\langle \eta_i \rangle - u^i V\langle \eta_i \rangle. \end{aligned}$$

On the other hand,

$$U \lrcorner d\eta = U \lrcorner (d\eta_i \wedge dx^i) = (U \lrcorner d\eta_i)\, dx^i - (U \lrcorner dx^i)\, d\eta_i$$

$$= U\langle \eta_i \rangle\, dx^i - u^i\, d\eta_i$$

so that

$$V \lrcorner U \lrcorner d\eta = U\langle \eta_i \rangle v^i - u^i V\langle \eta_i \rangle;$$

$$U\langle V \lrcorner \eta \rangle - V\langle U \lrcorner \eta \rangle = [U, V] \lrcorner \eta + V \lrcorner U \lrcorner d\eta. \qquad \square$$

Theorem 4-3.3. *An exterior system D_r that is defined by the r 1-forms $\{\omega^a\}$ is completely integrable if and only if its characteristic subspace $\mathscr{S}(D_r)$ is a Lie subalgebra of $T(E_n)$.*

Proof. Let $\{V_\alpha, \alpha = r+1, \dots, n\}$ be a basis for $\mathscr{S}(D_r)$, so that

$$V_\alpha \lrcorner \omega^a = 0, \qquad a = 1, \dots, r, \qquad \alpha = r+1, \dots, n \tag{4-3.9}$$

and let Greek indices have the range $r+1, \dots, n$. A direct application of (4-3.8) yields

$$V_\alpha \langle V_\beta \lrcorner \omega^a \rangle - V_\beta \langle V_\alpha \lrcorner \omega^a \rangle = [V_\alpha, V_\beta] \lrcorner \omega^a + V_\beta \lrcorner V_\alpha \lrcorner d\omega^a \tag{4-3.10}$$

and hence

$$[V_\alpha, V_\beta] \lrcorner \omega^a = V_\alpha \lrcorner V_\beta \lrcorner d\omega^a. \tag{4-3.11}$$

If D_r is completely integrable, then

$$d\omega^a = \Gamma^a_b \wedge \omega^b, \qquad V_\beta \lrcorner d\omega^a = (V_\beta \lrcorner \Gamma^a_b)\omega^b - \Gamma^a_b V_\beta \lrcorner \omega^b = (V_\beta \lrcorner \Gamma^a_b)\omega^b,$$

and

$$V_\alpha \lrcorner V_\beta \lrcorner d\omega^a = (V_\beta \lrcorner \Gamma^a_b) V_\alpha \lrcorner \omega^b = 0.$$

The right-hand side of (4-3.11) thus vanishes and hence $[V_\alpha, V_\beta] \lrcorner \omega^a = 0$ shows that $[V_\alpha, V_\beta] \in \mathscr{S}(D_r)$. Since $\{V_\alpha, \alpha = r+1, \dots, n\}$ is a basis for $\mathscr{S}(D_r)$, we have

$$[V_\alpha, V_\beta] = C^\gamma_{\alpha\beta} V_\gamma \tag{4-3.12}$$

and hence $\mathscr{S}(D_r)$ is a Lie subalgebra of $T(E_n)$. Conversely, if $\mathscr{S}(D_r)$ is a Lie subalgebra of $T(E_n)$, (4-3.12) holds and hence $[V_\alpha, V_\beta] \lrcorner \omega^a = C^\gamma_{\alpha\beta} V_\gamma \lrcorner \omega^a = 0$. In this event, (4-3.11) implies

$$V_\alpha \lrcorner V_\beta \lrcorner d\omega^a = 0. \tag{4-3.13}$$

Thus if we note that (4-3.6) implies

$$V_\alpha \lrcorner \omega^\beta = \delta^\beta_\alpha, \qquad V_\alpha \lrcorner \omega^a = 0,$$

$$d\omega^a = \xi^a_{bc}\omega^b \wedge \omega^c + 2\xi^a_{\alpha c}\omega^\alpha \wedge \omega^c + \xi^a_{\alpha\beta}\omega^\alpha \wedge \omega^\beta$$

then (4-3.13) can be satisfied only if $\xi^a_{\alpha\beta} = 0$; that is

$$d\omega^a = \left(\xi^a_{bc}\omega^b + 2\xi^a_{\alpha c}\omega^\alpha\right) \wedge \omega^c = \Gamma^a_c \wedge \omega^c. \tag{4-3.14}$$

The Frobenius theorem then shows that D_r is completely integrable. □

Corollary 4-3.1. *Any complete system*

$$V_\alpha\langle f \rangle = 0, \qquad \alpha = r+1, \dots, n \tag{4-3.15}$$

of linear partial differential equations in E_n defines and is defined by a completely integrable exterior system D_r whose 1-forms $\{\omega^a, a = 1, \dots, r\}$ satisfy the requirements

$$V_i \lrcorner \omega^j = \delta^j_i, \qquad i, j = 1, \dots, n, \tag{4-3.16}$$

where $\{V_i\}$ is a completion of $\{V_\alpha\}$ to a basis for $T(E_n)$. Thus the functions $\{g^a(x^j)\}$ that occur in

$$\omega^a = A^a_b\, dg^b \tag{4-3.17}$$

are r independent solutions of (4-3.15) *and the general solution of* (4-3.15) *is given by*

$$f = \Psi\left(g^1(x^j), \dots, g^r(x^j)\right). \tag{4-3.18}$$

Proof. The only thing that has not yet been proven is $V_\alpha\langle g^a \rangle = 0$. However, $V_\alpha \lrcorner \omega^a = A^a_b V_\alpha \lrcorner dg^b = A^a_b V_\alpha\langle g^b \rangle = 0$ because each V_α is a characteristic vector of D_r. Thus since $\det(A^a_b) \neq 0$ we indeed have $V_\alpha\langle g^a \rangle = 0$. It is then a trivial matter to see that (4-3.18) constitutes the general solution of (4-3.15); $V_\alpha\langle f \rangle = V_\alpha\langle \Psi \rangle = \partial\Psi/\partial g^a\, V_\alpha\langle g^a \rangle = 0$, $\alpha = r+1, \dots, n$. □

From the practical point of view, Corollary 4-3.1 is the important result, for it tells us how to find the independent integrals of a complete system of linear partial differential equations. The construction of the corresponding exterior system is also fundamental whenever we have to deal with mappings. This should be obvious since exterior forms behave very nicely under mappings while vector fields do not unless the mapping is invertible.

It is clear that not every exterior system D_r is completely integrable, for $dI\{D_r\}$ need not be contained in $I\{D_r\}$. The question naturally arises as to what can be said for not completely integrable exterior systems.

Definition. Let D_r be an exterior system with linearly independent 1-forms $\{\omega^a, a = 1, \ldots, r\}$ and let $\{V_\alpha, \alpha = r+1, \ldots, n\}$ be a basis for the characteristic subspace, $\mathscr{S}(D_r)$, of the exterior system. If $\bar{\mathscr{S}}(D_r)$ is the completion of $\mathscr{S}(D_r)$ to a Lie subalgebra of $T(E_n)$, then D_r is said to be *subintegrable* if and only if the dimension of $\bar{\mathscr{S}}(D_r)$ is less than n. If $\{V_\alpha, \alpha = r - s + 1, \ldots, n\}$ is a basis for $\bar{\mathscr{S}}(D_r)$, and $\{\gamma^1, \ldots, \gamma^{r-s}\}$ is a basis for the subspace of $\Lambda^1(E_n)$ whose characteristic subspace is $\bar{\mathscr{S}}(D_r)$, then $\{\gamma^1, \ldots, \gamma^{r-s}\}$ is the *maximal subexterior* system of the exterior system D_r that is completely integrable.

The concept of the maximal subexterior system of a subintegrable exterior system provides certain information that is quite often useful in the solution of explicit problems. Needless to say, subintegrable exterior systems are much harder to deal with than completely integrable exterior systems.

The closure of an ideal of $\Lambda(E_n)$ and the complete integrability of its system of characteristic vector fields is not restricted to the case where the ideal is generated by 1-forms. A demonstration of this will be given in Section 4-5.

4-4. THE DARBOUX CLASS OF A 1-FORM AND THE DARBOUX THEOREM

We now turn to the question of how to obtain an intrinsic characterization and a general representation of an arbitrary 1-form. As with most such problems the first task is to obtain the characterization.

Let ω be a given 1-form. Starting with ω we construct the following sequence of exterior forms of increasing degree:

(4-4.1) $$I_1 = \omega, \qquad I_2 = d\omega,$$

(4-4.2) $$I_{2k+1} = \omega \wedge I_{2k} = I_1 \wedge I_{2k},$$

(4-4.3) $$I_{2k+2} = dI_{2k+1} = I_2 \wedge I_{2k}.$$

This sequence is finite, for all forms of degree higher than n vanish identically and hence $I_{n+1} \equiv 0$. For each point $P:(x^i)$ in an n-dimensional subset $\mathscr{S}$ of E_n we can find an integer $K(P)$ such that

(4-4.4) $$I_{K(P)} \neq 0, \qquad I_{K(P)+m} = 0$$

for all $m > 0$.

Definition. The *Darboux class*, $K(\omega, \mathscr{S})$, of the 1-form ω relative to the set $\mathscr{S}$ is the integer

$$K(\omega, \mathscr{S}) = \max_{P \in \mathscr{S}} K(P). \tag{4-4.5}$$

Definition. If $K(P) = K(\omega, \mathscr{S})$ then $P:(x^j)$ is said to be a *regular point* of ω relative to $\mathscr{S}$. If $K(P) < K(\omega, \mathscr{S})$ then $P:(x^j)$ is said to be a *critical point* of ω relative to $\mathscr{S}$. The important thing to note here is that the class of a 1-form, its regular points and its critical points can be determined in a finite number of steps simply by constructing the sequence given by (4-4.1) through (4-4.3).

Definition. The *rank* of ω relative to $\mathscr{S}$ is the greatest even integer less than or equal to $K(\omega, \mathscr{S})$ and is written as $2p(\omega, \mathscr{S})$. The *index* of ω relative to $\mathscr{S}$ is given by

$$\varepsilon(\omega, \mathscr{S}) = K(\omega, \mathscr{S}) - 2p(\omega, \mathscr{S}). \tag{4-4.6}$$

The Darboux theorem gives the following representation.

Theorem 4-4.1. (Darboux) *Let ω be a 1-form with Darboux class $K(\omega, \mathscr{S})$ on an n-dimensional subset $\mathscr{S}$ of E_n. There exist $p(\omega, \mathscr{S})$ positive, scalar-valued functions $\{u_a(x^j)\}$ and $p(\omega, \mathscr{S}) + \varepsilon(\omega, \mathscr{S})$ scalar-valued functions $\{v_b(x^j)\}$ that constitute a system of $K(\omega, \mathscr{S})$ functionally independent functions at every regular point of $\mathscr{S}$ such that*

$$\omega = \sum_{a=1}^{p(\omega, \mathscr{S})} u_a \, dv_a + \varepsilon(\omega, \mathscr{S}) \, dv_{p(\omega, \mathscr{S})+1}. \tag{4-4.7}$$

A detailed proof of this theorem will not be given (see Schouten, pp. 89–94; Sternberg, pp. 137–141). There are, however, certain aspects of the proof that are worth mentioning since they give a direct method for obtaining the functions $\{u_a\}$ and $\{v_a\}$. Use of the Frobenius theorem shows that there exists a scalar-valued function η such that

$$I_{K(\omega)} = d\eta \wedge I_{K(\omega)-1} \tag{4-4.8}$$

at all regular points of $\omega(x^i)$. Two classes of transformations are then constructed. The first consists of similarity transformations. A 1-form ω is said to undergo a *similarity transformation* to an image form $`\omega$ if

$$`\omega = \sigma\omega, \qquad \sigma > 0. \tag{4-4.9}$$

If we construct the sequence ${}^{\backprime}I_k$ from the image form ${}^{\backprime}\omega$, it follows readily that

$$(4\text{-}4.10)\quad {}^{\backprime}I_{2k} = \sigma^k I_{2k} + k\sigma^{k-1} d\sigma \wedge I_{2k-1}, \qquad {}^{\backprime}I_{2k+1} = \sigma^{k+1} I_{k+1}.$$

Accordingly, if $K(\omega) = 2p(\omega)$, (4-4.8) can be used to obtain

$${}^{\backprime}I_{2p({}^{\backprime}\omega)} = \sigma^k d\eta \wedge I_{2k-1} + k\sigma^{k-1} d\sigma \wedge I_{2k-1}, \qquad k = p(\omega).$$

The choice $\sigma = e^{-\eta/k} = e^{-\eta/p(\omega)}$ then reduces the class of ${}^{\backprime}\omega$ to $2p(\omega) - 1$. The fact that all of the u's are positive follows from the observation $e^{-\eta/k} > 0$. A 1-form ω is said to undergo a *gradient transformation* to an image form ${}^{\backprime}\omega$ if

$$(4\text{-}4.11)\qquad {}^{\backprime}\omega = \omega + d\lambda.$$

This gives

$$(4\text{-}4.12)\qquad {}^{\backprime}I_{2k} = I_{2k}, \qquad {}^{\backprime}I_{2k+1} = I_{2k+1} + d\lambda \wedge I_{2k}.$$

Thus if $k(\omega)$ is an odd integer $= 2p(\omega) + 1$, then use of (4-4.8) gives

$${}^{\backprime}I_{K({}^{\backprime}\omega)} = I_{K(\omega)} + d\lambda \wedge I_{2p(\omega)} = d\eta \wedge I_{2p(\omega)} + d\lambda \wedge I_{2p(\omega)}.$$

The choice $\lambda = -\eta$ reduces the class of ${}^{\backprime}\omega$ to $2p(\omega)$. A succession of gradient transformations and similarity transformations will thus reduce the class of any form ω to one. The functions $\{u_a\}$ and $\{v_a\}$ are then constructed from the σ's and the λ's by inverting the sequence of similarity and gradient transformations, on noting that a 1-form of class one is an exact 1-form.

The Darboux theorem is very powerful. It tells us that any 1-form

$$\omega = w_i(x^j)\, dx^i$$

on E_n, which is necessarily a linear combination of the n exact forms $dx^1, dx^2, \ldots, dx^n$, can always be written as

$$\omega = \sum_{a=1}^{p(\omega,\mathscr{S})} u_a(x^j)\, dv_a(x^j) + \varepsilon(\omega, \mathscr{S})\, dv_{p(\omega,\mathscr{S})+1}(x^i).$$

There is a reduction from n exact 1-forms to $p(\omega, \mathscr{S}) + \varepsilon(\omega, \mathscr{S}) \leq [n/2] + 1$ exact 1-forms. The dimension reduction that is thus afforded is often instrumental in the solution of specific problems. For example, suppose that we have a 1-form ω on E_{12} that has Darboux class three. We may then pass from a linear combination of 12 exact 1-forms $dx^1, \ldots, dx^{12}$ to only 2, for we know that $\omega = u_1(x^j)\, dv_1(x^j) + dv_2(x^j)$ in this instance.

We note the lack of uniqueness in the representation given by the Darboux theorem. In particular, since $u_i\,dv_i = d(u_i v_i) - v_i\,du_i$, (4-4.7) can also be written as

(4-4.13) $$\omega = -\sum_{i=1}^{p(\omega)} v_i\,du_i + d\left\{\varepsilon(\omega)v_{2p(\omega)+1} + \sum_{i=1}^{p(\omega)} u_i v_i\right\}.$$

There are many different representations of $\omega(x)$ that involve the same total number of functions, and for each, the function whose exterior derivative occurs with unity coefficient will be different.

The Darboux theorem provides a standard or canonical form for closed 2-forms that is of importance in its own right.

Theorem 4-4.2. *If η is a closed 2-form on an n-dimensional region $\mathscr{S}$ of E_n that is starshaped, then η has the canonical form*

(4-4.14) $$\eta = \sum_{a=1}^{p(\omega,\mathscr{S})} du_a \wedge dv_a$$

at all regular points of ω in $\mathscr{S}$ and the functions $\{u_a, v_a;\, a = 1, \ldots, p(\omega, \mathscr{S})\}$ are independent.

Proof. Since η is closed and $\mathscr{S}$ is starshaped, the Poincaré lemma implies the existence of an $\omega \in \Lambda^1(\mathscr{S})$ such that $\eta = d\omega$. The Darboux theorem applied to ω then gives (4-4.7) and hence an exterior differentiation yields (4-4.14). □

The Darboux theorem provides explicit results in connection with the problem of solving $\omega = 0$. We simply substitute (4-4.7) for ω and equate the result to zero. Now observe that for $\varepsilon(\omega, \mathscr{S}) = 0$ we may always divide by u_1 since $u_1 \neq 0$ throughout $\mathscr{S}$. The problem of solving $\omega = 0$ can always be reduced to solving

(4-4.15) $$\omega = dv_1 + u_2\,dv_2 + \cdots + u_r\,dv_r = 0$$

with all of the u's positive. Solutions of (4-4.15) are obviously given by the intersection of the surfaces in $\mathscr{S}$ that are defined by

(4-4.16) $$v_1 = c_1, \quad v_2 = c_2, \ldots, v_r = c_r.$$

There are other solutions, but these obtain by relations between the u's and the v's. For example, suppose that we set

(4-4.17) $$v_1 = \phi(v_2, \ldots, v_r).$$

We then have

$$\bar{\omega} = (u_2 + \partial\phi/\partial v_2)\,dv_2 + \cdots + (u_r + \partial\phi/\partial v_r)\,dv_r$$

and hence $\bar{\omega} = 0$ can be satisfied by (4-4.17) and

$$u_2 = -\partial\phi/\partial v_2, \ldots, u_r = -\partial\phi/\partial v_r. \tag{4-4.18}$$

A concept that is important in applications is the accessibility of a 1-form originally introduced by Carathéodory.

Definition. A 1-form ω is said to have the *inaccessibility* property if and only if a sufficiently small neighborhood of any point P contains a point Q that cannot be reached by a path from P that satisfies $\omega = 0$. If ω does not have the inaccessibility property then it has the *accessibility* property.

Theorem 4-4.3. *A 1-form ω has the inaccessibility property on an n-dimensional set $\mathscr{S}$ of E_n if and only if the Darboux class of ω relative to $\mathscr{S}$ is less than three. If $K(\omega, \mathscr{S}) \geq 3$ then ω has the accessibility property on $\mathscr{S}$.*

Proof. If $K(\omega, \mathscr{S}) = 1$, then $\omega = dv_1$ and hence $\omega = 0$ holds only on $(n-1)$-dimensional surfaces $v_1 = c_1$. If P belongs to $v_1 = c_1$, then simply take Q to belong to $v_1 = c_1 + \delta$ for sufficiently small δ in order to obtain a point in the neighborhood of P that cannot be reached by any path in $\mathscr{S}$ for which $\omega = 0$. If $K(\omega, \mathscr{S}) = 2$, then $\omega = u_1\,dv_1$ and $\omega = 0$ is equivalent to $dv_1 = 0$ which is the previous case. For $K(\omega, \mathscr{S}) = 3$, $\omega = 0$ becomes

$$0 = dv_1 + u_2\,dv_2. \tag{4-4.19}$$

The cases with $K(\omega, \mathscr{S}) > 3$ proceed along the same lines as that for $K(\omega, \mathscr{S}) = 3$, so we will establish accessibility for (4-4.19) only. It is clear that $\{v_1 = \text{constant},\ v_2 = \text{constant}\}$ define surfaces of dimension $n-2$ of mutually accessible points. Now suppose that $P:(x_0^i)$ and $Q:(x_1^i)$ and set

$$v_1(x_0^i) = v_1^0, \qquad v_2(x_0^i) = v_2^0, \qquad v_1(x_1^i) = v_1^1, \qquad v_2(x_1^i) = v_2^1.$$

Accessibility will be established if we can find a path in E_n that connects the values (v_1^0, v_2^0) with (v_1^1, v_2^1). To this end, consider the path for which the induced functions

$$v_a(x^i(t)) = \bar{v}_a(t), \qquad u_2(x^i(t)) = \bar{u}_2(t)$$

are such that

$$\bar{v}_2(t) = v_2^0 + (v_2^1 - v_2^0)t, \qquad \bar{u}_2(t) = u_2^0 + ht. \tag{4-4.20}$$

Under these circumstances, (4-4.19) will be satisfied provided

(4-4.21) $$d\bar{v}_1/dt = (v_2^0 - v_2^1)(u_2^0 + ht)$$

and hence

(4-4.22) $$\bar{v}_1(t) = v_1^0 + (v_2^0 - v_2^1)(u_2^0 + ht/2)t.$$

Thus if we set $t = 1$, then $\bar{v}_2(1) = v_2^1$ and h can be chosen to secure

$$\bar{v}_1(1) = v_1^1 = v_1^0 + (v_2^0 - v_2^1)(u_2^0 + h/2),$$

provided $v_2^1 \neq v_2^0$. When $v_2^1 = v_2^0$, we first consider the path

$$\bar{v}_1(t) = v_1^0 + at, \qquad \bar{u}_2(t) = u_2^0, \qquad \bar{v}_2(t) = v_2^0 - at/u_2^0$$

for $u_2^0 \neq 0$. This path satisfies (4-4.19) and we have

$$\bar{v}_1(1) = v_1^0 + a, \qquad \bar{u}_2(1) = u_2^0, \qquad \bar{v}_2(1) = v_2^0 - a/u_2^0.$$

If $(\bar{v}_1(1), \bar{u}_2(1), \bar{v}_2(1))$ are used as a new starting point, the previous case shows that we can get to (v_1^1, u_2^1, v_2^1) by a path along which $\omega = 0$ since a can be so chosen that $v_2^1 \neq \bar{v}_2(1) = v_2^0 - a/u_2^0$ provided $u_2^0 \neq 0$. In case $u_2^0 = 0$, we simply start with the path

$$\bar{v}_1(t) = v_1^0, \qquad \bar{u}_2(t) = u_2^0 + bt, \qquad \bar{v}_2(t) = v_2^0$$

that satisfies (4-4.19) and obtain $\bar{u}_2(1) = u_2^0 + b \neq 0$ by an appropriate choice of b. □

Spaces of even dimension with a closed 2-form of maximal rank are now recognized as fundamental to the study of classical mechanics. The notion of the rank of a 2-form follows directly from the notion of the Darboux class of a 1-form and Theorem 4-4.2.

Definition. The *rank* of a closed 2-form ω on a region $\mathscr{S}$ is the number $2p(\omega, \mathscr{S})$ where $p(\omega, \mathscr{S})$ has the property

(4-4.23) $$\omega^{(p)} \neq 0, \qquad \omega^{(p+1)} = 0$$

at all points of $\mathscr{S}$. Here, $\omega^{(k)}$ means the k-fold exterior product of ω with itself.

Definition. A $2n$-dimensional starshaped region $\mathscr{S}$ of a $2n$-dimensional space E_{2n} is said to be a *symplectic region* if and only if there is a closed 2-form ω

defined on $\mathscr{S}$ that has maximal rank on $\mathscr{S}$ (i.e., $p(\omega, \mathscr{S}) = n$), and ω is said to be the *fundamental* or *symplectic* 2-form on $\mathscr{S}$. A symplectic region is said to admit a *symplectic structure*.

Theorem 4-4.4. *Let $\mathscr{S}$ be a symplectic region of E_{2n} with fundamental 2-form ω. There exists a coordinate cover $(q^1, \ldots, q^n; p_1, \ldots, p_n) = (q^\alpha; p_\alpha)$ of $\mathscr{S}$ for which*

$$\omega = dp_\alpha \wedge dq^\alpha \tag{4-4.24}$$

and we have

$$\omega = d\pi \tag{4-4.25}$$

where the symplectic potential 1-form π is determined by

$$\pi = p_\alpha \, dq^\alpha + d\rho \tag{4-4.26}$$

to within an arbitrary gradient (gauge) transformation for any gauge function $\rho(q^\alpha; p_\alpha) \in \Lambda^0(\mathscr{S})$.

Proof. By definition ω is a closed 2-form of maximal rank and hence $2p(\omega, \mathscr{S}) = 2n$. Accordingly, Theorem 4-4.3 shows that ω can be written on $\mathscr{S}$ as

$$\omega = \sum_{a=1}^{n} du_a \wedge dv_a \tag{4-4.27}$$

and that

$$du_1 \wedge du_2 \wedge \cdots \wedge du_n \wedge dv_1 \wedge dv_2 \wedge \cdots \wedge dv_n \neq 0 \tag{4-4.28}$$

throughout $\mathscr{S}$ since the u's and the v's are independent. In view of (4-4.28),

$$p_\alpha = u_\alpha(x^i), \qquad q^\alpha = v_\alpha(x^i), \qquad \alpha = 1, \ldots, n \tag{4-4.29}$$

defines a one-to-one regular transformation of coordinate covers of $\mathscr{S}$. The representation (4-4.24) then obtains for the coordinate cover $(q^\alpha; p_\alpha)$ by (4-4.27). Now (4-4.24) can be written as $\omega = d(p_\alpha \, dq^\alpha)$, and hence $\omega = d(p_\alpha \, dq^\alpha + d\rho)$ for any $\rho \in \Lambda^0(\mathscr{S})$. □

The relevance to classical mechanics comes about through the following consideration.

Definition. A vector field $V \in T(\mathscr{S})$, on a symplectic region $\mathscr{S}$ with fundamental 2-form ω, is a *Hamiltonian* vector field with generating function

$H(q^\alpha; p_\alpha) \in \Lambda^0(\mathscr{S})$ if and only if

$$dH + V \lrcorner \omega = 0. \tag{4-4.30}$$

Lemma 4-4.1. *A vector field $V \in T(\mathscr{S})$ is a Hamiltonian vector field with generating function $H(q^\alpha; p_\alpha)$ if and only if*

$$V = \frac{\partial H}{\partial p_\alpha}\frac{\partial}{\partial q^\alpha} - \frac{\partial H}{\partial q^\alpha}\frac{\partial}{\partial p_\alpha}. \tag{4-4.31}$$

Proof. When (4-4.30) is written out, we have

$$\frac{\partial H}{\partial q^\alpha} dq^\alpha + \frac{\partial H}{\partial p_\alpha} dp_\alpha + V\langle p_\alpha\rangle dq^\alpha - V\langle q^\alpha\rangle dp_\alpha = 0.$$

The representation (4-4.31) then follows directly on resolving this relation on the basis $\{dq^\alpha; dp_\alpha\}$. □

Theorem 4-4.5. *If V is a Hamiltonian vector field generated by $H(q^\alpha; p_\alpha) \in \Lambda^0(\mathscr{S})$ on a symplectic region $\mathscr{S}$ with fundamental 2-form ω, then the orbital equations for V are Hamilton's equations*

$$\frac{dq^\alpha}{ds} = \frac{\partial}{\partial p_\alpha} H(q^\beta; p_\beta), \qquad \frac{dp_\alpha}{ds} = -\frac{\partial}{\partial q^\alpha} H(q^\beta; p_\beta) \tag{4-4.32}$$

for a mechanical system with n degrees of freedom and Hamiltonian function $H(q^\alpha; p_\alpha)$.

Proof. The result is an immediate consequence of Lemma 4-4.1 and the definition of the orbital equations of a vector field. □

Suppose we need to solve the exterior equation

$$\omega = 0 \tag{4-4.33}$$

on a region $\mathscr{S}$ of E_n comprised solely of regular points of the 1-form ω. The Darboux theorem shows that ω can be written in the form

$$\omega = \sum_{a=1}^{p} u_a \, dv_a \tag{4-4.34}$$

if the Darboux class is even ($K(\omega, \mathscr{S}) = 2p$) and in the form

$$\omega = \sum_{a=1}^{p} u_a \, dv_a + dv_{p+1} \tag{4-4.35}$$

if the Darboux class is odd ($K(\omega, \mathscr{S}) = 2p + 1$). In both cases the p functions $\{u_a\}$ may be chosen to be strictly positive. Thus in particular, (4-4.34) can also be written as

$$\omega = \left(\sum_{a=1}^{p-1} \frac{u_a}{u_p} dv_a + dv_p \right) u_p. \tag{4-4.36}$$

The problem of solving $\omega = 0$ on $\mathscr{S}$ is thus equivalent to solving the problem

$$\omega' = \sum_{a=1}^{\bar{p}} u_a\, dv_a + dv_{\bar{p}+1} = 0, \tag{4-4.37}$$

with $\bar{p}$ a fixed integer on $\mathscr{S}$ such that

$$2\bar{p} < n \tag{4-4.38}$$

and $\{u_1, \ldots, u_{\bar{p}}; v_1, \ldots, v_{\bar{p}+1}\}$ are functionally independent on $\mathscr{S}$.

There are the obvious solutions of dimension $n - \bar{p} - 1$ that are given by

$$v_a(x^j) = c_a, \qquad a = 1, \ldots, \bar{p} + 1 \tag{4-4.39}$$

for each choice of the $\bar{p} + 1$ constants $\{c_a, a = 1, \ldots, \bar{p} + 1\}$. An inspection of (4-4.37) shows that any other solution must involve relations between the v's. Let r be any integer such that

$$1 \le r \le \bar{p} \tag{4-4.40}$$

for $\bar{p} > 0$, and set

$$v_1(x^j) = y^1, \ldots, v_r(x^j) = y^r, \tag{4-4.41}$$

$$v_{r+1}(x^j) = f^{r+1}(y^1, \ldots, y^r), \ldots, \tag{4-4.42}$$

$$v_{\bar{p}}(x^j) = f^{\bar{p}}(y^1, \ldots, y^r),$$

$$v_{\bar{p}+1}(x^j) = f^{\bar{p}+1}(y^1, \ldots, y^r), \tag{4-4.43}$$

where the $\bar{p} + 1 - r$ functions $f^{r+1}, \ldots, f^{\bar{p}+1}$ are arbitrary. When this is used in conjunction with (4-4.37), we have

$$\omega' = \sum_{\alpha=1}^{r} \left(u_\alpha + \frac{\partial f^{\bar{p}+1}}{\partial y^\alpha} + \sum_{k=1}^{\bar{p}-r} u_{r+k} \frac{\partial f^{r+k}}{\partial y^\alpha} \right) dy^\alpha. \tag{4-4.44}$$

It is then a simple matter to see that

(4-4.45) $$u_{r+1}(x^j) = y^{r+1}, \dots, u_{\bar{p}}(x^j) = y^{\bar{p}},$$

(4-4.46) $$u_\alpha(x^j) = -\frac{\partial f^{\bar{p}+1}}{\partial y^\alpha} - \sum_{k=1}^{\bar{p}-r} y^{r+k}\frac{\partial f^{r+k}}{\partial y^\alpha}, \qquad \alpha = 1, \dots, r,$$

gives $\omega = 0$ and hence solves the problem.

The $\bar{p}$ quantities $y^1, \dots, y^{\bar{p}}$ play the role of variables that can be eliminated. This elimination will result in $\bar{p} + 1 - r$ relations between the x's that come form (4-4.42), (4-4.43) and r relations between the x's that come from (4-4.46), for a total of $\bar{p} + 1$ relations between the x's. *The solution given by* (4-4.41) through (4-4.43), (4-4.45) and (4-4.46) *is thus a solution of dimension* $n - \bar{p} - 1$ *that involves* $\bar{p} + 1 - r$ *arbitrary functions* $f^{r+1}, \dots f^{\bar{p}+1}$ *where* r *is any number such that* $1 \leq r \leq \bar{p}$.

An inspection of (4-4.34) through (4-4.37) shows that the number $\bar{p} = \bar{p}(\omega, \mathscr{S})$ is determined by

(4-4.47) $$\bar{p}(\omega, \mathscr{S}) = p(\omega, \mathscr{S}) + \varepsilon(\omega, \mathscr{S}) - 1$$

and is thus known once the Darboux class is known. We summarize the results in Table 2 for a given value of n. The number of arbitrary functions that can occur in a solution of maximal dimension is of particular importance. If there are r arbitrary functions, then an r-fold infinity of solutions of maximal dimension will pass through any point P of $\mathscr{S}$. Thus for 1-forms of Darboux class one or two, there is one and only one solution of maximal dimension $(n - 1)$ that will pass through any point P of $\mathscr{S}$. We know, however, that any 1-form of Darboux class one or two has the inaccessibility property, and this may also be viewed as the confinement of all one-dimensional solutions through a point P to lie in the unique maximal solution that passes through P. On the other hand, for $K > 2$, the table shows that there will always be at least

TABLE 2. Solutions of $\omega = 0$

Darboux Class $\leq n$	$\bar{p}$	Maximal Dimension of Solutions	Number of Arbitrary Functions
1	0	$n - 1$	0
2	0	$n - 1$	0
3	1	$n - 2$	0, 1
4	1	$n - 2$	0, 1
5	2	$n - 3$	0, 1, 2
6	2	$n - 3$	0, 1, 2
7	3	$n - 4$	0, 1, 2, 3
8	3	$n - 4$	0, 1, 2, 3
9	4	$n - 5$	0, 1, 2, 3, 4
10	4	$n - 5$	0, 1, 2, 3, 4

a 1-fold infinity of solutions of maximal dimension that passes through a given point P and we know that any 1-form with $K > 2$ has the accessibility property. These two facts combine to show that the one-dimensional solutions for $K > 2$ are not necessarily contained in solutions of maximal dimension; rather, they can weave between solutions of maximal dimension by changing the selection of the arbitrary functions from one point to another. In any event, it should now be clear that $\omega = 0$ with $K(\omega, \mathscr{S}) > 2$ can have solutions that are complicated in the extreme.

All of the solutions obtained in the above discussion are solutions of *maximal dimension* $n - \bar{p} - 1$. Clearly, there are solutions of lower dimension and these are obtained by constructing mappings $\Phi: E_m \to E_n$ with $m < n - \bar{p} - 1$ for which $\Phi^*\omega' = 0$ is identically satisfied on the domain of Φ (see Section 3-7). It should be noted, however, that with the exception of one-dimensional solutions, the problem of finding mappings Φ such that $\Phi^*\omega' = 0$ is an even more difficult problem since there is no explicit construction similar to that given above for solutions of maximal dimension.

4-5. FINITE DEFORMATIONS AND LIE DERIVATIVES

If $V = v^i(x^j)\partial_i$ is a given element of $T(E_n)$, then the orbits of V are given by the solutions of

$$\frac{d\bar{x}^i(t)}{dt} = v^i(\bar{x}^j(t)) \tag{4-5.1}$$

subject to the initial data

$$\bar{x}^i(0) = x^i. \tag{4-5.2}$$

We have also seen that V may be considered as a derivation on $\Lambda^0(E_n)$. Thus since $\Lambda^0(E_n)$ is the first element in the sequence $\Lambda^0(E_n), \Lambda^1(E_n), \Lambda^2(E_n), \ldots, \Lambda^n(E_n)$, it is natural to ask whether V can be extended to a derivation on all of $\Lambda(E_n)$ in such a way that it agrees with V when applied to elements of Λ^0. This would indeed be a useful construct if we could also extend in such a fashion that $[U, V]\langle f\rangle = U\langle V\langle f\rangle\rangle - V\langle U\langle f\rangle\rangle$ is also extended to apply to all of $\Lambda(E_n)$.

The underlying process from which we can construct such an extension is the operator series representation of the solutions of (4-5.1), (4-5.2):

$$\bar{x}^i(t) = \exp(tV)\langle x^i\rangle. \tag{4-5.3}$$

These equations can be thought of as defining a 1-parameter family of maps

$$T_V(t): E_n \to E_n | \bar{x}^i(t) = \exp(tV)\langle x^i\rangle \tag{4-5.4}$$

that moves any point $P:(x^i)$ at $t = 0$ to the point $P_t:(\exp(tV)\langle x^i\rangle)$ at time t. It is, however, conceptually simpler to consider $T_V(t)$ as a map from the space E_n into a replica $E_{n,t}$ of E_n, where the replicas are parameterized by t; that is, we shift consideration to $\mathbb{R} \times E_n$. This alternative view gives the graphic situation shown in Fig. 13.

If we label points in $\mathbb{R} \times E_n$ by $(t; x^1, \ldots, x^n)$ there is the obvious projection map

$$\pi : \mathbb{R} \times E_n \to E_n | (t; x^1, \ldots, x^n) \mapsto (x^1, \ldots, x^n). \tag{4-5.5}$$

Let ω be a k-form on E_n. Since the map π induces the map π^* on k-forms that maps in the direction opposite to π, the k-form $\pi^*\omega$ is induced on every replica of E_n in $\mathbb{R} \times E_n$ and this collection of k-forms is independent of the value of the parameter t. From now on, we simply assume that ω is defined in this way on every replica $E_{n,t}$ of E_n in $\mathbb{R} \times E_n$ and drop the π^* map.

Let $Q:(\exp(\tau V)\langle x^i\rangle)$ be the image in $E_{n,\tau}$ of the point $P:(x^i)$ in $E_{n,0}$ under the map $T_V(\tau)$. We can then evaluate ω at Q in $E_{n,\tau}$ and at P in $E_{n,0}$, but we cannot compare them since forms at different points or forms at different points in different spaces cannot be added (we are unable to add an element of $T_P^*(E_n)$ to an element of $T_Q^*(E_n)$ for P different from Q). The form

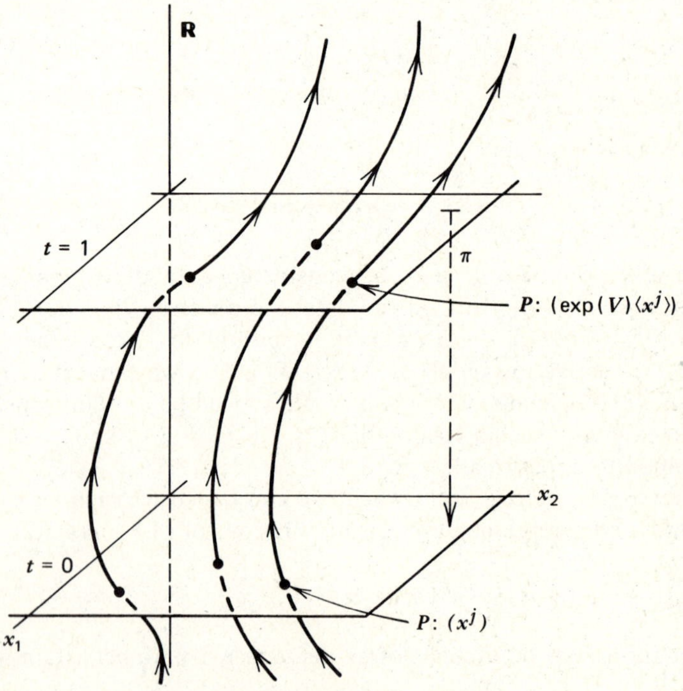

FIGURE 13. The family of maps Ψ_t

ω at Q in $E_{n,\tau}$ can be mapped into a form in $E_{n,0}$ at P by the action of $T_V(\tau)^*$, however. This will result in a new k-form

(4-5.6) $$\hat{\omega}(\tau) = T_V(\tau)^*\omega = \exp(\tau V)^*\omega$$

on $E_{n,0}$ that can be evaluated (at P) and compared with ω at P. The definition (4-5.6) is expressed in the following graphic way: *the new field* $\hat{\omega}(\tau)$ *at the old point P is equal to the old field at the new point Q pulled back to P by the map* $\exp(\tau V)^*$.

Definition. The *finite deformation*, $\Delta_V(\tau)\omega$, of a given $\omega \in \Lambda^k(E_n)$, that is induced by the vector field V, is given by

(4-5.7) $$\Delta_V(\tau)\omega = \exp(\tau V)^*\omega - \omega = \hat{\omega}(\tau) - \omega.$$

Definition. The *Lie derivative*, $£_V\omega$, of a given $\omega \in \Lambda^k(E_n)$ with respect to the vector field V is given by

(4-5.8) $$£_V\omega = \lim_{\tau \to 0}\left(\frac{\exp(\tau V)^*\omega - \omega}{\tau}\right).$$

Clearly $£_V\omega \in \Lambda^k(E_n)$ for $\omega \in \Lambda^k(E_n)$. The following theorem provides a direct parallel between $\bar{x}^i(t)$ and $\hat{\omega}(t)$.

Theorem 4-5.1. *If* $\omega \in \Lambda^k(E_n)$ *and* $V \in T(E_n)$, *then*

(4-5.9) $$\hat{\omega}(t) = \exp(t£_V)\omega.$$

Proof. The definition of the Lie derivative and (4-5.6) yield

$$\frac{d\hat{\omega}(t)}{dt} = \lim_{\tau \to 0}\left(\frac{\exp((t+\tau)V)^*\omega - \exp(tV)^*\omega}{\tau}\right).$$

But $\exp((t+\tau)V) = \exp(tV)\circ\exp(\tau V)$ because tV and τV commute, and hence

$$\begin{aligned}\exp((t+\tau)V)^* &= \{\exp(tV)\circ\exp(\tau V)\}^*\\ &= \exp(\tau V)^*\exp(tV)^*;\end{aligned}$$

that is

$$\frac{d\hat{\omega}(t)}{dt} = \lim_{\tau \to 0}\left(\frac{\exp(\tau V)^* - 1}{\tau}\right)\exp(tV)^*\omega.$$

However, $\exp(tV)^*\omega = \hat{\omega}(t)$ belongs to $\Lambda^k(E_n)$ and hence (4-5.8) gives

$$\frac{d\hat{\omega}(t)}{dt} = £_V\,\hat{\omega}(t). \tag{4-5.10}$$

Thus since (4-5.6) also implies

$$\hat{\omega}(0) = \omega,$$

the series solution of (4-5.10) subject to this initial data gives (4-5.9). □

The Lie derivative would seem to be a candidate for the extension of V to an operator on forms of arbitrary degree. The following lemma provides the explicit evaluation whereby the requisite properties are easily established.

Lemma 4-5.1. *The Lie derivative of a k-form* ω *with respect to* $V \in T(E_n)$ *has the explicit evaluation*

$$£_V\omega = V \lrcorner d\omega + d(V \lrcorner \omega). \tag{4-5.11}$$

Proof. Since the proof is the same for forms of any degree, we confine attention to forms of degree one in the interests of computational simplicity. If $\omega = \omega_i\,dx^i$, $V = v^i\partial_i$, then

$$\begin{aligned}\exp(\tau V)^*\omega &= \omega_i(\exp(\tau V)\langle x^j\rangle)\,d\{\exp(\tau V)\langle x^i\rangle\}\\ &= \omega + \tau\left(\frac{\partial\omega_i}{\partial x^j}v^j + \omega_j\frac{\partial v^j}{\partial x^i}\right)dx^i + o(\tau).\end{aligned}$$

The definition (4-5.8) thus gives

$$£_V\omega = \left(\frac{\partial\omega_i}{\partial x^j}v^j + \omega_j\frac{\partial v^j}{\partial x^i}\right)dx^i.$$

If we add and subtract $(\partial\omega_j/\partial x^i)v^j\,dx^i$, we obtain

$$£_V\omega = \left\{\left(\frac{\partial\omega_i}{\partial x^j} - \frac{\partial\omega_j}{\partial x^i}\right)v^j + \frac{\partial}{\partial x^i}\left(\omega_j v^j\right)\right\}dx^i$$

from which (4-5.11) follows directly. □

Theorem 4-5.2. *The Lie derivative is a map from* $\Lambda^k(E_n)$ *to* $\Lambda^k(E_n)$, $k = 0, 1, \ldots, n$, *that exhibits the properties*

L1. $\pounds_V f = V\langle f \rangle \, \forall f \in \Lambda^0(E_n)$.
L2. $\pounds_V(\alpha + \beta) = \pounds_V \alpha + \pounds_V \beta$.
L3. $\pounds_V(\alpha \wedge \beta) = (\pounds_V \alpha) \wedge \beta + \alpha \wedge \pounds_V \beta$.
L4. $\pounds_V \, d\alpha = d(\pounds_V \alpha)$.
L5. $\pounds_{f \cdot V} \alpha = f \cdot \pounds_V \alpha + df \wedge (V \lrcorner \alpha)$.
L6. $\pounds_{U+V} \alpha = \pounds_U \alpha + \pounds_V \alpha$.

Hence $\pounds_V$ *is a derivation on the graded algebra* $\Lambda(E_n)$ *that agrees with the derivation* V *on* $\Lambda^0(E_n)$.

Proof. That $\pounds_V$ is on Λ^k to Λ^k follows from

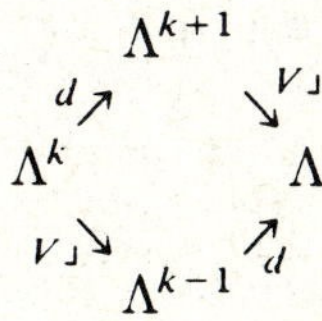

and (4-5.11). Properties **L2** and **L6** are immediate consequences of the linearity of $\pounds_V \alpha$ in V for fixed α and in α for fixed V, as shown by the Lemma 4-5.1. For **L4**, we note that

$$\pounds_V \, d\alpha = V \lrcorner d\, d\alpha + d(V \lrcorner d\alpha) = d(V \lrcorner d\alpha)$$

$$d\pounds_V \, \alpha = d\{V \lrcorner d\alpha + d(V \lrcorner \alpha)\} = d(V \lrcorner d\alpha),$$

while **L1** follows from

$$\pounds_V f = V \lrcorner df + d(V \lrcorner f) = V \lrcorner df = V\langle f \rangle.$$

By definition

$$\pounds_{f \cdot V} \alpha = f \cdot V \lrcorner d\alpha + d(f \cdot V \lrcorner \alpha) = f \cdot V \lrcorner d\alpha + f \cdot d(V \lrcorner \alpha) + df \wedge (V \lrcorner \alpha)$$

$$= f \cdot \pounds_V \alpha + df \wedge (V \lrcorner \alpha),$$

which is property **L5**. Again, by definition, with $a = \deg(\alpha)$,

$$\begin{aligned}
\pounds_V(\alpha \wedge \beta) &= V \lrcorner d(\alpha \wedge \beta) + d(V \lrcorner(\alpha \wedge \beta)) \\
&= V \lrcorner \big(d\alpha \wedge \beta + (-1)^a \alpha \wedge d\beta\big) \\
&\quad + d\big((V \lrcorner \alpha) \wedge \beta + (-1)^a \alpha \wedge (V \lrcorner \beta)\big) \\
&= (V \lrcorner d\alpha) \wedge \beta + (-1)^{a+1} d\alpha \wedge (V \lrcorner \beta) + (-1)^a (V \lrcorner \alpha) \wedge d\beta \\
&\quad + (-1)^{2a} \alpha \wedge (V \lrcorner d\beta) + d(V \lrcorner \alpha) \wedge \beta + (-1)^{a-1} (V \lrcorner \alpha) \wedge d\beta \\
&\quad + (-1)^a d\alpha \wedge (V \lrcorner \beta) + (-1)^{2a} \alpha \wedge d(V \lrcorner \beta) \\
&= \{(V \lrcorner d\alpha) + d(V \lrcorner \alpha)\} \wedge \beta + \alpha \wedge \{V \lrcorner d\beta + d(V \lrcorner \beta)\},
\end{aligned}$$

which is property **L3**. □

The conclusion that the Lie derivative acts as a derivation on $\Lambda(E_n)$ is a special case of a more general result that is of interest in its own right. For purposes of discussion, let

$$D_1 := d, \qquad D_2 := V \lrcorner,$$

then D_1 and D_2 are antiderivations on the graded algebra $\Lambda(E_n)$. Further D_1 acting on any element of the algebra increases the grade of that element by one while D_2 decreases the grade by one. Accordingly, D_1 and D_2 are said to have *odd parity* (they change the grading by an odd multiple of the grading unit). Finally, let us observe that

$$\pounds_V \omega = D_2 D_1 \omega + D_1 D_2 \omega = (D_2 D_1 + D_1 D_2)\omega,$$

and hence the Lie derivative is the *anticommutator* of the operators D_1 and D_2, $D_1 D_2 + D_2 D_1$. The following general result can now be established.

Lemma 4-5.2. *Let $\mathscr{G}$ be a graded algebra with multiplication $(\cdot \mid \cdot)$ and grading* $\mathrm{gr}(\cdot)$. *The anticommutator of two antiderivations of odd parity on $\mathscr{G}$ is a derivation on $\mathscr{G}$.*

Proof. Since D_1 is an antiderivation on $\mathscr{G}$, we have

$$D_1(\alpha|\beta) = (D_1\alpha|\beta) + (-1)^{\mathrm{gr}(\alpha)}(\alpha|D_1\beta),$$

and hence

$$D_2D_1(\alpha|\beta) = (D_2D_1\alpha|\beta) + (-1)^{\mathrm{gr}(D_1\alpha)}(D_1\alpha|D_2\beta) \\ + (-1)^{\mathrm{gr}(\alpha)}(D_2\alpha|D_1\beta) + (-1)^{2\,\mathrm{gr}(\alpha)}(\alpha|D_2D_1\beta).$$

An interchange of the order of application of D_1 and D_2 shows that

$$(D_2D_1 + D_1D_2)(\alpha|\beta) = (D_2D_1\alpha|\beta) + (D_1D_2\alpha|\beta) \\ + (-1)^{\mathrm{gr}(D_1\alpha)}(D_1\alpha|D_2\beta) + (-1)^{\mathrm{gr}(D_2\alpha)}(D_2\alpha|D_1\beta) \\ + (-1)^{\mathrm{gr}(\alpha)}(D_2\alpha|D_1\beta) + (-1)^{\mathrm{gr}(\alpha)}(D_1\alpha|D_2\beta) \\ + (\alpha|D_2D_1\beta) + (\alpha|D_1D_2\beta).$$

Because D_1 and D_2 have odd parity, we have

$$(-1)^{\mathrm{gr}(\alpha)} + (-1)^{\mathrm{gr}(D_1\alpha)} = (-1)^{\mathrm{gr}(\alpha)} + (-1)^{\mathrm{gr}(D_2\alpha)} = 0,$$

and hence

$$(D_2D_1 + D_1D_2)(\alpha|\beta) = ((D_2D_1 + D_1D_2)\alpha|\beta) + (\alpha|(D_2D_1 + D_1D_2)\beta).$$

□

Theorem 4-5.2 provides a full description of the action of $\pounds_V$ on exterior forms. The question now arises as to what happens when $\pounds_V$ acts on elements of $T(E_n)$. We note that $V \lrcorner \omega$ is a well defined exterior form for any given exterior form ω, and hence $\pounds_U(V \lrcorner \omega)$ is well defined. Now, property **L3** shows that $\pounds_U$ is a derivation on $\Lambda(E_n)$, so it is consistent to define the action of $\pounds_U$ on $V \in T(E_n)$ by requiring $\pounds_U$ to act as a derivation on $V \lrcorner \omega$ for all $\omega \in \Lambda(E_n)$, that is

$$\pounds_U(V \lrcorner \omega) = (\pounds_U V) \lrcorner \omega + V \lrcorner \pounds_U \omega.$$

Definition. The action of $\pounds_U$ on $T(E_n)$ is defined by the requirement that

$$(\pounds_U V) \lrcorner \omega = \pounds_U(V \lrcorner \omega) - V \lrcorner \pounds_U \omega \tag{4-5.12}$$

holds for all $\omega \in \Lambda(E_n)$.

Lemma 4-5.3. *If $U, V \in T(E_n)$ then the action of $\pounds_U$ on V has the explicit evaluation*

$$\pounds_U V = [U, V]. \tag{4-5.13}$$

Thus, $\pounds_U$ is a map from $T(E_n)$ to $T(E_n)$ and

(4-5.14) $$\pounds_U V = -\pounds_V U.$$

Proof. Since (4-5.12) must hold for all ω, putting $\omega = df \in \Lambda^1$, and noting that $V \lrcorner df = V\langle f \rangle$, give

$$\begin{aligned}(\pounds_U V)\langle f \rangle &= (\pounds_U V) \lrcorner df \\ &= \pounds_U(V\langle f \rangle) - V \lrcorner \pounds_U df = \pounds_U(V\langle f \rangle) - V \lrcorner d(U\langle f \rangle) \\ &= U \lrcorner d(V\langle f \rangle) - V \lrcorner d(U\langle f \rangle) = U\langle V\langle f \rangle\rangle - V\langle U\langle f \rangle\rangle.\end{aligned}$$

The definition of the Lie product thus gives

$$(\pounds_U V)\langle f \rangle = [U, V]\langle f \rangle$$

for all $f \in \Lambda^0(E_n)$ and hence (4-5.13) is established. The relation (4-5.14) is an immediate consequence of the properties of the Lie product. □

It is instructive to see that (4-5.13) agrees with a direct definition of the Lie derivative similar to (4-5.8). Let U and V be given elements of $T(E_n)$ and define $\bar{x}^i(s)$ by $\bar{x}^i(s) = \exp(sU)\langle x^i \rangle$ so that $\bar{P}:(\bar{x}^i(s))$ is the image of $P:(x^i)$ under transport along the orbits of the vector field U. Since $\exp(sU)_*$ maps in the same direction as $\exp(sU)$, the appropriate definition of the Lie derivative is

(4-5.15) $$\pounds_U V = \lim_{s \to 0} \left\{ \frac{\bar{V} - (\exp(sU))_* V}{s} \right\},$$

where $\bar{V}$ denotes the evaluation of the vector field V at $\bar{P}:(\bar{x}^i(s))$; that is, $\bar{V} = v^i(\bar{x}^j(s))\bar{\partial}_i$ with $\bar{\partial}_i = \partial/\partial \bar{x}^i$. It is now an easy matter to see that

$$\begin{aligned}\bar{V} - (\exp(sU))_* V &= v^i(\exp(sU)\langle x^j \rangle)\bar{\partial}_i - (\exp(sU))_*(v^k(x^j)\partial_k) \\ &= s\{u^j \partial_j v^i - v^j \partial_j u^i\}\bar{\partial}_i + o(s),\end{aligned}$$

and hence $\pounds_U V = [U, V]$. The respective definitions of the Lie derivative show that $\exp(s\pounds_U)\omega$ is similar to a description in terms of Eulerian variables while $\exp(s\pounds_U)V$ is similar to a description in terms of Lagrangian variables in fluid mechanics.

Now that we know that $\pounds_U V = [U, V]$, we turn to the computation of $\pounds_{[U,V]}$. By definition,

(4-5.16) $$\pounds_{[U,V]}\Omega = \pounds_{\pounds_U V}\Omega = (\pounds_U V) \lrcorner d\Omega + d((\pounds_U V) \lrcorner \Omega).$$

Use of (4-5.12) first with $\omega = d\Omega$ and then with $\omega = \Omega$, yields

$$(\pounds_U V)\lrcorner d\Omega = \pounds_U(V\lrcorner d\Omega) - V\lrcorner(\pounds_U d\Omega),$$

$$(\pounds_U V)\lrcorner\Omega = \pounds_U(V\lrcorner\Omega) - V\lrcorner(\pounds_U\Omega).$$

When these results are put back into (4-5.16), we have

$$\begin{aligned}\pounds_{[U,V]}\Omega &= \pounds_U(V\lrcorner d\Omega) - V\lrcorner(\pounds_U d\Omega) + d\left[\pounds_U(V\lrcorner\Omega) - V\lrcorner\pounds_U\Omega\right]\\ &= \pounds_U(V\lrcorner d\Omega) - V\lrcorner d\pounds_U\Omega + \pounds_U d(V\lrcorner\Omega) - d\{V\lrcorner\pounds_U\Omega\}\\ &= \pounds_U\{V\lrcorner d\Omega + d(V\lrcorner\Omega)\} - V\lrcorner d\pounds_U\Omega - d(V\lrcorner\pounds_U\Omega)\\ &= \pounds_U(\pounds_V\Omega) - \pounds_V(\pounds_U\Omega) = (\pounds_U\pounds_V - \pounds_V\pounds_U)\Omega.\end{aligned}$$

Theorem 4-5.3. *The Lie derivative satisfies*

(4-5.17) $$\pounds_{[U,V]}\omega = (\pounds_U\pounds_V - \pounds_V\pounds_U)\omega$$

and hence the Lie algebra of $T(E_n)$ as operators on $\Lambda^0(E_n)$ extends to a Lie algebra of operators on $\Lambda(E_n)$ under the correspondence

(4-5.18) $$[U,V] \to \pounds_U\pounds_V - \pounds_V\pounds_U.$$

If $\{U_a,\ a = 1,\ldots,r\}$ form a Lie subalgebra of $T(E_n)$ with

(4-5.19) $$[U_a, U_b] = C^e_{ab}U_e,$$

then

(4-5.20) $$(\pounds_{U_a}\pounds_{U_b} - \pounds_{U_b}\pounds_{U_a})\omega = C^e_{ab}\pounds_{U_e}\omega + dC^e_{ab}\wedge(U_e\lrcorner\omega).$$

Thus a Lie subalgebra of $T(E_n)$ extends to a Lie subalgebra of operators on $\Lambda(E_n)$ only if

(4-5.21) $$dC^e_{ab} = 0;$$

that is, the Lie subalgebra is the Lie subalgebra of a Lie group.

Proof. We have already established (4-5.17). If the collection $\{U_a\}$ satisfy (4-5.19), then (4-5.17) gives

$$\begin{aligned}\pounds_{[U_a,U_b]}\omega &= (\pounds_{U_a}\pounds_{U_b} - \pounds_{U_b}\pounds_{U_a})\omega\\ &= \pounds_{C^e_{ab}U_e}\omega = C^e_{ab}\pounds_{U_e}\omega + dC^e_{ab}\wedge(U_e\lrcorner\omega)\end{aligned}$$

which is just (4-5.20). The remaining conclusions are obvious. □

The definition of the Lie derivative with respect to a vector field V, (4-5.8), shows that this operator measures the change in an exterior form that would be observed when we move an infinitesimal amount along the orbits of a vector field and simultaneously transport the natural basis elements of the exterior algebra. Let $\mathscr{S}$ be a subspace of $\Lambda(E_n)$ and let ω be a generic element of $\mathscr{S}$. If we evaluate ω at a point on an orbit of V, say at $Q:(\exp(\tau V)\langle x^i\rangle)$, we can pull this evaluation back to $P:(\exp(0V)\langle x^i\rangle) = P:(x^i)$ by applying $\exp(\tau V)^*$. When (4-5.6) and (4-5.9) are combined, we then have

$$\exp(\tau V)^*\omega(\exp(\tau V)\langle x^i\rangle) = (\exp(\tau \pounds_V)\omega)(x^i).$$

Accordingly, $\omega \in \mathscr{S}$ at $P:(x^i)$ will imply $\omega \in \mathscr{S}$ at $Q:(\exp(\tau V)\langle x^i\rangle)$ if and only if $(\pounds_V\omega)$ at $P:(x^i)$ belongs to $\mathscr{S}$.

Definition. A subspace $\mathscr{S}$ of $\Lambda(E_n)$ is said to be *stable under Lie transport* along the orbits of a vector field V if $\pounds_V\omega$ belongs to $\mathscr{S}$ for every $\omega \in \mathscr{S}$; that is

(4-5.22) $$\pounds_V\mathscr{S} \subset \mathscr{S}.$$

Stability under Lie transport is a very important property of a subspace. If a subspace has this property, it is impossible to get outside the subspace by Lie differentiation. We have already met two fundamental subspaces of $\Lambda(E_n)$ with this property.

Theorem 4-5.4. *The subspace of all closed forms $\mathscr{C}(E_n)$, and the subspace of all exact forms $\mathscr{E}(E_n)$, are stable under Lie transport by any element of $T(E_n)$; that is,*

(4-5.23) $$\pounds_V\mathscr{C}(E_n) \subset \mathscr{C}(E_n) \qquad \forall V \in T(E_n),$$

(4-5.24) $$\pounds_V\mathscr{E}(E_n) \subset \mathscr{E}(E_n) \qquad \forall V \in T(E_n).$$

Proof. If $\omega \in \mathscr{C}(E_n)$, we have $d\omega = 0$. Thus since d and $\pounds_V$ commute for any $V \in T(E_n)$ by property **L4**, $d\pounds_V\omega = \pounds_V\, d\omega = 0$. Any exact element of $\Lambda(E_n)$ has the form $d\rho$ for some $\rho \in \Lambda(E_n)$. Accordingly, we have $\pounds_V\, d\rho = V \lrcorner d\, d\rho + d(V \lrcorner d\rho) = d(V \lrcorner d\rho)$ is again exact for any $V \in T(E_n)$. □

4-6. IDEALS AND ISOVECTOR FIELDS

The Lie derivative proves to be a particularly effective tool in the analysis of ideals of the graded exterior algebra. We saw in Section 4-3 that any closed ideal generated by 1-forms had a characteristic subspace of $T(E_n)$ that formed a Lie subalgebra of $T(E_n)$. The following theorem of Cartan shows that the ideal need not be generated by 1-forms in order that the result follow.

Theorem 4-6.1. (Cartan) *Let I be an ideal of $\Lambda(E_n)$ whose characteristic system forms a subspace $\mathscr{S}$ of $T(E_n)$ with constant dimension. If I is closed then $\mathscr{S}$ forms a Lie subalgebra of $T(E_n)$.*

Proof. By definition

$$\mathscr{S} = \{V \in T(E_n) | V \lrcorner I \subset I\}.$$

Since $\mathscr{S}$ is of fixed dimension, say r, there exist r linearly independent elements $\{V_\alpha,\ \alpha = 1, \dots, r\}$ of $T(E_n)$ that span $\mathscr{S}$. The definition of the Lie derivative yields

$$V \lrcorner \pounds_U \omega = V \lrcorner U \lrcorner d\omega + V \lrcorner d(U \lrcorner \omega),$$

$$\pounds_U(V \lrcorner \omega) = [U, V] \lrcorner \omega + V \lrcorner \pounds_U \omega$$

$$= U \lrcorner d(V \lrcorner \omega) + d(U \lrcorner V \lrcorner \omega).$$

When these relations are combined, we obtain the basic relation

$$[U, V] \lrcorner \omega = U \lrcorner d(V \lrcorner \omega) - V \lrcorner d(U \lrcorner \omega) - V \lrcorner U \lrcorner d\omega + d(U \lrcorner V \lrcorner \omega) \tag{4-6.1}$$

that is a generalization of (4-3.8). Thus if $\omega \in I$, we have

$$\begin{aligned} [V_\alpha, V_\beta] \lrcorner \omega = {} & V_\alpha \lrcorner d(V_\beta \lrcorner \omega) - V_\beta \lrcorner d(V_\alpha \lrcorner \omega) \\ & - V_\beta \lrcorner V_\alpha \lrcorner d\omega + d(V_\alpha \lrcorner V_\beta \lrcorner \omega). \end{aligned} \tag{4-6.2}$$

By hypothesis, $V_\alpha \lrcorner I \subset I$ for all $V_\alpha \in \mathscr{S}$ and $dI \subset I$. These conditions show that each term on the right-hand side of (4-6.2) is contained in I. Thus, $[V_\alpha, V_\beta] \lrcorner \omega$ is contained in I for every $\omega \in I$ so that $[V_\alpha, V_\beta]$ belongs to $\mathscr{S}$. Accordingly, $\mathscr{S}$ forms a Lie subalgebra of $T(E_n)$. □

The Cartan theorem provides the basis for a simple and direct proof of the Frobenius theorem.

Proof of the Frobenius Theorem. Let D_r denote the exterior system that is defined by r 1-forms $\{\omega^a\}$ and let $I\{D_r\}$ denote the ideal of $\Lambda(E_n)$ that is generated by D_r. We need to show that $dI\{D_r\} \subset I\{D_r\}$ implies that there exist a nonsingular r-by-r matrix $((A^a_b))$ and r independent elements $\{g^a(x^j)\}$ of $\Lambda^0(E_n)$ such that $\omega^a = A^a_b\, dg^b$. Let $\{V_\alpha,\ \alpha = r+1, \dots, n\}$ be a basis for the characteristic subspace $\mathscr{S}(D_r)$ of D_r. The Cartan theorem shows that $\mathscr{S}(D_r)$ forms a Lie subalgebra of $T(E_n)$ and hence there exist r independent elements $g^a(x^j)$, $a = 1, \dots, r$ of $\Lambda^0(E_n)$ such that

$$V_\alpha\langle g^a \rangle = V_\alpha \lrcorner dg^a = 0, \qquad a = 1, \dots, r, \qquad \alpha = r+1, \dots, n, \tag{4-6.3}$$

$$\Omega = dg^1 \wedge dg^2 \wedge \cdots \wedge dg^r \neq 0. \tag{4-6.4}$$

These equations show that $\mathscr{S}(D_r)$ is a characteristic system of the system of r 1-forms $\{dg^a\}$ that spans an r-dimensional subspace $\mathscr{S}^*$ of $\Lambda^1(E_n)$ characterized by the simple r-form Ω. However, $\mathscr{S}(D_r)$ is also the characteristic subspace of D_r, so that $V_\alpha \lrcorner \omega^a = 0$, and any $V \in T(E_n)$ that does not belong to $\mathscr{S}(D_r)$ gives $V \lrcorner \omega^a \neq 0$. Thus each of the ω's belongs to $\mathscr{S}^*$; that is, $\omega^a \wedge \Omega = 0$. It follows that $\omega^a = A^a_b\, dg^b$ for some matrix $((A^a_b))$, and we have

$$\omega^1 \wedge \omega^2 \wedge \cdots \wedge \omega^r = \det(A^a_b)\Omega.$$

The independence of the 1-forms $\{\omega^a\}$ then gives $\det(A^a_b) \neq 0$. □

Particular note should be taken of the fact that the A's and the g's are not unique. Simply observe that if $\{F^a(g^1, \ldots, g^r)\}$ is any system of functions such that $\det(\partial F^a/\partial g^b) \neq 0$, the dF^a is another system of r independent elements of $\Lambda^1(E_n)$ such that $V_\alpha \lrcorner dF^a = 0$. We could then write $\omega^a = B^a_b\, dF^b$, in which case $\omega^a = A^a_b\, dg^b$ yields the relations $A^a_b = B^a_c \partial F^c/\partial g^b$.

The Cartan theorem also leads to the following result that is often useful in specific problems.

Theorem 4-6.2. *If I is a closed ideal of $\Lambda(E_n)$ whose characteristic subspace is of dimension $n - r$, then there exist r independent elements $\{g^a(x^i),\ a = 1, \ldots, r\}$ of $\Lambda^0(E_n)$ such that I is contained in the closed ideal that is generated by the r elements $\{dg^1, \ldots, dg^r\}$.*

Proof. Since I is closed, its characteristic subspace $\mathscr{S}(I)$ forms a Lie subalgebra of $T(E_n)$ and this subalgebra is of dimension $n - r$ by hypothesis. There thus exist $n - r$ elements $\{V^\alpha\}$ of $T(E_n)$ that form a basis for $\mathscr{S}(I)$ and satisfy

$$[V^\alpha, V^\beta] = C^{\alpha\beta}_\gamma V^\gamma.$$

There are r independent first integrals $\{g^a,\ a = 1, \ldots, r\}$ of the system $V^\alpha\langle g\rangle = 0$, so that $V^\alpha \lrcorner dg^a = 0$. The ideal of $\Lambda(E_n)$ that is generated by $\{dg^1, \ldots, dg^r\}$ is a closed ideal that admits $\mathscr{S}(I)$ as its characteristic subspace. Thus since each of the generators $\{dg^1, \ldots, dg^r\}$ is of lowest possible degree, namely one, while some of the generators of I may be of degree greater than one, I is contained in the closed ideal generated by $\{dg^1, \ldots, dg^r\}$. □

A simple example is useful here. Consider the ideal I of $\Lambda(E_4)$ that is generated by

$$\omega^1 = dx^1 + x^2 dx^3 + dx^4, \qquad \omega^2 = x^2 dx^2 \wedge dx^3.$$

Since $\omega^2 = x^2 d\omega^1$ and $d\omega^2 = 0$, I is a closed ideal of $\Lambda(E_4)$. The characteristic subspace of I is one-dimensional and is spanned by $V^1 = \partial_1 - \partial_4$, and

$V^1\langle g \rangle = 0$ admits the three independent first integrals

$$g^1 = x^1 + x^4, \qquad g^2 = x^2, \qquad g^3 = x^3.$$

We thus consider the ideal J generated by the 1-forms $dg^1 = dx^1 + dx^4$, $dg^2 = dx^2$, $dg^3 = dx^3$. Since $\omega^1 = dg^1 + g^2 dg^2$, $\omega^2 = g^2 dg^2 \wedge dg^3$, I is contained in J. Further, I is a proper subideal of J for J contains terms of the form $f(x^i)\,dx^2 = f(x^i)\,dg^2$ while I does not.

Theorem 4-6.2 provides the means whereby subsets of E_n can be found on which a given ideal vanishes.

Theorem 4-6.3. *Let I be a given ideal of $\Lambda(E_n)$ and construct its closure $\bar{I} = I \cup dI$. If the dimension of the characteristic subspace, $\mathscr{S}(\bar{I})$, of $\bar{I}$ is $n - r > 0$, then there exist $(n - r)$-dimensional subsets*

$$(4\text{-}6.5) \qquad g^a(x^i) = k^a, \qquad a = 1, \ldots, r$$

of E_n on which every element of I and of $\bar{I}$ vanish. The r functions $\{g^a(x^j)\}$ are the primitive integrals of the characteristic system $V^\alpha\langle g \rangle = 0$, $\alpha = r + 1, \ldots, n$, where $\{V^\alpha, \alpha = r + 1, \ldots, n\}$ is a basis for the Lie subalgebra $\mathscr{S}(\bar{I})$ of $T(E_n)$.

Proof. Since $\bar{I}$ is a closed ideal of $\Lambda(E_n)$, Theorem 4-6.2 shows that $\bar{I}$ is contained in the ideal J that is generated by $\{dg^1, \ldots, dg^r\}$. Noting that every element of J vanishes on the $(n - r)$-dimensional subset (4-6.5) for any values of the constants k^a, the result is established. □

We observe that I is a proper subideal of $\bar{I}$, unless I is itself a closed ideal, and $\bar{I}$ is a proper subideal of J. It is clear that there will in general be subsets of E_n of dimension greater than $n - r$ on which each element of I will vanish. The problem of finding all subsets of E_n on which each element of I vanishes for dI not contained in I is a difficult one and is very sensitive to the structure of the elements of $\Lambda(E_n)$ that generate I. In most instances, Theorem 4-6.3 is the best that can be obtained without entering into the specifics of the generators of I.

We saw in the last section that a subspace $\mathscr{S}$ of $\Lambda(E_n)$ is stable under Lie transport by a vector field $V \in T(E_n)$ if and only if $\pounds_V \mathscr{S} \subset \mathscr{S}$; that is, you cannot get outside of $\mathscr{S}$ by transport of any element of $\mathscr{S}$ along the orbits of V. This concept is also applicable to ideals of $\Lambda(E_n)$.

Definition. An ideal I of $\Lambda(E_n)$ is said to be *stable under transport* along the orbits of a vector field $V \in T(E_n)$ if and only if

$$(4\text{-}6.6) \qquad \pounds_V I \subset I,$$

in which case V is said to be an *isovector* of I.

Theorem 4-6.4. *The isovectors of an ideal I of $\Lambda(E_n)$ form a Lie subalgebra of $T(E_n)$ that is the Lie subalgebra of a Lie group.*

Proof. By definition, V is an isovector of I if and only if

$$\pounds_V \omega \equiv 0 \bmod I \tag{4-6.7}$$

for all $\omega \in I$. Suppose U and V are isovectors of I. We then have

$$(\pounds_U \pounds_V - \pounds_V \pounds_U)\omega = \pounds_U(\pounds_V \omega) - \pounds_V(\pounds_U \omega) \equiv 0 \bmod I$$

and hence

$$\pounds_{[U,V]}\omega = (\pounds_U \pounds_V - \pounds_V \pounds_U)\omega \equiv 0 \bmod I.$$

Thus $[U, V]$ is an isovector of I whenever U and V are, and the set of all isovectors is closed under the Lie product. Further,

$$\pounds_{U+V}\omega = \pounds_U \omega + \pounds_V \omega \equiv 0 \bmod I$$

for all $\omega \in I$, but

$$\pounds_{f \cdot U}\omega = f \cdot \pounds_U \omega + df \wedge (U \lrcorner \omega) \equiv df \wedge (U \lrcorner \omega) \bmod I$$

for all $\omega \in I$ and all $f \in \Lambda^0(E_n)$. Hence all isovectors of I form a vector subspace of $T(E_n)$ over $\mathbb{R}$ but not over $\Lambda^0(E_n)$. The vector subspace property over $\mathbb{R}$ but not over $\Lambda^0(E_n)$ then implies

$$[U_a, U_b] = C^e_{ab} U_e, \qquad dC^e_{ab} = 0$$

if $\{U_a,\ a = 1, \ldots, r\}$ is a basis for the Lie subalgebra of isovectors of I. □

The following lemmas are often useful in specific problems.

Lemma 4-6.1. *Let $I\{\alpha^a\}$ be an ideal of $\Lambda(E_n)$ that is generated by $\{\alpha^a,\ a = 1, \ldots, r\}$ with all α's of the same degree $k > 0$, then V is an isovector of $I\{\alpha^a\}$ if and only if there exist elements $A^a_b(x^j) \in \Lambda^0(E_n)$ such that*

$$\pounds_V \alpha^a = A^a_b \alpha^b, \qquad a = 1, \ldots, r. \tag{4-6.8}$$

Proof. If $\rho \in I\{\alpha^a\}$, then $\rho = \gamma_a \wedge \alpha^a$ and hence

$$\pounds_V \rho = (\pounds_V \gamma_a) \wedge \alpha^a + \gamma_a \wedge \pounds_V \alpha^a \equiv \gamma_a \wedge \pounds_V \alpha^a \bmod I\{\alpha^a\}.$$

Accordingly, $\pounds_V I\{\alpha^a\} \subset I\{\alpha^a\}$ only if $\pounds_V \alpha^a \subset I\{\alpha^a\}$, $a = 1, \ldots, r$. Thus since $\deg(\alpha^a) = \deg(\pounds_V \alpha^a)$, (4-6.8) must hold. Conversely, (4-6.8) gives $\pounds_V \alpha^a \in I\{\alpha^a\}$ and hence $\pounds_V \rho \equiv 0 \bmod I\{\alpha^a\}$ for any $\rho \in I\{\alpha^a\}$. □

Lemma 4-6.2. *Let $I\{\alpha^a\}$ be an ideal of $\Lambda(E_n)$ that is generated by $\{\alpha^a, a = 1, \ldots, r\}$ with all α's of the same degree $k > 0$ and let*

$$\bar{I}\{\alpha^a\} = I\{\alpha^a; d\alpha^a\}$$

be the closure of $I\{\alpha^a\}$. The collection of all isovectors of the ideal $I\{\alpha^a\}$ coincides with the collection of all isovectors of the closure of $I\{\alpha^a\}$.

Proof. If V is an isovector of $I\{\alpha^a\}$, then (4-6.8) holds and we have

$$£_V \, d\alpha^a = d£_V \, \alpha^a = d\left(A^a_b \alpha^b\right) = dA^a_b \wedge \alpha^b + A^a_b \, d\alpha^b.$$

Hence $£_V \, d\alpha^a$ belongs to $\bar{I}(\alpha^a\}$. Thus if $\rho \in \bar{I}\{\alpha^a\}$, $\rho = \beta_a \wedge \alpha^a + \gamma_a \wedge d\alpha^a$ and hence $£_V I\{\alpha^a\} \subset I\{\alpha^a\}$ implies $£_V \bar{I}\{\alpha^a\} \subset \bar{I}\{\alpha^a\}$. In order to obtain the converse, we note that $\alpha^a \in \bar{I}\{\alpha^a\}$ and hence $£_V \bar{I}\{\alpha^a\} \subset \bar{I}\{\alpha^a\}$ only if $£_V \alpha^a \subset \bar{I}\{\alpha^a\}$. However, $\deg(£_V \alpha^a) = k$ while $\deg(d\alpha^a) = k + 1$, and hence $£_V \alpha^a \in \bar{I}\{\alpha^a\}$ only if there are functions A^a_b such that $£_V \alpha^a = A^a_b \alpha^b$. Lemma 4-6.1 then shows that V is an isovector of the ideal $I\{\alpha^a\}$. □

Lemma 4-6.3. *Let $I\{\alpha^a\}$ be an ideal of $\Lambda(E_n)$ that is generated by $\{\alpha^a, a = 1, \ldots, r\}$ with all of the α's of degree less than k, and let $\{\beta^b, b = 1, \ldots, s\}$ be given forms such that $\deg(\beta^b) \geq k$, $b = 1, \ldots, s$. A vector field V is an isovector of the ideal $I\{\alpha^a; \beta^b\}$ if and only if*

1 *V is an isovector of $I\{\alpha^a\}$ and*
2 *$£_V \beta^b \in I\{\alpha^a; \beta^b\}$.*

Proof. We first note that $I\{\alpha^a\} \subset I\{\alpha^a; \beta^b\}$, and hence $£_V I\{\alpha^a\} \subset I\{\alpha^a\}$ implies $£_V I\{\alpha^a\} \subset I\{\alpha^a; \beta^b\}$. Thus for any $\rho = \gamma_a \wedge \alpha^a + \omega_b \wedge \beta^b \in I\{\alpha^a; \beta^b\}$, satisfaction of the two conditions 1 and 2 implies that

$$£_V \rho = £_V \gamma_a \wedge \alpha^a + \gamma_a \wedge £_V \alpha^a + £_V \omega_b \wedge \beta^b + \omega_b \wedge £_V \beta^b$$

belongs to $I\{\alpha^a; \beta^b\}$. Conversely, $£_V \rho \equiv 0 \bmod I\{\alpha^a; \beta^b\}$ if and only if

$$\gamma_a \wedge £_V \alpha^a + \omega_b \wedge £_V \beta^b \equiv 0 \bmod I\{\alpha^a; \beta^b\}.$$

Thus if this result is to hold for all $\rho \in I\{\alpha^a; \beta^b\}$, we must have

$$£_V \alpha^a \equiv 0 \bmod I\{\alpha^a; \beta^b\}, \qquad £_V \beta^b \equiv 0 \bmod I\{\alpha^a; \beta^b\}.$$

However, $\deg(\alpha^a) < k$ while $\deg(\beta^b) \geq k$, so that $£_V \alpha^a \in I\{\alpha^a; \beta^b\}$ can be satisfied only if

$$£_V \alpha^a = \eta^a_c \wedge \alpha^c$$

for some $((\eta^a_c))$; that is $£_V \alpha^a \subset I\{\alpha^a\}$. This in turn implies $£_V(\sigma_a \wedge \alpha^a) =$

$(\pounds_V \sigma_a) \wedge \alpha^a + \sigma_a \wedge \pounds_V \alpha^a$ for all σ_a for which the sum is defined. We thus have the conditions

$$\pounds_V \alpha^a \subset I\{\alpha^a\}, \qquad \pounds_V \beta^b \subset I\{\alpha^a; \beta^b\}. \tag{4-6.9}$$

□

The result obtained in Lemma 4-6.3 is particularly useful whenever an ideal is generated by forms of different degrees. It tells us that the search for isovectors starts by finding all isovectors of the subideal of the ideal that is generated by generating exterior forms of the lowest degree. Calculation of isovectors of the original ideal is then simply a matter of walking up a ladder of generators of increasing degree as indicated by (4-6.9). In this regard, Lemma 4-6.2 is also useful since it allows us to disregard any generators that are exterior derivatives of other generators when looking for isovectors.

Suppose that we are given an exterior form $\alpha \in \Lambda(E_n)$ and we wish to solve the exterior equation $\alpha = 0$. We have seen that this is equivalent to the problem of finding a map Φ from E_m to E_n for which $\Phi^*\alpha = 0$ is identically satisfied on the domain of Φ in E_m. This concept is directly applicable to ideals.

Definition. A map

$$\Phi: E_m \to E_n | x^i = \phi^i(y^A), \qquad i = 1, \ldots, n \tag{4-6.10}$$

is said to *solve* an ideal I of $\Lambda(E_n)$ if and only if

$$\Phi^*\rho = 0 \tag{4-6.11}$$

is satisfied identically on the domain of Φ in E_m for every $\rho \in I$.

Lemma 4-6.4. *If $I\{\omega^a\}$ is an ideal of $\Lambda(E_n)$ that is generated by $\{\omega^a, a = 1, \ldots, r\}$, then a map Φ from E_m to E_n solves $I\{\omega^a\}$ if and only if*

$$\Phi^*\omega^a = 0, \qquad a = 1, \ldots, r \tag{4-6.12}$$

is satisfied identically on the domain of Φ in E_m.

Proof. If $\rho \in I\{\omega^a\}$, then $\rho = \gamma_a \wedge \omega^a$ for $\deg(\gamma_a) + \deg(\omega^a) = k \le n$. Accordingly, we have $\Phi^*\rho = \Phi^*\gamma_a \wedge \Phi^*\omega^a$. This shows that we can have $\Phi^*\rho = 0$ for all $\rho \in I\{\omega^a\}$ if and only if $\Phi^*\gamma_a \wedge \Phi^*\omega^a = 0$ for all $\{\gamma_a\}$ such that $\deg(\gamma_a) + \deg(\omega^a) = k$ for all possible $k \le n$, and this is the case if and only if (4-6.12) hold. □

Although Lemma 4-6.4 simplifies the problem of finding a map that solves a finitely generated ideal $I\{\omega^a\}$, this problem is still a complicated one in general. Suppose, therefore, that we have found one solving map Φ of $I\{\omega^a\}$. It is then natural to ask whether we can use this information to simplify the problem of finding other solving maps. Here is where isovectors of $I\{\omega^a\}$ come in as is shown by the following result.

Theorem 4-6.5. *If V is an isovector of $I\{\omega^a\}$, and Φ is a solving map of $I\{\omega^a\}$, then V imbeds Φ in a 1-parameter family of solving maps*

(4-6.13) $$\Phi_V(s) = T_V(s) \circ \Phi,$$

where $T_V(s): E_n \to E_n$ is the 1-parameter family of maps that is generated by the flow of V.

Proof. Let

(4-6.14) $$T_V(s): E_n \to E_n | \bar{x}^i = \exp(sV)\langle x^i \rangle$$

be the 1-parameter family of maps that is generated by the flow of V (i.e., solutions of $d\bar{x}^i/ds = V\langle \bar{x}^i \rangle$ subject to the initial data $\bar{x}^i(0) = x^i$). We then have

(4-6.15) $$\Phi_V(s)^* \omega^a = (T_V(s) \circ \Phi)^* \omega^a = \Phi^* \circ T_V(s)^* \omega^a$$
$$= \Phi^*(\exp(s£_V)\omega^a)$$

because $T_V(s)^*\omega^a = (\exp(sV))^*\omega^a = \exp(s£_V)\omega^a$. Since V is an isovector of $I\{\omega^a\}$, $£_V\omega^a$ belongs to $I\{\omega^a\}$ and hence $\exp(s£_V)\omega^a$ belongs to $I\{\alpha^a\}$. Thus since Φ^* annihilates $I\{\omega^a\}$ by hypothesis, $\Phi^*(\exp(s£_V)\omega^a) = 0$ identically on the domain of Φ and (4-6.15) shows that $\Phi_V(s)^*\omega^a = 0$, $a = 1, \ldots, r$ are satisfied identically on the domain of Φ. Lemma 4-6.4 then shows that $\Phi_V(s)$ is a 1-parameter family of maps that solves $I\{\omega^a\}$. Clearly, we have $\Phi_V(0) = \Phi$, and hence (4-6.13) defines an imbedding of Φ in the 1-parameter family. □

Theorem 4-6.5 is of fundamental importance in the study of nonlinear systems of second-order partial differential equations. The reason for this is that the system of second-order partial differential equations can be characterized by a finitely generated ideal of the exterior algebra on an appropriately chosen space following the work of E. Cartan. Solutions of the system of second-order partial differential equations are then realized by mappings that annihilate the ideal that characterizes the system. This system of concepts will be dealt with in detail in Part Two.

4-7. INTEGRATION OF EXTERIOR FORMS, STOKES' THEOREM, AND THE DIVERGENCE THEOREM

Let Y_r be a closed, r-dimensional region of r-dimensional number space E_r and let ∂Y_r denote the boundary of Y_r. We know that the dimension of $\Lambda^r(E_r)$

is one, so that any element ω of $\Lambda^r(E_r)$ can be written uniquely as

$$\omega = w(x^j)\mu, \tag{4-7.1}$$

where

$$\mu = dx^1 \wedge dx^2 \wedge \cdots \wedge dx^r \tag{4-7.2}$$

is the natural volume element of E_r with respect to the $(x^1, \ldots, x^r)$ coordinate cover. Further, any two elements of $\Lambda^r(E_r)$ can only differ in the choice of the function $w(x^j)$ and in the order in which $dx^1, dx^2, \ldots, dx^n$ are multiplied. Accordingly, the choice of the order given by the natural volume element μ fixes all simple elements of $\Lambda^r(E_r)$ to within a factor ± 1. Choice of the $+$ sign, as in (4-7.1), fixes things completely. This choice of the order in which the factors appear in the basis element μ is referred to as a choice of *orientation*, while the making of this choice is said to assign an orientation to Y_r. In three dimensions the choice $\mu = dx \wedge dy \wedge dz$ fixes the right-hand orientation on E_3, while the choice of $\eta = dy \wedge dx \wedge dz = -dx \wedge dy \wedge dz$ for the natural volume element would fix a left-hand orientation on E_3. If $\Phi : {}^{`}E_r \to E_r | x^i = \phi^i({}^{`}x^j)$ is an invertible map, then

$$\Phi^*\mu = \det(\partial\phi^i/\partial{}^{`}x^j)\,{}^{`}\mu, \qquad {}^{`}\mu = d{}^{`}x^1 \wedge \cdots \wedge d{}^{`}x^r,$$

in which case Φ is said to be *orientation preserving* if $\det(\partial\phi^i/\partial{}^{`}x^j) > 0$ and to be *orientation reversing* if $\det(\partial\phi^i/\partial{}^{`}x^j) < 0$. This shows that the choice of orientation with respect to any one coordinate cover (x^i) fixes the orientation for all coordinate covers that can be obtained from (x^i) by invertible coordinate transformations.

The choice of orientation for r-forms on E_r also fixes the orientation of $(r-1)$-forms on E_r. We have seen that $\Lambda^{r-1}(E_r)$ has a basis of the form

$$\mu_i = \partial_i \lrcorner \mu, \qquad i = 1, \ldots, r \tag{4-7.3}$$

such that

$$dx^i \wedge \mu_j = \delta^i_j \mu. \tag{4-7.4}$$

This shows that

$$dx^1 \wedge \mu_1 = dx^2 \wedge \mu_2 = \cdots = dx^r \wedge \mu_r = \mu, \tag{4-7.5}$$

and hence any change in orientation induces a like change in the signs of the basis elements $\{\mu_i\}$ of $\Lambda^{r-1}(E_r)$. We therefore say that the basis $\{\mu_i\}$ of $\Lambda^{r-1}(E_r)$ assigns an *orientation to* $\Lambda^{r-1}(E_r)$ *of equal parity.*

Let

$$\omega = w(x^j)\mu \tag{4-7.6}$$

be a given element of $\Lambda^r(E_r)$. We define the integral of ω over the set Y_r by

$$\int_{Y_r} \omega = \int_{Y_r} w(x^j)\, dx^1\, dx^2 \cdots dx^r, \tag{4-7.7}$$

where the integral on the right-hand side of (4-7.7) denotes r-dimensional Riemann integral over the region Y_r of r-dimensional Euclidean space. It is now a direct, but messy calculation to establish Stokes' theorem (see Fig. 14).

Theorem 4-7.1. (Stokes) *If $\omega \in \Lambda^{r-1}(E_r)$ is defined over a closed, arcwise connected, simply connected, r-dimensional region Y_r of E_r and Y_r has a smooth boundary, ∂Y_r, then*

$$\int_{Y_r} d\omega = \int_{\partial Y_r} j^*\omega, \tag{4-7.8}$$

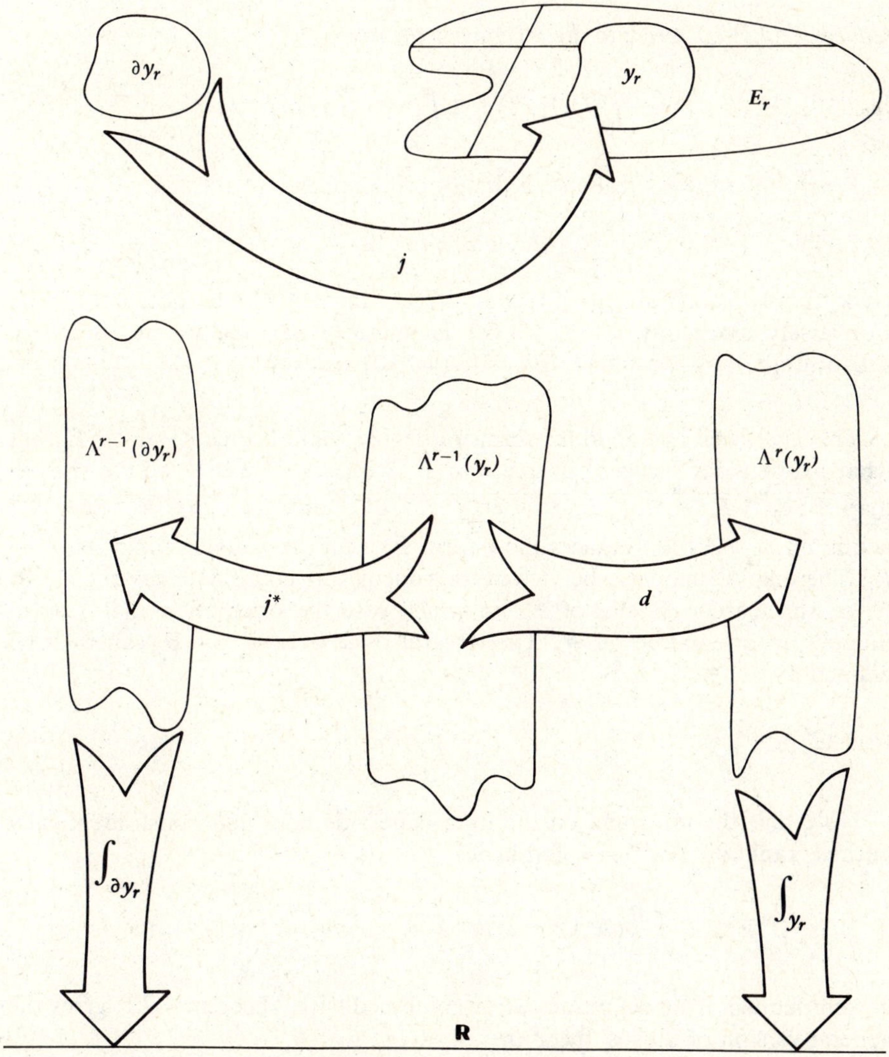

FIGURE 14. Stokes' theorem

when the orientation of ∂Y_r is of equal parity with the orientation of Y_r and j is the injection of ∂Y_r into E_r.

We will not give a proof of Stokes' theorem since there is a trivial reformulation that reduces Stokes' theorem to the divergence theorem, and the proof of the latter is simply a careful application of the integration by parts theorem.

Theorem 4-7.2. *If $\omega \in \Lambda^{r-1}$ is defined over a closed, arcwise connected, simply connected, r-dimensional region Y_r of E_r with a smooth boundary ∂Y_r, then the resolution of ω on the basis $\{\mu_i\}$ of $\Lambda^{r-1}(E_r)$ by*

$$\omega = w^i(x^j)\mu_i \tag{4-7.9}$$

reduces Stokes' theorem to the divergence theorem

$$\int_{Y_r} \partial_i w^i(x^j)\mu = \int_{\partial Y_r} j^* w^i(x^j)\mu_i. \tag{4-7.10}$$

Proof. We have already seen that (4-6.7) implies

$$d\omega = \partial_i w^i(x^j)\mu. \tag{4-7.11}$$

Now simply substitute (4-7.9) and (4-7.11) into (4-7.8) to obtain (4-7.10). Conversely since any $\omega \in \Lambda^{r-1}(E_r)$ is uniquely represented by (4-7.9), a substitution of (4-7.9) and (4-7.11) into (4-7.10) yields (4-7.8). □

This is all well and good for forms of degree r and degree $r-1$ on E_r, but what happens for forms of degree r in E_n with $r < n$? Let Ψ be the smooth map $\Psi: Y_r \subset E_r \xrightarrow{\text{onto}} \mathscr{Y}_r \subset E_n | x^i = \psi^i(u^1, \ldots, u^r)$, $\operatorname{rank}(\partial\psi^i/\partial u^\alpha) = r$ then Ψ^* acting on $\omega \in \Lambda^r(E_n)$ induces the r-form $\Psi^*\omega$ on the r-dimensional set Y_r in E_r. The map Ψ may also be viewed as inducing the coordinate cover $\{u^\alpha\}$ of E_r as surface coordinates of $\mathscr{Y}_r$, in which case the orientation $\mu(u^\alpha)$ of E_r induces an orientation on $\mathscr{Y}_r$. The integral of ω over $\mathscr{Y}_r$ in E_n can then be defined by

$$\int_{\mathscr{Y}_r} \omega = \int_{Y_r} \Psi^*\omega. \tag{4-7.12}$$

Note that the integrand on the right-hand side does not vanish identically because $\operatorname{rank}(\partial\psi^i/\partial u^\alpha) = r$ and hence

$$\int_{Y_r} \Psi^*\omega = \int_{Y_r} \hbar(u^\alpha)\, du^1 \wedge du^2 \wedge \cdots \wedge du^r = \int_{Y_r} \hbar(u^\alpha)\mu(u^\alpha)$$

is well defined. This definition, when combined with Theorem 4-7.1, gives the general version of Stokes' theorem.

Theorem 4-7.3. *If $\omega \in \Lambda^{r-1}(E_n)$, then*

(4-7.13)
$$\int_{\mathscr{Y}_r} d\omega = \int_{\partial \mathscr{Y}_r} j^*\omega$$

provided $\mathscr{Y}_r$ and $\partial\mathscr{Y}_r$ are such that there exists a map

(4-7.14)
$$\Psi : Y_r \subset E_r \overset{\text{onto}}{\to} \mathscr{Y}_r \subset E_n | x^i = \psi^i(u^\alpha),$$

$$\operatorname{rank}(\partial\psi^i/\partial u^\alpha) = r$$

for which Y_r is a closed, arcwise connected, simply connected, r-dimensional region with a smooth boundary.

A more useful way of writing this result is

(4-7.15)
$$\int_{Y_r} \Psi^* d\omega = \int_{Y_r} d\Psi^*\omega = \int_{\partial Y_r} \Psi^* j^*\omega$$

since $\int_{Y_r}\Psi^* d\omega$ and $\int_{\partial Y_r}\Psi^* j^*\omega$ are the actual values involved.

PROBLEMS

Let E_4 have the coordinate cover (x, y, z, t) and

$$\omega_1 = 3y^2 dx + 2te^{4z} dy - x^2y^3 dz + 5\sin(7y)\, dt,$$

$$\omega_2 = (xz - t^2)\, dx \wedge dy + 3e^{5t} dx \wedge dz - 12x^3 y\, dy \wedge dt,$$

$$\omega_3 = ze^{9t} dx \wedge dy \wedge dz + (6x^3 - 8yt^4)\, dx \wedge dy \wedge dt$$
$$+ e^{7x} dy \wedge dz \wedge dt,$$

$$\omega_4 = (3xz^2 + 5ye^{11zt})\, dx \wedge dy \wedge dz \wedge dt,$$

$$V = 2Ay\partial_x + 5Bt^2\partial_y + (6x^3y^2 - 7Ct^5)\partial_z + 2Fxy\partial_t.$$

4.1. Evaluate the following exterior derivatives: (a) $d\omega_1$; (b) $d\omega_2$; (c) $d\omega_3$; (d) $d\omega_4$; (e) $d(V \lrcorner \omega_1)$; (f) $d(V \lrcorner \omega_2)$; (g) $d(\omega_1 \wedge \omega_2)$.

Answers

(a) $(35\cos(7y) - 2e^{4z})\, dy \wedge dt - 2xy^3 dx \wedge dz$
$-(8te^{4z} + 3x^2y^2)\, dy \wedge dz - 6y\, dx \wedge dy,$

(b) $15e^{5t}\,dx \wedge dz \wedge dt - 2(t + 18x^2y)\,dx \wedge dy \wedge dt + x\,dx \wedge dy \wedge dz$,

(c) $(7e^{7x} - 9ze^{9t})\,dx \wedge dy \wedge dz \wedge dt$,

(d) 0,

(e) $5t^2(6Be^{4z} + 7Cx^2y^3t^2)\,dt + 40Bt^3e^{4z}\,dz$
$+(70\mathrm{Fxy}\cos(7\mathrm{y}) + 10\mathrm{Fx}\sin(7\mathrm{y}) + 18\mathrm{Ay}^2 + 21\mathrm{Cx}^2\mathrm{y}^2\mathrm{t}^5$
$-30x^5y^4)\,dy + 2y(5F\sin(7y) + 7Cxy^2t^5 - 15x^4y^4)\,dx$,

(f) $-30Aye^{5t}\,dz \wedge dt + 4t(Ay - 15Bx^3t)\,dy \wedge dt + 5Bxt^2\,dx \wedge dz$
$+2A(3e^{5t} - xy)\,dy \wedge dz + 2y(18x^3e^{5t} + 48Fx^3y + Az)\,dx \wedge dy$
$+5(18x^3y^2e^{5t} - 21Ct^5e^{5t} - 21Ct^4e^{5t} - 36Bx^2yt^2 + 2Bxzt$
$-4Bt^3)\,dx \wedge dt$,

(g) $(30te^{4z+5t} + 6e^{4z+5t} - 105e^{5t}\cos(7y)$
$+5x\sin(7y) - 60x^4y^4 - 2x^2y^3t)\,dx \wedge dy \wedge dz \wedge dt$.

4.2. Which quantities $\omega_1, \omega_2, \omega_3, \omega_4$ are closed?

4.3. Write out the conditions $d\alpha = 0$ for (a) $\alpha \in \Lambda^1(E_n)$; (b) $\alpha \in \Lambda^2(E_n)$; (c) $\alpha \in \Lambda^3(E_n)$; (d) $\alpha \in \Lambda^{n-1}(E_n)$.

4.4. Let γ^1, γ^2, be two 1-forms on E_4. Write out the conditions that $\{\gamma^1, \gamma^2\}$ form a completely integrable differential system.

4.5. Find all characteristic vectors of the following ideals of $\Lambda(E_4)$: (a) $I\{\omega_1\}$; (b)$I\{\omega_2\}$; (c) $I\{\omega_1, \omega_2\}$; (d) $I\{x\,dy + y\,dz, x^2\,dy - y^2\,dt\}$; (e) $I\{\omega_1, V \lrcorner \omega_2\}$.

4.6. Find the Darboux class, the rank, and the index of the following 1-forms on E_4: (a) $dx + x^2\,dy$; (b) $y\,dx + x^2y\,dt$; (c) $t\,dx - x\,dt + 5y\,dz$; (d) $(x - y^2)\,dx + (y^3 - z^2)\,dy + 5\,dt$.

4.7. Find all solutions of maximal dimension on E_4 of $\omega = 0$ where (a) $\omega = x^2\,dy - dt$; (b) $\omega = (1 - 3y^2)\,dz + e^{xt}\,dy$; (c) $\omega = (x^2 - 5t^3)\,d(y^2 + z) + zt\,d(xt + e^y)$; (d) $\omega = t\,dx - y\,dz$; (e) $\omega = (x^3 - 3yt)\,d(x + 4z)$.

4.8. Evaluate the following Lie derivatives: (a) $\pounds_V\omega_1$; (b) $\pounds_V\omega_2$; (c) $\pounds_V\omega_3$; (d) $\pounds_V\omega_4$.

Answers

(a) $5t^2(4Be^{4z} + 35B\cos(7y) + 7Cx^2y^3t^2)\,dt$
$-xy^2(4Ay^2 + 15Bxt^2)\,dz$
$+2(2Fxye^{4z} - 28Ct^6e^{4z} + 24x^3y^2te^{4z} + 5Fx\sin(7y)$
$+3Ay^2 - 6x^5y^4)\,dy$
$+2y(5F\sin(7y) + 15Bt^2 - 9x^4y^4)\,dx$,

(b) $-12x^2(6Ay^2 + 5Bxt^2)\,dy \wedge dt + 6Ae^{5t}\,dy \wedge dz$
$+5t(2Bxz - 2Bt^2 - 21Ct^3e^{5t})\,dx \wedge dt$
$+30Fxye^{5t}\,dx \wedge dz$
$+(36x^3ye^{5t} + 24Fx^3y^2 - 4Fxyt + 2Ayz - 7Cxt^5 + 6x^4y^2)\,dx \wedge dy$,

(c) $14Aye^{7x}\,dy \wedge dz \wedge dt - 10Bzte^{9t}\,dx \wedge dz \wedge dt$
$+(36Ax^2y - 18x^2y^2e^{7x} - 35Czt^4e^{9t} - 64Fxy^2t^3 - 40Bt^6)\,dx \wedge$
$dy \wedge dt$
$+(2Fye^{7x} + 18Fxyze^{9t} - 7Ct^5e^{9t} + 6x^3y^2e^{9t})\,dx \wedge dy \wedge dz,$

(d) $(110Fxy^2ze^{11zt} + 25Bt^2e^{11zt} - 385Cyt^6e^{11zt}$
$+330x^3y^3te^{11zt} + 6Ayz^2 - 42Cxzt^5 + 36x^4y^2z)\,dx \wedge dy \wedge dz \wedge$
$dt.$

4.9. Given $\alpha = x\,dy - y\,dz$ and $U = Cz\partial_y + Ax\partial_z$ on E_3, evaluate (a) $£_U\alpha$; (b) $£_U£_U\alpha$; (c) $£_U£_U£_U\alpha$; (d) $£_U£_U£_U£_U\alpha$; (e) $\exp(s£_U)\alpha$.

Answers

(a) $C(x - z)\,dz - Ay\,dx,$

(b) $AC(x - 2z)\,dx - ACx\,dz,$

(c) $-3A^2Cx\,dx,$

(d) $0,$

(e) $As(-y + \frac{1}{2}C(x - 2z)s - \frac{1}{2}ACxs^2)\,dx + x\,dy$
$+(-y + C(x - z)s - \frac{1}{2}ACxs^2)\,dz.$

4.10. Find all isovectors of the following ideals of $\Lambda(E_3)$: (a) $I\{x\,dy + y\,dz\}$; (b) $I\{dx + x\,dy, z^2\,dy\}$; (c) $I\{(1 + y^2)\,dx + dz, (1 + z^2)\,dx - dy\}$; (d) $I\{dx + 2\,dz, 3x\,dy \wedge dz\}$.

4.11. Let $g(x^j)$ be a given element of $\Lambda^0(E_n)$ such that $dg \neq 0$. Show that all $f \in \Lambda^0(E_n)$ that satisfy $df \wedge dg = 0$ are of the form $f(x^j) = F(g(x^j))$ for some smooth choice of the function F.

4.12. Let $g^1(x^j), \ldots, g^r(x^j)$ be r given elements of $\Lambda^0(E_n)$ such that $dg^1 \wedge dg^2 \wedge \cdots \wedge dg^r \neq 0$. Show that all functions $f \in \Lambda^0(E_n)$ that satisfy $df \wedge dg^1 \wedge \cdots \wedge dg^r = 0$ are of the form $f = F(g^1, g^2, \ldots, g^r)$ for some smooth choice of the function F.

4.13. Let $g^1, \ldots, g^r$ be r given elements of $\Lambda^0(E_n)$ such that $dg^1 \wedge \cdots \wedge dg^r \neq 0$. Determine all functions $f_1, f_2, \ldots, f_r$ on E_n such that $df_1 \wedge dg^1 + df_2 \wedge dg^2 + \cdots + df_r \wedge dg^r = 0$. (Hint: use the Cartan lemma.)

4.14. Show that $dg^1 \wedge dg^2 \wedge \cdots \wedge dg^r \neq 0$ and $dp = f_1\,dg^1 + f_2\,dg^2 + \cdots + f_r\,dg^r = f_a\,dg^a$ implies $p = p(g^1, \ldots, g^r)$ and $f_a = f_a(g^b) = \dfrac{\partial p}{\partial g^a}$.

CHAPTER FIVE

ANTIEXACT DIFFERENTIAL FORMS AND HOMOTOPY OPERATORS

5-1. PERSPECTIVE

The subspaces $\mathscr{C}(S)$ of all closed elements of $\Lambda(S)$ and $\mathscr{E}(S)$ of all exact elements of $\Lambda(S)$ considered in the preceding chapter assume particular importance in modern studies of geometry and topology. Studies of the structure of the relations between these two subspaces of $\Lambda(S)$ has led to a new level of understanding and to new mathematical disciplines through the works of de Rham, Hodge, and Cartan. There is an aspect of this problem that has been little touched, however. This is unfortunate, for it is of considerable practical importance and provides otherwise unavailable insights, even though the considerations are strictly local in nature. The presentation given here will follow that of the last chapter of the appendix of the book on isovector methods by Edelen, where this body of ideas was first discussed.

Let ω be a differential form on a starshaped region S of an n-dimensional space. It has been known, starting from the works of Poincaré, that a linear homotopy operator H can be defined on elements of $\Lambda(S)$ that verifies the identity $\omega = dH\omega + Hd\omega$. The customary use of this result is to establish the Poincaré lemma: a closed form on a starshaped region S is exact on S (i.e.,

$d\omega = 0$ implies $\omega = dH\omega$). Significantly more information is actually available, however. If ω is an arbitrary differential form on S, then $\omega = dH\omega + H\,d\omega$ allows us to associate an exact form $\omega_e = dH\omega$ with ω. It is thus meaningful to refer to ω_e as the exact part of ω. On the other hand, it can be shown that $\omega_a = \omega - \omega_e$ belongs to the kernel of the linear homotopy operator H. We accordingly refer to $\omega_a = \omega - \omega_e$ as the antiexact part of ω. The important thing about antiexact forms, in contrast with exact or closed forms, is that they form a graded submodule $\mathscr{A}(S)$, of the graded module $\Lambda(S)$ of exterior forms on S, and that *H inverts the exterior derivative operator*, d, *on this submodule.* It is then clear that the decomposition $\omega = \omega_e + \omega_a$ and the invertibility of d on $\mathscr{A}(S)$ allows the construction of general solutions to systems of exterior differential equations and provides a number of other useful results of both computational and structural natures. Thus in overview, the substance of this chapter is the study of the submodule $\mathscr{A}(S)$ of antiexact differential forms. The results obtained in this way will be instrumental in the discussions of electrodynamics and gauge theories given in Part Three.

5-2. STARSHAPED REGIONS

Let S be an open region of an n-dimensional space E_n. Such a region is said to be *starshaped* with respect to one of its points $P_0 \in S$ if the following conditions are met:

1. S is contained in a coordinate neighborhood U of P_0.
2. The coordinate functions of U assign coordinates (x_0^i) to the point P_0.
3. If P is any point in S and if the coordinate functions of U assign coordinates (x^i) to P, then the set of points with coordinates $(x_0^i + \lambda(x^i - x_0^i))$ for all $\lambda \in [0, 1]$ belongs to S.

It is clear that the notion of a starshaped region is a local one. Moreover, it is dependent upon the existence of a specific coordinate neighborhood with specific coordinate functions. We refer to the coordinate neighborhood U and the coordinate functions that verify that S is starshaped, as a *preferred coordinate neighborhood* and *preferred coordinate functions* of S. For the time being, we assume that a starshaped region S is referred to its preferred coordinate neighborhood with its preferred coordinate functions since the analysis will be of a local nature. Thus if $P \in S$, we write $P : (x^i)$, where (x^i) are the coordinates of P in the preferred coordinate system, and of course, $P_0 : (x_0^i)$. Thus for now, we fix the coordinate cover of S. Later we will consider what happens when the coordinate cover is changed.

The local nature of starshaped regions notwithstanding, there is certainly no scarcity of starshaped regions. For instance, one can generate uncountably many that are starshaped with respect to a given point, say P_0. Let $'S$ be a starshaped region with respect to the point $'P_0$ of an n-dimensional space $'M$ and let $\mathscr{K}$ denote the collection of all regular maps of $'M$ to the n-dimensional space E_n that map $'P_0$ onto P_0. It is then clear that the image of $'S$ under any map $\Phi \in \mathscr{K}$ yields a region in E_n that is starshaped with respect to P_0. Further, the Φ-image of the preferred coordinate cover of $'S$ yields a preferred coordinate cover of $\Phi('S)$.

The fixed nature of the preferred coordinate cover of S allows us to take advantage of certain notational conveniences. If $P:(x^i)$ belongs to S, then each point with coordinates $(x_0^i + \lambda(x^i - x_0^i))$ belongs to S. Thus if we have an exterior differential form

$$\omega = \omega_{i_1 \cdots i_k}(x^j)\, dx^{i_1} \wedge \cdots \wedge dx^{i_k} \tag{5-2.1}$$

defined on S, the exterior differential form

$$\tilde{\omega}(\lambda) = \omega_{i_1 \cdots i_k}\left(x_0^j + \lambda\left(x^j - x_0^j\right)\right) dx^{i_1} \wedge \cdots \wedge dx^{i_k} \tag{5-2.2}$$

is well defined on S for every $\lambda \in [0,1]$ and $\tilde{\omega}(1) = \omega$. In particular, we have

$$d(\tilde{\omega}(\lambda)) = \lambda(\widetilde{d\omega})(\lambda) \tag{5-2.3}$$

for every $\lambda \in [0,1]$, as follows directly from the definition of the exterior derivative.

A starshaped region S also has a naturally associated vector field that has many of the properties of the classical "radius vector." This vector field is defined relative to the preferred coordinate system on S by

$$\mathscr{X}(x^i) = \left(x^i - x_0^i\right)\partial_i = \left\{\frac{d}{d\lambda}\left(x_0^i + \lambda\left(x^i - x_0^i\right)\right)\right\}\partial_i \tag{5-2.4}$$

and satisfies the relation

$$\tilde{\mathscr{X}}(\lambda) = \mathscr{X}\left(x_0^i + \lambda\left(x^i - x_0^i\right)\right) = \lambda\mathscr{X} \tag{5-2.5}$$

for all $\lambda \in [0,1]$.

5-3. THE HOMOTOPY OPERATOR H

Let ω be a differential form of degree k on a starshaped region S. Since $\tilde{\omega}(\lambda)$ and $\mathscr{X}$ are well defined on S, the linear homotopy operator

$$H\omega = \int_0^1 \mathscr{X} \lrcorner\, \tilde{\omega}(\lambda)\lambda^{k-1}\, d\lambda, \qquad k = \deg(\omega), \tag{5-3.1}$$

is well defined on S as a Riemann-Graves integral. H is thus well defined on

$\Lambda(S)$. Its specific representation in terms of the preferred coordinate cover of S follows directly from (5-2.2) and (5-2.4):

$$(5\text{-}3.2) \qquad (H\omega)(x^i) = \int_0^1 \mathscr{X}(x^i) \lrcorner\, \omega\big(x_0^i + \lambda(x^i - x_0^i)\big)\lambda^{k-1}\, d\lambda.$$

Although some of the properties of H given in the following theorem are trivial consequences of others in the list, they are given explicitly in order to simplify proofs later on.

Theorem 5-3.1. *The operator H is a linear operator on $\Lambda(S)$ with the following properties*:

$\mathbf{H_1}$. *H maps $\Lambda^k(S)$ into $\Lambda^{k-1}(S)$ for $k \geq 1$ and maps $\Lambda^0(S)$ into zero.*
$\mathbf{H_2}$. *$dH + Hd =$ identity for $k \geq 1$, and $(H\,df)(x^i) = f(x^i) - f(x_0^i)$ for $k = 0$.*
$\mathbf{H_3}$. *$(HH\omega)(x^i) = 0$, $(H\omega)(x_0^i) = 0$.*
$\mathbf{H_4}$. *$H\,dH = H$, $dH\,d = d$.*
$\mathbf{H_5}$. *$H\,dH\,d = Hd$, $dH\,dH = dH$, $(dH)(Hd) = 0$, $(Hd)(dH) = 0$.*
$\mathbf{H_6}$. *$\mathscr{X}\lrcorner H = 0$, $H\mathscr{X}\lrcorner = 0$.*

Proof. The linearity of H on $\Lambda(S)$ is clear from the defining equation (5-3.1). Property $\mathbf{H_1}$ also follows directly from (5-3.1) since $\mathscr{X}\lrcorner$ maps $\Lambda^k(S)$ into $\Lambda^{k-1}(S)$ for $k \geq 1$ and maps $\Lambda^0(S)$ into the zero function on S. When (5-3.1) is used to evaluate $H\omega$ and $H\,d\omega$ for $\omega \in \Lambda^k(S)$, we obtain

$$(5\text{-}3.3) \quad (dH + Hd)\omega = \int_0^1 \big\{ d(\mathscr{X}\lrcorner\tilde{\omega}(\lambda))\lambda^{k-1} + \mathscr{X}\lrcorner(\widetilde{d\omega})(\lambda)\lambda^k \big\}\, d\lambda.$$

However, (5-2.3) gives $(\widetilde{d\omega})(\lambda) = \lambda^{-1} d(\tilde{\omega}(\lambda))$, so (5-3.3) can be written

$$(5\text{-}3.4) \qquad (Hd + dH)\omega = \int_0^1 L_\lambda \tilde{\omega}(\lambda)\, d\lambda,$$

where the linear operator L_λ is defined on $\Lambda(S)$ for all $\lambda \in [0,1]$ by

$$(5\text{-}3.5) \qquad L_\lambda \tilde{\omega}(\lambda) = \{ d(\mathscr{X}\lrcorner\tilde{\omega}(\lambda)) + \mathscr{X}\lrcorner d(\tilde{\omega}(\lambda))\}\lambda^{k-1}.$$

It might appear that there is some difficulty with (5-3.5) if $k = 0$. This is not the case, however, for $\mathscr{X}\lrcorner\tilde{f}(\lambda) = 0$ and $d(\tilde{f}(\lambda)) = \lambda(\widetilde{df})(\lambda)$ for $f \in \Lambda^0(S)$. Now, (5-3.5) and the definition of the Lie derivative give $\lambda^{1-k} L_\lambda \tilde{\omega}(\lambda) =$

$£_{\mathscr{X}}\tilde{\omega}(\lambda)$. Thus since £ is a derivation on $\Lambda(S)$, $\alpha \in \Lambda^k(S)$, $\beta \in \Lambda^m(S)$ imply

$$\begin{aligned}\lambda^{1-k-m}L_\lambda(\widetilde{\alpha \wedge \beta})(\lambda) &= \lambda^{1-k-m}L_\lambda(\tilde{\alpha}(\lambda) \wedge \tilde{\beta}(\lambda))\\ &= \lambda^{1-k}L_\lambda\tilde{\alpha}(\lambda) \wedge \tilde{\beta}(\lambda) + \tilde{\alpha}(\lambda) \wedge \lambda^{1-m}L_\lambda\tilde{\beta}(\lambda),\end{aligned}$$

so that

$$L_\lambda(\widetilde{\alpha \wedge \beta})(\lambda) = (L_\lambda\tilde{\alpha}(\lambda)) \wedge \lambda^m\tilde{\beta}(\lambda) + \lambda^k\tilde{\alpha}(\lambda) \wedge L_\lambda\tilde{\beta}(\lambda). \tag{5-3.6}$$

Accordingly, since any element of $\Lambda(S)$ can be written as a sum of simple monomials and an elementary calculation gives $L_\lambda\tilde{\alpha}(\lambda) = (d/d\lambda)(\lambda\tilde{\alpha}(\lambda))$ for $\alpha \in \Lambda^1(S)$, we obtain the following evaluation of $L_\lambda\tilde{\omega}(\lambda)$ for $\omega \in \Lambda^k(S)$:

$$L_\lambda\tilde{\omega}(\lambda) = \frac{d}{d\lambda}(\lambda^k\tilde{\omega}(\lambda)). \tag{5-3.7}$$

A substitution of this result back into (5-3.4) gives $(Hd + dH)\omega = \int_0^1(d/d\lambda)(\lambda^k\tilde{\omega}(\lambda))\,d\lambda$. Thus since $\tilde{\omega}(1) = \omega$, property $\mathbf{H}_2$ is established. Since $\mathscr{X}(x_0^i) = 0$ from (5-2.4), (5-3.2) yields $(H\omega)(x_0^i) = 0$. It also follows from (5-2.5) and (5-3.1) that

$$\begin{aligned}HH\omega &= \int_0^1 \mathscr{X}\lrcorner\left\{\int_0^1 \widetilde{\mathscr{X}\lrcorner\tilde{\omega}(\lambda)\lambda^{k-1}}\,d\lambda\right\}(\mu)\mu^{k-2}\,d\mu\\ &= \int_0^1\int_0^1 \mathscr{X}\lrcorner\tilde{\mathscr{X}}(\mu)\lrcorner(\widetilde{\tilde{\omega}(\lambda)})(\mu)\lambda^{k-1}\mu^{k-2}\,d\lambda\,d\mu\\ &= \int_0^1\int_0^1 \mathscr{X}\lrcorner\mu\mathscr{X}\lrcorner\widetilde{\tilde{\omega}(\lambda)}(\mu)\lambda^{k-1}\mu^{k-2}\,d\lambda\,d\mu = 0,\end{aligned}$$

which establishes $\mathbf{H}_3$. Property $\mathbf{H}_4$ follows directly by allowing H and d to act on $dH + Hd$ = identity and using $\mathbf{H}_3$; i.e., $H = H(\text{identity}) = H(dH + Hd) = H\,dH + HHd = H\,dH$. Allowing H and d to act on properties $\mathbf{H}_4$ establishes $\mathbf{H}_5$. Properties $\mathbf{H}_6$ are established in exactly the same way as that used to establish $\mathbf{H}_3$; i.e., (5-2.5) can be used to obtain a factor of the form $\mathscr{X}\lrcorner\mathscr{X}\lrcorner$ under the integral sign, which of course vanishes. □

The operator H on a starshaped region is of particular importance in many applications. It is not an unfamiliar operator in the exterior calculus, for it is customarily constructed in order to establish the Poincaré lemma; *every closed form on S is exact*. This result follows directly from $\mathbf{H}_2$ since ω closed implies $\omega = dH\omega + H\,d\omega = dH\omega$. In this vein, it may be remarked that the proof of property $\mathbf{H}_2$ given above is somewhat simpler than given in most texts, for it requires explicit calculation only at the point where $L_\lambda\tilde{\omega}(\lambda) = (d/d\lambda)(\lambda\tilde{\omega}(\lambda))$

is established for 1-forms. There is, however, significantly more information that can be gleaned from properties $\mathbf{H}_1$ through $\mathbf{H}_6$, and it is this additional body of information that will prove to be instrumental.

We consider a few elementary examples as an assist to the reader. Take two-dimensional Euclidean space with its global Cartesian coordinate cover (x, y) and let the 1-form ω be defined by

$$\omega = yU(3 - x)\,dx, \tag{5-3.8}$$

where $U(z) = \{1 \text{ for } z > 0, 0 \text{ for } z < 0\}$ is the unit function. Since $d\omega = U(3 - x)\,dy \wedge dx$ on $x \neq 3$, the definition of H given by (5-3.1) yields

$$H\omega = \int_0^1 x\lambda\, yU(3 - \lambda x)\,d\lambda = \begin{cases} xy/2 & \text{for} \quad x < 3 \\ 3^2 y/2x & \text{for} \quad x > 3, \end{cases} \tag{5-3.9}$$

$$H\,d\omega = \int_0^1 U(3 - \lambda x)\lambda(y\,dx - x\,dy)\,d\lambda \tag{5-3.10}$$

$$= (y\,dx - x\,dy)\begin{cases} 1/2 & \text{for} \quad x < 3 \\ 3^2/2x^2 & \text{for} \quad x > 3. \end{cases}$$

Thus although both ω and $d\omega$ vanish for $x > 3$, $H\omega$ and $H\,d\omega$ are nonzero for $x > 3$, $y \neq 0$. In fact, this "smearing out" that results from the action of H is characteristic of this operator. It now remains only to check that $dH + Hd$ = identity. However, (5-3.9) yields

$$dH\omega = \tfrac{1}{2}\begin{cases} y\,dx + x\,dy & \text{for} \quad x < 3 \\ 3^2(-y\,dx + x\,dy)/x^2 & \text{for} \quad x > 3, \end{cases} \tag{5-3.11}$$

so that (5-3.10) and (5-3.11) yield

$$dH\omega + H\,d\omega = \begin{cases} y\,dx & \text{for} \quad x < 3 \\ 0 & \text{for} \quad x > 3. \end{cases} \tag{5-3.12}$$

For an example of an exact form, we take

$$\omega = x^r U(3 - x)\,dx, \qquad r > 0, \tag{5-3.13}$$

in which case (5-3.1) gives

$$H\omega = \frac{1}{r + 1}\begin{cases} x^{r+1} & \text{for} \quad x < 3 \\ 3^{r+1} & \text{for} \quad x > 3. \end{cases} \tag{5-3.14}$$

Notice that in both examples, $H\omega$ is continuous at $x = 3$ even though ω has a jump discontinuity there. It is clear from $dH + Hd$ = identity, that the action of H on a form will result in a form with an improved continuity class.

5-4. THE EXACT PART OF A FORM AND THE VECTOR SUBSPACE OF EXACT FORMS

Let $\mathscr{E}^k(S)$ denote the collection of all exact elements of $\Lambda^k(S)$. Since the set $\mathscr{E}^k(S)$ is closed under addition and multiplication by numbers, but is not closed under multiplication by functions, $\mathscr{E}^k(S)$ forms a linear subspace of $\Lambda^k(S)$ but not a submodule of $\Lambda^k(S)$. Further, $\mathscr{E}^0(S)$ *is empty*. Since the exterior product of two exact forms is again an exact form, we can construct the graded algebra $\mathscr{E}(S)$ of exact forms on S, and $\mathscr{E}(S)$ is a subspace of $\Lambda(S)$ but not a submodule of $\Lambda(S)$. In fact, it is precisely because $\mathscr{E}^0(S)$ is empty that $\mathscr{E}(S)$ is not a submodule of $\Lambda(S)$. The following result is now an almost immediate consequence of properties $\mathbf{H}_1$ through $\mathbf{H}_6$.

Lemma 5-4.1. *The operator dH maps $\mathscr{E}^k(S)$ onto $\mathscr{E}^k(S)$ and $\Lambda(S)$ onto $\mathscr{E}(S)$.*

Proof. Since $dH\omega$ is exact for any $\omega \in \Lambda^k(S)$, dH maps $\Lambda^k(S)$ into $\mathscr{E}^k(S)$. The result then follows upon showing that this mapping is onto. We do this with the following lemma that is of importance in its own right. □

Lemma 5-4.2. *The operator d is the inverse of the operator H when the domain of H is restricted to $\mathscr{E}^k(S)$.*

Proof. Let ω be an arbitrary element of $\mathscr{E}^k(S)$, then $\omega = d\alpha$ for some $\alpha \in \Lambda^{k-1}(S)$, and the restriction of H to $\mathscr{E}^k(S)$ yields quantities of the form $H\omega = H\,d\alpha$. Allowing d to act on both sides of this relation yields $dH\omega = dH\,d\alpha = d\alpha = \omega$ when $\mathbf{H}_4$ is used. Thus dH restricted to $\mathscr{E}^k(S)$ is the identity and the result follows. It also follows that dH is an onto map of $\Lambda^k(S)$ to $\mathscr{E}^k(S)$ and hence Lemma 5-4.1 is established. □

We now go back to property $\mathbf{H}_2$ and note that any $\omega \in \Lambda^k(S)$ satisfies

$$\omega = dH\omega + H\,d\omega. \tag{5-4.1}$$

Since dH is the identity map of $\mathscr{E}^k(S)$ and maps $\Lambda^k(S)$ onto $\mathscr{E}^k(S)$, every element ω of $\Lambda^k(S)$ has a uniquely associated exact part, $dH\omega$. This motivates the following definition.

Definition. Let $\omega \in \Lambda^k(S)$ with $k \geq 1$. The element of $\mathscr{E}^k(S)$, defined by

$$\omega_e = dH\omega, \tag{5-4.2}$$

is the *exact part* of ω. Elements of $\Lambda^0(S)$ have no exact part.

If we use property $\mathbf{H}_2$, then for any $\omega \in \Lambda^k(S)$ we have $\omega = dH\omega + H\,d\omega$, the definition of ω_e yields $\omega = \omega_e + H\,d\omega$. Accordingly, allowing H to act on both sides of this relation and making use of property $\mathbf{H}_3$ shows that

$H(\omega - \omega_e) = 0$. We thus have the following result that acts as motivation for the considerations of the next section.

Lemma 5-4.3. *Let ω_e be the exact part of a form ω, then $\omega - \omega_e$ belongs to the kernel of the linear operator H.*

5-5. THE MODULE OF ANTIEXACT FORMS

The results of Lemma 5-4.3 and the fact that any ω can be written as $\omega = \omega_e + H\,d\omega$, by $\mathbf{H}_2$, suggests that we study quantities of the form $H\,d\alpha$. In this regard the following definition is the natural complement of the definition of ω_e.

Definition. Let $\omega \in \Lambda^k(S)$ for any $k \geq 0$. The element

$$\omega_a = H\,d\omega = \omega - \omega_e \tag{5-5.1}$$

of $\Lambda^k(S)$ is the *antiexact part* of ω.

The collection of all antiexact parts of $\Lambda^k(S)$ is denoted by $\mathscr{A}^k(S)$, for $k \geq 1$. Since $\mathscr{E}^0(S)$ is empty and any form satisfies $\omega = dH\omega + H\,d\omega = \omega_e + \omega_a$, we agree that any scalar function on S is its own antiexact part:

$$\mathscr{A}^0(S) = \Lambda^0(S). \tag{5-5.2}$$

It follows immediately from $\mathbf{H}_3$ and $\mathbf{H}_6$ that the antiexact part of any $\omega \in \Lambda^k(S)$, for $k \geq 1$, satisfies

$$\mathscr{X} \lrcorner \omega_a = 0, \qquad \omega_a(x_0^i) = 0. \tag{5-5.3}$$

Conversely, if $\alpha \in \Lambda^k(S)$ satisfies $\mathscr{X} \lrcorner \alpha = 0$, define an element ω of $\Lambda^k(S)$ by $\omega = d\beta + \alpha$ for $\beta \in \Lambda^{k-1}(S)$. Since $\widetilde{(\mathscr{X} \lrcorner \alpha)}(\lambda) = \tilde{\mathscr{X}}(\lambda) \lrcorner \tilde{\alpha}(\lambda) = \lambda \mathscr{X} \lrcorner \tilde{\alpha}(\lambda)$, (5-3.1) shows that $H\alpha = 0$. Allowing H to act on both sides of $\omega = d\beta + \alpha$ then gives $H\omega = H\,d\beta$ and we see that $\omega_e = dH\omega = dH\,d\beta = d\beta$ by $\mathbf{H}_4$. This in turn gives $\omega = \omega_e + \alpha$, from which we conclude that $\alpha = \omega_a$. However, if $\alpha = \omega_a$, then $\alpha = H\,d\omega$ and we must have $\alpha(x_0^i) = 0$ for $k > 0$. These considerations establish the following lemma.

Lemma 5-5.1. $\mathscr{A}^k(S) = \{\alpha \in \Lambda^k(S) | \mathscr{X} \lrcorner \alpha = 0,\ \alpha(x_0^i) = 0$ *for* $k > 0\}$.

Lemma 5-5.2. *The operator Hd maps $\Lambda^k(S)$ onto $\mathscr{A}^k(S)$ and $\mathscr{A}^n(S) = 0$ for S of dimension n.*

Proof. That Hd is into $\mathscr{A}^k(S)$ follows directly from $\omega_a = H\,d\omega$. On the other hand, $H\,d\omega_a = H\,dH\,d\omega = H\,d\omega = \omega_a$ by $\mathbf{H}_5$, and hence Hd is onto

$\mathscr{A}^k(S)$. The latter result follows on noting that $\Lambda^k \xrightarrow{d} \Lambda^{k+1} \xrightarrow{H} \Lambda^k$ and $\Lambda^{n+1}(S) = 0$ for S of dimension n. □

Definition. Any form $\alpha \in \Lambda^k(S)$ that satisfies $\mathscr{X} \lrcorner \alpha = 0$ and $\alpha(x_0^i) = 0$ for $k \geq 1$ is an *antiexact* form. $\mathscr{A}^k(S)$ is the vector space of antiexact forms of degree k on S, and the antiexact part of any form is an antiexact form.

We are now in possession of the necessary groundwork in order to establish the theorem underlying this discussion.

Theorem 5-5.1. *Antiexact forms possess the following properties*:

A$_1$. $\mathscr{A}^k(S) \subset \ker(H)$ *for all* k.
A$_2$. $\alpha \in \mathscr{A}^k(S)$, $\beta \in \mathscr{A}^m(S)$ *implies* $\alpha \wedge \beta \in \mathscr{A}^{k+m}(S)$.
A$_3$. $\mathscr{A}^k(S)$ *is a* C^∞*-module for* $k \geq 1$.
A$_4$. H *is the inverse of* d *on* $\mathscr{A}^k(S)$ *for* $k \geq 1$.

Proof. Property **A$_1$** follows directly from Lemma 5-4.3 since any ω can be written as $\omega = \omega_e + \omega_a$ and $\omega - \omega_e$ belongs to $\ker(H)$. Clearly, **A$_2$** holds for elements of $\mathscr{A}^0(S) = \Lambda^0(S)$, and hence it suffices to establish **A$_2$** for $\max(k, m) \geq 1$. Under the hypotheses of **A$_2$**, we know that $\alpha \wedge \beta \in \Lambda^{k+m}(S)$. Since $\max(k, m) \geq 1$, Lemma 5-5.1 shows that $(\alpha \wedge \beta)(x_0^i) = 0$ because at least one of the factors vanishes at (x_0^i). Further, $\mathscr{X} \lrcorner(\alpha \wedge \beta) = (\mathscr{X} \lrcorner \alpha) \wedge \beta + (-1)^k \alpha \wedge (\mathscr{X} \lrcorner \beta) = 0$ since $\mathscr{X} \lrcorner \alpha = 0$ and $\mathscr{X} \lrcorner \beta = 0$ by Lemma 5-5.1. Thus $\mathscr{X} \lrcorner(\alpha \wedge \beta) = 0$, $(\alpha \wedge \beta)(x_0^i) = 0$ and Lemma 5-5.1 implies $\alpha \wedge \beta \in \mathscr{A}^{k+m}(S)$. **A$_3$** then follows directly from **A$_2$** since the set $\mathscr{A}^k(S)$ is closed under addition and under exterior multiplication by all elements of $\mathscr{A}^0(S) = \Lambda^0(S)$. Property **H$_2$** gives for $\omega \in \Lambda^k(S)$, $k \geq 1$, $\omega = H\,d\omega + dH\omega$. However, **A$_1$** shows that $H\omega = 0$ for $\omega \in \mathscr{A}^k(S)$, and hence $\omega = H\,d\omega$ for $\omega \in \mathscr{A}^k(S)$ with $k \geq 1$. □

Remark. For forms of degree zero, property **H$_2$** gives $f(x^i) = f(x_0^i) + (H\,df)(x^i)$. Accordingly, the operator H can be used to invert the operator d on $\mathscr{A}^k(S)$ for all values of k.

Corollary 5-5.1. *Let* $\mathscr{A}(S)$ *be the graded associative algebra of antiexact forms with the operations of addition and exterior multiplication taken from* $\Lambda(S)$, *then* $\mathscr{A}(S)$ *is a subalgebra of* $\Lambda(S)$ *that is given by* $\mathscr{A}(S) = H\,d(\Lambda(S))$, *and* $\mathscr{A}$ *is the Hd-projection of* Λ.

Proof. The result is a direct consequence of Theorem 5-5.1 and Lemma 5-5.2: the latter showing that Hd maps $\Lambda(S)$ onto $\mathscr{A}(S)$. That Hd is a projection follows from property **H$_5$** and Lemma 5-5.2. □

Corollary 5-5.2. *The linear operator* H *maps* $\Lambda^{k+1}(S)$ *onto* $\mathscr{A}^k(S)$ *for* $k \geq 1$ *and hence we have* $\mathscr{A}^k(S) = H(\Lambda^{k+1}(S))$ *for* $k > 0$.

Proof. If $\omega \in \Lambda^{k+1}(S)$, then $H\omega \in \Lambda^k(S)$ by $\mathbf{H}_1$. Now, $(H\omega)(x_0^i) = 0$ by $\mathbf{H}_3$ and $\mathscr{X}\lrcorner H\omega = 0$ by $\mathbf{H}_6$. Thus, $H\omega \in \mathscr{A}^k(S)$ by Lemma 5-5.1. Conversely, if $\alpha \in \mathscr{A}^k(S)$, then $\alpha = H\,d\alpha$ by $\mathbf{A}_4$. There thus exists a $\beta = d\alpha \in \Lambda^{k+1}$ such that $\alpha = H\beta$ and $H(\Lambda^{k+1}(S))$ covers $\mathscr{A}^k(S)$. In contrast with Corollary 5-5.1, we note that H is not a projection because property $\mathbf{H}_3$ gives $HH\omega = 0$ for all $\omega \in \Lambda(S)$. □

The fact that $\mathscr{A}(S)$ is a submodule of $\Lambda(S)$ is central in the applicability of antiexact exterior forms. This fact seems surprising on first reading since Corollary 5-5.1 shows that there exists a $\gamma \in \Lambda(E_n)$ for each pair α, β of $\Lambda(E_n)$ such that $H\,d\gamma = H\,d\alpha \wedge H\,d\beta$ while it is an elementary exercise to see that $H(d\alpha \wedge d\beta) = H\,d(\alpha \wedge d\beta) \neq H\,d\alpha \wedge H\,d\beta$. In fact, the definition of the operator H shows that $H(\rho \wedge \eta)$ is not expressible in terms of $H\rho$, $H\eta$ and the operations $(\wedge, +, d)$. The following lemma provides the resolution of this seeming paradox.

Lemma 5-5.3. *If* $\alpha \in \mathscr{A}^k(S)$, *then there exists an* $\hat{\alpha} \in \Lambda^{k+1}(S)$ *such that*

$$\alpha = \mathscr{X}\lrcorner\hat{\alpha}. \tag{5-5.4}$$

Proof. Corollary 5-5.2 shows that there exists a $\beta \in \Lambda^{k+1}(S)$ for a given $\alpha \in \mathscr{A}^k(S)$ such that $\alpha = H\beta$. Since β is the sum of the basis elements of $\Lambda^{k+1}(S)$ with coefficients from $\Lambda^0(S)$, both $\Lambda^{k+1}(S)$ and $\mathscr{A}^k(S)$ are closed under addition, and $H(\rho + \eta) = H\rho + H\eta$, it is sufficient to establish the result for a monomial of the form $f(x^j)\,dx^{i_1} \wedge dx^{i_2} \wedge \cdots \wedge dx^{i_{k+1}}$. We now simply observe that

$$\begin{aligned} H\big(f(x^j)\,dx^{i_1} \cdots dx^{i_{k+1}}\big) &= \int_0^1 \lambda^k f\big(x_0^j + \lambda(x^j - x_0^j)\big)\,d\lambda\,\mathscr{X}\lrcorner\big(dx^{i_1} \wedge \cdots \wedge dx^{i_{k+1}}\big) \\ &= \mathscr{X}\lrcorner\big(\hat{f}(x^j)\,dx^{i_1} \wedge \cdots \wedge dx^{i_{k+1}}\big), \end{aligned}$$

where we have set

$$\hat{f}(x^j) = \int_0^1 \lambda^k f\big(x_0^j + \lambda(x^j - x_0^j)\big)\,d\lambda.$$

Thus since (5-5.4) gives $\mathscr{X}\lrcorner\alpha = 0$ and $\alpha|_{x=x_0} = 0$ because $\mathscr{X}|_{x=x_0} = 0$, the result is established. □

If α and β are antiexact, Lemma 5-5.3 gives

$$\alpha = \mathscr{X}\lrcorner\hat{\alpha}, \qquad \beta = \mathscr{X}\lrcorner\hat{\beta}$$

and hence

$$\alpha \wedge \beta = (\mathscr{X} \lrcorner \hat{\alpha}) \wedge (\mathscr{X} \lrcorner \hat{\beta}) = \mathscr{X} \lrcorner (\hat{\alpha} \wedge \mathscr{X} \lrcorner \hat{\beta}).$$

Thus since

$$Hd(\mathscr{X} \lrcorner (\hat{\alpha} \wedge \mathscr{X} \lrcorner \hat{\beta}) = \mathscr{X} \lrcorner (\hat{\alpha} \wedge \mathscr{X} \lrcorner \hat{\beta}) - dH(\mathscr{X} \lrcorner (\hat{\alpha} \wedge \mathscr{X} \lrcorner \hat{\beta})$$

$$= \mathscr{X} \lrcorner (\hat{\alpha} \wedge \mathscr{X} \lrcorner \hat{\beta}),$$

$$Hd(\mathscr{X} \lrcorner \hat{\alpha}) = \mathscr{X} \lrcorner \hat{\alpha} - dH(\mathscr{X} \lrcorner \hat{\alpha}) = \mathscr{X} \lrcorner \hat{\alpha},$$

we have

$$\alpha \wedge \beta = Hd(\mathscr{X} \lrcorner \hat{\alpha}) \wedge Hd(\mathscr{X} \lrcorner \hat{\beta}) = (\mathscr{X} \lrcorner \hat{\alpha}) \wedge (\mathscr{X} \lrcorner \hat{\beta})$$

$$= \mathscr{X} \lrcorner (\hat{\alpha} \wedge \mathscr{X} \lrcorner \hat{\beta}) = Hd(\mathscr{X} \lrcorner (\hat{\alpha} \wedge \mathscr{X} \lrcorner \hat{\beta})) = Hd\gamma$$

with $\gamma = \mathscr{X} \lrcorner (\hat{\alpha} \wedge \mathscr{X} \lrcorner \hat{\beta}) \in \mathscr{A}(S)$. Further if ρ and η are elements of $\mathscr{A}(S)$, Lemma 5-5.3 gives

$$H\rho = \mathscr{X} \lrcorner \alpha, \qquad H\eta = \mathscr{X} \lrcorner \beta$$

for some α and $\beta \in \Lambda(S)$, in which case we have

$$H\rho \wedge H\eta = (\mathscr{X} \lrcorner \alpha) \wedge (\mathscr{X} \lrcorner \beta) = \mathscr{X} \lrcorner (\alpha \wedge \mathscr{X} \lrcorner \beta) = H\gamma.$$

Since H annihilates the antiexact part of γ it is sufficient to exhibit an exact γ for which the above relation holds:

$$\gamma = d(\mathscr{X} \lrcorner (\alpha \wedge \mathscr{X} \lrcorner \beta)).$$

5-6. REPRESENTATIONS

Property $\mathbf{H_2}$ has shown us that any $\omega \in \Lambda^k(S)$ with $k \geq 1$ can be written in the form $\omega = dH\omega + Hd\omega$. Thus since $dH\omega = \omega_e \in \mathscr{E}^k(S)$ and $Hd\omega = \omega_a \in \mathscr{A}^k(S)$, we obtain the representation $\omega = \alpha + \beta$ with $\alpha \in \mathscr{E}^k(S)$ and $\beta \in \mathscr{A}^k(S)$. It would thus appear that we have to work with the two sets $\mathscr{E}^k(S) = dH(\Lambda^k(S))$ and $\mathscr{A}^k(S) = Hd(\Lambda^k(S))$ in order to represent any $\omega \in \Lambda^k(S)$. It turns out, however, that any element of $\Lambda^k(S)$ can be reconstructed from elements of $\mathscr{A}(S)$ alone, as we now proceed to show.

Theorem 5-6.1. *Any $\omega \in \Lambda^k(S)$, $k \geq 1$, has the unique representation $\omega = d\alpha_1 + \alpha_2$ under the conditions $\alpha_1 \in \mathscr{A}^{k-1}(S)$, $\alpha_2 \in \mathscr{A}^k(s)$. If these conditions* *are satisfied, then $\alpha_2 = H(d\omega)$ is unique, while $\alpha_1 = H(\omega)$ is unique for $k > 1$ and $\alpha_1 = H(\omega) +$ constant for $k = 1$.*

Proof. Property $\mathbf{H}_3$ gives $\omega = dH\omega + Hd\omega$ for any $\omega \in \Lambda^k(S)$ with $k \geq 1$. Corollary 5-5.2 shows that $H\omega \in \mathscr{A}^{k-1}(S)$ and $Hd\omega \in \mathscr{A}^k(S)$. Thus, $\omega \in \Lambda^k(S)$ admits the representation $\omega = d\alpha_1 + \alpha_2$ with $\alpha_1 = H\omega \in \mathscr{A}^{k-1}(S)$ and $\alpha_2 = Hd\omega \in \mathscr{A}^k(S)$. It remains only to establish uniqueness under the conditions $\alpha_1 \in \mathscr{A}^{k-1}(S)$, $\alpha_2 \in \mathscr{A}^k(S)$. Suppose therefore that $\omega = d\alpha_1 + \alpha_2 = d\beta_1 + \beta_2$ with α_1 and β_1 in $\mathscr{A}^{k-1}(S)$ and α_2 and β_2 in $\mathscr{A}^k(S)$. This gives $d(\alpha_1 - \beta_1) = \beta_2 - \alpha_2$ and $0 = d(\beta_2 - \alpha_2)$. Since H inverts d on $\mathscr{A}^{k+1}(S)$, we have $\beta_2 - \alpha_2 = Hd(\beta_2 - \alpha_2) = 0$. Thus $d(\alpha_1 - \beta_1) = \beta_2 - \alpha_2 = 0$, and the Poincaré lemma yields $\alpha_1 - \beta_1 = d\rho$ for some $\rho \in \Lambda^{k-2}(S)$. Since $\alpha_1 - \beta_1 \subset \ker(H)$ by $\mathbf{A}_1$, allowing H to act on both sides of $\alpha_1 - \beta_1 = d\rho$ gives $0 = Hd\rho$. However, $\rho = dH\rho + Hd\rho$ by $\mathbf{H}_2$, and so we conclude that $\rho = dH\rho$. This in turn gives $\alpha_1 - \beta_1 = d^2H\rho = 0$, and uniqueness is established. □

This result can be paraphrased in the following direct form:

(5-6.1) $$\Lambda^k(S) = d(\mathscr{A}^{k-1}(S)) \oplus \mathscr{A}^k(S), \qquad k \geq 1.$$

Similarly, $\mathbf{H}_2$ can be used to obtain the direct sum representation

(5-6.2) $$\Lambda^k(S) = \mathscr{E}^k(S) \oplus \mathscr{A}^k(S),$$

although this will turn out to be of significantly less use than (5-6.1). In this context it is of interest to note that Corollary 5-5.2 follows from (5-6.1) by direct calculation:

$$H(\Lambda^{k+1}(S)) = H(d\mathscr{A}^k(S) + \mathscr{A}^{k+1}(S)) = Hd(\mathscr{A}^k(S)) = \mathscr{A}^k(S)$$

since H inverts d on $\mathscr{A}^k(S)$. The following chart summarizes these findings in a convenient format:

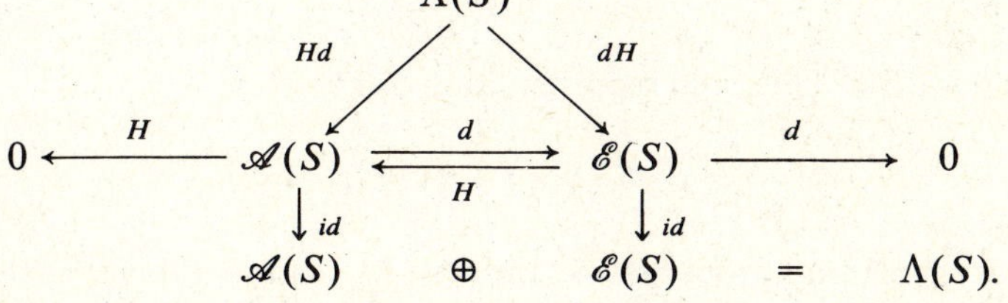

Corollary 5-6.1. *The decomposition*

(5-6.3) $$\omega = \omega_e + \omega_a, \qquad \omega_e = dH(\omega), \qquad \omega_a = H(d\omega)$$

of any form into the sum of an exact part and an antiexact part is unique.

Proof. The result follows directly from Theorem 5-6.1 for $\omega = d\alpha_1 + \alpha_2$ with $\alpha_1 = H(\omega)$, $\alpha_2 = H(d\omega)$ yields $\omega_e = d\alpha_1 = dH(\omega)$ and $\alpha_2 \in \mathscr{A}(S)$. □

Corollary 5-6.2. *An exterior form vanishes if and only if its exact part and its antiexact part vanish separately.*

Proof. Clearly, $\omega_e = 0$, $\omega_a = 0$ give $\omega = \omega_e + \omega_a = 0$. Conversely, $\omega = 0$ implies $H(\omega) = 0$ and $H(d\omega) = 0$ so that $\omega_e = dH(\omega) = 0$ and $\omega_a = H(d\omega) = 0$. □

Corollary 5-6.3. *We have*

(5-6.4) $$\mathscr{E}^k(S) \cap \mathscr{A}^k(S) = 0\text{-element of } \Lambda^k(S).$$

Proof. If $\omega = dH(\rho)$ then $\omega_a = H\,d^2H(\rho) = 0$, and if $\omega = H(d\rho)$ then $\omega_e = dH^2(d\omega) = 0$. Thus $\mathscr{E}^k(S)$ and $\mathscr{A}^k(S)$ have only the 0-element of $\Lambda^k(S)$ in common. □

Consider the case where

$$\omega = 2x\,dx - y^2\,dz$$

on E_3 and take $\{x_0^i\} = \{0, 0, 0\}$. We then have

$$H(\omega) = \int_0^1 (2\lambda x^2 - y^2\lambda^2 z)\,d\lambda = x^2 - \tfrac{1}{3}y^2 z$$

with $H(\omega) \in \mathscr{A}^0(E_3)$ and

$$\omega_e = dH(\omega) = d\left(x^2 - \tfrac{1}{3}y^2 z\right).$$

Use of this result and $\omega_a = \omega - \omega_e$ yields

$$\omega_a = \tfrac{2}{3}y(z\,dy - y\,dz)$$

with $\omega_a \in \mathscr{A}^1(E_3)$. Thus, the representation 5-6.1 gives

$$\omega = d\left(x^2 - \tfrac{1}{3}y^2 z\right) + \tfrac{2}{3}y(z\,dy - y\,dz).$$

Corollary 5-6.1 and properties $\mathbf{A}_2$ through $\mathbf{A}_4$ lead to a direct generalization of the process $\int_0^z f(y)\,dy$ of definite integration to forms and antiexact forms. In order to see this, we note that (5-6.3) yields

$$\omega(x) = d(H\omega)(x) + H(d\omega)(x),$$

for any $\omega \in \Lambda(S)$, and the term $d(H\omega)(x)$ plays the role of a "constant of

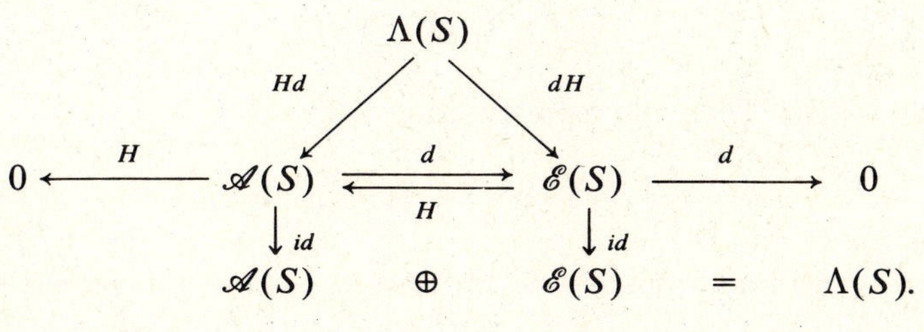

integration" since $d(d(H\omega)(x)) \equiv 0$. There is also a direct analogue of *integration by parts* as the following lemma shows.

Lemma 5-6.1. *If* $\alpha \in \mathscr{A}^k(S)$ *and* $\beta \in \mathscr{A}^m(S)$, *then*

$$H(d\alpha \wedge \beta) = \alpha \wedge \beta + (-1)^{k+1} H(\alpha \wedge d\beta). \tag{5-6.5}$$

Proof. Since $\alpha \in \mathscr{A}^k(S)$, $\beta \in \mathscr{A}^m(S)$, we have $\alpha \wedge \beta \in \mathscr{A}^{k+m}(S)$ by $\mathbf{A}_2$. Property $\mathbf{A}_4$ thus gives $\alpha \wedge \beta = Hd(\alpha \wedge \beta)$, and this leads directly to (5-6.5). □

As one might anticipate, use of H to invert d on $\mathscr{A}(S)$, and the integration by parts formula (5-6.5) provide a natural and direct access to the study of exterior differential equations. We take up this topic beginning at Section 5-9.

5-7. CHANGE OF CENTER

If S is a starshaped region with respect to the point $P \in S$, it is convenient to refer to P as a *center* of S. The homotopy operator H that is constructed on S by use of the center P will now be referred to as the *homotopy operator with center P*. Suppose that S is starshaped with respect to several points P_1, P_2, etc. We can then construct homotopy operators $H_1, H_2, \ldots,$ on S with centers $P_1, P_2, \ldots,$ respectively. The question thus arises as to the relation between homotopy operators on S with different centers.

Let P_1 and P_2 be two centers of a starshaped region S, and let H_1 and H_2 be the homotopy operators on S with centers P_1 and P_2, respectively. If ω is any element of $\Lambda^k(S)$, property $\mathbf{H}_2$ gives

$$\omega = dH_1\omega + H_1 d\omega = dH_2\omega + H_2 d\omega,$$

and we obtain the relations

$$d(H_1 - H_2)\omega = (H_2 - H_1)\, d\omega, \tag{5-7.1}$$

$$d(H_2 - H_1)(d\omega) = 0. \tag{5-7.2}$$

Since the Poincaré lemma holds in S, (5-7.2) yields

$$(H_2 - H_1)\, d\omega = d\beta \tag{5-7.3}$$

for some $\beta \in \Lambda^{k-1}(S)$. If we restrict β to belong to $\mathscr{A}_2^{k-1}(S)$, then allowing H_2 to act on both sides of (5-7.3) yields $-H_2H_1\, d\omega = H_2\, d\beta = \beta$ since H_2 inverts d on $\mathscr{A}_2(S)$. When this result together with (5-7.3) is substituted back into (5-7.1), we obtain $d(H_1 - H_2)\omega = d\beta$, and the Poincaré lemma gives

$$(H_1 - H_2)\omega = \beta + d\alpha \tag{5-7.4}$$

for some $\alpha \in \Lambda^{k-2}(S)$. Restricting α to $\mathscr{A}_2^{k-2}(S)$, application of H_2 to both sides of (5-7.4) gives $H_2 H_1 \omega = \alpha$. We thus have

$$H_1\omega = H_2\omega + \beta + d\alpha \tag{5-7.5}$$

with

$$\beta = -H_2 H_1 \, d\omega \in \mathscr{A}_2^{k-1}(S), \qquad \alpha = H_2 H_1 \omega \in \mathscr{A}_2^{k-2}(S). \tag{5-7.6}$$

It now only remains to show that the restriction of α and β to $\mathscr{A}_2(S)$ is inconsequential. For this result, it is sufficient to substitute (5-7.6) into (5-7.5) and to use the fact that $dH_2 = -H_2 d + id$ to obtain

$$\begin{aligned} H_2\omega + \beta + d\alpha &= H_2\omega - H_2 H_1 \, d\omega + dH_2 \, H_1\omega \\ &= H_2\omega - H_2 H_1 \, d\omega + H_1\omega - H_2 \, dH_1 \, \omega \\ &= H_1\omega + H_2(\omega - H_1 \, d\omega - dH_1 \, \omega) = H_1\omega. \end{aligned}$$

These considerations establish the following result.

Theorem 5-7.1. *Let S be a starshaped region and let H_1 and H_2 be two homotopy operators on S with centers P_1 and P_2, respectively. The operators H_1 and H_2 then stand in the relation*

$$H_1\omega = H_2\omega + \beta + d\alpha,$$

with α and $\beta = -H_2 H_1 \, d\omega$ belonging to $\mathscr{A}_2(S)$, and $\alpha =$ constant if $\omega \in \Lambda^1(S)$, $\alpha = H_2 H_1 \omega$ if $\omega \in \Lambda^k(S)$, $k > 1$.

This result is important, for it allows us to shift from any one center of S to any other center of S without essential changes in the results. Granted, for $\omega = dH_1 \, \omega + H_1 \, d\omega$ and $\omega = dH_2 \, \omega + H_2 \, d\omega$, the exact element $dH_1 \, \omega$ changes to the exact element $dH_2 \omega$ and the antiexact element $H_1 \, d\omega$ changes to the antiexact element $H_2 \, d\omega$ on change from the center P_1 to the center P_2, but the representation in terms of elements of $\mathscr{A}_1(S)$ goes over exactly into the same representation in terms of elements of $\mathscr{A}_2(S)$ and H_2 inverts d on $\mathscr{A}_2(S)$ for exactly the same reasons that H_1 inverts d on $\mathscr{A}_1(S)$.

For example, with $\omega = 2x \, dx - y^2 \, dz$, we can construct H_1 with center $(0, 0, 0)$ and H_2 with center $(1, 0, 1)$. We then have

$$H_1\omega = x^2 - \tfrac{1}{3}y^2 z, \qquad H_2\omega = x^2 - \tfrac{1}{3}y^2 z - 1 + \tfrac{1}{3}y^2.$$

However, $H_1\, d\omega = \frac{2}{3}y(z\,dy - y\,dz)$, and hence

$$\beta = -H_2 H_1 \omega = -\tfrac{1}{3}y^2.$$

A combination of these results shows that

$$H_1\omega = H_2\omega + \beta + 1,$$

and hence $\alpha = 1 =$ constant is admissible since $\omega \in \Lambda^1(E_3)$.

5-8. BEHAVIOR UNDER MAPPINGS

Let S be a starshaped region of an m-dimensional space E_m with center P and with homotopy operator H with center P. Let Φ be a differentiable mapping from E_m to a space E_n of dimension $n \geq m$, that is regular on S. Finally, let $\Lambda(\Phi(S))$ denote the restriction of $\Lambda(E_n)$ to the range of Φ in the sense that $\Lambda(\Phi(S)) = \Phi^{-1*}\Lambda(S)$, where Φ^{-1} is the relation inverse of Φ restricted to the range of Φ. If E_m and E_n have the same dimension m, then Φ^{-1} is the mapping inverse of Φ since Φ is assumed to be regular on S (i.e., Φ^* maps at least one nonzero m-form onto a nonzero m-form).

If α is any k-form on $\Phi(S)$, then $\Phi^*\alpha$ is a k-form on S and $H\Phi^*\alpha$ belongs to $\mathscr{A}(S)$. Since Φ^{-1*} maps elements of $\Lambda(S)$ onto elements of $\Lambda(\Phi(S))$, the quantity $\Phi^{-1*}H\Phi^*\alpha$ belongs to $\Lambda(\Phi(S))$. Thus if we restrict α to $\Lambda(\Phi(S))$, then $\Phi^{-1*}H\Phi^* = H^*$ induces a linear map of $\Lambda(\Phi(S))$ to $\Lambda(\Phi(S))$. The linear operator H^* obtained in this way is referred to as the *homotopy operator induced by* Φ on $\Lambda(\Phi(S))$.

Theorem 5-8.1. *Let Φ be a map of a space E_m into a space E_n with $m \leq n$, and let Φ be regular on a starshaped region S of E_m with center P and homotopy operator H. Then $\Phi(S)$ is a starshaped region in $\Phi(E_m)$ with center $\Phi(P)$, Φ induces the homotopy operator H^* on $\Lambda(\Phi(S))$ that is defined by*

$$\Phi^*H^* = H\Phi^*, \tag{5-8.1}$$

and

$$dH^* + H^*d = \text{identity} \tag{5-8.2}$$

on $\Lambda(\Phi(S))$.

Proof. Since Φ is regular on S, $\Phi(S)$ is a regular region of E_n of dimension equal to that of E_m. The preferred coordinate system on S can thus be lifted by Φ to a coordinate cover of $\Phi(S)$ from which we conclude that $\Phi(S)$ is starshaped with center $\Phi(P)$. If we cut down $\Lambda(E_n)$ to $\Lambda(\Phi(S))$, then the invertibility of Φ on $\Phi(S)$ yields a well defined map Φ^{-1*} of $\Lambda(S)$ onto

$\Lambda(\Phi(S))$. Thus pulling (5-8.1) back to $\Lambda(\Phi(S))$ by Φ^{-1*} yields the operator $H^* = \Phi^{-1*}H\Phi^*$ that we obtained previously. It remains only to verify the relation (5-8.2). If $\alpha \in \Lambda(\Phi(S))$, then $\beta = \Phi^*\alpha \in \Lambda(S)$ and $\mathbf{H_2}$ gives $\Phi^*\alpha = dH\beta + H\,d\beta = dH\Phi^*\alpha + H\,d\Phi^*\alpha = dH\Phi^*\alpha + H\Phi^*\,d\alpha$. Thus (5-8.1) yields $\Phi^*\alpha = \Phi^*(dH^*\alpha + H^*d\alpha)$. Since all forms that occur belong to $\Lambda(\Phi(S))$, action by Φ^{-1*} yields (5-8.2). □

This theorem is of particular importance if we allow Φ to be a map of E_m to E_m. In this event every image of a starshaped region S in E_m under a regular map of E_m to E_m is a starshaped region and Φ induces a homotopy operator H^* on the image of S. Since such maps Φ can also be thought of as inducing changes in the coordinate cover of the underlying coordinate neighborhood U that contains S if Φ is regular on U, the relation $\Phi^*H^* = H\Phi^*$ gives a complete accounting of how homotopy operators behave under coordinate maps. It is also clear that $H_2 = \Phi^{-1*}H_1\Phi^*$ yields the change of center formula if $\Phi: S \to S$ is the translation that maps the center P_1 onto the center P_2.

Consider the map

$$\Phi: E_2 \to E_3 | x = 1 + \alpha, \qquad y = \alpha + e^{-\beta}, \qquad z = e^{\beta},$$

then $\Phi(E_2)$ is the surface $y = x + 1/z - 1$, $z > 0$ in E_3 since $\Phi^{-1}(\Phi(E_2))$ is given by $\alpha = x - 1$, $\beta = \ln z$. If $\omega = 2x\,dx - y\,dz$, then

$$\Phi^*\omega = 2(1 + \alpha)\,d\alpha - (\alpha e^{\beta} + 1)\,d\beta$$

is well defined on E_2. Let H be the homotopy operator on E_2 with center $(0, 0)$, then

$$H\Phi^*\omega = \int_0^1 \{2(1 + \lambda\alpha)\alpha - (\lambda\alpha e^{\lambda\beta} + 1)\beta\}\,d\lambda$$

$$= \alpha(2 + \alpha) - \frac{\alpha}{\beta}(\beta e^{\beta} - e^{\beta} + 1) - \beta.$$

However, ω restricted to $\Phi(E_2)$ (i.e., $y = x + 1/z - 1$) gives

$$\omega|_{\Phi(E_2)} = 2x\,dx - \left(x + \frac{1}{z} - 1\right)dz$$

while

$$H^*\omega|_{\Phi(E_2)} = \Phi^{-1*}(H\Phi^*\omega) = (x - 1)(x + 1) - \ln z$$

$$- \frac{(x - 1)}{\ln z}\{z \ln z - z + 1\}.$$

We note, in this regard, that $\Phi:(0,0) \to (1,1,1)$, so that the center of H^* on $y = x + \dfrac{1}{z} - 1$, $z > 0$ is $(1,1,1)$.

It is essential in these considerations to note that *the homotopy operator H on $\Lambda(E_n)$ does not commute with restrictions to subregions*. Thus for example, if $\Phi: E_{n-1} \to E_n$ is regular, and the center, P, of H on E_n is not contained in $\Phi(E_{n-1})$, then $H(\omega|_{\Phi(E_{n-1})})$ is not well defined. Here $\omega|_{\Phi(E_{n-1})}$ denotes the restriction of $\omega \in \Lambda(E_n)$ to the hypersurface $\Phi(E_{n-1})$.

5-9. AN INTRODUCTORY PROBLEM

The purpose of this section is to show how the homotopy operator, H, and the unique decomposition of forms into sums of exact and antiexact parts allow us to obtain explicit solutions of exterior differential equations. To this end, we confine attention to the simplest problem in exterior differential equations: Find all k-forms Ω on S that verify

$$d\Omega = \Gamma \wedge \Omega + \Sigma$$

for $\Gamma \in \Lambda^1(S)$ and $\Sigma \in \Lambda^{k+1}(S)$.

Clearly, Γ and Σ cannot be assigned arbitrarily if there is to be a solution of $d\Omega = \Gamma \wedge \Omega + \Sigma$, for the closure condition, $d^2\Omega = 0$, of this equation must be satisfied. Exterior differentiation and elimination give us the complete resulting system

$$d\Omega = \Gamma \wedge \Omega + \Sigma, \tag{5-9.1}$$

$$d\Sigma = \Gamma \wedge \Sigma - \Theta \wedge \Omega, \tag{5-9.2}$$

$$d\Gamma = \Theta, \qquad d\Theta = 0. \tag{5-9.3}$$

We thus have the problem of finding all $\Omega \in \Lambda^k(S)$, $\Gamma \in \Lambda^1(S)$, $\Sigma \in \Lambda^{k+1}(S)$, $\Theta \in \Lambda^2(S)$ that satisfy the system (5-9.1) through (5-9.3).

The system consisting of the two exterior equations (5-9.3) is particularly simple, so we start with it. Since the Poincaré lemma holds on S, $d\Theta = 0$ integrates to give

$$\Theta = d\theta \tag{5-9.4}$$

and property $\mathbf{H_2}$ and the first of (5-9.3) give

$$\theta = H(\Theta) = H(d\Gamma), \tag{5-9.5}$$

so that $\theta \in \mathscr{A}^1(S)$. Use of property $\mathbf{H_2}$ and the first of (5-9.3) shows that

$\Gamma = dH(\Gamma) + H(d\Gamma) = dH(\Gamma) + \theta$. We thus have

$$\Gamma = d\gamma + \theta, \tag{5-9.6}$$

with

$$\gamma = H(\Gamma), \qquad \theta = H(d\Gamma) \tag{5-9.7}$$

belonging to $\mathscr{A}(S)$.

We now turn to the system (5-9.1), (5-9.2) and make the similarity transformations

$$\Omega = e^{\rho}\omega, \qquad \Sigma = e^{\rho}\sigma \tag{5-9.8}$$

with $\rho \in \Lambda^0(S)$. When (5-9.6) is used, this gives

$$d\omega = (d\gamma - d\rho + \theta) \wedge \omega + \sigma,$$

$$d\sigma = (d\gamma - d\rho + \theta) \wedge \sigma - d\theta \wedge \omega.$$

If ρ is given by

$$\rho = \gamma = H(\Gamma), \tag{5-9.9}$$

then the system reduces to

$$d\omega = \theta \wedge \omega + \sigma, \qquad d\sigma = \theta \wedge \sigma - d\theta \wedge \omega.$$

The further substitution

$$\beta = \theta \wedge \omega, \tag{5-9.10}$$

yields the final reduced system

$$d\omega = \beta + \sigma, \qquad d\sigma = -d\beta \tag{5-9.11}$$

since $d\beta = d\theta \wedge \omega - \theta \wedge d\omega = d\theta \wedge \omega - \theta \wedge (\theta \wedge \omega + \sigma) = d\theta \wedge \omega - \theta \wedge \sigma$.

The integration of the system (5-9.11) is direct since property $\mathbf{H}_2$ gives $\omega = dH(\omega) + H(d\omega)$ and $\sigma = dH(\sigma) + H(d\sigma)$. Thus $\sigma = dH(\sigma) - H(d\beta) = dH(\sigma) - \beta + dH(\beta)$, since $H(d\beta) + dH(\beta) = \beta$, and $\omega = dH(\omega) + H(\beta + \sigma)$. Accordingly if we define $\eta \in \mathscr{A}^k(S)$ and $\phi \in \mathscr{A}^{k-1}(S)$ by

$$\eta = H(\sigma), \qquad \phi = H(\omega), \tag{5-9.12}$$

we obtain

$$\sigma = d\eta - \beta + dH(\beta), \qquad \omega = d\phi + H(\beta) + \eta \tag{5-9.13}$$

because $H(\sigma) = H(d\eta - \beta + dH(\beta)) = H(d\eta - H(d\beta)) = H\,d\eta = \eta$, where the last equality follows from the fact that H inverts d on $\mathscr{A}(S)$ and $\eta \in \mathscr{A}^k(S)$.

It now remains only to eliminate the subsidiary form β. Substituting the second of (5-9.13) into (5-9.10) gives us

$$\beta = \theta \wedge d\phi + \theta \wedge H(\beta) + \theta \wedge \eta,$$

and hence

$$H(\beta) = H(\theta \wedge d\phi + \theta \wedge H(\beta) + \theta \wedge \eta).$$

However $\theta \in \mathscr{A}(S)$, $\eta \in \mathscr{A}(S)$ imply $\theta \wedge H(\beta) \in \mathscr{A}(S)$, $\theta \wedge \eta \in \mathscr{A}(S)$ by property $\mathbf{A}_2$ and hence property $\mathbf{A}_1$ gives $H(\theta \wedge H(\beta) + \theta \wedge \eta) = 0$. We thus obtain

$$\beta = \theta \wedge d\phi + \theta \wedge H(\theta \wedge d\phi) + \theta \wedge \eta, \qquad H(\beta) = H(\theta \wedge d\phi),$$

and (5-9.13) give

$$\omega = d\phi + \eta + H(\theta \wedge d\phi),$$

$$\sigma = d\eta - \theta \wedge \eta - \theta \wedge H(\theta \wedge d\phi) - Hd(\theta \wedge d\phi).$$

Thus since $\theta \wedge d\phi = dH(\theta \wedge d\phi) + Hd(\theta \wedge d\phi)$, σ can be written in the equivalent form

$$\begin{aligned}\sigma &= d\eta - \theta \wedge \eta - \theta \wedge H(\theta \wedge d\phi) - \theta \wedge d\phi + dH(\theta \wedge d\phi)\\ &= d(d\phi + \eta + H(\theta \wedge d\phi)) - \theta \wedge (d\phi + \eta + H(\theta \wedge d\phi)).\end{aligned}$$

It is then a simple but lengthy calculation to verify that substituting ω and σ back into (5-9.8) together with $\rho = \gamma$ and (5-9.6) lead to the identical satisfaction of the system (5-9.1) through (5-9.3) for any γ and ϕ and any antiexact forms θ and η. We have thus established the following result.

Theorem 5-9.1. *The general solution of the differential system*

(5-9.14) $$d\Omega = \Gamma \wedge \Omega + \Sigma, \qquad d\Sigma = \Gamma \wedge \Sigma - \Theta \wedge \Omega,$$

(5-9.15) $$d\Gamma = \Theta, \qquad d\Theta = 0,$$

on a starshaped region S is given by

(5-9.16) $$\Omega = e^{\gamma}\{d\phi + \eta + H(\theta \wedge d\phi)\},$$

(5-9.17) $$\Sigma = e^{\gamma}\{d(\eta + H(\theta \wedge d\phi)) - \theta \wedge (d\phi + \eta + H(\theta \wedge d\phi))\},$$

(5-9.18) $$\Gamma = d\gamma + \theta, \qquad \Theta = d\theta$$

for any choice of the scalar-valued function γ and any choice of the antiexact *forms* ϕ, η, θ, *in which case*

$$\gamma = H(\Gamma), \qquad \phi = H(e^{-\gamma}\Omega), \qquad \eta = H(e^{-\gamma}\Sigma), \qquad \theta = H\,d\Gamma. \tag{5-9.19}$$

We noted at the beginning of this section that Γ and Σ that occur in $d\Omega = \Gamma \wedge \Omega + \Sigma$ could not be arbitrary in view of the necessity of satisfying the integrability conditions (5-9.2) and (5-9.3). This fact is clearly borne out by Theorem 5-9.1; simply note that Ω, Γ and Σ depend on the quantities γ, ϕ, η and θ and their derivatives in an essentially inseparable manner. Thus since $(\gamma, \phi, \eta, \theta)$ serve to parameterize the general solution, a change in any one of these functions or forms changes the quantities Ω, Σ, Γ, Θ in an intrinsic way. It should be noted, however, that we may independently specify Γ and then determine $\gamma = H\Gamma$ and $\theta = H\,d\Gamma$. The set of all compatible Ω, Σ are then determined by (5-9.16) and (5-9.17) in terms of a choice of the quantities ϕ and η.

It is of interest to inquire as to what amount of information is required in order to determine the antiexact $(k-1)$-form ϕ. Since ϕ is given by the second of (5-9.19), (5-3.1) and (5-2.5) yield

$$\phi = \int_0^1 \mathscr{X} \lrcorner e^{-\tilde{\gamma}(\lambda)} \tilde{\Omega}(\lambda) \lambda^{k-1}\, d\lambda$$

$$= \int_0^1 e^{-\tilde{\gamma}(\lambda)} (\widetilde{\mathscr{X} \lrcorner \Omega})(\lambda) \lambda^{k-2}\, d\lambda.$$

It thus follows that specification of the $(k-1)$-form

$$\xi = \mathscr{X} \lrcorner \Omega$$

on S serves to determine ϕ uniquely; that is,

$$\phi = \int_0^1 e^{-\tilde{\gamma}(\lambda)} \tilde{\xi}(\lambda) \lambda^{k-2}\, d\lambda.$$

Let us now go back to the starting point, namely

$$d\Omega = \Gamma \wedge \Omega + \Sigma. \tag{5-9.20}$$

Theorem 5-9.1 shows that for given $\Gamma = d\gamma + \theta$, $\gamma = H(\Gamma)$, $\theta = H(d\Gamma)$, we can determine families of quantities Ω and Σ that satisfy (5-9.20). There is, however, a significant simplification that can be effected by noting that the sum $\Gamma \wedge \Omega + \Sigma$ does not determine either Γ or Σ uniquely. Thus if we substitute $\Gamma = d\gamma + \theta$ into (5-9.20), we obtain

$$d\Omega = d\gamma \wedge \Omega + \Sigma' \tag{5-9.21}$$

with

$$\Sigma' = \Sigma + \theta \wedge \Omega. \tag{5-9.22}$$

Now (5-9.17) and (5-9.18) show that (5-9.22) yields

$$\Sigma' = e^{\gamma} d(\eta + H(\theta \wedge d\phi)), \tag{5-9.23}$$

and hence

$$d\Sigma' = d\gamma \wedge \Sigma'. \tag{5-9.24}$$

The original system

$$d\Omega = \Gamma \wedge \Omega + \Sigma, \qquad d\Sigma = \Gamma \wedge \Sigma - \Theta \wedge \Omega,$$

$$d\Gamma = \Theta, \qquad d\Theta = 0,$$

is thus equivalent to the system

$$d\Omega = d\gamma \wedge \Omega + \Sigma', \qquad d\Sigma' = d\gamma \wedge \Sigma', \qquad \gamma = H(\Gamma).$$

Application of Theorem 5-9.1 to this equivalent system gives the following result.

Theorem 5-9.2. *Any system of exterior equations*

$$d\Omega = \Gamma \wedge \Omega + \Sigma, \qquad d\Sigma = \Gamma \wedge \Sigma - \Theta \wedge \Omega, \tag{5-9.25}$$

$$d\Gamma = \Theta, \qquad d\Theta = 0, \tag{5-9.26}$$

on S is equivalent to the system

$$d\Omega = d\gamma \wedge \Omega + \Sigma', \qquad d\Sigma' = d\gamma \wedge \Sigma', \qquad \gamma = H(\Gamma) \tag{5-9.27}$$

and this system has the general solution

$$\Omega = e^{\gamma}(d\phi + \eta), \qquad \Sigma' = e^{\gamma} d\eta \tag{5-9.28}$$

with $\Sigma' = \Sigma + H(d\Gamma) \wedge \Omega$

$$\phi = H(e^{-\gamma}\Omega), \qquad \eta = H(e^{-\gamma}\Sigma') = H d(e^{-\gamma}\Omega) \tag{5-9.29}$$

belonging to $\mathscr{A}(S)$.

It may have seemed to be a waste of time to solve the original system, when the equivalent system (5-9.27) is so much simpler. There are instances, however, in which the solution to the original system is required in terms of the original forms Γ and Σ. This is particularly true when dealing with several Ω's, as is the case in later sections.

A class of exterior forms that figures heavily in applications consists of those for which $d\Omega = \Gamma \wedge \Omega$.

Definition. An exterior form Ω of degree k is said to be *recursive* with *coefficient form* Γ if Ω satisfies

$$d\Omega = \Gamma \wedge \Omega \qquad (\Theta \wedge \Omega = 0,\ d\Gamma = \Theta). \tag{5-9.30}$$

Theorem 5-9.1 then leads to the following immediate conclusion.

Theorem 5-9.3. *If Ω is a recursive exterior form with coefficient form Γ then*

$$\Omega = e^{\gamma}\{d\phi + H(\theta \wedge d\phi)\} \tag{5-9.31}$$

with

$$\gamma = H(\Gamma), \qquad \theta = H(d\Gamma), \qquad \phi = H(e^{-\gamma}\Omega), \tag{5-9.32}$$

and θ must satisfy

$$d\theta \wedge \{d\phi + H(\theta \wedge d\phi)\} = 0. \tag{5-9.33}$$

This result is awkward in view of the condition (5-9.33) that results from the requirement $\Theta \wedge \Omega = d\Gamma \wedge \Omega = 0$. This motivates consideration of a special class of recursive forms.

Definition. A recursive form Ω is said to be *gradient recursive* if the coefficient Γ is exact.

Theorem 5-9.3 then gives the following immediate result.

Theorem 5-9.4. *If Ω is a gradient recursive k-form with coefficient form $d\gamma$, then*

$$\Omega = e^{\gamma}\, d\phi \tag{5-9.34}$$

with $\phi = H(e^{-\gamma}\Omega)$.

We observe from this result that $d\phi = e^{-\gamma}\Omega$, from which we have the following conclusion.

Corollary 5-9.1. *A k-form Ω is gradient recursive if and only if there exists a scalar-valued function γ such that $e^{-\gamma}\Omega$ is exact.*

5-10. REPRESENTATIONS FOR 2-FORMS THAT ARE NOT CLOSED

We have seen that the classic theorem of Darboux provides a canonical form for any closed 2-form. Specifically, let $\pi(x)$ be a closed 2-form on an open region S of E_n, and define the *rank* of $\pi(x)$ on S to be the even integer $2p$ such that $(\pi(x))^{(p)} \neq 0$ for some x in S and $(\pi(x))^{(p+1)} = 0$ for all x in S. Here, $(\pi(x))^{(p)}$ denotes the pth exterior power of $\pi(x)$. The points $x \in S$ for which $(\pi(x))^{(p)} \neq 0$ are *regular points* of $\pi(x)$, while those points for which $(\pi(x))^{(p)} = 0$ are referred to as *critical points* of $\pi(x)$. The Darboux theorem established the existence of $2p$ scalar-valued functions $\{u_i(x) | 1 \leq i \leq 2p\}$ that are independent at all regular points of $\pi(x)$ (i.e., $du_1(x) \wedge du_2(x) \wedge \cdots \wedge du_{2p}(x) \neq 0$) and such that

$$\pi(x) = \sum_{j=1}^{p} du_j(x) \wedge du_{j+p}(x) = d\left(\sum_{j=1}^{p} u_j(x)\, du_{j+p}(x)\right)$$

on S. The question thus arises as to what happens when the given 2-form $\pi(x)$ is not closed.

One answer to this question can be provided directly by the fundamental decomposition

$$\text{(5-10.1)} \qquad \pi = \pi_e + \pi_a, \qquad \pi_e = dH\pi, \qquad \pi_a = H\,d\pi$$

on any starshaped region S of E_n with homotopy operator H. Since the 2-form π_e is exact, it is closed and hence the Darboux representation of the previous paragraph is applicable. We therefore define the *exact rank* of a 2-form $\pi(x)$ to be the even integer $2p_e$ such that $(\pi_e(x))^{(p_e)} \neq 0$ for some x in S and $(\pi_e(x))^{(p_e+1)} = 0$ for all x in S. A point x is then an *exact regular* point of $\pi(x)$ if $(\pi_e(x))^{(p_e)} \neq 0$. The reader should note that the exact rank of a 2-form will usually be less than the rank as the following example shows,

$$\pi = \pi_e + \pi_a, \qquad \pi_e = dx \wedge dy,$$

$$\pi_a = x\,dz \wedge dt - z\,dx \wedge dt + t\,dx \wedge dz,$$

$$(\pi_e)^{(2)} = 0, \qquad (\pi)^{(2)} = 2x\,dx \wedge dy \wedge dz \wedge dt.$$

Be this as it may, the Darboux representation gives

$$\pi_e = \sum_{i=1}^{p_e} du_i \wedge du_{p_e+i},$$

at all exact regular points of $dH\pi$ in S.

Theorem 5-10.1. *Let π be a 2-form on a starshaped region S of E_n that has exact rank $2p_e$ on S then π has a representation*

$$\pi(x) = \sum_{i=1}^{p_e} du_i(x) \wedge du_{p_e+i}(x) + H\,d\pi(x) \tag{5-10.2}$$

on S and the $2p_e$ functions $\{u_j(x) | 1 \leq j \leq 2p_e\}$ are independent at all exact regular points of $\pi(x)$.

Another answer is available through use of Theorem 5-9.1. Let Ω be a given 2-form and assume that $d\Omega \neq 0$ on S. Starting with this given Ω, we can always construct the differential system

$$d\Omega = \Gamma \wedge \Omega + \Sigma, \qquad d\Sigma = \Gamma \wedge \Sigma - \Theta \wedge \Omega, \tag{5-10.3}$$
$$d\Gamma = \Theta, \qquad d\Theta = 0,$$

and thereby determine a 1-form Γ and the 3-form Σ, although not necessarily uniquely. Theorem 5-9.1 then gives

$$\Omega = e^{\gamma}(d\phi + \eta + H(\theta \wedge d\phi)) \tag{5-10.4}$$

where $\gamma \in \mathscr{A}^0(S)$, $\phi \in \mathscr{A}^1(S)$, $\eta \in \mathscr{A}^2(S)$, $\theta \in \mathscr{A}^1(S)$ are determined by

$$\gamma = H(\Gamma), \qquad \phi = H(e^{-\gamma}\Omega), \qquad \eta = H(e^{-\gamma}\Sigma), \qquad \theta = H(d\Gamma). \tag{5-10.5}$$

Now $d\phi$ is a closed 2-form to which the Darboux theorem is applicable, and we have the following result.

Theorem 5-10.2. *Let Ω be a given 2-form on S with the differential system*

$$d\Omega = \Gamma \wedge \Omega + \Sigma, \qquad d\Sigma = \Gamma \wedge \Sigma - \Theta \wedge \Omega, \tag{5-10.6}$$

and $d\Gamma = \Theta$, $d\Theta = 0$. The 2-form Ω admits the canonical representation

$$\Omega = e^{\gamma}(\pi + \eta + H(\theta \wedge \pi)), \tag{5-10.7}$$

with

$$\gamma = H(\Gamma), \qquad \theta = H(d\Gamma), \qquad \eta = H(e^{-\gamma}\Sigma), \tag{5-10.8}$$

and

$$\pi = \sum_{j=1}^{p} du_j \wedge du_{j+p} \tag{5-10.9}$$

where $2p$ is the rank of $dH(e^{-\gamma}\Omega)$ on S. Further the $2p$ functions $\{u_i\}$ are linearly independent at all regular points of $dH(e^{-\gamma}\Omega)$.

Further results can be obtained from Theorem 5-10.1 by particularization of the structures of Γ and Σ.

Corollary 5-10.1. *If Ω is a 2-form such that*

$$d\Omega = d\gamma \wedge \Omega + \Sigma, \tag{5-10.10}$$

then

$$\Omega = e^{\gamma}(\pi + \eta), \qquad \Sigma = e^{\gamma}\, d\eta \tag{5-10.11}$$

with

$$\eta = H(e^{-\gamma}\Sigma) \tag{5-10.12}$$

and

$$\pi = \sum_{j=1}^{p} du_j \wedge du_{j+p}, \tag{5-10.13}$$

where $2p$ is the rank of $dH(e^{-\gamma}\Omega)$ on S. Further, Σ is a gradient recursive 3-form with coefficient form $d\gamma$.

5-11. DIFFERENTIAL SYSTEMS OF DEGREE k AND CLASS r

The next several sections study problems associated with solving systems of simultaneous exterior differential equations. The purpose of this section is to obtain a formulation of these problems.

A significant simplification can be obtained through introduction of matrix notation. Let $\Lambda^k_{r,s}(E_n)$ denote the collection of all r-by-s matrices of elements of $\Lambda^k(E_n)$. Elements of $\Lambda^k_{r,s}(E_n)$ may also be viewed as k-forms that take their values in the collection of all r-by-s matrices. If $\boldsymbol{\Gamma} = ((\Gamma^p_q)) \in \Lambda^k_{a,b}$ and $\boldsymbol{\Omega} = ((\Omega^r_s)) \in \Lambda^m_{b,c}$, then $\boldsymbol{\Sigma} = \boldsymbol{\Gamma} \wedge \boldsymbol{\Omega} = ((\Sigma^p_s)) \in \Lambda^{k+m}_{a,c}$ is defined by

$$\Sigma^p_s = \sum_{q=1}^{b} \Gamma^p_q \wedge \Omega^q_s;$$

that is, the matrix product with exterior multiplication. It follows directly from this definition that

$$(\boldsymbol{\Gamma} \wedge \boldsymbol{\Omega})^T = (-1)^{\deg(\boldsymbol{\Gamma})\deg(\boldsymbol{\Omega})}(\boldsymbol{\Omega})^T \wedge (\boldsymbol{\Gamma})^T$$

where the superscript T denotes the transpose operation.

Definition. A collection of exterior forms Ω, Γ, Σ, with $\Omega \in \Lambda^k_{r,1}$, $\Gamma \in \Lambda^1_{r,r}$, $\Sigma \in \Lambda^{k+1}_{r,1}$ is said to form *a differential system of degree k and class r on S* if and only if

$$d\Omega = -\Gamma \wedge \Omega + \Sigma \tag{5-11.1}$$

is satisfied throughout S.

A comparison of (5-11.1) with (5-9.1) shows that a differential system of degree k and class r is a natural generalization of the considerations of Section 9 to systems of r simultaneous exterior differential equations. Such systems of exterior differential equations arise in a natural fashion in many problems of both practical and theoretical importance. A specific example will be given in full detail in a later section.

An arbitrary collection $\{\Omega, \Gamma, \Sigma\}$ cannot form a differential system on S, for the closure of the system (5-11.1) must likewise be satisfied throughout S. Exterior differentiation of (5-11.1) gives

$$\mathbf{0} = -d\Gamma \wedge \Omega + \Gamma \wedge d\Omega + d\Sigma.$$

When (5-11.1) is used to eliminate $d\Omega$, we then have

$$\begin{aligned}\mathbf{0} &= -d\Gamma \wedge \Omega + \Gamma \wedge (-\Gamma \wedge \Omega + \Sigma) + d\Sigma \\ &= -(d\Gamma + \Gamma \wedge \Gamma) \wedge \Omega + \Gamma \wedge \Sigma + d\Sigma.\end{aligned}$$

Thus if we define $\Theta \in \Lambda^2_{r,r}$ by

$$\Theta = d\Gamma + \Gamma \wedge \Gamma,$$

we obtain

$$d\Sigma = -\Gamma \wedge \Sigma + \Theta \wedge \Omega.$$

The subsidiary conditions $d\Gamma = \Theta - \Gamma \wedge \Gamma$ yield the further closure conditions

$$\mathbf{0} = d\Theta - d\Gamma \wedge \Gamma + \Gamma \wedge d\Gamma.$$

When $d\Gamma = \Theta - \Gamma \wedge \Gamma$ is used to eliminate $d\Gamma$, we obtain

$$\begin{aligned}\mathbf{0} &= d\Theta - (\Theta - \Gamma \wedge \Gamma) \wedge \Gamma + \Gamma \wedge (\Theta - \Gamma \wedge \Gamma) \\ &= d\Theta - \Theta \wedge \Gamma + \Gamma \wedge \Theta;\end{aligned}$$

that is

$$d\Theta = -\Gamma \wedge \Theta + \Theta \wedge \Gamma.$$

Theorem 5-11.1. *A collection* $\Omega \in \Lambda^k_{r,1}$, $\Gamma \in \Lambda^1_{r,r}$, $\Sigma \in \Lambda^{k+1}_{r,1}$ *forms a differential system of degree k and class r on S if and only if*

$$d\Omega = -\Gamma \wedge \Omega + \Sigma, \tag{5-11.2}$$

$$d\Sigma = -\Gamma \wedge \Sigma + \Theta \wedge \Omega, \tag{5-11.3}$$

$$d\Gamma = -\Gamma \wedge \Gamma + \Theta, \tag{5-11.4}$$

$$d\Theta = -\Gamma \wedge \Theta + \Theta \wedge \Gamma \tag{5-11.5}$$

are satisfied throughout S for some $\Theta \in \Lambda^2_{r,r}$.

For $r = n = \dim(S)$ and $k = 1$, the equations (5-11.2) through (5-11.5) are the equations of structure of Cartan and their closure equations if the n entries of Ω are linearly independent. Further, if $r = n$, the system (5-11.4) and (5-11.5) constitutes the second half of the Cartan structure equations irrespective of the value of k. Differential systems are thus of intrinsic interest in regard to the structure of the region S. On the other hand, we take particular note of the fact that a differential system makes no demand concerning the linear independence of the entries that comprise Ω. There is also no demand concerning the degrees of the entries that comprise Ω except that they are all the same for any given differential system. The study of differential systems of degree k will thus afford a foundation for the structural analysis of a region S relative to its forms of degree k, for general $k < n$.

The analogy with the Cartan structure equations is, however, a useful one in the general context of differential systems. We accordingly give the following definitions in relation to equations (5-11.2) through (5-11.5) which we refer to as the "differential system (*)."

Definition. The entries of Γ are the *connection* 1-forms of the differential system (*) and the equations

$$d\Gamma = -\Gamma \wedge \Gamma + \Theta$$

are the *connection equations.*

Definition. The entries of Θ are the *curvature* 2-forms of the differential system (*) and the equations

$$d\Theta = -\Gamma \wedge \Theta + \Theta \wedge \Gamma$$

are the *curvature equations.*

Definition. The entries of Σ are the *torsion* $(k + 1)$-forms of the differential system (*) and the equations

$$d\Sigma = -\Gamma \wedge \Sigma + \Theta \wedge \Omega$$

are the *torsion equations.*

It is essential to realize at the outset that a differential system admits several different interpretations depending on what is assumed to be given and what is to be determined by "solving" the system. In the first instance if $\mathbf{\Omega}$ and $\mathbf{\Gamma}$ are given, then (5-11.2) serves to determine $\mathbf{\Sigma}$ while (5-11.4) determines $\mathbf{\Theta}$. Under these circumstances (5-11.3) and (5-11.5) are satisfied identically. Next suppose that we are given $\mathbf{\Gamma}$ and $\mathbf{\Sigma}$. Here (5-11.4) serves to determine $\mathbf{\Theta}$, which will satisfy (5-11.5) identically. We are then left with the exterior differential equation (5-11.2) for the determination of $\mathbf{\Omega}$ and any solution of (5-11.2) must be such as to secure satisfaction of the algebraic conditions (5-11.3). The difficulty in this case revolves around satisfaction of the additional algebraic conditions that (5-11.3) impose. Next suppose that we are given $\mathbf{\Sigma}$ and $\mathbf{\Theta}$, then $\mathbf{\Omega}$ and $\mathbf{\Gamma}$ have to be determined by simultaneous integration of the full system of exterior equations (5-11.2) through (5-11.5). Finally we may view the system (5-11.2) through (5-11.5) as a system for the simultaneous determination of the quantities $(\mathbf{\Omega}, \mathbf{\Gamma}, \mathbf{\Sigma})$, in which case the curvature quantities $\mathbf{\Theta}$ must be determined in such a fashion that the full system is consistent. In this latter instance, it is often preferable to write the differential system (*) in the equivalent but more symmetric representation:

$$d\mathbf{\Omega} + \mathbf{\Gamma} \wedge \mathbf{\Omega} = \mathbf{\Sigma},$$

$$d\mathbf{\Sigma} + \mathbf{\Gamma} \wedge \mathbf{\Sigma} = \mathbf{\Theta} \wedge \mathbf{\Omega},$$

$$d\mathbf{\Gamma} + \mathbf{\Gamma} \wedge \mathbf{\Gamma} = \mathbf{\Theta},$$

$$d\mathbf{\Theta} + \mathbf{\Gamma} \wedge \mathbf{\Theta} = \mathbf{\Theta} \wedge \mathbf{\Gamma}.$$

5-12. INTEGRATION OF THE CONNECTION EQUATIONS

We start the analysis of the differential system (*) by studying the connection equation

(5-12.1) $$d\mathbf{\Gamma} = -\mathbf{\Gamma} \wedge \mathbf{\Gamma} + \mathbf{\Theta}.$$

Use of the representation Theorem 5-6.1 allows us to write

(5-12.2) $$\mathbf{\Gamma} = \mathbf{\Gamma}_e + \mathbf{\Gamma}_a, \qquad \mathbf{\Gamma}_e = d\boldsymbol{\gamma},$$

where

(5-12.3) $$\boldsymbol{\gamma} = H(\mathbf{\Gamma})$$

belongs to $\mathscr{A}^0(S) = \Lambda^0(S)$ with $\boldsymbol{\gamma}(x_0^i) = \mathbf{0}$. A substitution of (5-12.2) into (5-12.1) gives us

(5-12.4) $$d\mathbf{\Gamma}_a = -d\boldsymbol{\gamma} \wedge d\boldsymbol{\gamma} - d\boldsymbol{\gamma} \wedge \mathbf{\Gamma}_a - \mathbf{\Gamma}_a \wedge d\boldsymbol{\gamma} + \mathbf{\Theta} - \mathbf{\Gamma}_a \wedge \mathbf{\Gamma}_a.$$

If $\mathbf{A} \in \Lambda^0_{r,r}$ is a nonsingular matrix on S, we can change variables by the substitution

$$(5\text{-}12.5) \qquad \boldsymbol{\Gamma}_a = \mathbf{A}\bar{\boldsymbol{\Gamma}}\mathbf{A}^{-1}, \qquad \boldsymbol{\Theta} = \mathbf{A}\bar{\boldsymbol{\Theta}}\mathbf{A}^{-1}$$

with $\bar{\boldsymbol{\Gamma}} \in \mathscr{A}^1_{r,r}$, due to the module property of antiexact forms. In this event, $d\boldsymbol{\Gamma}_a = d\mathbf{A} \wedge \bar{\boldsymbol{\Gamma}}\mathbf{A}^{-1} + \mathbf{A}\, d\bar{\boldsymbol{\Gamma}}\mathbf{A}^{-1} - \mathbf{A}\bar{\boldsymbol{\Gamma}} \wedge d\mathbf{A}^{-1} = d\mathbf{A} \wedge \bar{\boldsymbol{\Gamma}}\mathbf{A}^{-1} + \mathbf{A}\, d\bar{\boldsymbol{\Gamma}}\mathbf{A}^{-1} + \mathbf{A}\bar{\boldsymbol{\Gamma}}\mathbf{A}^{-1} \wedge d\mathbf{A}\,\mathbf{A}^{-1}$, because $d\mathbf{A}^{-1} = -\mathbf{A}^{-1} d\mathbf{A}\,\mathbf{A}^{-1}$, and (5-12.4) becomes

$$(5\text{-}12.6) \qquad d\bar{\boldsymbol{\Gamma}} = -\mathbf{A}^{-1}(d\boldsymbol{\gamma}\,\mathbf{A} + d\mathbf{A}) \wedge \bar{\boldsymbol{\Gamma}} - \bar{\boldsymbol{\Gamma}} \wedge \mathbf{A}^{-1}(d\boldsymbol{\gamma}\,\mathbf{A} + d\mathbf{A})$$
$$-\mathbf{A}^{-1} d\boldsymbol{\gamma} \wedge d\boldsymbol{\gamma}\,\mathbf{A} + \bar{\boldsymbol{\Theta}} - \bar{\boldsymbol{\Gamma}} \wedge \bar{\boldsymbol{\Gamma}}.$$

Since $\bar{\boldsymbol{\Gamma}} \in \mathscr{A}^1_{r,r}$ is antiexact, $\bar{\boldsymbol{\Gamma}} = H(d\bar{\boldsymbol{\Gamma}})$ by property $\mathbf{A}_4$. A significant simplification can thus be achieved in the system (5-12.6) if we can choose $\mathbf{A}$ so that $d\boldsymbol{\gamma}\,\mathbf{A} + d\mathbf{A}$ belongs to $\mathscr{A}^1_{r,r}$, for in that event $(d\boldsymbol{\gamma}\,\mathbf{A} + d\mathbf{A}) \wedge \bar{\boldsymbol{\Gamma}}$ and $\bar{\boldsymbol{\Gamma}} \wedge (d\boldsymbol{\gamma}\,\mathbf{A} + d\mathbf{A})$ belong to $\mathscr{A}^2_{r,r}$ by property $\mathbf{A}_2$. We therefore consider the system

$$(5\text{-}12.7) \qquad d\mathbf{A} = -d\boldsymbol{\gamma}\mathbf{A} - \mathbf{A}\boldsymbol{\mu}, \qquad \boldsymbol{\mu} \in \mathscr{A}^1_{r,r}.$$

Exterior differentiation of (5-12.7) yields the closure conditions

$$(5\text{-}12.8) \qquad d\boldsymbol{\mu} = -\mathbf{A}^{-1}\, d\boldsymbol{\gamma} \wedge d\boldsymbol{\gamma}\mathbf{A} + \boldsymbol{\mu} \wedge \boldsymbol{\mu}.$$

Since $\boldsymbol{\mu} \in \mathscr{A}^1_{r,r}$, we have $\boldsymbol{\mu} = H(d\boldsymbol{\mu})$ and $H(\boldsymbol{\mu} \wedge \boldsymbol{\mu}) = \mathbf{0}$. Thus, (5-12.8) yields

$$(5\text{-}12.9) \qquad \boldsymbol{\mu} = -H\big(\mathbf{A}^{-1}\, d\boldsymbol{\gamma} \wedge d\boldsymbol{\gamma}\,\mathbf{A}\big).$$

When this is substituted into (5-12.7), we obtain

$$(5\text{-}12.10) \qquad d\mathbf{A} = -d\boldsymbol{\gamma}\,\mathbf{A} + \mathbf{A}H\big(\mathbf{A}^{-1}\, d\boldsymbol{\gamma} \wedge d\boldsymbol{\gamma}\,\mathbf{A}\big)$$

and (5-12.6), (5-12.7) and (5-12.8) yield

$$(5\text{-}12.11) \qquad d\bar{\boldsymbol{\Gamma}} = \boldsymbol{\mu} \wedge \bar{\boldsymbol{\Gamma}} + \bar{\boldsymbol{\Gamma}} \wedge \boldsymbol{\mu} + d\boldsymbol{\mu} - \boldsymbol{\mu} \wedge \boldsymbol{\mu} + \bar{\boldsymbol{\Theta}} - \bar{\boldsymbol{\Gamma}} \wedge \bar{\boldsymbol{\Gamma}}.$$

If we now use the facts that H inverts d on $\mathscr{A}(S)$ and that every element of $\mathscr{A}(S)$ belongs to the kernel of H, (5-12.11) yields $\bar{\boldsymbol{\Gamma}} = \boldsymbol{\mu} + H(\bar{\boldsymbol{\Theta}})$. Thus (5-12.2), (5-12.5) and (5-12.11) give $\boldsymbol{\Gamma} = d\boldsymbol{\gamma} + \mathbf{A}\bar{\boldsymbol{\Gamma}}\mathbf{A}^{-1} = d\boldsymbol{\gamma} + \mathbf{A}\boldsymbol{\mu}\mathbf{A}^{-1} + \mathbf{A}H(\bar{\boldsymbol{\Theta}})\mathbf{A}^{-1}$. An elimination of the term $\mathbf{A}\boldsymbol{\mu}$ between this result and equation (5-12.7) then results in $\boldsymbol{\Gamma} = (-d\mathbf{A} + \mathbf{A}H(\bar{\boldsymbol{\Theta}}))\mathbf{A}^{-1}$. Since $\mathbf{A} \in \mathscr{A}^0_{r,r}$, $\mathbf{A}_0 = \mathbf{A}(x^i_0) + H(d\mathbf{A})$, and (5-12.10) yields $\mathbf{A} = \mathbf{A}_0 - H(d\boldsymbol{\gamma}\,\mathbf{A})$. We have thus established the following results.

Theorem 5-12.1. *The general solution of the connection equations,*

$$d\Gamma = -\Gamma \wedge \Gamma + \Theta,$$

is given in terms of Θ *and* $\gamma = H(\Gamma)$ *by*

$$\Gamma = (\mathbf{A}H(\bar{\Theta}) - d\mathbf{A})\mathbf{A}^{-1}, \qquad \bar{\Theta} = \mathbf{A}^{-1}\Theta\mathbf{A} \tag{5-12.12}$$

in the region $D \subset S$ *where the matrix integral equation*

$$\mathbf{A} = \mathbf{A}_0 - H(d\gamma\,\mathbf{A}), \qquad \gamma = H(\Gamma) \tag{5-12.13}$$

possesses a nonsingular solution.

5-13. THE ATTITUDE MATRIX OF A DIFFERENTIAL SYSTEM

The result established Theorem 5-12.1 shows that everything hinges on the question of the solvability of the matrix integral equation $\mathbf{A} = \mathbf{A}_0 - H(d\gamma\mathbf{A})$ and the domain D over which the solution is nonsingular. We give the following definition in direct analogy with the underlying theory of the Cartan equations of structure.

Definition. If a matrix $\mathbf{A} \in \mathscr{A}^0_{r,r}$ satisfies the matrix integral equation

$$\mathbf{A}(x) = \mathbf{A}_0 - H(d\gamma\mathbf{A})(x), \qquad \gamma(x) = H(\Gamma)(x) \tag{5-13.1}$$

on a domain $D \subset S$ and $\det(\mathbf{A})(x) \neq 0$ throughout D, then $\mathbf{A}$ is an *attitude matrix* of the connection Γ and D is a *domain of regularity of* $\mathbf{A}$.

We first note that the homotopy operator H, is a linear operator, and hence $H(d\gamma\mathbf{A})$ is a linear operator on the space of all matrices $\mathbf{A}(x) \in \mathscr{A}^0_{r,r}$ for any given $\gamma(x) \in \mathscr{A}^0_{r,r}$. Noting that elements of $\Lambda^0(E_n)$ are C^∞ functions on E_n, it is then an easy matter to establish local existence of a solution $\mathbf{A}(x)$ of (5-12.1) (simply apply the contraction mapping theorem given in the appendix). Further since $H(d\gamma\mathbf{A})$ is antiexact, it follows that $\mathbf{A}(x_0^i) = \mathbf{A}_0$. We need to establish under what conditions is this solution nonsingular on D.

Lemma 5-13.1. *Let* $\mathbf{A}_0$ *be any constant element of* $\mathscr{A}^0_{r,r}$ *such that* $\det(\mathbf{A}_0) \neq 0$, *and let* $\mathbf{A}(x^i)$ *satisfy the matrix integral equation* $\mathbf{A}(x^i) = \mathbf{A}_0 - H(d\gamma\mathbf{A})(x^i)$. *Each connection matrix* Γ *has a unique attitude matrix* $\mathbf{A}(x^i)$ *such that*

$$\det(\mathbf{A}(x^i)) = \det(\mathbf{A}_0)\exp(-\operatorname{tr}(\gamma)(x^i)), \qquad \gamma = H(\Gamma). \tag{5-13.2}$$

Proof. Set $a(x^i) = a = \det(\mathbf{A}(x^i))$ and let $\mathrm{tr}(\mathbf{B})$ denote the trace of the matrix $\mathbf{B}$. If $\mathbf{C}$ denotes the matrix of co-factors of the matrix $\mathbf{A}$, we have

$$da = \mathrm{tr}(d\mathbf{A}\,\mathbf{C}) = \mathrm{tr}(d\mathbf{A}\,\mathbf{A}^{-1}\mathbf{A}\mathbf{C}).$$

Thus, since $\mathbf{AC} = a\mathbf{E}$, where $\mathbf{E}$ is the identity matrix (5-12.7) may be used to obtain

$$da = a\,\mathrm{tr}(d\mathbf{A}\,\mathbf{A}^{-1}) = -a\,\mathrm{tr}(d\gamma + \mathbf{A}\mu\mathbf{A}^{-1});$$

that is

$$d(\ln a) = -d\,\mathrm{tr}(\gamma) - \mathrm{tr}(\mu).$$

Now, $a \in \Lambda^0(E_n)$ implies $\ln a = \ln a_0 + Hd(\ln a)$ while $\mu \in \mathscr{A}^1_{r,r}$ gives $H(\mathrm{tr}(\mu)) = 0$. Accordingly, we have

$$a(x^i) = a_0 \exp(-\mathrm{tr}(\gamma))(x^i).$$

Finally since $\gamma = H(\Gamma)$, $\gamma(x_0^i) = \mathbf{0}$ and $a_0 = a(x_0^i) = \det(\mathbf{A}(x_0^i)) = \det(\mathbf{A}_0)$. □

An inspection of (5-13.2) shows that there are as many attitude matrices of a connection Γ as there are matrices $\mathbf{A}_0$ such that $\det(\mathbf{A}_0) \neq 0$. This many to one relation between attitude matrices and connections can be bothersome. If we set $\mathbf{A} = \mathbf{BA}_0$, (5-13.1) gives $\mathbf{BA}_0 = \mathbf{A}_0 - H(d\gamma\,\mathbf{BA}_0)$. Accordingly since $\mathbf{A}_0$ is a constant matrix, $H(d\gamma\,\mathbf{BA}_0) = H(d\gamma\,\mathbf{B})\mathbf{A}_0$ and we obtain

(5-13.3) $$\mathbf{B} = \mathbf{E} - H(d\gamma\,\mathbf{B}).$$

Further if $\mathbf{A} = \mathbf{BA}_0$ is substituted into (5-12.12), we have

$$\Gamma = \left(\mathbf{BA}_0 H(\mathbf{A}_0^{-1}\mathbf{B}^{-1}\Theta\mathbf{BA}_0) - d\mathbf{BA}_0\right)\mathbf{A}_0^{-1}\mathbf{B}^{-1}$$

$$= (\mathbf{B}H(\mathbf{B}^{-1}\Theta\mathbf{B}) - d\mathbf{B})\mathbf{B}^{-1}.$$

The connection Γ is thus independent of the choice of the constant matrix $\mathbf{A}_0$ such that $\det(\mathbf{A}_0) \neq 0$. We may thus assign $\mathbf{A}_0$ without loss of generality.

Lemma 5-13.2. *A one-to-one correspondence can be established between connection matrices* Γ *and attitude matrices* $\mathbf{A}$ *by the assignment* $\mathbf{A}_0 = \mathbf{E}$, *and* Γ *is invariant under the transformation* $\mathbf{A} \to \mathbf{AC}_0$ *for any constant matrix* $\mathbf{C}_0$ *such that* $\det(\mathbf{C}_0) \neq 0$.

There is a further simplification that can be effected.

Lemma 5-13.3. *Any connection matrix* Γ *has a unique attitude matrix* $\mathbf{A}$ *that satisfies the matrix integral equation*

$$\mathbf{A} = \mathbf{E} - H(\Gamma\mathbf{A}). \tag{5-13.4}$$

Proof. The analysis given in the previous section started with the decomposition $\Gamma = d\gamma + \Gamma_a$ with $\gamma = H(\Gamma)$. Accordingly, $H(d\gamma\mathbf{A}) = H(\Gamma\mathbf{A} - \Gamma_a\mathbf{A}) = H(\Gamma\mathbf{A})$ because Γ_a antiexact implies that $\Gamma_a\mathbf{A}$ is also antiexact and hence belongs to $\ker H$. □

The results established up to this point can be placed in a more useful context in the following manner. The attitude matrix $\mathbf{A}$ of a connection matrix Γ is a nonsingular matrix-valued function of position on S that is generated by $\mathbf{A} = \mathbf{E} - H(\Gamma\mathbf{A})$. Thus if we take the standard representation of the general linear group $GL(r, \mathbb{R})$ in terms of nonsingular r-by-r matrices, we may view $\mathbf{A}$ as a mapping ρ_Γ of S into $GL(r, \mathbb{R})$:

$$\rho_\Gamma : S \to GL(r, \mathbb{R}) \,|\, \mathbf{A} = \mathbf{E} - H(\Gamma\mathbf{A}). \tag{5-13.5}$$

Further, if $P_0 : (x_0^i)$ is the center of the starshaped region S, then ρ_Γ maps P_0 into $\mathbf{E}$ because $H(\Gamma\mathbf{A})(x_0^i) = \mathbf{0}$ by property $\mathbf{H}_3$.

There is a deep and important relation between connection matrices that take their values in matrix representations of Lie algebras and matrix representations of Lie groups of matrices. Let g be a k-dimensional Lie algebra with constants of structure C_{ab}^e and let $\{\sigma_a, a = 1, \ldots, k\}$ be a basis for a representation of g as a subspace of $gl(r, \mathbb{R})$. By this we mean that each σ_a is an r-by-r constant valued matrix such that

$$[\sigma_a, \sigma_b] = \sigma_a\sigma_b - \sigma_b\sigma_a = C_{ab}^e\sigma_e. \tag{5-13.6}$$

It is then well known that g is the Lie algebra of a matrix Lie group G that is continuously connected to the identity matrix and that any element of G has the form $\mathbf{B} = \exp(u^a\sigma_a)$ for some choice of $\{u^a\} \in \mathbb{R}^k$.

Theorem 5-13.1. *Let* $\Gamma \in \Lambda^1_{r,r}(E_n)$ *be a connection matrix of a differential system that takes its values in the finite dimensional matrix Lie algebra of an r-by-r matrix Lie group G, then the attitude matrix* $\mathbf{A}(x^i)$ *of* Γ *belongs to the matrix Lie group G for each* $P : (x^i) \in E_n$.

The proof of Theorem 5-13.1 in the general case is lengthy and involves concepts and constructs from the theory of Lie groups whose foundations have not been prepared. We therefore restrict attention to a special case that is of interest in its own right. If $\mathbf{A}$ is the attitude matrix of Γ, then $\mathbf{A}$ satisfies $\mathbf{A} = \mathbf{E} - H(\Gamma\mathbf{A})$. Since $\Gamma \in \Lambda^1_{r,r}(E_n)$, we have $\Gamma = -d\gamma + \Gamma_a$ with Γ_a antiexact and $\gamma = -H(\Gamma)$. Thus $\mathbf{A}$ satisfies $\mathbf{A} = \mathbf{E} + H(d\gamma\,\mathbf{A})$ because Γ_a, and hence $\Gamma_a \mathbf{A}$, is antiexact and belongs to $\ker H$. By hypothesis, Γ takes its values in the matrix Lie algebra of G. We therefore consider the situation in which $\Gamma = -du\,\sigma + \Gamma_a$ with $u \in \Lambda^0(E_n)$ and σ belongs to the matrix Lie algebra of G. In this case, the attitude matrix $\mathbf{A}$ has to satisfy the Riemann-Graves matrix integral equation

$$\mathbf{A} = \mathbf{E} + H(du\,\sigma\mathbf{A}) = \mathbf{E} + \sigma H(du\,\mathbf{A}). \tag{5-13.7}$$

We proceed to solve this equation by the iteration algorithm

$$\mathbf{A}_{(k+1)} = \mathbf{E} + \sigma H\big(du\,\mathbf{A}_{(k)}\big). \tag{5-13.8}$$

Starting with $\mathbf{A}_{(0)} = \mathbf{E}$, we have $\mathbf{A}_{(1)} = \mathbf{E} + \sigma H(du\,\mathbf{E}) = \mathbf{E} + u\sigma$, and $\mathbf{A}_{(2)} = \mathbf{A}_{(1)} + \sigma H(du\,u\sigma) = \mathbf{A}_{(1)} + \frac{1}{2}u^2\sigma^2$. It is then a simple matter to verify the general result

$$\mathbf{A}_{(n)} = \mathbf{A}_{(n-1)} + \frac{1}{n!}u^n\sigma^n \tag{5-13.9}$$

by induction on n. We thus have

$$\mathbf{A}_{(\infty)} = \lim_{n\to\infty} \mathbf{A}_{(n)} = \exp(u\sigma). \tag{5-13.10}$$

On the other hand, since $d(u\sigma) = du\,\sigma$, it follows that $\mathbf{E} + \sigma H(du \exp(u\sigma)) = \mathbf{E} + H(d(u\sigma)\exp(u\sigma)) = \mathbf{E} + Hd(\exp(u\sigma)) = \mathbf{E} + \exp(u\sigma) - \exp(u\sigma)|_{P_0}$, where P_0 is the center for the homotopy operator H. Thus since $u\sigma = -H(\Gamma)$, $(u\sigma)|_{P_0} = \mathbf{0}$, and we have $\mathbf{E} + \sigma H(du\exp(u\sigma)) = \exp(u\sigma)$; that is, $\exp(u\sigma)$ satisfies (5-13.7). The result then follows on noting that $\mathbf{A} = \exp(u\sigma)$ necessarily belongs to the matrix Lie group G for each $P:(x^j)$ whenever σ belongs to the matrix Lie algebra g of G.

Direct verification of Theorem 5-13.1 is straightforward whenever we have a specific matrix Lie group with which to deal. Suppose, therefore, that Γ satisfies

$$\Gamma + \Gamma^T = \mathbf{0} \tag{5-13.11}$$

so that Γ takes its values in the Lie algebra of $SO(r)$. The attitude matrix $\mathbf{A}$ of Γ thus satisfies—see (5-12.7)—

$$d\mathbf{A} = -d\gamma\,\mathbf{A} - \mathbf{A}\mu \tag{5-13.12}$$

with $\gamma = H(\Gamma)$. (5-13.11) then shows that γ likewise satisfies

$$\gamma + \gamma^T = 0. \tag{5-13.13}$$

Thus since (5-13.12) gives

$$d\mathbf{A}^T = -\mathbf{A}^T d\gamma^T - \mu^T\mathbf{A}^T,$$

it follows that

$$d(\mathbf{A}^T\mathbf{A}) = (d\mathbf{A}^T)\mathbf{A} + \mathbf{A}^T d\mathbf{A} = -\mathbf{A}^T(d\boldsymbol{\gamma}^T + d\boldsymbol{\gamma})\mathbf{A} - \boldsymbol{\mu}^T\mathbf{A}^T\mathbf{A} - \mathbf{A}^T\mathbf{A}\boldsymbol{\mu}$$

$$= -\boldsymbol{\mu}^T\mathbf{A}^T\mathbf{A} - \mathbf{A}^T\mathbf{A}\boldsymbol{\mu}$$

when (5-13.13) is used. Since $\boldsymbol{\mu} \in \mathscr{A}^1_{r,r}$ the module property of antiexact forms shows that $\boldsymbol{\mu}^T\mathbf{A}^T\mathbf{A} + \mathbf{A}^T\mathbf{A}\boldsymbol{\mu}$ belongs to $\ker H$. Thus, applying H to both sides of the above relation and using property $\mathbf{H}_2$ with $\mathbf{A}|_{P_0} = \mathbf{E}$, we have

$$\mathbf{A}^T\mathbf{A} = \mathbf{E}.$$

The following equivalent formulation of Theorem 5-13.1 often proves useful. *Let G be a finite dimensional matrix Lie group with Lie algebra g. If a connection matrix $\boldsymbol{\Gamma}$ takes its values in g then the mapping ρ_{Γ} given by* (5-13.5) *maps S into G.*

5-14. INTEGRATION OF THE CURVATURE EQUATIONS

Theorem 5-12.1 gave the general solution of the connection equations as

$$\boldsymbol{\Gamma} = (\mathbf{A}H(\overline{\boldsymbol{\Theta}}) - d\mathbf{A})\mathbf{A}^{-1} \tag{5-14.1}$$

where $\mathbf{A}$ is a solution of

$$\mathbf{A} = \mathbf{E} - H(\boldsymbol{\Gamma}\mathbf{A}) \tag{5-14.2}$$

and $\overline{\boldsymbol{\Theta}}$ is defined in terms of $\boldsymbol{\Theta}$ and $\mathbf{A}$ by

$$\overline{\boldsymbol{\Theta}} = \mathbf{A}^{-1}\boldsymbol{\Theta}\mathbf{A}. \tag{5-14.3}$$

The quantities $\boldsymbol{\Theta}$ still remain to be determined, for they must satisfy the curvature equations

$$d\boldsymbol{\Theta} = -\boldsymbol{\Gamma} \wedge \boldsymbol{\Theta} + \boldsymbol{\Theta} \wedge \boldsymbol{\Gamma}. \tag{5-14.4}$$

We now proceed to solve (5-14.4).

Under the substitution $\boldsymbol{\Theta} = \mathbf{A}\overline{\boldsymbol{\Theta}}\mathbf{A}^{-1}$ that is the inverse of (5-14.3), the system (5-14.4) becomes

$$\begin{aligned} d\overline{\boldsymbol{\Theta}} &= -\mathbf{A}^{-1}d\mathbf{A} \wedge \overline{\boldsymbol{\Theta}} + \overline{\boldsymbol{\Theta}} \wedge \mathbf{A}^{-1}d\mathbf{A} - \mathbf{A}^{-1}(\boldsymbol{\Gamma} \wedge \overline{\boldsymbol{\Theta}} - \overline{\boldsymbol{\Theta}} \wedge \boldsymbol{\Gamma})\mathbf{A} \\ &= -H(\overline{\boldsymbol{\Theta}}) \wedge \overline{\boldsymbol{\Theta}} + \overline{\boldsymbol{\Theta}} \wedge H(\overline{\boldsymbol{\Theta}}), \end{aligned} \tag{5-14.5}$$

where the last equality follows from (5-14.1). We now use the decomposition of forms into exact and antiexact parts to write

(5-14.6) $$\overline{\mathbf{\Theta}} = d\boldsymbol{\theta} + \overline{\mathbf{\Theta}}_a, \qquad \boldsymbol{\theta} = H(\overline{\mathbf{\Theta}}) \in \mathscr{A}^1_{r,r}$$

in which case (5-14.5) becomes

(5-14.7) $$d\overline{\mathbf{\Theta}}_a = +d(\boldsymbol{\theta} \wedge \boldsymbol{\theta}) - \boldsymbol{\theta} \wedge \overline{\mathbf{\Theta}}_a + \overline{\mathbf{\Theta}}_a \wedge \boldsymbol{\theta}.$$

Since $\boldsymbol{\theta} \wedge \overline{\mathbf{\Theta}}_a$ and $\overline{\mathbf{\Theta}}_a \wedge \boldsymbol{\theta}$ belong to $\mathscr{A}^3_{r,r}$ by property $\mathbf{A}_2$, and hence to the kernel of H, while H inverts d on $\mathscr{A}(S)$ by property $\mathbf{A}_4$, application of H to (5-14.7) yields

(5-14.8) $$\overline{\mathbf{\Theta}}_a = \boldsymbol{\theta} \wedge \boldsymbol{\theta}.$$

Accordingly (5-14.6) yields

(5-14.9) $$\overline{\mathbf{\Theta}} = d\boldsymbol{\theta} + \boldsymbol{\theta} \wedge \boldsymbol{\theta}.$$

Combination of the above results now yields the following theorem.

Theorem 5-14.1. *The general solution of the connection and the curvature equations of a differential system is given by*

(5-14.10) $$\mathbf{\Gamma} = (\mathbf{A}\boldsymbol{\theta} - d\mathbf{A})\mathbf{A}^{-1},$$

(5-14.11) $$\mathbf{\Theta} = \mathbf{A}(d\boldsymbol{\theta} + \boldsymbol{\theta} \wedge \boldsymbol{\theta})\mathbf{A}^{-1},$$

where the matrix $\mathbf{A}$ *and the matrix* $\boldsymbol{\theta}$ *of antiexact* 1*-forms are given by*

(5-14.12) $$\mathbf{A} = \mathbf{A}_0 - H(\mathbf{\Gamma}\mathbf{A}), \qquad d\mathbf{A}_0 = \mathbf{0}, \qquad \det(\mathbf{A}_0) \neq 0,$$

(5-14.13) $$\boldsymbol{\theta} = H(\mathbf{A}^{-1}\mathbf{\Theta}\mathbf{A})$$

and $\mathbf{\Gamma}$ *and* $\mathbf{\Theta}$ *are independent of the choice of* $\mathbf{A}_0$.

It is clear from these results that $\mathbf{\Gamma}$ determines $\mathbf{\Theta}$ uniquely, for $\mathbf{\Theta} = d\mathbf{\Gamma} + \mathbf{\Gamma} \wedge \mathbf{\Gamma}$ from the connection equations and (5-14.10) gives $\boldsymbol{\theta} = \mathbf{A}^{-1}(\mathbf{\Gamma}\mathbf{A} + d\mathbf{A})$. On the other hand, a given curvature matrix $\mathbf{\Theta}$ can result from many connections.

As an example, suppose that $\mathbf{U} \in \Lambda^k_{2,1}$ is to satisfy $d\mathbf{U} = -\mathbf{\Gamma} \wedge \mathbf{U}$ with

$$\mathbf{\Gamma} = -\begin{pmatrix} 0 & a \\ b & 0 \end{pmatrix} d\xi, \qquad \xi \in \Lambda^0_{1,1}, \qquad da = db = 0,$$

then $d\mathbf{\Gamma} = 0$, $\mathbf{\Gamma} \wedge \mathbf{\Gamma} = 0$, and hence $\mathbf{\Theta} = \mathbf{0}$, $\mathbf{\Sigma} = \mathbf{0}$. In this case

$$\mathbf{A} = \exp\left[\begin{pmatrix} 0 & a \\ b & 0 \end{pmatrix} \xi\right]$$

satisfies $\mathbf{A} = \mathbf{E} - H(\mathbf{\Gamma A})$; that is

$$\mathbf{\Gamma} = -(d\mathbf{A})\mathbf{A}^{-1}.$$

Thus

$$\mathbf{U} = \exp\left[\begin{pmatrix} 0 & a \\ b & 0 \end{pmatrix} \xi\right] d\boldsymbol{\phi}$$

with arbitrary $\boldsymbol{\phi} \in \Lambda_{2,1}^{k-1}$; that is,

$$\mathbf{U} = \begin{pmatrix} \cosh(\xi\sqrt{ab}) & \dfrac{a}{\sqrt{ab}} \sinh(\xi\sqrt{ab}) \\ \dfrac{b}{\sqrt{ab}} \sinh(\xi\sqrt{ab}) & \cosh(\xi\sqrt{ab}) \end{pmatrix} d\boldsymbol{\phi}.$$

5-15. INTEGRATION OF A DIFFERENTIAL SYSTEM

Now that we have solved the connection equations and the curvature equations,

(5-15.1) $$\mathbf{\Gamma} = (\mathbf{A}\boldsymbol{\theta} - d\mathbf{A})\mathbf{A}^{-1}, \qquad \mathbf{A} = \mathbf{E} - H(\mathbf{\Gamma A}),$$

(5-15.2) $$\mathbf{\Theta} = \mathbf{A}(d\boldsymbol{\theta} + \boldsymbol{\theta} \wedge \boldsymbol{\theta})\mathbf{A}^{-1}, \qquad \boldsymbol{\theta} = H(\mathbf{A}^{-1}\mathbf{\Theta A}),$$

the task outstanding consists of integrating the remaining equations

(5-15.3) $$d\mathbf{\Omega} = -\mathbf{\Gamma} \wedge \mathbf{\Omega} + \mathbf{\Sigma}, \qquad d\mathbf{\Sigma} = -\mathbf{\Gamma} \wedge \mathbf{\Sigma} + \mathbf{\Theta} \wedge \mathbf{\Omega}$$

of the differential system (5-11.2) through (5-11.5). Under the substitution

(5-15.4) $$\mathbf{\Omega} = \mathbf{A}\boldsymbol{\omega}, \qquad \mathbf{\Sigma} = \mathbf{A}\boldsymbol{\sigma},$$

equations (5-15.1) through (5-15.3) combine to yield

(5-15.5) $$d\boldsymbol{\omega} = -\boldsymbol{\theta} \wedge \boldsymbol{\omega} + \boldsymbol{\sigma}, \qquad d\boldsymbol{\sigma} = -\boldsymbol{\theta} \wedge \boldsymbol{\sigma} + (d\boldsymbol{\theta} + \boldsymbol{\theta} \wedge \boldsymbol{\theta}) \wedge \boldsymbol{\omega}.$$

If we put

(5-15.6) $$\boldsymbol{\rho} = \boldsymbol{\theta} \wedge \boldsymbol{\omega},$$

and note that this implies $d\boldsymbol{\rho} = d\boldsymbol{\theta} \wedge \boldsymbol{\omega} + \boldsymbol{\theta} \wedge (\boldsymbol{\theta} \wedge \boldsymbol{\omega} - \boldsymbol{\sigma})$, we arrive at the system

(5-15.7) $$d\boldsymbol{\omega} = -\boldsymbol{\rho} + \boldsymbol{\sigma}, \qquad d\boldsymbol{\sigma} = d\boldsymbol{\rho}.$$

We now set

$$\sigma = d\eta + \sigma_a, \qquad \eta = H(\sigma) \tag{5-15.8}$$

$$\omega = d\phi + \omega_a, \qquad \phi = H(\omega) \tag{5-15.9}$$

and obtain

$$d\omega_a = -\rho + d\eta + \sigma_a, \qquad d\sigma_a = d\rho. \tag{5-15.10}$$

Since H inverts d on $\mathscr{A}(S)$, we obtain $\omega_a = H(-\rho + d\eta + \sigma_a) = -H(\rho) + \eta$ and $\sigma_a = H(d\rho) = -dH(\rho) + \rho$ because $dH + Hd =$ identity. Equations (5-15.8) and (5-15.9) thus yield

$$\omega = d\phi - H(\rho) + \eta, \qquad \sigma = d\eta - dH(\rho) + \rho. \tag{5-15.11}$$

It now remains only to eliminate the subsidiary variable ρ. A direct combination of (5-15.6) and (5-15.11) yields

$$\rho = \theta \wedge d\phi - \theta \wedge H(\rho) + \theta \wedge \eta$$

so that

$$H(\rho) = H(\theta \wedge d\phi - \theta \wedge H(\rho) + \theta \wedge \eta) = H(\theta \wedge d\phi)$$

because $\theta \wedge H(\rho)$ and $\theta \wedge \eta$ belong to $\mathscr{A}(S)$. Thus we obtain

$$\rho = \theta \wedge d\phi - \theta \wedge H(\theta \wedge d\phi) + \theta \wedge \eta,$$

and hence

$$\omega = d\phi + \eta - H(\theta \wedge d\phi),$$

$$\sigma = d\eta - dH(\theta \wedge d\phi) + \theta \wedge d\phi - \theta \wedge H(\theta \wedge d\phi) + \theta \wedge \eta.$$

A combination of the above substitutions now establishes the following result.

Theorem 5-15.1. *The general solution of the differential system*

$$d\Omega = -\Gamma \wedge \Omega + \Sigma, \qquad d\Sigma = -\Gamma \wedge \Sigma + \Theta \wedge \Omega, \tag{5-15.12}$$

$$d\Gamma = -\Gamma \wedge \Gamma + \Theta, \qquad d\Theta = -\Gamma \wedge \Theta + \Theta \wedge \Gamma \tag{5-15.13}$$

is given in terms of the matrices of antiexact forms

$$\phi = H(\mathbf{A}^{-1}\Omega) \tag{5-15.14}$$

$$\eta = H(\mathbf{A}^{-1}\Sigma) \tag{5-15.15}$$

$$\theta = H(\mathbf{A}^{-1}\Theta\mathbf{A}) \tag{5-15.16}$$

and the attitude matrix **A**, *satisfying*

$$\mathbf{A} = \mathbf{E} - H(\boldsymbol{\Gamma}\mathbf{A}), \tag{5-15.17}$$

by

$$\boldsymbol{\Omega} = \mathbf{A}\{d\boldsymbol{\phi} + \boldsymbol{\eta} - H(\boldsymbol{\theta} \wedge d\boldsymbol{\phi})\}, \tag{5-15.18}$$

$$\boldsymbol{\Sigma} = \mathbf{A}\{d\boldsymbol{\eta} + \boldsymbol{\theta} \wedge \boldsymbol{\eta} + H(d\boldsymbol{\theta} \wedge d\boldsymbol{\phi}) - \boldsymbol{\theta} \wedge H(\boldsymbol{\theta} \wedge d\boldsymbol{\phi})\}, \tag{5-15.19}$$

$$\boldsymbol{\Gamma} = (\mathbf{A}\boldsymbol{\theta} - d\mathbf{A})\mathbf{A}^{-1}, \tag{5-15.20}$$

$$\boldsymbol{\Theta} = \mathbf{A}(d\boldsymbol{\theta} + \boldsymbol{\theta} \wedge \boldsymbol{\theta})\mathbf{A}^{-1}. \tag{5-15.21}$$

The following corollary is now immediate.

Corollary 5-15.1. *We have*

$$(\mathbf{A}^{-1}\boldsymbol{\Omega})_e = d\boldsymbol{\phi}, \qquad (\mathbf{A}^{-1}\boldsymbol{\Omega})_a = \boldsymbol{\eta} - H(\boldsymbol{\theta} \wedge d\boldsymbol{\phi}), \tag{5-15.22}$$

$$(\mathbf{A}^{-1}\boldsymbol{\Sigma})_e = d\boldsymbol{\eta}, \tag{5-15.23}$$

$$(\mathbf{A}^{-1}\boldsymbol{\Sigma})_a = +\boldsymbol{\theta} \wedge \boldsymbol{\eta} + H(\boldsymbol{\theta} \wedge d\boldsymbol{\phi}) - \boldsymbol{\theta} \wedge H(\boldsymbol{\theta} \wedge d\boldsymbol{\phi}),$$

$$(\boldsymbol{\Gamma}\mathbf{A})_e = d\mathbf{A}, \qquad (\boldsymbol{\Gamma}\mathbf{A})_a = \mathbf{A}\boldsymbol{\theta}, \tag{5-15.24}$$

$$(\mathbf{A}^{-1}\boldsymbol{\Theta}\mathbf{A})_e = d\boldsymbol{\theta}, \qquad (\mathbf{A}^{-1}\boldsymbol{\Theta}\mathbf{A})_a = \boldsymbol{\theta} \wedge \boldsymbol{\theta}. \tag{5-15.25}$$

5-16. EQUIVALENT DIFFERENTIAL SYSTEMS

The discussion given in Section 5-11 showed that a differential system

$$d\boldsymbol{\Omega} = -\boldsymbol{\Gamma} \wedge \boldsymbol{\Omega} + \boldsymbol{\Sigma}, \qquad d\boldsymbol{\Sigma} = -\boldsymbol{\Gamma} \wedge \boldsymbol{\Sigma} + \boldsymbol{\Theta} \wedge \boldsymbol{\Omega} \tag{5-16.1}$$

$$d\boldsymbol{\Gamma} = -\boldsymbol{\Gamma} \wedge \boldsymbol{\Gamma} + \boldsymbol{\Theta}, \qquad d\boldsymbol{\Theta} = -\boldsymbol{\Gamma} \wedge \boldsymbol{\Theta} + \boldsymbol{\Theta} \wedge \boldsymbol{\Gamma} \tag{5-16.2}$$

arises from the system of equations

$$d\boldsymbol{\Omega} = -\boldsymbol{\Gamma} \wedge \boldsymbol{\Omega} + \boldsymbol{\Sigma} \tag{5-16.3}$$

by adjoining to this system the equations that obtain from it by demanding closure. Now for given $\boldsymbol{\Omega}$, *the quantities* $\boldsymbol{\Gamma}$ *and* $\boldsymbol{\Sigma}$ *that occur in* (5-16.3) *are not uniquely determined*. This leads to the following natural definition of equivalence.

Definition. Two differential systems,

$$d\Omega = -\Gamma \wedge \Omega + \Sigma, \qquad d\Sigma = -\Gamma \wedge \Sigma + \Theta \wedge \Omega,$$

$$d\Gamma = -\Gamma \wedge \Gamma + \Theta, \qquad d\Theta = -\Gamma \wedge \Theta + \Theta \wedge \Gamma,$$

and

$$d\bar{\Omega} = -\bar{\Gamma} \wedge \bar{\Omega} + \bar{\Sigma}, \qquad d\bar{\Sigma} = -\bar{\Gamma} \wedge \bar{\Sigma} + \bar{\Theta} \wedge \bar{\Omega},$$

$$d\bar{\Omega} = -\bar{\Gamma} \wedge \bar{\Gamma} + \bar{\Theta}, \qquad d\bar{\Theta} = -\bar{\Gamma} \wedge \bar{\Theta} + \bar{\Theta} \wedge \bar{\Gamma}$$

are said to be *equivalent* on S if and only if $\bar{\Omega} = \Omega$ at all points of S.

There are obviously many differential systems that are equivalent to a given differential system. They are obtained by the requirement that

$$d\Omega = -\Gamma \wedge \Omega + \Sigma = -\bar{\Gamma} \wedge \Omega + \bar{\Sigma},$$

from which we obtain

$$\bar{\Sigma} = \Sigma + (\bar{\Gamma} - \Gamma) \wedge \Omega. \tag{5-16.4}$$

Theorem 5-15.1 can then be used to obtain the solutions of the two equivalent systems whereby the various antiexact forms that determine these solutions can be related. The results are rather complex and not altogether illuminating with the exception of one very important case.

Let us start with a given differential system with connection matrix Γ, in which case,

$$\Gamma = (\mathbf{A}\boldsymbol{\theta} - d\mathbf{A})\mathbf{A}^{-1}, \qquad \mathbf{A} = \mathbf{E} - H(\Gamma\mathbf{A}). \tag{5-16.5}$$

If we now take

$$\bar{\Gamma} = -d\mathbf{A}\,\mathbf{A}^{-1} \tag{5-16.6}$$

then ($\boldsymbol{\theta} \in \mathscr{A}^1_{r,r}$)

$$\begin{aligned} \bar{\gamma} = H(\bar{\Gamma}) &= -H(d\mathbf{A}\,\mathbf{A}^{-1}) = -H(d\mathbf{A}\,\mathbf{A}^{-1} - \mathbf{A}\boldsymbol{\theta}\mathbf{A}^{-1}) \\ &= H(\Gamma) = \gamma \end{aligned} \tag{5-16.7}$$

and the attitude matrix $\bar{\mathbf{A}}$ of the connection $\bar{\Gamma}$ can be taken to be the same as the attitude matrix $\mathbf{A}$ of the connection Γ, for $\mathbf{A}$ and $\bar{\mathbf{A}}$ satisfy the same matrix integral equation $\mathbf{A} = \mathbf{E}_0 - H(d\gamma\mathbf{A})$. In this event (5-16.4) yields $\bar{\boldsymbol{\Sigma}} = \boldsymbol{\Sigma} - (\boldsymbol{\Gamma} - \bar{\boldsymbol{\Gamma}}) \wedge \boldsymbol{\Omega}$ and hence (5-15.18), (5-15.19), (5-15.20), and (5.16.6) yield

(5-16.8) $$\begin{aligned}\bar{\boldsymbol{\Sigma}} &= \mathbf{A}\{d\boldsymbol{\eta} + \boldsymbol{\theta} \wedge \boldsymbol{\eta} + H(d\boldsymbol{\theta} \wedge d\boldsymbol{\phi}) - \boldsymbol{\theta} \wedge H(\boldsymbol{\theta} \wedge d\boldsymbol{\phi}) \\ &\quad - \boldsymbol{\theta} \wedge (d\boldsymbol{\phi} + \boldsymbol{\eta} - H(\boldsymbol{\theta} \wedge d\boldsymbol{\phi}))\} \\ &= A\{d\boldsymbol{\eta} + H(d\boldsymbol{\theta} \wedge d\boldsymbol{\phi}) - \theta \wedge d\boldsymbol{\phi}\} = \mathbf{A}d(\boldsymbol{\eta} - H(\boldsymbol{\theta} \wedge d\boldsymbol{\phi})).\end{aligned}$$

Accordingly, (5-15.14) and (5-15.15) yield

(5-16.9) $$\bar{\boldsymbol{\phi}} = H(\mathbf{A}^{-1}\boldsymbol{\Omega}) = H(\mathbf{A}^{-1}\boldsymbol{\Omega}) = \boldsymbol{\phi},$$

(5-16.10) $$\bar{\boldsymbol{\eta}} = H(\mathbf{A}^{-1}\bar{\boldsymbol{\Sigma}}) = Hd(\boldsymbol{\eta} - H(\boldsymbol{\theta} \wedge d\boldsymbol{\phi})) = \boldsymbol{\eta} - H(\boldsymbol{\theta} \wedge d\boldsymbol{\phi}),$$

while $d\bar{\boldsymbol{\Gamma}} = d\bar{\boldsymbol{\Gamma}} \wedge d\bar{\boldsymbol{\Gamma}} + \bar{\boldsymbol{\Theta}}$ and (5-16.6) give

(5-16.11) $$\bar{\boldsymbol{\Theta}} = \mathbf{0}, \qquad \bar{\boldsymbol{\theta}} = \mathbf{0}.$$

We have thus established the following result.

Theorem 5-16.1. *Any differential system*

(5-16.12) $$d\boldsymbol{\Omega} = -\boldsymbol{\Gamma} \wedge \boldsymbol{\Omega} + \boldsymbol{\Sigma}, \qquad d\boldsymbol{\Sigma} = -\boldsymbol{\Gamma} \wedge \boldsymbol{\Sigma} + \boldsymbol{\Theta} \wedge \boldsymbol{\Omega}$$

(5-16.13) $$d\boldsymbol{\Gamma} = -\boldsymbol{\Gamma} \wedge \boldsymbol{\Gamma} + \boldsymbol{\Theta}, \qquad d\boldsymbol{\Theta} = -\boldsymbol{\Gamma} \wedge \boldsymbol{\Theta} + \boldsymbol{\Theta} \wedge \boldsymbol{\Gamma}$$

is the equivalent to a curvature-free differential system

(5-16.14) $$d\boldsymbol{\Omega} = -\bar{\boldsymbol{\Gamma}} \wedge \bar{\boldsymbol{\Omega}} + \bar{\boldsymbol{\Sigma}}, \qquad d\bar{\boldsymbol{\Sigma}} = -\bar{\boldsymbol{\Gamma}} \wedge \bar{\boldsymbol{\Sigma}},$$

(5-16.15) $$d\bar{\boldsymbol{\Gamma}} = -\bar{\boldsymbol{\Gamma}} \wedge \bar{\boldsymbol{\Gamma}}, \qquad \bar{\boldsymbol{\Theta}} = \mathbf{0}$$

whose general solution is given by

(5-16.16) $$\boldsymbol{\Omega} = \mathbf{A}(d\boldsymbol{\phi} + \bar{\boldsymbol{\eta}}),$$

(5-16.17) $$\bar{\boldsymbol{\Sigma}} = \mathbf{A}\, d\bar{\boldsymbol{\eta}},$$

(5-16.18) $$\bar{\boldsymbol{\Gamma}} = -d\mathbf{A}\,\mathbf{A}^{-1},$$

with

(5-16.19) $$\mathbf{A} = \mathbf{E} - H(\bar{\boldsymbol{\Gamma}}\mathbf{A}) = \mathbf{E} - H(\boldsymbol{\Gamma}\mathbf{A})$$

(5-16.20) $$\bar{\boldsymbol{\eta}} = H(\mathbf{A}^{-1}\bar{\boldsymbol{\Sigma}}), \qquad \boldsymbol{\phi} = H(\mathbf{A}^{-1}\boldsymbol{\Omega}),$$

and

$$\bar{\Sigma} = \Sigma - (\Gamma - \bar{\Gamma}) \wedge \Omega, \qquad \bar{\eta} = \eta - H(\theta \wedge d\phi).$$

Theorem 5-16.1 provides access to results similar to the Frobenius theorem (Section 4-3) for systems of differential forms of degree greater than one.

Theorem 5-16.2. *Let Ω be a column matrix of k-forms on a starshaped region S of E_n. Ω admits the representation*

$$\Omega = \mathbf{A}\, d\rho \tag{5-16.21}$$

for a column matrix of $(k-1)$-forms ρ if and only if Ω can be completed to a complete differential system that is both curvature free and torsion free.

Proof. Starting with Ω, we complete to obtain a complete differential system (5-16.12), (5-16.13). Application of Theorem 5-16.1 then converts this system of a curvature free differential system (5-16.14), (5-16.15) for which

$$\Omega = \mathbf{A}(d\phi + \bar{\eta}), \qquad \bar{\Sigma} = \mathbf{A}\, d\bar{\eta}, \tag{5-16.22}$$

where $\mathbf{A}$ satisfies $\mathbf{A} = \mathbf{E} - H(\Gamma\mathbf{A})$ and ϕ and $\bar{\eta}$ are antiexact ($\phi = H(A^{-1}\Omega)$, $\bar{\eta} = H(\mathbf{A}^{-1}\bar{\Sigma})$). The first of (5-16.22) shows that $\Omega = \mathbf{A}\, d\phi$ if and only if $\bar{\eta} = \mathbf{0}$. Thus since $\bar{\eta} = H(\mathbf{A}^{-1}\bar{\Sigma})$, we must have $\mathbf{A}^{-1}\bar{\Sigma} \in \ker H$; that is $\mathbf{A}^{-1}\bar{\Sigma} \in \mathscr{A}_{r,1}^{k+1}$. However, the second of (5-16.22) gives $\mathbf{A}^{-1}\bar{\Sigma} = d\bar{\eta}$ and hence $\mathbf{A}^{-1}\bar{\Sigma} \in \mathscr{E}_{r,1}^{k+1}$. Since exact and antiexact forms intersect only in the zero element by Corollary 5-6.3, it follows that $\bar{\Sigma} = \mathbf{0}$ must hold throughout S; the curvature free differential system must also be torsion free. Under these circumstances, we have $\Omega = \mathbf{A}\, d\phi = \mathbf{A}\, d(\phi + d\beta) = \mathbf{A}\, d\rho$ with $\rho = \phi + d\beta$. □

5.17. HORIZONTAL AND VERTICAL IDEALS OF A DISTRIBUTION

A situation at the other extreme is where both the matrices of connection 1-forms and the torsion 2-forms are restricted so that they do not belong to the ideal generated by the entries of Ω. Here it is customary to assume that Ω is a column matrix of 1-forms. Further it is also convenient to replace the underlying space, E_n, by a larger space as will become evident in a moment. The general structure that results from these considerations finds applications in a wide range of both mathematical and physical situations.

Let B be an $(N+n)$-dimensional space that is locally of the form $E_n \times \mathbb{R}_N$. The space B can be given local coordinates $(x^i, q^\alpha | 1 \leq i \leq n, 1 \leq \alpha \leq N)$ for which the natural projection operator π has the representation

$$\pi : B \to E_n | (x^i, q^\alpha) \to (x^i). \tag{5-17.1}$$

A space B with this structure is referred to as a *space of fibers* with fibers $\pi^{-1}(x^i)$ that are diffeomorphic to $\mathbb{R}_N$. The simplest situation in which a space of fibers naturally arises is in the study of systems of partial differential equations. There E_n is the space of independent variables and $\mathbb{R}_N$ is the range space of the dependent variables $(q^1, \dots, q^N)$; that is, B is the graph space of the solutions to the given system of partial differential equations (see Chapter Six).

Let $T(B)$ denote the tangent space of the space of fibers B and consider the N vector fields

$$V_\alpha = \partial_\alpha, \qquad \alpha = 1, \dots, N, \tag{5-17.2}$$

where $\partial_\alpha := \partial/\partial q^\alpha$. The *vertical subspace* of $T(B)$ is defined by

$$\mathscr{V}(B) = \operatorname{span}(V_\alpha | 1 \le \alpha \le N), \tag{5-17.3}$$

since the orbit of any $U \in \mathscr{V}(B)$ passing through (x^i, q^α) is contained in $\pi^{-1}(x^i)$ and is thus tangent to the fibers of B. Let $\{Y_i^\alpha(x^j, q^\beta) | 1 \le i \le n, 1 \le \alpha \le N\}$ be a given system of nN elements of $\Lambda^0(B)$. They give rise to n independent vector fields

$$H_i = \partial_i + Y_i^\alpha(x, q)\partial_\alpha = \partial_i + Y_i^\alpha(x, q)V_\alpha \tag{5-17.4}$$

with $\partial_i := \partial/\partial x^i$. These may be taken as a basis a *horizontal distribution* on B:

$$\mathscr{H}(B) = \operatorname{span}(H_i | 1 \le i \le n). \tag{5-17.5}$$

It is then an elementary calculation to see that $T(B)$ admits the direct sum decomposition

$$T(B) = \mathscr{V}(B) \oplus \mathscr{H}(B)$$

into vertical and horizontal subspaces for any assignment of the Y's.

As with most problems significant simplifications obtain by introducing the natural duals of $\mathscr{V}(B)$ and $\mathscr{H}(B)$. We take the *vertical dual* to be defined by

$$\mathscr{V}^*(B) = \operatorname{span}(V^i | 1 \le i \le n) \tag{5-17.6}$$

with the *vertical* 1-forms

$$V^i = dx^i, \qquad i = 1, \dots, n \tag{5-17.7}$$

(i.e., $V^i = 0 \Leftrightarrow x^i =$ constant). The *horizontal dual*,

$$\mathscr{H}^*(B) = \operatorname{span}(H^\alpha | 1 \le \alpha \le N) \tag{5-17.8}$$

is defined in terms of the *horizontal* 1-forms

$$H^{\alpha} = dq^{\alpha} - Y_i^{\alpha}(x,q)\,dx^i = dq^{\alpha} - Y_i^{\alpha}(x,q)V^i. \tag{5-17.9}$$

The reason for these assignments is that the Y's are always associated with horizontal quantities and we have the fundamental relations

$$V_{\alpha} \lrcorner V^i = 0, \qquad H_j \lrcorner V^i = \delta_j^i, \tag{5-17.10}$$

$$V_{\alpha} \lrcorner H^{\beta} = \delta_{\alpha}^{\beta}, \qquad H_j \lrcorner H^{\beta} = 0, \tag{5-17.11}$$

$$\Lambda^1(B) = \mathscr{V}^*(B) \oplus \mathscr{H}^*(B). \tag{5-17.12}$$

The homogeneous ideal

$$H = I\{H^1, H^2, \dots, H^N\} \tag{5-17.13}$$

of $\Lambda(B)$ that is generated by the horizontal 1-forms is referred to as the *horizontal ideal*. The *vertical ideal* is defined by

$$V = I\{V^1, V^2, \dots, V^n\}.$$

The first of (5-17.10) shows that $\mathscr{V}(B)$ is the characteristic subspace of the vertical ideal, V, while the second of (5-17.11) shows that $\mathscr{H}(B)$ is the characteristic subspace of the horizontal ideal. The remaining equations in the system (5-17.10), (5-17.11) are simply a convenient system of normalization conditions.

Let Ψ be any smooth curve in E_n,

$$\Psi: (a,b) \subset \mathbb{R} \to E_n | x^i = \psi^i(s), \qquad a < s < b,$$

then

$$\Psi_H: (a,b) \to B | x^i = \psi^i(s), \qquad q^{\alpha} = \psi^{\alpha}(s), \qquad a < s < b$$

is the *horizontal lift* of Ψ to B if and only if

$$\Psi^*\mathscr{H}^*(B) = 0 \qquad (\text{i.e., } \Psi^*H = 0).$$

Noting that $\mathscr{H}^*(B)$ is generated by the 1-forms H^{α}, $\Psi^*\mathscr{H}^*(B) = 0$ if and only if $\Psi^*H^{\alpha} = 0$, $1 \le \alpha \le N$; that is,

$$\frac{d\psi^{\alpha}}{ds} = Y_i^{\alpha}(\Psi^j, \Psi^{\beta})\frac{d\psi^i}{ds}, \qquad \alpha = 1, \dots, N.$$

A space of fibers has well defined horizontal lifts, for any sufficiently smooth horizontal distribution assigned by the Y's, by the standard existence and uniqueness theorems for systems of first order ordinary differential equations. Thus any sufficiently smooth curve in the base space E_n can be horizontally lifted to a family of curves in the space of fibers B (simply choose different initial data for the functions $\psi^\alpha(s)$).

The system of defining equations (5-17.10), (5-17.11) leads to some remarkable simplifications.

Lemma 5-17.1. *Any* $\omega \in \Lambda^1(B)$ *has the representation*

$$\omega = (H_i \lrcorner \omega) V^i + (V_\alpha \lrcorner \omega) H^\alpha. \tag{5-17.14}$$

Any $\eta \in \Lambda^2(B)$ *has the representation*

$$\eta = \tfrac{1}{2}(H_j \lrcorner H_i \lrcorner \eta) V^i \wedge V^j + (V_\alpha \lrcorner H_i \lrcorner \eta) V^i \wedge H^\alpha + \tfrac{1}{2}(V_\beta \lrcorner V_\alpha \lrcorner \eta) H^\alpha \wedge H^\beta. \tag{5-17.15}$$

Proof. The direct sum decomposition (5-17.12) shows that any $\omega \in \Lambda^1(B)$ can be written as

$$\omega = \omega_i V^i + \omega_\alpha H^\alpha, \tag{5-17.16}$$

and hence

$$V_\beta \lrcorner \omega = \omega_i V_\beta \lrcorner V^i + \omega_\alpha V_\beta \lrcorner H^\alpha = \omega_\beta,$$

$$H_j \lrcorner \omega = \omega_i H_j \lrcorner V^i + \omega_\alpha H_j \lrcorner H^\alpha = \omega_j$$

when (5-17.10), (5-17.11) are used. Similarly, any $\eta \in \Lambda^2(B)$ can be written as

$$\eta = \tfrac{1}{2}\eta_{ij} V^i \wedge V^j + \eta_{\alpha i} V^i \wedge H^\alpha + \tfrac{1}{2}\eta_{\alpha\beta} H^\alpha \wedge H^\beta$$

with $\eta_{ij} = -\eta_{ji}$ and $\eta_{\alpha\beta} = -\eta_{\beta\alpha}$. The explicit expression (5-17.15) then follows upon computing the indicated inner multiplications and using (5-17.11). □

Starting with the horizontal 1-forms $H^\alpha = dq^\alpha - Y_i^\alpha \, dx^i$, exterior differentiation yields $dH^\alpha = -dY_i^\alpha \wedge dx^i = -dY_i^\alpha \wedge V^i$. However, Lemma 5-17.1 shows that

$$\begin{aligned} dY_i^\alpha &= (H_k \lrcorner dY_i^\alpha) V^k + (V_\beta \lrcorner dY_i^\alpha) H^\beta \\ &= H_k \langle Y_i^\alpha \rangle V^k + V_\beta \langle Y_i^\alpha \rangle H^\beta, \end{aligned}$$

and hence

$$dH^\alpha = -H_k\langle Y_i^\alpha\rangle V^k \wedge V^i - V_\beta\langle Y_i^\alpha\rangle H^\beta \wedge V^i$$

$$= V_\beta\langle Y_i^\alpha\rangle V^i \wedge H^\beta + \tfrac{1}{2}\left(H_i\langle Y_k^\alpha\rangle - H_k\langle Y_i^\alpha\rangle\right)V^k \wedge V^i.$$

Theorem 5-17.1. *A horizontal system*

$$H^\alpha = dq^\alpha - Y_i^\alpha(x,q)\,dx^i \tag{5-17.17}$$

on a space of fibers, B, generates a unique complete differential system if the matrices of connection 1-forms Γ_β^α and torsion 2-forms Σ^α belong to the vertical ideal of $\Lambda(B)$.

Proof. The computation given above shows that

$$dH^\alpha = -\Gamma_\beta^\alpha \wedge H^\beta + \Sigma^\alpha \tag{5-17.18}$$

with

$$\Gamma_\beta^\alpha = -V_\beta\langle Y_i^\alpha\rangle V^i \in \mathscr{V}^*(B), \tag{5-17.19}$$

$$\Sigma^\alpha = \tfrac{1}{2}\left(H_i\langle Y_j^\alpha\rangle - H_j\langle Y_i^\alpha\rangle\right)V^j \wedge V^i \in \mathscr{V}^*(B). \tag{5-17.20}$$

The results then follow upon noting that $\Gamma_\beta^\alpha \wedge H^\beta$ belongs to the horizontal ideal while Σ^α belongs solely to the vertical ideal and hence $\Gamma_\beta^\alpha \mapsto \Gamma_\beta^\alpha + \rho_\beta^\alpha$, $\rho_\beta^\alpha \in \mathscr{H}^*(B)$ will require $\Sigma^\alpha \mapsto \Sigma^\alpha - \rho_\beta^\alpha \wedge H^\beta \notin \mathscr{V}^*(B)$.

The requirements $\Gamma_\beta^\alpha \in \mathscr{V}^*(B)$, $\Sigma^\alpha \in \mathscr{V}^*(B)$ are in fact natural. The torsion 2-forms then represent the part of $\mathscr{H}^*(B)$ that is carried outside of $\mathscr{H}^*(B)$ by exterior differentiation, while $\Gamma_\beta^\alpha \wedge H^\beta$ represents the part of $\mathscr{H}^*(B)$ that remains inside $\mathscr{H}^*(B)$ with coefficient 1-forms Γ_β^α that live only outside $\mathscr{H}^*(B)$. Clearly these circumstances give maximal differential separation between $\mathscr{H}^*(B)$ and $\mathscr{V}^*(B)$; particularly in view of the fact that $dV^i = d\,dx^i = 0$ give $d\mathscr{V}^*(B) \subset \mathscr{V}^*(B)$.

The curvature 2-forms associated with the horizontal ideal $\mathscr{H}^*(B)$ are defined by

$$\theta_\beta^\alpha = d\Gamma_\beta^\alpha + \Gamma_\gamma^\alpha \wedge \Gamma_\beta^\gamma. \tag{5-17.21}$$

It is natural to inquire as to how the curvature 2-forms distribute with respect to the horizontal ideal. A direct analogy with (5-17.14) allows us to write

$$d = dx^i \wedge \partial_i + dq^\alpha \wedge \partial_\alpha = V^i \wedge H_i\langle\ \rangle + H^\alpha \wedge V_\alpha\langle\ \rangle$$

where the operators H_i and V_α are interpreted as acting on the scalar coefficients of the exterior forms involved in the differentiation process. It is then easily seen that (5-17.19) and (5-17.21) yield

$$\theta_\beta^\alpha = -H_k\langle V_\beta\langle Y_i^\alpha\rangle\rangle V^k \wedge V^i - V_\gamma\langle V_\beta\langle Y_i^\alpha\rangle\rangle H^\gamma \wedge V^i$$

$$+ V_\gamma\langle Y_k^\alpha\rangle V_\beta\langle Y_i^\gamma\rangle V^k \wedge V^i.$$

An interchange of the order of action of the operators H_k and V_β in the first term (i.e., $H_k\langle V_\beta\langle \cdot \rangle\rangle = V_\beta\langle H_k\langle \cdot \rangle\rangle + [H_k, V_\beta]\langle \cdot \rangle$) gives

$$\text{(5-17.22)} \qquad \theta_\beta^\alpha = V_\gamma\langle V_\beta\langle Y_i^\alpha\rangle\rangle V^i \wedge H^\gamma + V_\beta\langle H_i\langle Y_k^\alpha\rangle\rangle V^k \wedge V^i.$$

When (5-17.19) and (5-17.20) are used, the following result is obtained.

Lemma 5-17.2. *Let*

$$\text{(5-17.23)} \qquad \Gamma_\beta^\alpha = -V_\beta\langle Y_i^\alpha\rangle V^i, \qquad 2\Sigma^\alpha = \left(H_i\langle Y_j^\alpha\rangle - H_j\langle Y_i^\alpha\rangle\right) V^j \wedge V^i$$

be the unique vertical-valued connection 1*-forms and torsion* 2*-forms of the horizontal ideal* $\mathscr{H}^*(B)$*. The curvature* 2*-forms have the evaluation*

$$\text{(5-17.24)} \qquad \theta_\beta^\alpha = -V_\gamma\langle \Gamma_\beta^\alpha\rangle \wedge H^\gamma + V_\beta\langle \Sigma^\alpha\rangle$$

and

$$\text{(5-17.25)} \qquad \theta_\beta^\alpha \wedge H^\beta = V_\beta\langle \Sigma^\alpha\rangle \wedge H^\beta.$$

Proof. The only result not already verified is (5-17.25). However, (5-17.24) gives

$$\theta_\beta^\alpha \wedge H^\beta = -V_\gamma\langle \Gamma_\beta^\alpha\rangle \wedge H^\gamma \wedge H^\beta + V_\beta\langle \Sigma^\alpha\rangle \wedge H^\beta.$$

Thus since $[V_\alpha, V_\beta] = 0$ and $\Gamma_\beta^\alpha = -V_\beta\langle Y_i^\alpha\rangle V^i$, we have $-V_\gamma\langle \Gamma_\beta^\alpha\rangle \wedge H^\gamma \wedge H^\beta = V_\gamma\langle V_\beta\langle Y_i^\alpha\rangle\rangle V^i \wedge H^\gamma \wedge H^\beta = 0$. □

The Frobenius theorem shows that the horizontal 1-forms $\{H^\alpha\}$ are completely integrable if and only if $d\mathscr{H}^*(B) \subset \mathscr{H}^*(B)$. In view of (5-17.18) we have $dH^\alpha \equiv \Sigma^\alpha \bmod \mathscr{H}^*(B)$. Thus since $\{\Sigma^\alpha\}$ do not belong to $\mathscr{H}^*(B)$, *the horizontal* 1*-forms* $\{H^\alpha\}$ *are completely integrable if and only if the* $\mathscr{V}^*(B)$*-valued torsion* 2*-forms vanish identically*. In this event Lemma 5-17.2 shows that the curvature 2-forms have the evaluation $\theta_\beta^\alpha = -V_\gamma\langle \Gamma_\beta^\alpha\rangle \wedge H^\gamma$ and accordingly vanish only if $V_\gamma\langle V_\beta\langle Y_i^\alpha\rangle\rangle = \partial_\gamma \partial_\beta Y_i^\alpha = 0$. Further it follows directly from (5-17.20) that $\Sigma^\alpha = 0$ if and only if $H_i\langle Y_j^\alpha\rangle = H_j\langle Y_i^\alpha\rangle$, and this is the case if and only if $[H_i, H_j] = 0$ (i.e., simply note that $H_i = \partial_i + Y_i^\alpha \partial_\alpha$). When these equations are written out in full, we obtain

$$\partial_i Y_j^\alpha + Y_i^\beta \partial_\beta Y_j^\alpha = \partial_j Y_i^\alpha + Y_j^\beta \partial_\beta Y_i^\alpha. \tag{5-17.26}$$

5-18. n-FORMS AND INTEGRATION

Let S be a starshaped region of an n-dimensional space M with center P_0 and let $\{x^i\}$ be a preferred coordinate cover of S for which $P_0 : \{x_0^i = 0\}$. The natural volume element (n-form) of S with respect to the coordinate cover $\{x^i\}$ is denoted by

$$\mu = dx^1 \wedge dx^2 \wedge \cdots \wedge dx^n. \tag{5-18.1}$$

Now μ is closed, and hence Theorem 5-6.1 gives the unique representation

$$\mu = dH(\mu). \tag{5-18.2}$$

It also follows from (5-2.2) that $\tilde{\mu}(\lambda) = \mu$, and hence (5-3.1) shows that

$$H(\mu) = \int_0^1 \mathscr{X} \lrcorner \tilde{\mu}(\lambda) \lambda^{n-1}\, d\lambda = \left(\int_0^1 \lambda^{n-1}\, d\lambda \right) \mathscr{X} \lrcorner \mu = \frac{1}{n} \mathscr{X} \lrcorner \mu.$$

However, since $x_0^i = 0$,

$$\mathscr{X} \lrcorner \mu = \left(x^i - x_0^i \right) \partial_i \lrcorner \mu = x^i \mu_i. \tag{5-18.3}$$

It thus follows that $\rho = H(\mu) = (1/n) x^i \mu_i$. Further since $\rho \in \mathscr{A}^{n-1}(S)$, we have $\mathscr{X} \lrcorner \rho = 0$ and $\rho(x_0^i) = 0$. These considerations establish the following results.

Lemma 5-18.1. *Let S be a starshaped region of an n-dimensional space whose center P_0 has the coordinates $x_0^i = 0$ in a preferred coordinate cover $\{x^i\}$ of S, and let μ be the natural volume element of S with respect to $\{x^i\}$. There exists a unique $(n - 1)$-form*

$$\rho = \frac{1}{n} x^i \mu_i \tag{5-18.4}$$

on S such that

$$\mu = d\rho, \tag{5-18.5}$$

$$\mathscr{X} \lrcorner \rho = 0, \qquad \rho\left(x_0^i\right) = 0. \tag{5-18.6}$$

Theorem 5-18.1. *Let S be a starshaped region of an n-dimensional space whose center P_0 has coordinates $x_0^i = 0$ in a preferred coordinate cover $\{x^i\}$ of S, and let*

$$\omega(x) = f(x)\mu \tag{5-18.7}$$

be any n-form on S where μ is the natural volume element of S with respect to the coordinate cover $\{x^i\}$. There exists a unique, antiexact, $(n-1)$-form

$$\rho = F^i\mu_i \tag{5-18.8}$$

on S such that

$$\omega = d\rho, \tag{5-18.9}$$

$$\mathscr{X} \lrcorner \rho = 0, \qquad \rho(x_0^i) = 0, \tag{5-18.10}$$

and the functions $F^i(x)$ are given by

$$F^i = x^i h(f) \tag{5-18.11}$$

where

$$h(f)(x) = \int_0^1 \lambda^{n-1}\tilde{f}(\lambda)\, d\lambda \tag{5-18.12}$$

satisfies

$$nh(f) + x^i\partial_i h(f) = f \tag{5-18.13}$$

identically on S.

Proof. Since ω is an n-form on an n-dimensional region, $d\omega = 0$, Theorem 5-6.1 gives the unique representation $\omega = dH(\omega)$. We set $\rho = H(\omega)$, in which case (5-18.4) and (5-18.5) hold. Now

$$\rho = H(\omega) = H(f\mu) = \int_0^1 \mathscr{X} \lrcorner \widetilde{(f\mu)}(\lambda)\lambda^{n-1}\, d\lambda$$

$$= \mathscr{X} \lrcorner \mu \int_0^1 \tilde{f}(\lambda)\lambda^{n-1}\, d\lambda = h(f)\mathscr{X} \lrcorner \mu$$

since $\tilde{\mu}(\lambda) = \mu$. Thus $\rho = F^i\mu_i$, where $F^i = x^i h(f)$. This establishes (5-18.8) through (5-18.12). Finally we have

$$\omega = f\mu = d\rho = d\big(h(f)x^i\mu_i\big) = d\big(h(f)x^i\big) \wedge \mu_i$$

$$= \partial_j\big(h(f)x^i\big)\, dx^j \wedge \mu_i = \big(x^i\partial_i h(f) + nh(f)\big)\mu.$$

Thus since μ constitutes a basis for $\Lambda^n(S)$, we obtain the identical satisfaction of (5-18.13) on S. □

Theorem 5-18.2. *Let B_n be an open, connected, arc-wise simply connected set whose closure, $\bar{B}_n$, is compact and contained in a starshaped region S of an n-dimensional space. Let the boundary of $\bar{B}_n$ be locally smooth with the exception of a finite number of edges and vertices and let x^i be a preferred coordinate system on S such that the center of S has coordinates $x_0^i = 0$. There exists a unique, antiexact $(n-1)$-form ρ on S, for given $\omega \in \Lambda^n(S)$, such that*

$$\int_{\bar{B}_n} \omega = \int_{\partial \bar{B}_n} \rho \tag{5-18.14}$$

$$\mathscr{X} \lrcorner \rho = 0, \qquad \rho(0) = 0. \tag{5-18.15}$$

This form ρ is determined by

$$\rho = F^i \mu_i, \qquad F^i = h(f) x^i, \qquad h(f)(x) = \int_0^1 \lambda^{n-1} f(\lambda x^i)\, d\lambda \tag{5-18.16}$$

for $\omega = f\mu$.

Proof. Under the given hypotheses Theorem 5-18.1 gives $\omega = d\rho$ where ρ verifies (5-18.15) and (5-18.16) for $\omega = f\mu$, and is unique. The result then follows directly from Stokes' theorem. □

The importance of this result is in its showing that any volume integral over a "nice" n-dimensional region B_n of E_n can be reduced to a surface integral over the boundary of that region. Further the $(n-1)$-form $\rho = F^i \mu_i$ is universal in the sense that its values do not depend on the particulars of either the region or the boundary; that is ρ is defined throughout the region $\bar{B}_n$. Thus if b_n is any "nice" subregion of B_n, we likewise have

$$\int_{\bar{b}_n} \omega = \int_{\partial \bar{b}_n} \rho$$

with the same ρ.

Particular stress has been placed on the fact that ρ is an antiexact $(n-1)$-form. This condition can be relaxed, however. Let us define $\bar{\rho}$ by

$$\bar{\rho} = \rho + d\eta \tag{5-18.17}$$

where η is an arbitrary $(n-2)$-form. Noting that $\partial\partial \bar{B}_n \equiv 0$, Stokes' theorem shows that

$$\int_{\partial \bar{B}_n} (\rho + d\eta) = \int_{\partial \bar{B}_n} \rho + \int_{\partial\partial \bar{B}_n} \eta = \int_{\partial \bar{B}_n} \rho.$$

Thus for any $\omega \in \Lambda^n(E_n)$, we have

$$\int_{\bar{B}_n} \omega = \int_{\partial \bar{B}_n} \bar{\rho}$$

where the exact part of $\bar{\rho}$ is arbitrary and the antiexact part is determined by

$$\bar{\rho}_a = F^i \mu_i .$$

5-19. THE ADJOINT OF A LINEAR OPERATOR ON $\Lambda(E_n)$

There is one further property of the linear homotopy operator that is often of great use, particularly in problems that arise in the calculus of variations. The underlying idea here rests on the linearity property of the operator H, and hence the basic definition will be given in terms of an arbitrary linear operator whose domain and range are contained in the exterior algebra $\Lambda(E_n)$.

Definition. Let L be a linear operator whose domain and range are contained in $\Lambda(E_n)$. If a linear operator L^+ exists such that the domain and range of L^+ are contained in $\Lambda(E_n)$ and

$$\alpha \wedge L(\beta) - L^+(\alpha) \wedge \beta = dP(\alpha, \beta) \tag{5-19.1}$$

for all α and β such that $\alpha \wedge L(\beta) \in \Lambda^n(E_n)$, then L^+ is said to be the *generalized adjoint* of L with bilinear concomitant $P(\alpha, \beta) \in \Lambda^{n-1}(E_n)$. If $P(\alpha, \beta) = 0$, then L^+ is said to be the *adjoint* of L.

The reason why forms of degree n on E_n are involved is as follows. Let B_n be a region of E_n for which Stokes' theorem holds (Theorem 4-7.1), and let L^+ be the generalized adjoint of L with bilinear concomitant $P(\cdot, \cdot)$. An integration of (5-19.1) over B_n and Stokes' theorem gives

$$\int_{B_n} \{\alpha \wedge L(\beta) - L^+(\alpha) \wedge \beta\} = \int_{\partial B_n} P(\alpha, \beta), \tag{5-19.2}$$

while a relaxation of the condition $\alpha \wedge L(\beta) \in \Lambda^n(E_n)$ would not provide the integral formula (5-19.2) that is easily recognized as the "standard" definition of the generalized adjoint.

Generalized adjoints of some of the linear operators we have already encountered follow directly from the definition.

Lemma 5-19.1. *The generalized adjoint of the Lie derivative and the bilinear concomitant are given by*

$$\text{(5-19.3)} \qquad \pounds_V^+ = -\pounds_V, \qquad P(\alpha,\beta) = V \lrcorner (\alpha \wedge \beta).$$

Proof. For any α, β such that $\deg(\alpha \wedge \beta) = n$, we have

$$\pounds_V(\alpha \wedge \beta) = (\pounds_V\alpha) \wedge \beta + \alpha \wedge \pounds_V\beta = V \lrcorner d(\alpha \wedge \beta) + d\{V \lrcorner (\alpha \wedge \beta)\}.$$

Noting that $d(\alpha \wedge \beta) \in \Lambda^{n+1}(E_n)$, $d(\alpha \wedge \beta) = 0$ and we have

$$\alpha \wedge \pounds_V\beta - (-\pounds_V\alpha) \wedge \beta = d\{V \lrcorner (\alpha \wedge \beta)\}$$

for all α, β such that $\alpha \wedge \pounds_V\beta \in \Lambda^n(E_n)$. □

Lemma 5-19.2. *The generalized adjoint of the exterior derivative and the bilinear concomitant are given by*

$$\text{(5-19.4)} \qquad d^+ = (-1)^{\deg(\alpha)+1} d, \qquad P(\alpha,\beta) = (-1)^{\deg(\alpha)} \alpha \wedge \beta.$$

Proof. The proof follows the same lines as that given above on noting that $d(\alpha \wedge \beta) = (d\alpha) \wedge \beta + (-1)^{\deg(\alpha)}\alpha \wedge d\beta$ for all α, β such that $\deg(\alpha \wedge d\beta) = n$. □

The adjoint of the linear homotopy operator can be obtained by straightforward but messy explicit calculations that are not in the least instructive. We therefore simply state the result and refer the reader to the account given by Edelen.

Theorem 5-19.1. *Let S be a starshaped region of n-dimensional space with center P_0 and let $\bar{S}$ denote the closure of S. The homotopy operator H with center P_0 has an adjoint operator H^+ that is defined for any $\alpha(x) \in \Lambda^k(S)$ by*

$$\text{(5-19.5)} \qquad H^+(\alpha) = (-1)^{k+1} \int_1^\infty \mathscr{X} \lrcorner \widetilde{(e\alpha)}(\lambda) \lambda^{k-1}\, d\lambda$$

where e is the characteristic function of $\bar{S}$. Thus

$$\text{(5-19.6)} \qquad H^+(\alpha)(x) = (-1)^{k+1} \int_1^\infty e(\lambda x) \mathscr{X}(x) \lrcorner \alpha(\lambda x) \lambda^{k-1}\, d\lambda$$

with $x_0^i = 0$ and

$$\text{(5-19.7)} \qquad e(\lambda x) = \begin{cases} 1 & \text{for } \lambda x \in \bar{S} \\ 0 & \text{for } \lambda x \notin \bar{S}. \end{cases}$$

Corollary 5-19.1. *The operator H^+ is a linear operator that maps $\Lambda^k(S)$ into $\mathscr{A}^{k-1}(S)$.*

Proof. That H^+ maps $\Lambda^k(S)$ into $\Lambda^{k-1}(S)$ follows directly from (5-19.5) or (5-19.6), and the linearity of H^+ is likewise immediate. Since $\mathscr{X}(x)$ does not depend on λ, it can be taken outside the integral in (5-19.6) so as to obtain

$$H^+(\alpha)(x) = (-1)^{k+1}\mathscr{X}(x)\lrcorner\int_1^\infty e(\lambda x)\alpha(\lambda x)\lambda^{k-1}\,d\lambda. \tag{5-19.8}$$

We thus have

$$\mathscr{X}(x)\lrcorner H^+(\alpha)(x) = 0, \qquad H^+(\alpha)(x_0) = 0, \tag{5-19.9}$$

where the last equality follows from $\mathscr{X}(x_0) = 0$ and the fact that the integral is finite due to the continuity of α and the presence of the characteristic function e of $\bar{S}$. Thus $H^+(\alpha) \in \mathscr{A}^{k-1}(S)$. □

Since the image of any α under H^+ is antiexact, we have the following immediate result.

Corollary 5-19.2. *The operator H^+ satisfies*

$$HH^+\alpha \equiv 0, \qquad H^+H\alpha \equiv 0. \tag{5-19.10}$$

Let α and β be forms such that $\alpha \wedge \beta \in \Lambda^n(S)$. If we use property $\mathbf{H}_2$, it follows that $\alpha \wedge \beta = \alpha \wedge dH(\beta) + \alpha \wedge H(d\beta)$, and hence

$$\int_{\bar{S}}\alpha \wedge \beta = \int_{\bar{S}}\alpha \wedge dH(\beta) + \int_{\bar{S}}\alpha \wedge H(d\beta). \tag{5-19.11}$$

Now let $\alpha \in \Lambda^k(S)$. In this event, $\alpha \wedge dH(\beta) = (-1)^k d(\alpha \wedge H(\beta)) + (-1)^{k+1} d\alpha \wedge H(\beta)$, so that

$$\begin{aligned}\int_{\bar{S}}\alpha \wedge dH(\beta) &= (-1)^k\int_{\partial\bar{S}}\alpha \wedge H(\beta) + (-1)^{k+1}\int_{\bar{S}}d\alpha \wedge H(\beta)\\ &= (-1)^k\int_{\partial\bar{S}}\alpha \wedge H(\beta) + (-1)^{k+1}\int_{\bar{S}}H^+(d\alpha) \wedge \beta.\end{aligned}$$

Similarly,

$$\begin{aligned}\int_{\bar{S}}\alpha \wedge H(d\beta) &= \int_{\bar{S}}H^+(\alpha) \wedge d\beta\\ &= (-1)^{k-1}\int_{\partial\bar{S}}H^+(\alpha) \wedge \beta + (-1)^k\int_{\bar{S}}dH^+(\alpha) \wedge \beta.\end{aligned}$$

A combination of these equalities yields

$$\text{(5-19.12)}\quad \int_{\bar{S}} \alpha \wedge \beta = \int_{\bar{S}} \alpha \wedge (dH + Hd)(\beta)$$

$$= \int_{\bar{S}} \{(-1)^k (dH^+ - H^+ d)(\alpha)\} \wedge \beta$$

$$+ \int_{\partial \bar{S}} (-1)^k \{\alpha \wedge H(\beta) - H^+(\alpha) \wedge \beta\}.$$

This yields the following useful result.

Lemma 5-19.3. *The generalized adjoint of* $dH + Hd$ *is given, for any* $\alpha \in \Lambda^k(S)$, *by*

$$\text{(5-19.13)}\quad (dH + Hd)^+(\alpha) = (-1)^k (dH^+ - H^+ d)(\alpha)$$

with the bilinear concomitant

$$\text{(5-19.14)}\quad P(\alpha, \beta) = (-1)^n \{\alpha \wedge H(\beta) - H^+(\alpha) \wedge \beta\}$$

for $\alpha \wedge \beta \in \Lambda^n(S)$.

This result may seem strange on first examination, for $dH + Hd =$ identity on S. However, it is easily seen that

$$\text{(5-19.15)}\quad \int_{\bar{S}} \alpha \wedge d\beta = \int_{\bar{S}} \{(-1)^{k+1} d\alpha\} \wedge \beta + \int_{\partial \bar{S}} (-1)^k \alpha \wedge \beta$$

for any α and β such that $\alpha \wedge d\beta \in \Lambda^n(S)$. The occurrence of a nonzero bilinear concomitant thus comes about from the presence of the operator d (see Lemma 5-19.2).

PROBLEMS

5.1. Let E_4 have the coordinate cover (x, y, z, t) and let

$$\omega^1 = y^3\, dx + xz\, dy + (1 - t^2)\, dz + 6y^5\, dt,$$

$$\omega^2 = 6A\, dx \wedge dy + 3y^2 dx \wedge dz + (z + 5xt^3)\, dy \wedge dt,$$

$$\omega^3 = te^{2x}\, dx \wedge dy \wedge dz,$$

$$\omega^4 = (A + 6x + y^3 + t^4)\, dx \wedge dy \wedge dz \wedge dt,$$

and let ω_e denote the exact part of ω; ω_a denote the antiexact part of ω. Find

(a) $H\omega^1, \omega_e^1, \omega_a^1$; (b) $H\omega^2, \omega_e^2, \omega_a^2$,

(c) $H\omega^3, \omega_e^3, \omega_a^3$; (d) $H\omega^4, \omega_e^4, \omega_a^4$.

Answers

(a)
$$H\omega^1 = \frac{1}{12}(3xy^3 + 4xyz + 12y^5t - 4zt^2 + 12z),$$

$$\omega_e^1 = \frac{y}{12}(3y^2 + 4z)\,dx + \frac{1}{12}(9xy^2 + 4xz + 60y^4t)\,dy + \frac{1}{3}(xy - t^2 + 3)\,dz + \frac{1}{3}(3y^5 - 2zt)\,dt,$$

$$\omega_a^1 = \frac{1}{12}(9y^3 - 4yz)\,dx + \frac{1}{12}(12xy - 9xy^2 - 4xz - 60y^4t)\,dy - \frac{1}{3}(xy + 2t^2)\,dz + \frac{1}{3}(15y^5 + 2zt)\,dt.$$

(b)
$$H\omega^2 = -\frac{3y}{4}(4A + yz)\,dx + \frac{1}{6}(18Ax - 5xt^4 - 2zt)\,dy + \frac{3}{4}xy^2\,dz + \frac{y}{6}(5xt^3 + 2z)\,dz,$$

$$\omega_e^2 = \frac{1}{3}y\,dz \wedge dt + \frac{1}{6}(25xt^3 + 4z)\,dy \wedge dt + \frac{3}{2}y^2\,dx \wedge dz + \frac{1}{6}(9xy + 2t)\,dy \wedge dz + \frac{5}{6}yt^3\,dx \wedge dt + \frac{1}{6}(36A + 9yz - 5t^4)\,dx \wedge dy,$$

$$\omega_a^2 = -\frac{1}{3}y\,dz \wedge dt + \frac{1}{6}(5xt^3 + 2z)\,dy \wedge dt - \frac{1}{6}(9xy + 2t)\,dy \wedge dz - \frac{5}{6}yt^3\,dx \wedge dt + \frac{3}{2}y^2\,dx \wedge dz + \frac{1}{6}(5t^4 - 9yz)\,dx \wedge dy.$$

(c)
$$H\omega^3 = \frac{te^{2x}}{8x^3}(4x^3 - 6x^2 + 6x - 3 + 3e^{-2x})\,dy \wedge dz + \frac{yte^{2x}}{8x^4}(-4x^3 + 6x^2 - 6x + 3 - 3e^{-2x})\,dx \wedge dz + \frac{zte^{2x}}{8x^4}(4x^3 - 6x^2 + 6x - 3 + 3e^{-2x})\,dx \wedge dy,$$

$$\begin{aligned}\omega_e^3 =& \frac{e^{2x}}{8x^3}(4x^3 - 6x^2 + 6x - 3 + 3e^{-2x})\, dy \wedge dz \wedge dt\\ &+ \frac{ye^{2x}}{8x^4}(-4x^3 + 6x^2 - 6x + 3 - 3e^{-2x})\, dx \wedge dz \wedge dt\\ &+ \frac{ze^{2x}}{8x^4}(-4x^3 - 6x^2 + 6x - 3 + 3e^{-2x})\, dx \wedge dy \wedge dt\\ &+ \frac{te^{2x}}{8x^4}(8x^4 - 4x^3 + 6x^2 - 6x + 3 - 3e^{-2x})\, dx \wedge dy \wedge dz,\end{aligned}$$

$$\begin{aligned}\omega_a^2 =& \frac{e^{2x}}{8x^3}(-4x^3 + 6x^2 - 6x + 3 - 3e^{-2x})\, dy \wedge dz \wedge dt\\ &+ \frac{ye^{2x}}{8x^4}(4x^3 - 6x^2 + 6x - 3 + 3e^{-2x})\, dx \wedge dz \wedge dt\\ &+ \frac{ze^{2x}}{8x^4}(-4x^3 + 6x^2 - 6x + 3 - 3e^{-2x})\, dx \wedge dy \wedge dt\\ &+ \frac{e^{2x}}{8x^4}(8x^5 - 8x^4t + 4x^3t - 6x^2t + 6xt - 3t\\ &+ 3te^{-2x})\, dx \wedge dy \wedge dz.\end{aligned}$$

(d)
$$\begin{aligned}H\omega^4 =& \frac{1}{280}\{x(70A + 336x + 40y^3 + 35t^4)\, dy \wedge dz \wedge dt\\ &- y(70A + 336x + 40y^3 + 35t^4)\, dx \wedge dz \wedge dt\\ &+ z(70A + 336x + 40y^3 + 35t^4)\, dx \wedge dy \wedge dt\\ &- t(70A + 336x + 40y^3 + 35t^4)\, dx \wedge dy \wedge dt\},\end{aligned}$$

$$\omega_e^4 = (A + 6x + y^3 + t^4)\, dx \wedge dy \wedge dz \wedge dt,$$

$$\omega_a^4 = 0.$$

5.2. Show that

$$\begin{aligned}H\left(\frac{1}{z+1}(x^2\, dz + t^3\, dy + dx)\right) =& \frac{1}{6z^4}(6x^2zt\ln(z+1) + 6xz^3\ln(z+1)\\ &- 6yt^3\ln(z+1) + 3x^2z^3t - 6x^2z^2t\\ &+ 2yz^3t^3 - 3yz^2t^3 + 6yzt^3).\end{aligned}$$

5.3. Show that

$$\begin{aligned}H\omega &= H(\cos(3t)\, dy + \sin(z)\, dx)\\ &= \frac{1}{3zt}(yz\sin(3t) - 3xt\cos(z)),\end{aligned}$$

and that

$$\omega_e = \frac{1}{3z^2t^2}\{yz^2(3t\cos(3t) - \sin(3t))\,dt + 3xt^2(\cos(z) - z\sin(z) - 1)\,dz + z^2t\sin(3t)\,dy + 3zt^2(1 - \cos(z))\,dx\},$$

$$\omega_a = \frac{1}{3z^2t^2}\{yz^2(\sin(3t) - 3t\cos(3t))\,dt + 3xt^2(1 - \cos(z) - z\sin(z))\,dz + z^2t(3t\cos(3t) - \sin(3t))\,dy + 3zt^2(\cos(z) + z\sin(z) - 1)\,dx\}.$$

5.4. Show that

$$H\left(\frac{x^3t}{1+y}\,dx\wedge dy\wedge dz + ye^{6z}\,dx\wedge dz\wedge dt\right)$$

$$= \frac{1}{1080y^7z^4}\{5xy^8e^{6z}(36z^3 - 18z^2 + 6z - 1 + e^{-6z})\,dz\wedge dt + 18x^4z^4t(60\ln(y+1) + 10y^6 - 12y^5 + 15y^4 - 20y^3 + 30y^2 - 60y)\,dy\wedge dz + 5y^8ze^{6z}(-36z^3 + 18z^2 - 6z + 1 - e^{-6z})\,dx\wedge dt + yt(180e^{6z}y^7z^3 - 90e^{6z}y^7z^2 + 30e^{6z}y^7z - 5e^{6z}y^7 - 1080x^3y^4\ln(y+1) - 180x^3y^6z^4 + 216x^3y^5z^4 - 270x^3y^4z^4 + 360x^3y^3z^4 - 540x^3y^2z^4 + 1080x^3yz^4 + 5y^7)\,dx\wedge dz + 18x^3z^5t(60\ln(y+1) + 10y^6 - 12y^5 + 15y^4 - 20y^3 + 30y^2 - 60y)\,dx\wedge dy\}.$$

5.5. Use $H\,dH\,\omega = H\omega$ and $dH\,d\omega = d\omega$ to show that

$$(\exp(t\,dH))\,\omega = H\,d\omega + e^t dH\,\omega,$$

$$(\exp(tHd))\,\omega = dH\,\omega + e^t H\,d\omega.$$

5.6. Show that a function f that satisfies the Riemann-Graves integral equation $f = 1 + H(f\,du)$, where u is a given function that vanishes at the center of the homotopy operator H, is given by $f = \exp(u)$.

What is the solution of $f = k + H(f\,du)$, where $k \neq 0$ is a given constant and u vanishes at the center of H? Investigate what happens when u is not required to vanish at the center of H for $f = k + H(f\,du)$.

REFERENCES FOR FURTHER STUDY

Bishop, R. L. and R. J. Crittenden, *Geometry of Manifolds*. Academic Press, New York, 1964.

Cartan, H., *Differential Calculus*. Houghton Mifflin, Boston, 1971.

Cartan, H., *Differential Forms*. Houghton Mifflin, Boston, 1970.

Choquet-Bruhat, Y., C. Dewitt-Morette and M. Dillard-Bleick, *Analysis, Manifolds and Physics*. North-Holland, Amsterdam, 1977.

Edelen, D. G. B., *Isovector Methods for Equations of Balance*. Sijthoff and Noordhoff, Alphen aan den Rijn, The Netherlands, 1980.

Flanders, H., *Differential Forms with Applications to Physical Sciences*. Academic Press, New York, 1963.

Lang, S., *Introduction to Differentiable Manifolds*. John Wiley, New York, 1962.

Loomis, L. H. and S. Sternberg, *Advanced Calculus*. Addison-Wesley, Reading, Massachusetts, 1968.

Lovelock, D. and H. Rund, *Tensors, Differential Forms, and Variational Principles*. Wiley-Interscience, New York, 1975.

Schouten, J. A., *Ricci-Calculus*, sec. ed. Springer-Verlag, Berlin, 1954.

Ślebodziński, W., *Exterior Forms and their Applications*. Polish Scientific Publishers, Warsaw, 1970.

Sternberg, S., *Lectures on Differential Geometry*. Prentice-Hall, Englewood Cliffs, 1964.

von Westenholz, C., *Differential Forms in Mathematical Physics*. North-Holland, Amsterdam, 1981.

Warner, F. W., *Foundations of Differentiable Manifolds and Lie Groups*. Scott-Foresman, Dallas, 1971.

PART TWO

APPLICATIONS TO MATHEMATICS

CHAPTER SIX

ISOVECTOR METHODS FOR SECOND-ORDER PARTIAL DIFFERENTIAL EQUATIONS

6-1. THE GRAPH SPACE OF SOLUTIONS TO PARTIAL DIFFERENTIAL EQUATIONS

The topics of this chapter are the algebraic and geometric structures that are naturally associated with solutions of systems of nonlinear partial differential equations of the second order. Since it is the structures of the solutions rather than the solutions themselves that is of interest, realization of the solutions in terms of geometric and algebraic constructs proves to be most useful. Now, these constructs should represent the intrinsic structure of the solutions of the system of partial differential equations under study, rather than a geometric structure that is induced by the intrinsic geometry of the space of independent variables. Accordingly, it is sufficient for our purposes to take the space of independent variables to be an n-dimensional number space E_n with a fixed Cartesian coordinate cover (x^i). Other coordinate covers and specific intrinsic geometric structures of the independent variable space can be introduced subsequently by standard techniques of modern differential geometry.

Suppose that we have N functions $\{\phi^\alpha(x^j), 1 \leq \alpha \leq N\}$ that constitute a solution of the system of partial differential equations under study. The simplest geometric structure that can be associated with these functions is their graph. To this end, we introduce a *graph space* $\mathscr{G} = E_n \times \mathbb{R}_N$ with a global Cartesian coordinate cover (x^i, q^α), $i = 1, \ldots, n$, $\alpha = 1, \ldots, N$, and realize the solution functions in terms of a mapping from E_n into $\mathcal{G}$. Clearly some care must be exercised here, for the mapping of E_n into $\mathcal{G}$ should characterize a subset of $\mathcal{G}$ that is n-dimensional.

Let Φ denote a map from an n-dimensional point set B_n of E_n into $\mathcal{G}$. Such a map is said to be *regular* if

$$\Phi^*\mu = \Phi^*(dx^1 \wedge dx^2 \wedge \cdots \wedge dx^n) \neq 0 \tag{6-1.1}$$

throughout B_n. The collection of all regular maps of B_n into $\mathcal{G}$ is denoted by $R(B_n)$,

$$R(B_n) = \{\Phi : B_n \to \mathcal{G} | \Phi^*\mu \neq 0\}. \tag{6-1.2}$$

Regular maps are realized by

$$\Phi | x^i = \phi^i(\tau^i), \qquad q^\alpha = \bar{\phi}^\alpha(\tau^j) \tag{6-1.3}$$

where (τ^j) denotes the coordinates of B_n relative to a fixed coordinate cover of E_n. Since Φ is regular,

$$\Phi^*\mu = \frac{\partial(\phi^1, \phi^2, \ldots, \phi^n)}{\partial(\tau^1, \tau^2, \ldots, \tau^n)} d\tau^1 \wedge d\tau^2 \wedge \cdots \wedge d\tau^n \neq 0,$$

we may solve for the parameters (τ^j) in terms of the x's, at least locally, to obtain $\tau^j = m^j(x^i)$. Composition of this with the second of (6-1.3) then yields

$$q^\alpha = \bar{\phi}^\alpha(m^j(x^k)) = \phi^\alpha(x^k). \tag{6-1.4}$$

Thus every regular map Φ of B_n into $\mathcal{G}$ yields the graph $(x^i = \phi^i(\tau^j)$, $q^\alpha = \bar{\phi}^\alpha(\tau^j))$ that constitutes an n-dimensional surface in $\mathcal{G}$. Of course, if we are given a solution set $\{\phi^\alpha(x^j)\}$, then the admissible choice $x^i = \phi^i(\tau^j) = \tau^i$ $q^\alpha = \bar{\phi}^\alpha(\tau^j) = \phi^\alpha(x^j)$ gives the graph of such a solution set directly as a map from B_n to $\mathcal{G}$. From this point of view, the parameters (τ^i) are superfluous. They are introduced, however, in order to keep the distinction between E_n and $\mathcal{G}$ quite clear. On the other hand, the parameters (τ^j) often prove to be useful, for (6-1.3) can be viewed as a parametric description that serves to define the solution set implicitly. Such implicitly defined solution sets are often the rule rather than the exception when dealing with nonlinear second-order partial differential equations.

If $f = f(x^i, q^\alpha)$ is a function defined on $\mathcal{G}$, then composition with a regular map

$$\Phi : B_n \to \mathcal{G} | x^i = \tau^i, \qquad q^\alpha = \phi^\alpha(\tau^j)$$

yields

$$F(x^j) = f \circ \Phi = f(x^j, \phi^\alpha(x^j)).$$

The total x^i-wise derivative of $F = f \circ \Phi$ is denoted by D_i so that

(6-1.5) $$D_i F = \partial_i f\left(x^j, \phi^\alpha(x^j)\right)\big|_{\phi^\alpha} + \frac{\partial f\left(x^j, \phi^\alpha(x^j)\right)}{\partial \phi^\beta(x^j)} \partial_i \phi^\beta(x^j).$$

The operator ∂_i is thus used throughout these discussions to denote *explicit* x^i-wise differentiation; i.e., differentiation of functions of several variables with all variables but the explicit occurrences of the x's held fixed.

6-2. KINEMATIC SPACE AND THE CONTACT 1-FORMS

Characterization and study of systems of second order partial differential equations demand a direct access to the first derivatives of the dependent variables with respect to the independent variables. This is most easily accomplished by imbedding graph space $\mathcal{G}$ in a larger space whose coordinate cover will contain "place holders" for the first derivatives. To this end, we introduce $(n + N + nN)$-dimensional *kinematic space* $K = \mathcal{G} \times \mathbb{R}_{nN}$ with the global coordinate cover $(x^i, q^\alpha, y_i^\alpha)$, $i = 1, \ldots, n$, $\alpha = 1, \ldots, N$. Since K has the product structure $\mathcal{G} \times \mathbb{R}_{nN}$, there is a natural projection

(6-2.1) $$\pi : K \to \mathcal{G} \,|\, \left(x^i, q^\alpha, y_i^\alpha\right) \mapsto \left(x^i, q^\alpha\right).$$

The induced map π^* thus pulls any exterior form on $\mathcal{G}$ up to an exterior form on K, and hence K inherits any and all exterior differential structures of $\mathcal{G}$. On the other hand, a differential structure on K that contains any of the natural basis elements $\{dy_i^\alpha\}$ cannot arise from pulling a differential structure from $\mathcal{G}$ up to K. This situation is further compounded by noting that $\pi^*\Lambda^0(\mathcal{G})$ yields only functions on K that do not depend on the coordinates (y_i^α). Accordingly, differential forms on K that are linear combinations of the base elements $\{dx^i, dq^\alpha\}$ and their exterior products with coefficients that depend on the y's cannot be obtained by pulling up exterior structures from $\mathcal{G}$ to K!

The new variables (y_i^α) of K allow us to introduce N nontrivial 1-forms

(6-2.2) $$C^\alpha = dq^\alpha - y_i^\alpha \, dx^i, \qquad \alpha = 1, \ldots, N,$$

the *contact forms* of K. Since

(6-2.3) $$C^1 \wedge C^2 \wedge \cdots \wedge C^N = dq^1 \wedge dq^2 \wedge \cdots \wedge dq^N + \cdots \neq 0,$$

the N contact forms are independent. Further $dC^\alpha = -dy_i^\alpha \wedge dx^i$ implies

(6-2.4) $$C^1 \wedge C^2 \wedge \cdots \wedge C^N \wedge dC^\alpha \neq 0, \qquad \alpha = 1, \ldots, N.$$

and

(6-2.5) $$C^\alpha \wedge (dC^\alpha)^{(n)} \neq 0, \qquad (dC^\alpha)^{(n+1)} = 0,$$

where $(\cdot)^{(k)}$ denotes the kth exterior power. Thus, each of the N contact forms had Darboux class $2n+1$.

Irrespective of the motivation for introduction of the new coordinates (y_i^α), the coordinates (y_i^α) are as yet quite arbitrary. We now give them a precise identification. Let $\Phi: B_n \to \mathscr{G} | x^i = \phi^i(\tau^j), q^\alpha = \bar{\phi}^\alpha(\tau^j)$ be a regular map. We extend Φ to a map of B_n into K by the requirements

$$\text{(6-2.6)} \qquad \Phi^* C^\alpha = 0, \qquad \alpha = 1, \dots, N;$$

that is, *the section of K that is generated by any regular section of $\mathscr{G}$ annihilates the contact forms of K*. A direct calculation based on (6-2.2) gives

$$\text{(6-2.7)} \quad \Phi^* C^\alpha = \Phi^*\left(dq^\alpha - y_i^\alpha\, dx^i\right) = d\bar{\phi}^\alpha(\tau^j) - \Phi^*(y_i^\alpha)\, d\phi^i(\tau^j).$$

Since $\Phi: B_n \to \mathscr{G}$ is regular, the lift of Φ to K yields

$$\text{(6-2.8)} \qquad \Phi^*(y_i^\alpha) = \partial\bar{\phi}^\alpha(\tau^j)/\partial\phi^i(\tau^j) = \partial\phi^\alpha(x^k)/\partial x^i$$

when (6-1.4) is used. Although the coordinates (y_i^α) of K are arbitrary and independent of the other coordinates (x^i, q^α), the values of the y's on any section Φ of K such that $\Phi^*\mu \neq 0$, $\Phi^* C^\alpha = 0$, $\alpha = 1, \dots, N$ are given by $\Phi^*(y_i^\alpha) = \partial\phi^\alpha(x^k)/\partial x^i$. This, however, is exactly what is required in order that the y's shall serve as place holders that assign the first derivatives of the dependent variables on any regular section of K.

Definition. A map $\Phi: B_n \to K$ is *regular* if and only if

$$\text{(6-2.9)} \qquad \Phi^*\mu \neq 0; \qquad \Phi^* C^\alpha = 0,\ \alpha = 1, \dots, N.$$

The collection of all regular maps of B_n to K is denoted by

$$\text{(6-2.10)} \quad R(B_n) = \left\{\Phi: B_n \to K | \Phi^*\mu \neq 0;\ \Phi^* C^\alpha = 0,\ 1 \leq \alpha \leq N\right\}.$$

The following view now emerges. The space K is an arbitrary $(n + N + nN)$-dimensional number space with coordinate cover $(x^i, q^\alpha, y_i^\alpha)$, so that the x's, the q's and the y's can take arbitrary values and can be incremented independently of one another. The restriction of maps from B_n to K to be regular maps then defines all possible n-dimensional sections of K for which the y's become the derivatives of the q's with respect to the x's.

6-3. THE CONTACT IDEAL AND ITS ISOVECTORS

A regular map $\Phi: B_n \to K$ annihilates all N contact 1-forms of K. It thus follows that $\Phi^*\rho = 0$ for any $\rho = \gamma_\alpha \wedge C^\alpha$, and hence a regular map annihilates (solves) the ideal

$$\text{(6-3.1)} \qquad C = I\{C^1, C^2, \dots, C^N\}$$

of $\Lambda(K)$ that is generated by the contact 1-forms.

Definition. The ideal C that is defined by (6-3.1) is referred to as the *contact ideal* of $\Lambda(K)$.

We note that $dC^\alpha = -dy_i^\alpha \wedge dx^i$ and $\Phi^*(y_i^\alpha) = \partial_i \phi^\alpha(x^k)$ give

$$\Phi^*(dC^\alpha) = -d(\partial_i \phi^\alpha(x^k)) \wedge dx^i = -(\partial_j \partial_i \phi^\alpha(x^k))\, dx^j \wedge dx^i = 0 \tag{6-3.2}$$

for any regular map. Thus any regular map $\Phi: B_n \to K$ annihilates the closed ideal

$$\bar{C} = I\{C^1, \ldots, C^N, dC^1, \ldots, dC^N\} \tag{6-3.3}$$

of $\Lambda(K)$ that is generated by the contact 1-forms. We thus have the following result.

Lemma 6-3.1. *A map $\Phi: B_n \to K$ is regular if and only if*

$$\Phi^*\mu \neq 0, \qquad \Phi^*\bar{C} = 0. \tag{6-3.4}$$

Thus $\Phi: B_n \to K$ is regular if and only if $\Phi^*\mu \neq 0$ and Φ solves the contact ideal C.

Proof. The definition given just before Lemma 4-6.1 shows that Φ solves the ideal C if and only if Φ^* annihilates all exterior forms in C. We have just seen, however, that any regular map annihilates any $\rho \in C$, in which case it also annihilates any $\eta \in \bar{C}$. Conversely, if Φ^* solves $\bar{C}$ then Lemma 4-6.4 shows that we must have

$$\Phi^* C^\alpha = 0, \qquad \Phi^* \, dC^\alpha = 0, \qquad \alpha = 1, \ldots, N.$$

However, $\Phi^* \, dC^\alpha = d\Phi^* C^\alpha$ and hence $\Phi^* C^\alpha = 0$ implies $\Phi^* \, dC^\alpha = 0$. □

The reason for replacing the conditions $\Phi^* C^\alpha = 0$ by the equivalent statement $\Phi^*\bar{C} = 0$ is that all of the results that were established in Section 4-6 on transport properties along orbits of vector fields become immediately accessible. Let $V \in T(K)$, then V has the representation

$$V = v^i \partial_i + v^\alpha \partial_\alpha + v_i^\alpha \partial_\alpha^i \tag{6-3.5}$$

where v^i, v^α, v_i^α are functions on K (i.e., elements of $\Lambda^0(K)$) and we have used the notation

$$\partial_i := \frac{\partial}{\partial x^i}, \qquad \partial_\alpha := \frac{\partial}{\partial q^\alpha}, \qquad \partial_\alpha^i := \frac{\partial}{\partial y_i^\alpha}.$$

By definition the contact ideal C is stable under transport along the orbits of $V \in T(K)$ if and only if

$$\pounds_V C \subset C, \tag{6-3.6}$$

in which case V is an *isovector* of the contact ideal. The collection of all isovectors of the contact ideal is denoted by

$$TC(K) = \{V \in T(K) | \pounds_V C \subset C\}. \tag{6-3.7}$$

Theorem 4-6.4 then leads directly to the following results.

Theorem 6-3.1. *The isovectors of the contact ideal C of $\Lambda(K)$ form a Lie subalgebra of $T(K)$ that is the Lie algebra of a Lie group, the isogroup of the contact ideal.*

The situation is actually stronger than this since $\bar{C}$ is the closure of C. Lemma 4-6.2 then shows that *the closed ideal $\bar{C}$ is also stable under transport along the orbits of any $V \in TC(K)$* and that $TC(K)$ is the largest Lie subalgebra of $T(K)$ for which this is true.

Now that isovectors of the contact ideal have emerged, we have some immediate and very useful information concerning the collection $R(B_n)$ of all regular maps (see Fig. 15).

Theorem 6-3.2. *Let V be an isovector field of the contact ideal and let Φ be a regular map of B_n into K. Then V embeds Φ in a 1-parameter family of regular maps*

$$\Phi_V(s) = T_V(s) \circ \Phi \tag{6-3.8}$$

for s in a sufficiently small neighborhood of $s = 0$. Here $T_V(s): K \to K$ is the 1-parameter family of maps that is generated by the flow of V.

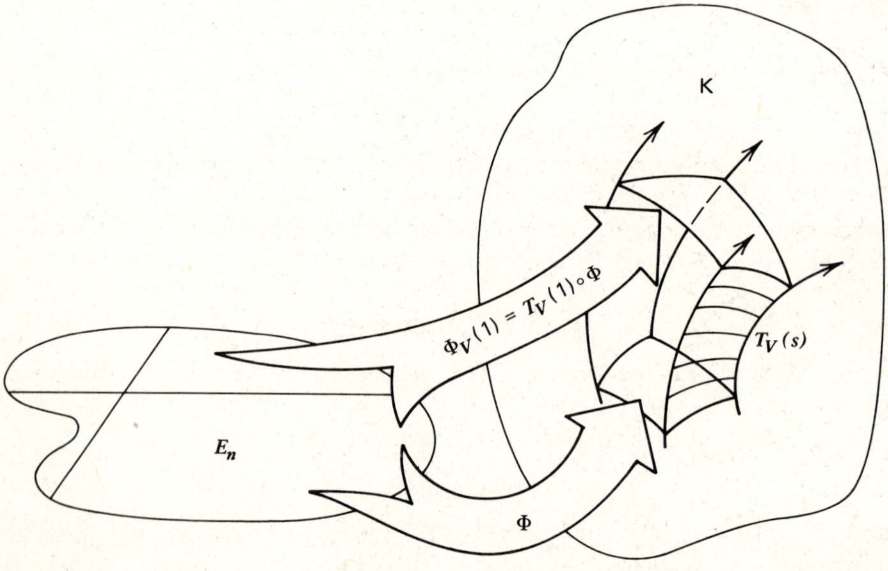

FIGURE 15. Composition of a regular map $\Phi :_n \to K$ with the flow $T_V(s)$ of a vector field V

Proof. Lemma 6-3.1 shows that Φ is a regular map if and only if

$$\Phi^*\mu \neq 0 \tag{6-3.9}$$

and Φ solves the contact ideal, $\Phi^*C = 0$. Theorem 4-6.5 then shows that $\Phi_V(s)$ solves C; that is, $\Phi_V(s)^*C = 0$. Further

$$\Phi_V(s)^*\mu = (T_V(s)\circ\Phi)^*\mu = \Phi^*\big(T_V(s)^*\mu\big) = \Phi^*(\exp(s£_V)\mu)$$

$$= \Phi^*\big(\mu + s£_V\mu + \tfrac{1}{2}s^2£_V£_V\mu + \cdots\big)$$

which is nonzero for s in a sufficiently small neighborhood of $s = 0$ because $\Phi^*\mu \neq 0$. □

We shall see in the next section that $TC(K)$ is not empty. Theorem 6-3.2 then shows that any regular map Φ is thereby a member of at least a 1-parameter family of regular maps. Accordingly, *$R(B_n)$ is at least arcwise locally connected*. Explicit realization of this connectedness property obtains directly through the study of the mapping $T_V(s)$ that is generated by the flow of $V \in TC(K)$. Because any $V \in TC(K)$ is an element of $T(K)$, we have

$$V = v^i(x,q,y)\,\partial_i + v^\alpha(x,q,y)\,\partial_\alpha + v_i^\alpha(x,q,y)\,\partial_\alpha^i$$

and hence the orbital equations for V are of the form

$$\frac{d\bar{x}^i}{ds} = v^i\big(\bar{x}^j,\bar{q}^\beta,\bar{y}_j^\beta\big), \qquad \frac{d\bar{q}^\alpha}{ds} = v^\alpha\big(\bar{x}^j,\bar{q}^\beta,\bar{y}_j^\beta\big), \tag{6-3.10}$$

$$\frac{d\bar{y}_i^\alpha}{ds} = v_i^\alpha\big(\bar{x}^j,\bar{q}^\beta,\bar{y}_j^\beta\big),$$

subject to the initial data

$$\bar{x}^i(0) = x^i, \qquad \bar{q}^\alpha(0) = q^\alpha, \qquad \bar{y}_i^\alpha(0) = y_i^\alpha. \tag{6-3.11}$$

If for the moment we borrow from the next section the fact that the functions v^i and v^α cannot depend on the $\bar{y}$'s when V belongs to $TC(K)$, the solution of the above initial value problem takes the form

$$\bar{x}^i = X^i\big(x^j,q^\beta;s\big) = \exp(sV)\langle x^i\rangle = \exp(s£_V)x^i, \tag{6-3.12}$$

$$\bar{q}^\alpha = Q^\alpha\big(x^j,q^\beta;s\big) = \exp(s£_V)q^\alpha,$$

$$\bar{y}_i^\alpha = Y_i^\alpha\big(x^j,q^\beta;y_j^\beta;s\big) = \exp(s£_V)y_i^\alpha.$$

These equations describe the 1-parameter family of maps $T_V(s)$ of K to K that results from transport along the orbits of V. For any regular map Φ from B_n to K, the parameters (τ^j) can be eliminated from (6-1.2) because $\Phi^*\mu \neq 0$, in which case we have

$$\text{(6-3.13)} \qquad \Phi|x^i = x^i, \qquad q^\alpha = \phi^\alpha(x^j), \qquad y_i^\alpha = \partial_i \phi^\alpha(x^j).$$

The composition map $\Phi_V(s) = T_V(s) \circ \Phi$ is then explicitly obtained by using (6-3.13) for the initial data:

$$\text{(6-3.14)} \qquad \bar{x}^i = X^i(x^j, \phi^\alpha(x^j); s), \qquad q^\alpha = Q^\alpha(x^j, \phi^\alpha(x^j); s),$$

$$\bar{y}_i^\alpha = Y_i^\alpha(x^j, \phi^\alpha(x^j), \partial_k \phi^\alpha(x^j); s).$$

There are two things that must be noted about the representation (6-3.14). First, it is an implicit representation due to the occurrence of the x's. However, since $\Phi_V(s)^*\mu \neq 0$ for all s in a sufficiently small neighborhood of $s = 0$, the relations $\bar{x}^i = X^i(x^j, \phi^\alpha(x^j); s)$ can be solved for the x's in terms of the $\bar{x}$'s and s to obtain

$$x^i = \hat{X}^i(\bar{x}^j; s).$$

We now use this to eliminate the variables (x^j) in the second and third of (6-3.14). When this is done, we obtain the explicit representation

$$\text{(6-3.15)} \quad \Phi_V(s)|\bar{x}^i = \bar{x}^i, \qquad \bar{q}^\alpha = \bar{Q}^\alpha(\bar{x}^j; s), \qquad \bar{y}_i^\alpha = \bar{Y}_i^\alpha(\bar{x}^j; s),$$

where $\bar{Q}^\alpha(\bar{x}^j; s) = Q^\alpha(\hat{X}^j(\bar{x}^k; s), \phi^\alpha(\hat{X}^j(\bar{x}^k; s)); s)$ etc. Second, we know that $\Phi_V(s)^*C = 0$ for any $V \in TC(K)$. Thus in particular, the functions $\bar{Y}_i^\alpha$ that occur in the explicit representation (6-3.15) actually have the evaluation

$$\text{(6-3.16)} \qquad \bar{Y}_i^\alpha(\bar{x}^j; s) = \frac{\partial}{\partial \bar{x}^i} \bar{Q}^\alpha(\bar{x}^j; s).$$

Transport of a regular map along the orbits of an isovector field of the contact ideal does exactly what it is supposed to; it preserves the relation that the $\bar{y}$'s are the derivatives on the $\bar{q}$'s with respect to the $\bar{x}$'s.

This situation seems complicated in the extreme on first reading due to the fact that we are only able to represent solutions of the orbital equations in the symbolic form (6-3.12) without an explicit $V \in TC(K)$ to work with. However, once an explicit $V \in TC(K)$ is given, direct integration of the orbital equations is often possible and the problem falls into manageable proportions. The reader will see this in later sections where specific examples will be worked in full.

6-4. EXPLICIT CHARACTERIZATION OF TC(K)

We now turn to the problem of obtaining explicit characterizations of isovectors of the contact ideal. By definition, $V \in TC(K)$ if and only if $£_V C \subset C$. Thus since $C = I\{C^1, \dots, C^N\}$ and all of the contact forms are of the same degree, Lemma 4-6.1 is applicable. This gives the following preliminary result.

Lemma 6-4.1. *A vector field $V \in T(K)$ is an isovector field of the contact ideal if and only if there exist N^2 functions $\{A^{\alpha}_{\beta}(x^j, q^{\gamma}, y^{\gamma}_j)\}$ such that*

$$£_V C^{\alpha} = A^{\alpha}_{\beta} C^{\beta} \tag{6-4.1}$$

We can now state the basic result.

Theorem 6-4.1. *A vector field*

$$V = v^i \partial_i + \bar{v}^{\alpha} \partial_{\alpha} + v^{\alpha}_i \partial^i_{\alpha} \tag{6-4.2}$$

is an isovector field of the contact ideal if and only if

$$v^{\alpha}_i = Z_i \langle V \lrcorner C^{\alpha} \rangle = Z_i \langle \bar{v}^{\alpha} - y^{\alpha}_j v^j \rangle, \tag{6-4.3}$$

$$£_V C^{\alpha} = \partial_{\beta}(V \lrcorner C^{\alpha}) C^{\beta}, \tag{6-4.4}$$

where Z_i is the linear differential operator

$$Z_i = \partial_i + y^{\beta}_i \partial_{\beta}, \tag{6-4.5}$$

and for $N > 1$,

$$v^i = f^i(x^j, q^{\beta}), \qquad \bar{v}^{\alpha} = f^{\alpha}(x^j, q^{\beta}), \tag{6-4.6}$$

$\{f^i, f^{\alpha}\}$ arbitrary functions of their indicated arguments, while for $N = 1$,

$$v^i = \partial^i_1 \xi(x^j, q^1, y^1_j), \tag{6-4.7}$$

$$\bar{v}^1 = (y^1_i \partial^i_1 - 1) \xi(x^j, q^1, y^1_j), \tag{6-4.8}$$

ξ an arbitrary function of its indicated arguments.

Proof. Computation of the Lie derivative of $C^{\alpha} = dq^{\alpha} - y^{\alpha}_i\, dx^i$, for any V of the form (6-4.2) and substitution into (6-4.1) lead to the conditions

$$d\bar{v}^{\alpha} - y^{\alpha}_i\, dv^i - v^{\alpha}_i\, dx^i = A^{\alpha}_{\beta}(dq^{\beta} - y^{\beta}_i\, dx^i). \tag{6-4.9}$$

Thus since v^i and $\bar{v}^\alpha$ are functions of the arguments $(x^j, q^\beta, y_j^\beta)$, explicit computation of dv^i and $d\bar{v}^\alpha$ and resolution on the basis $(dx^i, dq^\beta, dy_i^\beta)$ of $\Lambda^1(K)$ shows that (6-4.9) can be satisfied if and only if

$$\partial_i \bar{v}^\alpha - y_j^\alpha \partial_i v^j - v_i^\alpha = -A_\beta^\alpha y_i^\beta, \tag{6-4.10}$$

$$\partial_\beta \bar{v}^\alpha - y_j^\alpha \partial_\beta v^j = A_\beta^\alpha, \tag{6-4.11}$$

$$\partial_\beta^k \bar{v}^\alpha - y_j^\alpha \partial_\beta^k v^j = 0. \tag{6-4.12}$$

If we note that $V \lrcorner C^\alpha = \bar{v}^\alpha - y_j^\alpha v^j$, then (6-4.11) yields

$$A_\beta^\alpha = \partial_\beta (V \lrcorner C^\alpha), \tag{6-4.13}$$

and (6-4.1) and (6-4.13) give (6-4.4). Now that we know A_β^α, (6-4.10) may be solved for v_i^α:

$$v_i^\alpha = Z_i \langle V \lrcorner C^\alpha \rangle \tag{6-4.14}$$

for Z_i defined by (6-4.5). (Note that $Z_i \lrcorner C^\alpha = 0$ so that Z_i are characteristic vector fields of the contact ideal.) The only conditions that remain to be satisfied are those given by (6-4.12), which we now write as

$$\partial_\beta^k \bar{v}^\alpha = y_j^\alpha \partial_\beta^k v^j. \tag{6-4.15}$$

Because the left-hand sides of (6-4.15) are the components of the gradient of $\bar{v}^\alpha$ with respect to the variables y_k^β, satisfaction of (6-4.15) entails the integrability conditions $(\partial_\gamma^m \partial_\beta^k - \partial_\beta^k \partial_\gamma^m)\bar{v}^\alpha = 0$; that is,

$$\delta_\gamma^\alpha \partial_\beta^k v^m = \delta_\beta^\alpha \partial_\gamma^m v^k. \tag{6-4.16}$$

These conditions must be satisfied for all values of the indices $(\alpha, \beta, \gamma, k, m)$, so they must also hold when we set $\alpha = \gamma$ and perform the indicated summation, $N \partial_\beta^k v^m = \partial_\beta^m v^k$. When this result is put back into (6-4.16), we obtain

$$\frac{1}{N} \delta_\gamma^\alpha \partial_\beta^m v^k = \delta_\beta^\alpha \partial_\gamma^m v^k.$$

If we now set $\alpha = \beta$ and sum again on the repeated index, we see that

$$(1 - N^2)\, \partial_\gamma^m v^k = 0. \tag{6-4.17}$$

There are thus two cases according to whether $N = 1$ or $N > 1$. For the case $N > 1$ we must have $\partial_\gamma^m v^k = 0$, and hence $v^k = f^k(x^j, q^\beta)$ where $\{ f^k(x^j, q^\beta)\}$

are arbitrary functions of their indicated arguments. A substitution of $\partial^m_\gamma v^k = 0$ into (6-4.15) gives $\partial^k_\beta \bar{v}^\alpha = 0$ and hence $\bar{v}^\alpha = f^\alpha(x^j, q^\beta)$, where $\{f^\alpha(x^j, q^\beta)\}$ are arbitrary functions of their indicated arguments. This establishes (6-4.6). When $N = 1$, (6-4.16) reduces to

$$\partial^k_1 v^m = \partial^m_1 v^k$$

from which we infer the existence of a function $\xi(x^j, q^1, y^1_j)$ such that

$$v^m = \partial^m_1 \xi.$$

A substitution of this result back into (6-4.15) gives

$$\partial^k_1 \bar{v}^1 = y^1_j \partial^k_1 \partial^j_1 \xi = \partial^k_1 \left(y^1_j \partial^j_1 \xi - \xi \right),$$

and hence an integration yields

$$\bar{v}^1 = h(x^k, q^1) + \left(y^1_j \partial^j_1 - 1 \right) \xi(x^k, q^1, y^1_k).$$

It is then a simple matter to note that $\xi \mapsto \xi + h$ can be used to absorb the function of integration, $h(x^k, q^1)$, into ξ. □

There is an alternative view that is sometimes helpful in understanding what is going on here. Consider an invertible map

$$\rho : \mathscr{G} \to {}'\mathscr{G} \,|\, {}'x^i = \rho^i(x^j, q^\beta), \qquad {}'q^\alpha = \rho^\alpha(x^j, q^\beta).$$

between graph spaces, where ${}'\mathscr{G}$ is a replica of $\mathscr{G}$ with the coordinate cover $({}'x^i, 8j7q^\beta)$. If we extend $\mathscr{G}$ to K and ${}'\mathscr{G}$ to ${}'K$, we have the contact forms $C^\alpha = dq^\alpha - y^\alpha_i\, dx^i$ on K and ${}'C^\alpha = d\,{}'q^\alpha - {}'y^\alpha_i\, d\,{}'x^i$ on ${}'K$. A direct substitution of the transformation equations into these latter expressions gives

$$\rho^*\,{}'C^\alpha = \left(\partial_\beta \rho^\alpha - {}'y^\alpha_i \partial_\beta \rho^i \right) C^\beta + \left\{ Z_i\langle \rho^\alpha \rangle - {}'y^\alpha_j Z_i \langle \rho^j \rangle \right\} dx^i.$$

The following result is then immediate.

Lemma 6-4.2. *Let*

(6-4.18) $$\rho : \mathscr{G} \to {}'\mathscr{G} \,|\, {}'x^i = \rho^i(x^j, q^\beta), \qquad {}'q^\alpha = \rho^\alpha(x^j, q^\beta)$$

be an invertible map between graph spaces. The map ρ can be extended to a map between the kinematic spaces K and ${}'K$ such that ρ^{-1*} maps the contact ideal C

into the contact ideal 'C if and only if the transformation equations for the y's satisfy

$$Z_i\langle \rho^j \rangle' y_j^\alpha = Z_i\langle \rho^\alpha \rangle. \tag{6-4.19}$$

When the system of relations (6-4.19) is written out in full by use of (6-4.5), we have

$$'y_j^\alpha\left(\partial_i \rho^j + y_i^\beta \partial_\beta \rho^j\right) = \partial_i \rho^\alpha + y_i^\beta \partial_\beta \rho^\alpha. \tag{6-4.20}$$

The induced transformation law for the y's is thus seen to be very complicated.

In order to obtain a correlation with the results given in Theorem 6-4.1, consider a 1-parameter family of mappings of $\mathscr{G}$ to $'\mathscr{G}$ that are close to the identity mapping:

$$'x^i = x^i + sf^i(x^j, q^\beta) + o(s), \tag{6-4.21}$$

$$'q^\alpha = q^\alpha + sf^\alpha(x^j, q^\beta) + o(s).$$

Extension to 1-parameter family of mappings from K to $'K$ will then result in (6-4.21) and

$$'y_i^\alpha = y_i^\alpha + s\eta_i^\alpha\left(x^j, q^\beta, y_j^\beta\right) + o(s) \tag{6-4.22}$$

provided (6-4.20) is satisfied. When (6-4.21) and (6-4.22) are substituted into (6-4.20) and we neglect all terms that are $o(s)$, it is seen that the functions η_i^α are given by

$$\eta_i^\alpha = Z_i\langle f^\alpha - y_j^\alpha f^j \rangle. \tag{6-4.23}$$

We now identify $f^i(x^j, q^\beta)$ with v^i and $f^\alpha(x^j, q^\beta)$ with $\bar{v}^\alpha$ in accordance with (6-4.6), in which case (6-4.23) shows that η_i^α agrees with v_i^α. In this event (6-4.21) and (6-4.22) combine to give

$$'x^i = x^i + sv^i + o(s), \qquad 'q^\alpha = q^\alpha + s\bar{v}^\alpha + o(s), \tag{6-4.24}$$

$$'y_i^\alpha = y_i^\alpha + sv_i^\alpha + o(s),$$

where $V = v^i\partial_i + \bar{v}^\alpha\partial_\alpha + v_i^\alpha\partial_\alpha^i$ is an isovector of the contact ideal by Theorem 6-4.1. Accordingly, since (6-4.24) are a representation of an infinitesimal transformation of a 1-parameter group $\exp(sV): K \to 'K$, an identification of

K with $'K$ shows that

$$\exp(s\pounds_V)'C \subset C.$$

What both of these points of view say in one form or another, is that the isogroup of the contract ideal always maps $y_i^\alpha = \partial_i\phi^\alpha$ onto new y's such that $'y_i^\alpha = \partial'\phi^\alpha/\partial'x^i$.

The isogroup of the contact ideal is a group of automorphisms of the collection $R(B_n)$ of all regular maps $\Phi: B_n \to K$.

There is another aspect of isovectors of the contact ideal that is often useful in calculations and in obtaining simple and direct interpretations of results.

Theorem 6-4.2. *Any vector field*

$$V_g = f^i(x^j, q^\beta)\,\partial_i + \bar{f}^\alpha(x^j, q^\beta)\,\partial_\alpha \tag{6-4.25}$$

on graph space lifts to a unique isovector field

$$V = f^i(x^j, q^\beta)\,\partial_i + \bar{f}^\alpha(x^j, q^\beta)\,\partial_\alpha + Z_i\langle \bar{f}^\alpha - y_j^\alpha f^j\rangle\,\partial_{u\alpha}^i \tag{6-4.26}$$

of the contact ideal on kinematic space, and

$$V = V_g + Z_i\langle V_g\langle q^\alpha\rangle - y_j^\alpha V_g\langle x^j\rangle\rangle\,\partial_\alpha^i. \tag{6-4.27}$$

For $N > 1$, any isovector field of the contact ideal on kinematic space projects onto a unique vector field on graph space,

$$\pi_*: TC(K) \to T(\mathscr{G})\,|\,V \mapsto V_g, \qquad N > 1. \tag{6-4.28}$$

Proof. Theorem 6-4.1 shows that any isovector field of the contact ideal on K has the representation (6-4.26) for $N > 1$, while for $N = 1$ the choice $\xi = y_i^1 f^i(x^j, q^\beta) - \bar{f}^\alpha(x^j, q^\beta)$ and (6-4.8), (6-4.9) give $v^i = f^i(x^j, q^\beta)$, $\bar{v}^1 = \bar{f}^1(x^j, q^\beta)$. We then have $f^i = V_g\langle x^i\rangle$, $\bar{f}^\alpha = V_g\langle q^\alpha\rangle$ and hence (6-4.27) follows from (6-4.26). The results are then immediate upon comparing (6-4.25) and (6-4.26).

This result provided the basis for a simplification in specifying isovector fields of the contact ideal.

Definition. An isovector field

$$V = f^i(x^j, q^\beta)\,\partial_i + \bar{f}^\alpha(x^j, q^\beta)\,\partial_\alpha + Z_i\langle \bar{f}^\alpha - y_j^\alpha f^j\rangle\,\partial_\alpha^i \tag{6-4.29}$$

of the contact ideal is said to be *generated* by the vector field

$$V_g = f^i(x^j, q^\beta)\,\partial_i + \bar{f}^\alpha(x^j, q^\beta)\,\partial_\alpha \tag{6-4.30}$$

on graph space;

$$V = V_g + Z_i\langle V_g \lrcorner C^\alpha\rangle \partial^i_\alpha. \tag{6-4.31}$$

6-5. SECOND-ORDER PARTIAL DIFFERENTIAL EQUATIONS AND BALANCE FORMS

The results obtained in previous sections provide the necessary foundation for a detailed study of the structure of the solution set of second-order partial differential equations. Let us start the discussion with systems of quasilinear partial differential equations in general position. To this end let (x^i), $1 \le i \le n$ be the independent variables, and let $\phi^\alpha(x^j)$, $1 \le \alpha \le N$ be the dependent variables. The latter are to be obtained by solving the system of second-order partial differential equations

$$A^{ij}_{\alpha\beta}\big(x^k, \phi^\gamma, \partial_k\phi^\gamma\big)\,\partial_i\partial_j\phi^\beta + B_\alpha\big(x^k, \phi^\gamma, \partial_k\phi^\gamma\big) = 0, \tag{6-5.1}$$

where $(A^{ij}_{\alpha\beta}, B_\alpha)$ are given functions of their indicated arguments.

The original idea here is due to Cartan: reformulate the problem in terms of an ideal of the exterior algebra on a properly chosen underlying space. The underlying space to be used here is kinematic space K with the coordinate cover $(x^i, q^\alpha, y_i^\alpha)$. If $\Phi: B_n \to K$ is a regular map, $\Phi^*C = 0$, $\Phi^*\mu \neq 0$ and Φ can be represented by

$$\Phi|\,x^i = x^i, \qquad q^\alpha = \phi^\alpha(x^j), \qquad y_i^\alpha = \partial_i\phi^\alpha(x^j). \tag{6-5.2}$$

Let the system of n-forms E_α be defined by

$$E_\alpha = A^{ij}_{\alpha\beta}\big(x^k, q^\gamma, y_k^\gamma\big)\,dy_i^\beta \wedge \mu_j + B_\alpha\big(x^k, q^\gamma, y_k^\gamma\big)\mu, \tag{6-5.3}$$

where $\mu = dx^1 \wedge dx^2 \wedge \cdots \wedge dx^n$ is the volume element of E_n and $\mu_i = \partial_i \lrcorner \mu$. When (6-5.2) is used, we see that

$$\Phi^*E_\alpha = A^{ij}_{\alpha\beta}\big(x^k, \phi^\gamma, \partial_k\phi^\gamma\big)\,d\big(\partial_i\phi^\beta\big) \wedge \mu_j$$
$$+ B_\alpha\big(x^k, \phi^\gamma, \partial_k\phi^\gamma\big)\mu.$$

However, $d(\partial_i\phi^\beta) \wedge \mu_j = (\partial_k\partial_i\phi^\beta)\,dx^k \wedge \mu_j = \partial_j\partial_i\phi^\beta\mu$ because $dx^k \wedge \mu_j = \delta_j^k\mu$, and thus

$$\Phi^*E_\alpha = \big\{A^{ij}_{\alpha\beta}\big(x^k, \phi^\gamma, \partial_k\phi^\gamma\big)\,\partial_i\partial_j\phi^\beta + B_\alpha\big(x_k, \phi^\gamma, \partial_k\phi^\gamma\big)\big\}\mu. \tag{6-5.4}$$

It is thus clear that *a regular map* $\Phi: B_n \to K$ *solves the system of quasilinear partial differential equations* (6-5.1) *if and only if*

$$\Phi^* E_\alpha = 0. \tag{6-5.5}$$

A more restrictive class of quasilinear partial differential equations, but of particular practical significance, are those referred to as *balance equations*. These are obtained starting from the *balance n-forms*

$$B_\alpha = W_\alpha(x^k, q^\gamma, y_k^\gamma)\mu - dW_\alpha^i(x^k, q^\gamma, y_k^\gamma) \wedge \mu_i. \tag{6-5.6}$$

For these systems, we have

$$\Phi^* B_\alpha = \{W_\alpha(x^k, \phi^\gamma, \partial_k\phi^\gamma) - D_i W_\alpha^i(x^k, \phi^\gamma, \partial_k\phi^\gamma)\}\mu. \tag{6-5.7}$$

The reason why the n-forms given by (6-5.6) are referred to as balance forms is that an integration of (6-5.7) over B_n and Stokes' theorem yield

$$\int_{B_n} \Phi^*(W_\alpha \mu) = \int_{\partial B_n} \Phi^*(W_\alpha^i \mu_i), \tag{6-5.8}$$

which have the form of integral laws of balance (i.e., $\Phi^* W_\alpha$ is the source term and $\Phi^* W_\alpha^i$ are the boundary flux terms).

Explicit examples of systems of second-order partial differential equations that arise in well defined physical contexts are given in Section 6-7. A quick look at these examples may do more at this point than anything else in gaining an understanding of the translation of partial differential equations into exterior differential forms. For the time being, however, we go the other way and consider an arbitrary system of N exterior forms $\{E_\alpha\}$ of degree n.

As an example of what can arise consider an arbitrary 2-form $E_1 \in \Lambda^2(K)$ with $n = 2$ and $N = 1$,

$$\begin{aligned} E_1 = {} & A_{12}\, dx^1 \wedge dx^2 + A_{13}\, dx^1 \wedge dy_1 + A_{14}\, dx^1 \wedge dy_2 \\ & + A_{15}\, dx^1 \wedge dq + A_{23}\, dx^2 \wedge dy_1 + A_{24}\, dx^2 \wedge dy_2 \\ & + A_{25}\, dx^2 \wedge dq + A_{34}\, dy_1 \wedge dy_2 + A_{35}\, dy_1 \wedge dq \\ & + A_{45}\, dy_2 \wedge dq, \end{aligned}$$

where the A's are arbitrary functions of (x^1, x^2, q, y_1, y_2). If Φ is a regular map from B_2 to K,

$$\begin{aligned} \Phi | x^1 = x, \qquad x^2 = t, \qquad q = g(x,t), \qquad y_1 = \partial_x g, \\ y_2 = \partial_t g, \end{aligned}$$

we have $\Phi^* E_1 = \eta\, dx \wedge dt$ with

$$\eta = A_{12} + A_{15}\partial_t g - A_{25}\partial_x g - A_{23}\partial_x \partial_x g + (A_{13} + A_{24})\partial_x \partial_t g$$

$$+ A_{34}(\partial_x \partial_x g\, \partial_t \partial_t g - \partial_x \partial_t g\, \partial_x \partial_t g) + A_{14}\partial_t \partial_t g$$

$$+ A_{35}(\partial_x \partial_x g\, \partial_t g - \partial_x \partial_t g\, \partial_x g)$$

$$+ A_{45}(\partial_x \partial_t g\, \partial_t g - \partial_t \partial_t g\, \partial_x g).$$

The reader will have realized by now that there are many ways that a given system of second-order quasilinear partial differential equations can be transcribed into a system of n-forms on kinematic space. To cite just one instance, we have $\Phi^*(dq \wedge dx^2) = \partial_x g\, dx \wedge dt$ and $\Phi^*(y_1\, dx^1 \wedge dx^2) = \partial_x g\, dx \wedge dt$ in the notation of the previous example. However, $C = dq - y_1\, dx^1 - y_2\, dx^2$, and hence $dq \wedge dx^2 = (C + y_1\, dx^1 + y_2\, dx^2) \wedge dx^2 \equiv y_1\, dx^1 \wedge dx^2 \bmod C$. Thus since any regular map Φ annihilates the contact ideal, $dq \wedge dx^2$ and $y_1\, dx^1 \wedge dx^2$ are equivalent.

Definition. Two systems of n-forms $\{E_\alpha\}$ and $\{'E_\alpha\}$ are said to be *equivalent* if and only if there exist $\{A^\alpha_\beta\} \in \Lambda^0(K)$, $\{F_{\alpha\beta}\} \in \Lambda^{n-1}(K)$, $\{G_{\alpha\beta}\} \in \Lambda^{n-2}(K)$ such that

$$'E_\alpha = A^\beta_\alpha E_\beta + F_{\alpha\beta} \wedge C^\beta + G_{\alpha\beta} \wedge dC^\beta, \qquad \det(A^\beta_\alpha) \neq 0, \tag{6-5.9}$$

in which case we have

$$'E_\alpha = A^\beta_\alpha E_\beta \bmod \bar{C}, \qquad \det(A^\beta_\alpha) \neq 0. \tag{6-5.10}$$

It is now essential that we restrict attention to the collection of regular maps from $B_n \to K$ that have the further property of representing solutions to the given system of quasilinear partial differential equations.

Definition. A map $\Phi: B_n \to K$ is a *solution* map of a given system $\{E_\alpha\}$ of n-forms on K if and only if Φ is a regular map from $B_n \to K$ such that

$$\Phi^* E_\alpha = 0, \qquad \alpha = 1, \dots, N. \tag{6-5.11}$$

The collection of all solution maps is denoted by

$$S(E_\alpha) = \{\Phi: B_n \to K \mid \Phi \in R(B_n); \Phi^* E_\alpha = 0, 1 \le \alpha \le N\}. \tag{6-5.12}$$

Lemma 6-5.1. *A map $\Phi: B_n \to K$ is a solution map if and only if $\Phi^*\mu \neq 0$ and Φ solves the balance ideal*

(6-5.13) $$B = I\{C^1,\ldots,C^N, dC^1,\ldots, dC^N, E_1,\ldots, E_N\}$$

of $\Lambda(K)$.

Proof. Φ is regular if and only if $\Phi^*\mu = 0$, $\Phi^*C = 0$, $\Phi^*dC^\alpha = 0$ by Lemma 6-5.1. Thus since the balance ideal B is generated by $\{C^\alpha, dC^\alpha, E_\alpha\}$, Φ solves B if and only if Φ^* is regular and $\Phi^*E_\alpha = 0$. □

We can now show that equivalent systems of n-forms have the same solution maps and thus put to rest any questions concerning the specific choice of the n-forms that represent the given system of quasilinear partial differential equations.

Lemma 6-5.2. *If $\{E_\alpha\}$ and $\{'E_\alpha\}$ are two equivalent systems of n-forms, then any solution map of $\{E_\alpha\}$ is a solution map of $\{'E_\alpha\}$ and conversely.*

Proof. The two systems of n-forms E_α and $'E_\alpha$ are equivalent if and only if $'E_\alpha \equiv A_\alpha^\beta E_\beta \bmod \overline{C}$, $\det(A_\alpha^\beta) \neq 0$. Thus

$$'B = I\{C^\alpha, dC^\alpha, A_\alpha^\beta E_\beta \bmod \overline{C}\} = I\{C^\alpha, dC^\alpha, E_\alpha\} = B$$

since $\det(A_\alpha^\beta) \neq 0$. Accordingly, any Φ for which $\Phi^*\mu \neq 0$, $\Phi^*B = 0$ is such that $\Phi^*{}'B = 0$ and conversely. The result then follows by Lemma 6-5.1. □

The reader may wish to note in this context that equivalence of systems of n-forms is obviously symmetric, reflexive and transitive; it is a well defined equivalence relation of $\Lambda^n(K)$.

6-6. ISOVECTORS OF THE BALANCE IDEAL AND GENERATION OF SOLUTIONS

Solving a system of simultaneous, nonlinear, second-order partial differential equations is not an easy task. It is thus natural to try to make as much out of any known solutions as possible by using it to generate new solutions. This is where isovector methods genuinely shine, for they are tailor made to embed a given solution in a Lie group of solutions.

Definition. A vector field $V \in T(K)$ is an *isovector of the balance ideal* if and only if

(6-6.1) $$\pounds_V B \subset B.$$

The collection of all isovectors of the balance ideal is denoted by

(6-6.2) $$TB(E_\alpha) = \{V \in T(K) | \pounds_V B \subset B\}.$$

It is now a simple matter to combine this definition with Theorems 4-6.4 and 4-6.5 in order to obtain the following results.

Theorem 6-6.1. *The isovectors of the balance ideal form a Lie subalgebra of $T(K)$ that is the Lie algebra of a Lie group. If $V \in TB(E_\alpha)$ and Φ is a solution map of the system of partial differential equations that is characterized by the n-forms $\{E_\alpha\}$, then V embeds Φ in a 1-parameter family of solution maps*

(6-6.3) $$\Phi_V(s) = T_V(s) \circ \Phi,$$

where $T_V(s): K \to K$ is the 1-parameter family of maps that is generated by the flow of V. Thus $\exp(TB(E_\alpha))$ *generates a Lie group of automorphisms of the solution set $S(E_\alpha)$ by composition.*

All of the remarks given in the discussion following Theorem 6-3.2 are directly applicable here. The full content is best realized, however, in context of explicit problems. Detailed comment will thus be postponed until the next section.

There is still a question outstanding, for we have not as yet obtained explicit conditions for V to belong to $TB(E_\alpha)$.

Theorem 6-6.2. *A vector field V is an isovector field of the balance ideal if and only if V is an isovector field of the contact ideal and*

(6-6.4) $$\pounds_V E_\alpha \subset B;$$

that is,

(6-6.5) $$TB(E_\alpha) = \{V \in TC(K) | \pounds_V E_\alpha \subset B, 1 \leq \alpha \leq N\}.$$

Thus $V \in TC(K)$ is an isovector field of the balance ideal if and only if there exist quantities $A^\alpha_\beta \in \Lambda^0(K)$, $F_{\alpha\beta} \in \Lambda^{n-1}(K)$, $G_{\alpha\beta} \in \Lambda^{n-2}(K)$ such that

(6-6.6) $$\pounds_V E_\alpha = A^\beta_\alpha E_\beta + F_{\alpha\beta} \wedge C^\beta + G_{\alpha\beta} \wedge dC^\beta.$$

Proof. Examination of the balance ideal $B = I\{C^\alpha, dC^\alpha, E_\alpha\}$ shows that B is generated by the 1-forms C^α and the n-forms E_α. Thus identifying $\{\alpha^a\}$ of Lemma 4-6.3 with $\{C^\alpha, dC^\alpha\}$ and $\{\beta^b\}$ of the same lemma with $\{E_\alpha\}$, we see that any isovector of $I\{C^\alpha, dC^\alpha, E_\alpha\}$ is an isovector of the ideal C that satisfies (6-6.4). An application of Lemma 4-6.1 then shows that this latter set of conditions is satisfied if and only if there exist quantities $(A^\alpha_\beta, F_{\alpha\beta}, G_{\alpha\beta})$ such that (6-6.6) holds. □

The importance of Theorem 6-6.2 is twofold. First, it allows us to use all of the previous results on isovector fields of the contact ideal. In particular, we have the characterization

$$V = f^i(x,q)\,\partial_i + \bar{f}^\alpha(x,q)\,\partial_\alpha + Z_i\langle f^\alpha - y^\alpha_j \bar{f}^j\rangle\,\partial^i_\alpha \tag{6-6.7}$$

of $TC(K)$ and this accomplishes a significant part of the work. Second, starting with this characterization of $TC(K)$, it then remains only to determine the functions $\{f^i(x^j, q^\beta), \bar{f}^\alpha(x^j, q^\beta)\}$ so as to secure satisfaction of the conditions (6-6.6). At this juncture, we must take particular note of the fact that *the functions* $\{f^i, \bar{f}^\alpha\}$ *do not depend on the variables* $\{y^\alpha_i\}$. Accordingly, when (6-6.7) is used to compute $\pounds_V E_\alpha$ and the results are put into (6-6.6), we obtain a system of n-forms whose coefficients have to vanish *identically* in the nN arguments $\{y^\alpha_i\}$. This identical satisfaction in the variables $\{y^\alpha_i\}$ is what leads to the large number of equations that often plague calculation of isovectors of balance ideals.

6-7. EXAMPLES

One of the simplest nontrivial problems is that associated with linear diffusion in one spatial dimension. Here, we have $n = 2$, $N = 1$, K has the coordinate cover (x^1, x^2, q, y_1, y_2) and solution maps are represented by

$$\Phi | x^1 = x, \qquad x^2 = t, \qquad q = T(x,t), \qquad y_1 = \partial_x T, \qquad y_2 = \partial_t T, \tag{6-7.1}$$

where $T(x, t)$ is the temperature at station x at time t. The contact 1-form is

$$C = dq - y_1\,dx^1 - y_2\,dx^2 \tag{6-7.2}$$

and the balance 2-form can be written as

$$E_1 = d\big(q\,dx^1 + y_1\,dx^2\big). \tag{6-7.3}$$

It is then a simple matter to see that $\Phi^*B = \Phi^*I\{C, dC, E_1\} = 0$ if and only if

$$\Phi^*E_1 = (\partial_t T - \partial_x\partial_x T)\,dx \wedge dt = 0.$$

A lengthy but straightforward calculation shows that all isovectors of the balance ideal are generated by

$$V_g = v^1\partial_1 + v^2\partial_2 + \bar{v}\partial_q \tag{6-7.4}$$

where

$$\begin{aligned} v^1 &= k_1 + k_4x^1 - 2k_5x^2 + 2k_6x^1x^2, \\ v^2 &= k_2 + 2k_4x^2 + 2k_6(x^2)^2, \\ \bar{v} &= k_3q + k_5x^1q - k_6\left(x^2 + \tfrac{1}{2}(x^1)^2\right)q + g(x^1, x^2), \end{aligned} \tag{6-7.5}$$

the k's are constants (parameters of the isogroup), and $g(x^1, x^2)$ is any solution of $\partial_2 g = \partial_1\partial_1 g$. The isovector $V_g = g(x^1, x^2)\partial_q$ arises from the fact that the linear diffusion equation is indeed linear and hence satisfies the superposition principle (i.e., the translation of any solution by a solution gives a solution).

For the moment let us concentrate on the isovector that is generated by

$$V_g = -2x^2\partial_1 + x^1q\partial_q \tag{6-7.6}$$

(i.e., $g(x^1, x^2) = 0$, $k_5 = 1$ and all other k's vanish). When (6-4.29) and (6-4.30) are used, the isovector V that is generated by V_g is seen to be

$$V = -2x^2\partial_1 + x^1q\partial_q + \left(q + x^1y_1\right)\partial^1 + \left(2y_1 + x^1y_2\right)\partial^2 \tag{6-7.7}$$

($\partial^1 = \partial/\partial y_1$, $\partial^2 = \partial/\partial y_2$), and hence the orbital equations are

$$\frac{d\bar{x}^1}{ds} = -2\bar{x}^2, \qquad \frac{d\bar{x}^2}{ds} = 0, \qquad \frac{d\bar{q}}{ds} = \bar{x}^1\bar{q}, \tag{6-7.8}$$

$$\frac{d\bar{y}_1}{ds} = \bar{q} + \bar{x}^1\bar{y}_1, \qquad \frac{d\bar{y}_2}{ds} = 2\bar{y}_1 + \bar{x}^1\bar{y}_2. \tag{6-7.9}$$

The general solution of these differential equations subject to the initial data $\bar{x}^1(0) = x^1$, $\bar{x}^2(0) = x^2$, $\bar{q}(0) = q$, $\bar{y}_1(0) = y_1$, $\bar{y}_2(0) = y_2$ is given by

$$\bar{x}^1 = x^1 - 2x^2s, \qquad \bar{x}^2 = x^2, \tag{6-7.10}$$

$$\begin{aligned} \bar{q} &= q\exp(x^1s - x^2s^2), \\ \bar{y}_1 &= (y_1 + qs)\exp(x^1s - x^2s^2), \end{aligned} \tag{6-7.11}$$

$$\bar{y}_2 = \left(y_2 + 2y_1s + qs^2\right)\exp(x^1s - x^2s^2). \tag{6-7.12}$$

Isovector methods do not give solutions out of nowhere, for we must know at least one solution before we can imbed in a 1-parameter family of solutions. To this end let the function $T(x, t)$ that occurs in (6-7.1) satisfy $\partial_t T = \partial_x \partial_x T$. We then know $\Phi_V(s) = T_V(s) \circ \Phi$ and that

$$\Phi_V(s)^* y_1 = \Phi^* \bar{y}_1 = \partial\bar{T}/\partial\bar{x}, \qquad \Phi_V(s)^* y_2 = \Phi^* \bar{y}_2 = \partial\bar{T}/\partial\bar{t}.$$

Hence (6-7.12) give

(6-7.13)
$$\frac{\partial\bar{T}}{\partial\bar{x}} = \left(\frac{\partial T}{\partial x} + Ts\right)\exp(xs - ts^2),$$

$$\frac{\partial\bar{T}}{\partial\bar{t}} = \left(\frac{\partial T}{\partial t} + 2\frac{\partial T}{\partial x}s + Ts^2\right)\exp(xs - ts^2),$$

which are reasonably complicated relations between the old and the new partial derivatives. Further, (6-7.10) and (6-7.11) yield the relations

(6-7.14)
$$\bar{x} = x - 2ts, \qquad \bar{t} = t,$$

(6-7.15)
$$\bar{T} = T(x, t)\exp(xs - ts^2).$$

It is then a simple matter to see that (6-7.14) and (6-7.15) give

(6-7.16)
$$\frac{\partial\bar{T}}{\partial\bar{t}} - \frac{\partial^2\bar{T}}{\partial\bar{x}^2} = \left(\frac{\partial T}{\partial t} - \frac{\partial^2 T}{\partial x^2}\right)\exp(xs - ts^2),$$

(6-7.17)
$$\frac{\partial(\bar{x}, \bar{t})}{\partial(x, t)} = 1,$$

and hence $\bar{T} = \bar{T}(\bar{x}, \bar{t}; s)$ will be a solution of the linear diffusion equations for all values of the parameter s. The function $\bar{T}(\bar{x}, \bar{t}; s)$ is only defined implicitly by (6-7.14) and (6-7.15). It is, however, a simple matter to solve (6-7.14) for (x, t) in terms of $(\bar{x}, \bar{t}; s)$;

(6-7.18)
$$x = \bar{x} + 2\bar{t}s, \qquad t = \bar{t}.$$

When these are now substituted into (6-7.15), we obtain *the* 1-*parameter family*

(6-7.19)
$$\bar{T}(\bar{x}, \bar{t}; s) = T(\bar{x} + 2\bar{t}s, \bar{t})\exp(\bar{x}s + \bar{t}s^2)$$

of new solutions of the linear diffusion equation that arise from any known solution $T(x, t)$.

Next consider the isovector that is generated by

(6-7.20)
$$V_g = x^1\partial_1 + 2x^2\partial_2;$$

that is, $g(x^1, x^2) = 0$, $k_4 = 1$ and all other k's equal to zero in (6-7.5). The orbital equations on graph space are

$$\frac{d\bar{x}^1}{ds} = \bar{x}^1, \qquad \frac{d\bar{x}^2}{ds} = 2\bar{x}^2, \qquad \frac{d\bar{q}}{ds} = 0, \tag{6-7.21}$$

subject to the initial data $\bar{x}^1(0) = x^1$, $\bar{x}^2(0) = x^2$, $\bar{q}(0) = q$. We thus have

$$\bar{x}^1 = x^1 e^s, \qquad \bar{x}^2 = x^2 e^{2s}, \qquad \bar{q} = q, \tag{6-7.22}$$

and hence $\Phi_V(s)$ is given by

$$\bar{x} = xe^s, \qquad \bar{t} = te^{2s}, \qquad \bar{T} = T(x, t). \tag{6-7.23}$$

An elimination of the variables (x, t) between these equations gives the new solution

$$\bar{T}(\bar{x}, \bar{t}; s) = T(\bar{x}e^{-s}, \bar{t}e^{-2s}) \tag{6-7.24}$$

for any old solution $T(x, t)$ of the linear diffusion equation.

Finally, the isovector generated with $k_6 = 1$ and everything else zero in (6-7.5) is

$$V_g = 2x^1x^2\partial_1 + 2(x^2)^2\partial_2 - \left(x^2 + \tfrac{1}{2}(x^1)^2\right)\partial_q, \tag{6-7.25}$$

which is the isovector most characteristic of linear diffusion processes. The general solution of the orbital equations of V_g on graph space subject to the standard initial data is given by

$$\bar{x}^1 = x^1/(1 - 2x^2s), \qquad \bar{x}^2 = x^2/(1 - 2x^2s), \tag{6-7.26}$$

$$\bar{q} = q(1 - 2x^2s)^{1/2}\exp\left(-s(x^1)^2/2(1 - 2x^2s)\right). \tag{6-7.27}$$

Accordingly, composition with Φ gives $\Phi_V(s)$ with

$$\bar{x} = x/(1 - 2ts), \qquad \bar{t} = t/(1 - 2ts), \tag{6-7.28}$$

$$\bar{q} = T(x, t)(1 - 2ts)^{1/2}\exp(-sx^2/2(1 - 2ts)). \tag{6-7.29}$$

It is now only necessary to solve (6-7.28) for (x, t) in terms of $(\bar{x}, \bar{t}; s)$ to obtain

$$x = \bar{x}/(1 + 2s\bar{t}), \qquad t = \bar{t}/(1 + 2s\bar{t}) \tag{6-7.30}$$

and then substitute into (6-7.29). The result is the 1-parameter family of solutions

$$(6\text{-}7.31)\qquad \bar{T}(\bar{x}, \bar{t}; s) = \frac{T\left(\dfrac{\bar{x}}{1+2s\bar{t}}, \dfrac{\bar{t}}{1+2s\bar{t}}\right)}{\sqrt{1+2s\bar{t}}} \exp\left(\frac{-s\bar{x}^2}{2(1+2s\bar{t})}\right).$$

Linear diffusion in two spatial dimensions is described in kinematic space with the coordinate cover $(x^1, x^2, x^3, q, y_1, y_2, y_3)$ and by regular maps that are the lift of maps

$$(6\text{-}7.32)\qquad \Phi | x^1 = x, \qquad x^2 = y, \qquad x^3 = t, \qquad q = T(x, y, t).$$

Here, the contact form is

$$C = dq - y_1\,dx^1 - y_2\,dx^2 - y_3\,dx^3$$

and the balance form is

$$(6\text{-}7.33)\qquad E_1 = d\left(Rq\,dx^1 \wedge dx^2 - y_1\,dx^2 \wedge dx^3 + y_2\,dx^1 \wedge dx^3\right),$$

with $R =$ constant, since

$$\Phi^* E_1 = \left(R\partial_t T - \partial_x \partial_x T - \partial_y \partial_y T\right) dx \wedge dy \wedge dt.$$

The isovectors for this balance system are generated by

$$(6\text{-}7.34)\qquad V_g = v^1\partial_1 + v^2\partial_2 + v^3\partial_3 + \bar{v}\partial_q$$

where

$$(6\text{-}7.35)\qquad \begin{aligned} v^1 &= k_2 + k_5 x^2 + k_6 x^1 + k_7 x^3 + k_9 x^1 x^3, \\ v^2 &= k_3 - k_5 x^1 + k_6 x^2 + k_8 x^3 + k_9 x^2 x^3, \\ v^3 &= k_1 + 2k_6 x^3 + k_9 (x^3)^2, \\ \bar{v} &= k_4 q - k_7 \frac{R}{2} x^1 q - k_8 \frac{R}{2} x^2 q + r(x^1, x^2, x^3) \\ &\quad - k_9\left(x^3 + \frac{R}{4}(x^1)^2 + \frac{R}{4}(x^2)^2\right) q, \end{aligned}$$

where $r(x, y, t)$ is any solution of $R\partial_t r = \partial_x \partial_x r + \partial_y \partial_y r$.

From now on, it will save both time and space if we indicate the independent and the dependent variables explicitly rather than through an ensueing map. For example, the previous example would then be written with the coordinate cover $(x, y, t, T, y_1, y_2, y_3)$ for kinematic space. Solution maps are then designated by writing $T(x, y, t)$ explicitly instead of just the place holder T.

Let Ψ denote the stream function and ζ the y component of the vorticity of a steady, incompressible flow of a viscous fluid in the (x, z)-plane. The governing equations for such a system may be written in the form

$$(6\text{-}7.36) \qquad \nabla^2\zeta = \frac{\partial\Psi}{\partial z}\frac{\partial\zeta}{\partial x} - \frac{\partial\Psi}{\partial x}\frac{\partial\zeta}{\partial z}$$

$$\nabla^2\Psi = \zeta.$$

The contact 1-forms for this system are

$$C^1 = d\Psi - y_1^1\,dx - y_2^1\,dz, \qquad C^2 = d\zeta - y_1^2\,dx - y_2^2\,dz$$

on a kinematic space with coordinate cover $(x, z, \Psi, \zeta, y_1^1, y_2^1, y_1^2, y_2^2)$. Since the 2-forms

$$(6\text{-}7.37) \qquad E_1 = dy_1^2 \wedge dz - dy_2^2 \wedge dx + d\zeta \wedge d\Psi,$$

$$E_2 = dy_1^1 \wedge dz - dy_2^1 \wedge dx - \zeta\,dx \wedge dz$$

have the evaluations

$$E_1 = \left(\nabla^2\zeta - \frac{\partial\Psi}{\partial z}\frac{\partial\zeta}{\partial x} + \frac{\partial\Psi}{\partial x}\frac{\partial\zeta}{\partial z}\right) dx \wedge dz$$

$$E_2 = (\nabla^2\Psi - \zeta)\,dx \wedge dz$$

for $C^1 = 0$, $C^2 = 0$ (i.e., for $y_1^1 = \partial_x\Psi, \ldots$), the system (6-7.36) can be written as

$$(6\text{-}7.38) \qquad C^1 = 0, \qquad C^2 = 0, \qquad E_1 = 0, \qquad E_2 = 0.$$

If V is an isovector of the balance ideal generated by the C's and the E's, then V is generated by

$$(6\text{-}7.39) \qquad V_g = J_1\partial_x + J_2\partial_z + J_3\partial_\Psi + J_4\partial_\zeta.$$

A lengthy calculation shows that the J's are given by

$$(6\text{-}7.40) \qquad J_1 = k_1x - k_2z + k_3, \qquad J_2 = k_1z + k_2x + k_4,$$

$$J_3 = k_5, \qquad J_4 = -2k_1\zeta.$$

The equations governing thermal boundary layer flows in porous media are

$$(6\text{-}7.41)\qquad \frac{\partial \Psi}{\partial x} = f(z) - T, \qquad \frac{\partial^2 T}{\partial x^2} = \frac{\partial \Psi}{\partial z}\frac{\partial T}{\partial x} - \frac{\partial \Psi}{\partial x}\frac{\partial T}{\partial z},$$

where $f(z)$ is a given function of the indicated argument, Ψ is the stream function, T is the temperature distribution and the force of gravity acts in the direction of negative z. Noting that there are no second derivatives of Ψ, we need only the single contact form

$$C^1 = dT - y_1\,dx - y_2\,dz.$$

It is then a simple matter to see that (6-7.41) are given by

$$(6\text{-}7.42)\qquad C^1 = 0, \qquad E_1 = 0, \qquad E_2 = 0$$

on the kinematic space with coordinates $(x, z, T, \Psi, y_1, y_2)$, where the 2-forms E_1, E_2 are

$$(6\text{-}7.43)\qquad E_1 = dz \wedge d\Psi + (T - f(z))\,dz \wedge dx,$$

$$E_2 = d\Psi \wedge dT - dz \wedge dy_1.$$

If V is an isovector field of the balance ideal generated by C^1 and the E's, then V is generated by

$$(6\text{-}7.44)\qquad V_g = J_1\partial_x + J_2\partial_z + J_3\partial_T + J_4\partial_\Psi$$

with

$$(6\text{-}7.45)\qquad J_1 = \alpha(z) + (k_3 - k_2)x, \qquad J_2 = k_4 + (2k_3 - k_2)z,$$

$$J_3 = k_1 + k_2 T, \qquad J_4 = k_5 + k_3\Psi,$$

provided the function $f(z)$ has the form

$$(6\text{-}7.46)\qquad f(z) = -k_1/k_2 + b(k_4 + (2k_3 - k_2)z)^{k_2/(2k_3 - k_2)}.$$

Here k's and b are constants and $\alpha(z)$ is an arbitrary function of the indicated argument.

This example points up a very useful aspect of isovector theory. The function $f(z)$ was left arbitrary in the formulation of the problem and then solved for in order to have as large a collection of isovectors as possible. Now that we have (6-7.46), there are necessarily relations between the k's that become fixed. For instance, suppose that

$$f(z) = h + b\left(c + \frac{2}{1 + r}z\right)^r;$$

then

$$J_1 = \alpha(z) + k_3\frac{1-r}{1+r}x, \qquad J_2 = k_3\left(c + \frac{2}{1+r}z\right),$$

$$J_3 = k_3\frac{2r}{1+r}(T-b), \qquad J_4 = k_5 + k_3\Psi.$$

Next, we consider incompressible, unsteady viscous flows in three spatial dimensions. Introducing the parameters

$$P = p/\rho, \qquad M = \bar{\mu}/\rho,$$

we have the governing equations

$$\text{(6-7.47)} \qquad \partial_t\mathbf{U} + (\mathbf{U}\cdot\nabla)\mathbf{U} + \nabla P - M\nabla^2\mathbf{U} = \mathbf{0}, \qquad \nabla\cdot\mathbf{U} = 0,$$

where $\mathbf{U} = U\mathbf{i} + V\mathbf{j} + W\mathbf{k}$ denotes the velocity vector field of the flow referred to the standard Cartesian coordinate cover (x, y, z) and t is the time variable. Since the dependent variables of the problem are (U, V, W, P), the contact 1-forms are

$$\text{(6-7.48)} \qquad C^1 = dU - y_1^1\,dx - y_2^1\,dy - y_3^1\,dz - y_4^1\,dt,$$

$$C^2 = dV - y_1^2\,dx - y_2^2\,dy - y_3^2\,dz - y_4^2\,dt,$$

$$C^3 = dW - y_1^3\,dx - y_2^3\,dy - y_3^3\,dz - y_4^3\,dt,$$

$$C^4 = dP - y_1^4\,dx - y_2^4\,dy - y_3^4\,dz - y_4^4\,dt.$$

Noting that (6-7.47) have the form of laws of balance, it is a simple matter to construct the corresponding balance forms

$$\text{(6-7.49)} \qquad B_1 = \left(y_4^1 + Uy_1^1 + Vy_2^1 + Wy_3^1 + y_1^4\right)\mu - dF_1,$$

$$B_2 = \left(y_4^2 + Uy_1^2 + Vy_2^2 + Wy_3^2 + y_2^4\right)\mu - dF_2,$$

$$B_3 = \left(y_4^3 + Uy_1^3 + Vy_2^3 + Wy_3^3 + y_3^4\right)\mu - dF_3,$$

$$B_4 = \left(y_1^1 + y_2^2 + y_3^3\right)\mu,$$

where $\mu = dx \wedge dy \wedge dz$, $\mu_1 = \partial_x \lrcorner \mu$, $\mu_2 = \partial_y \lrcorner \mu$, $\mu_3 = \partial_z \lrcorner \mu$, and

$$F_1 = \left(y_1^1\mu_1 + y_2^1\mu_2 + y_3^1\mu_3\right)M, \tag{6-7.50}$$

$$F_2 = \left(y_1^2\mu_1 + y_2^2\mu_2 + y_3^2\mu_3\right)M,$$

$$F_3 = \left(y_1^3\mu_1 + y_2^3\mu_2 + y_3^3\mu_3\right)M$$

are the flux 3-forms that generate $M\nabla^2\mathbf{U}$ from $d\mathbf{F}$ (recall that $C^\alpha = 0$ replace the y's by the first partial derivatives of (U, V, W, P) with respect to (x, y, z, t)).

For simplicity let us write the generating vector of an isovector field V for the balance ideal as

$$V_g = J_X\frac{\partial}{\partial x} + J_Y\frac{\partial}{\partial y} + J_Z\frac{\partial}{\partial z} + J_T\frac{\partial}{\partial t} + J_U\frac{\partial}{\partial U} \tag{6-7.51}$$

$$+J_V\frac{\partial}{\partial V} + J_W\frac{\partial}{\partial W} + J_P\frac{\partial}{\partial P},$$

where $(J_X, J_Y, J_Z, J_T, J_U, J_V, J_W, J_P)$ are eight functions of the eight arguments (x, y, z, t, U, V, W, P). An exhausting calculation shows that

$$J_X = A_1(t) + bx + g_1y - g_2z, \tag{6-7.52}$$

$$J_Y = A_2(t) - g_1x + by - g_3z,$$

$$J_Z = A_3(t) + g_2x + g_3y + bz,$$

$$J_T = \bar{a} + 2bt,$$

$$J_U = dA_1/dt - bU + g_1V - g_2W,$$

$$J_V = dA_2/dt - g_1U - bV - g_3W,$$

$$J_W = dA_3/dt + g_2U + g_3V - bW,$$

$$J_P = -2bP - x\,d^2A_1/dt^2 - y\,d^2A_2/dt^2 - z\,d^2A_3/dt^2 + m(t),$$

where $\bar{a}$, b, g_1, g_2 and g_3 are arbitrary constants and $A_1(t)$, $A_2(t)$, $A_3(t)$ and $m(t)$ are arbitrary functions of the time variable.

Let λ and $\bar{\mu}$ be the Lamé constants, ρ the density, β the coefficient of thermal expansion, c the specific heat at constant volume, T_0 a reference thermodynamic temperature, k the thermal diffusion coefficient, and introduce the following parameters

$$E = \frac{\lambda + \bar{\mu}}{\rho}, \qquad F = \beta/\rho, \qquad G = \bar{\mu}/\rho, \qquad H = k/c, \qquad M = \beta T_0/c.$$

The governing equations of linear thermoelasticity are then given by

$$(6\text{-}7.53)\qquad \partial_m\{(E\partial_j U^j - FT)\delta^{mk} + G\partial_m U^k\} - \partial_t\partial_t U^k = 0,$$

$$\partial_m\{H\partial_m T - M\partial_t U^m\} - \partial_t T = 0,$$

where $(U^k) = (U, V, W)$ are the three elastic displacement functions and T is the thermodynamic temperature. Since the field variables are (U, V, W, T), the contact 1-forms are

$$C^1 = dU - y_1^1\,dx - y_2^1\,dy - y_3^1\,dz - y_4^1\,dt,$$

$$C^2 = dV - y_1^2\,dx - y_2^2\,dy - y_3^2\,dz - y_4^2\,dt,$$

$$C^3 = dW - y_1^3\,dx - y_2^3\,dy - y_3^3\,dz - y_4^3\,dt,$$

$$C^4 = dT - y_1^4\,dx - y_2^4\,dy - y_3^4\,dz - y_4^4\,dt.$$

An examination of the field equations (6-7.53) shows that they are laws of balance. It is then a simple matter to construct the corresponding balance forms

$$(6\text{-}7.54)\qquad B_1 = dR_1 - Fy_1^4\mu, \qquad B_2 = dR_2 - Fy_2^4\mu,$$

$$B_3 = dR_3 - Fy_3^4\mu, \qquad B_4 = dR_4 - y_4^4\mu,$$

where

$$(6\text{-}7.55)\qquad \theta = y_1^1 + y_2^2 + y_3^3,$$

$$R_1 = (E\theta + Gy_1^1)\mu_1 + Gy_2^1\mu_2 + Gy_3^1\mu_3 - y_4^1\mu_4,$$

$$R_2 = Gy_1^2\mu_1 + (E\theta + Gy_2^2)\mu_2 + Gy_3^2\mu_3 - y_4^2\mu_4,$$

$$R_3 = Gy_1^3\mu_1 + Gy_2^3\mu_2 + (E\theta + Gy_3^3)\mu_3 - y_4^3\mu_4,$$

$$R_4 = (Hy_1^4 - My_4^1)\mu_1 + (Hy_2^4 - My_4^2)\mu_2 + (Hy_3^4 - My_4^3)\mu_3.$$

For simplicity, we write the generating vector of an isovector field V for the system as

$$(6\text{-}7.56)\qquad V_g = J_x\partial/\partial x + J_y\partial/\partial y + J_z\partial/\partial z + J_t\partial/\partial t + J_U\partial/\partial U$$

$$+J_V\partial/\partial V + J_W\partial/\partial W + J_T\partial/\partial T,$$

where $(J_x, J_y, J_z, J_t, J_U, J_V, J_W, J_T)$ are eight functions of the eight arguments (x, y, z, t, U, V, W, T). We then have

$$
\begin{aligned}
J_x &= S_3 y + S_2 z + k_1, \\
J_y &= -S_3 x + S_1 z + k_2, \\
J_z &= -S_2 x - S_1 y + k_3, \\
J_t &= k_4, \\
J_U &= k_5 U + S_3 V + S_2 W + u(x, y, z, t), \\
J_V &= -S_3 U + k_5 V + S_1 W + v(x, y, z, t), \\
J_W &= -S_2 U - S_1 V + k_5 W + w(x, y, z, t), \\
J_T &= k_5 T + \tau(x, y, z, t).
\end{aligned}
\tag{6-7.57}
$$

Here, k_1, k_2, k_3, k_4, k_5, S_1, S_2 and S_3 are arbitrary constants and $u(x, y, z, t)$, $v(x, y, z, t)$, $w(x, y, z, t)$ and $\tau(x, y, z, t)$ are any four functions of the arguments (x, y, z, t) that constitute a solution of the given equations of linear thermoelasticity.

Clearly, k_1, k_2, k_3 and k_4 generate homogeneous translations of the space and time variables. Similarly, S_1, S_2 and S_3 generate homogeneous rigid body rotations of the reference frame of the independent spatial variables and the concomitant homogeneous rigid body rotations of the displacement functions. These seven transformations are those that are well known from a number of papers in the current literature. The parameter k_5 generates homogeneous scalings of (U, V, W, T). In addition to these, there is the family of transformations that is unbounded in number; namely, those generated by all solutions u, v, w and τ of the given equations of linear thermoelasticity. These arise from the superposition principle associated with linear equations, for (6-7.57) shows that translations of solutions by solutions gives solutions.

Isovector methods can also be applied directly to systems of nonlinear or quasilinear partial differential equations. Quasilinear systems are particularly simple since they do not require contact 1-forms. All that is required is for the system to be written in terms of balance n-forms $E_1, E_2, \ldots,$ and the isovectors are then given by solving

$$£_V E_\alpha = A_\alpha^\beta E_\beta.$$

Isotropic power-law creep in the antiplane strain case is described by the two nonvanishing stress components σ_{xz}, σ_{yz} and the out-of-plane displacement rate W. The governing equations are (see Delph, T. J.)

$$\partial\sigma_{xz}/\partial x + \partial\sigma_{yz}/\partial y = 0, \tag{6-7.58}$$

$$\partial W/\partial x = 4\mu\,\partial\sigma_{xz}/\partial t + 2L\left(\sigma_{xz}^2 + \sigma_{yz}^2\right)^{(n-1)/2}\sigma_{xz},$$

$$\partial W/\partial y = 4\mu\,\partial\sigma_{yz}/\partial t + 2L\left(\sigma_{xz}^2 + \sigma_{yz}^2\right)^{(n-1)/2}\sigma_{yz},$$

where μ is the shear modulus and L and n are material constants. It is now a simple matter to introduce a kinematic space with coordinate cover $(x, y, t, \sigma_{xz}, \sigma_{yz}, W)$, in which case (6-7.58) can be written in terms of the balance 3-forms

$$E_1 = d\sigma_{xz} \wedge dy \wedge dt - d\sigma_{yz} \wedge dx \wedge dt, \tag{6-7.59}$$

$$E_2 = dW \wedge dy \wedge dt - 4\mu\, d\sigma_{xz} \wedge dx \wedge dy$$
$$-2L\left(\sigma_{xz}^2 + \sigma_{yz}^2\right)^{\frac{n-1}{2}}\sigma_{xz}\, dx \wedge dy \wedge dt$$

$$E_3 = -dW \wedge dx \wedge dt - 4\mu\, d\sigma_{yz} \wedge dx \wedge dy$$
$$-2L\left(\sigma_{xz}^2 + \sigma_{yz}^2\right)^{\frac{n-1}{2}}\sigma_{yz}\, dx \wedge dy \wedge dt.$$

All isovectors of this system are given by

$$V = J_1\partial_x + J_2\partial_y + J_3\partial_t + J_4\partial_{\sigma_{xz}} + J_5\partial_{\sigma_{yz}} + J_6\partial_W \tag{6-7.60}$$

where

$$J_1 = k_1 + k_4\lambda x + k_5 y, \qquad J_2 = k_2 + k_4\lambda y - k_5 x,$$

$$J_3 = k_3 + (1-n)k_4 t, \qquad J_4 = k_4\sigma_{xz} + k_5\sigma_{yz},$$

$$J_5 = k_4\sigma_{yz} - k_5\sigma_{xz}, \qquad J_6 = k_4(n+\lambda)W$$

and the k's and λ are arbitrary constants. This problem has the remarkable feature that there is a five-dimensional Lie algebra of isovector fields for each choice of the constant λ (i.e., the five k's are the constants that arise from forming arbitrary linear combinations over $\mathbb{R}$ of the five infinitesimal generating vectors).

6-8. SIMILARITY VARIABLES AND SIMILARITY SOLUTIONS

Similarity solutions of systems of second order partial differential equations were first constructed by noting that the governing partial differential equations admitted homogeneous scaling transformations of the independent and the dependent variables. The drastic simplifications that were often achieved through use of this relatively simple technique served as motivation for deeper and more penetrating analyses of increased scope and generality. As it turns out, similarity solutions and the associated similarity variables are intimately related to and illuminated by the theory of isovector fields of the associated balance ideal.

Let K be a kinematic space with coordinate cover $(x^i, q^\alpha, y_i^\alpha)$ and contact forms

$$C^\alpha = dq^\alpha - y_i^\alpha\, dx^i, \qquad \alpha = 1, \ldots, N, \tag{6-8.1}$$

and let $\{E_\alpha\}$ be the collection of n-forms that characterizes the given system of quasilinear, second order partial differential equations. We then have the balance ideal

$$B = I\{C^\alpha, dC^\alpha, E_\alpha\} \tag{6-8.2}$$

and all solutions of the given system of partial differential equations are realized by maps $\Phi: B_n \to K$ such that

$$\Phi^* B = 0, \qquad \Phi^* \mu \neq 0. \tag{6-8.3}$$

Since B contains the contact ideal as a subideal, Φ^* annihilates C and any solution map is necessarily of the form

$$\Phi: B_n \to K | x^i = x^i, \qquad q^\alpha = \phi^\alpha(x^j), \qquad y_i^\alpha = \partial_i \phi^\alpha(x^j). \tag{6-8.4}$$

Similarity solutions are solution maps that satisfy a system of first order differential constraints that are generated by an isovector field of the balance ideal.

Definition. A map $\Phi: B_n \to K$ is a *similarity solution* that is generated by an isovector field V of the balance ideal if and only if Φ is a solution map of the balance ideal that satisfies the first-order differential constraints

$$\Phi^*(V \lrcorner C^\alpha) = 0, \qquad \alpha = 1, \ldots, N. \tag{6-8.5}$$

A similarity solution generated by $V \in TB(E_\alpha)$ thus satisfies

$$\Phi^* B = 0, \qquad \Phi^* \mu \neq 0, \qquad \Phi^*(V \lrcorner C^\alpha) = 0. \tag{6-8.6}$$

Lemma 6-8.1. *Let $N > 1$ and let $V \in TB(E_\alpha)$ be generated by*

(6-8.7) $$V_g = v^i(x^j, q^\beta)\,\partial_i + \bar{v}^\alpha(x^j, q^\beta)\,\partial_\alpha.$$

A map $\Phi: B \to K$ is a similarity generated by V if and only if Φ is a solution map of the balance ideal that satisfies the system of first order differential constraints

(6-8.8) $$\Phi^*(V_g \lrcorner C^\alpha) = 0, \qquad \alpha = 1, \ldots, N.$$

Proof. When $N > 1$, any $V \in TB(E_\alpha)$ is generated by its projection onto $T(\mathcal{G})$. We have $V = V_g + Z_i\langle V_g\langle q^\alpha\rangle - y_j^\alpha V_g\langle x^j\rangle\rangle\,\partial_\alpha^i$ and $C^\alpha = dq^\alpha - y_i^\alpha\,dx^i$, and hence $V \lrcorner C^\alpha = V_g \lrcorner C^\alpha$. □

Lemma 6-8.1 allows us to ignore the components of elements of $TB(E_\alpha)$ in the $\partial/\partial y_i^\alpha$ directions; that is, a similarity generated by $V \in TB(E_\alpha)$ is generated by its projection onto $T(\mathcal{G})$.

Lemma 6-8.2. *A map*

(6-8.9) $$\Phi | x^i = x^i, \qquad q^\alpha = \phi^\alpha(x^j), \qquad y_i^\alpha = \partial_i \phi^\alpha(x^j)$$

is a similarity solution of a given balance system that is generated by an isovector field V with

(6-8.10) $$V_g = v^i(x^j, q^\beta)\,\partial_i + \bar{v}^\alpha(x^j, q^\beta)\,\partial_\alpha$$

and $N > 1$ if and only if Φ is a solution of the balance ideal and the functions $\{\phi^\alpha(x)\}$ satisfy the system of N quasilinear first-order partial differential equations

(6-8.11) $$v^i(x^j, \phi^\beta(x^j))\,\partial_i \phi^\alpha(x^j) = \bar{v}^\alpha(x^j, \phi^\beta(x^j)), \qquad \alpha = 1, \ldots, N$$

with the same principal part $v^i(x^j, \phi^\beta(x^j))\,\partial_i$.

Proof. A direct calculation based on (6-8.1) and (6-8.10) gives $V_g \lrcorner C^\alpha = \bar{v}^\alpha(x^j, q^\beta) - y_i^\alpha v^i(x^j, q^\beta)$, and hence (6-8.9) shows that

$$\Phi^*(V_g \lrcorner C^\alpha) = \bar{v}^\alpha(x^j, \phi^\beta(x^j)) - v^i(x^j, \phi^\beta(x^j))\,\partial_i \phi^\alpha(x^j).$$

The result then follows from Lemma 6-8.1. □

Theorem 6-8.1. *Let $V \in TB(E_\alpha)$ be an isovector field of the balance ideal that is generated by*

(6-8.12) $$V_g = v^i(x^j, q^\beta)\,\partial_i + \bar{v}^\alpha(x^j, q^\beta)\,\partial_\alpha,$$

and let $N > 1$. A map Φ that solves the balance ideal is a similarity solution generated by V if and only if the functions $\phi^\alpha(x^j)$ satisfy

$$(6\text{-}8.13) \quad F^\alpha\big(g_1(x^j, \phi^\beta), \dots, g_{n+N-1}(x^j, \phi^\beta)\big) = k^\alpha, \qquad \alpha = 1, \dots, N,$$

$$(6\text{-}8.14) \qquad \frac{\partial(F^1, F^2, \dots, F^N)}{\partial(\phi^1, \phi^2, \dots, \phi^N)} \neq 0,$$

where the $n + N - 1$ functions $g_1(x^j, q^\beta), \dots, g_{n+N-1}(x^j, q^\beta)$ are a system of primitive integrals of the linear partial differential equation

$$(6\text{-}8.15) \qquad V_g\langle F(x^j, q^\beta)\rangle = 0.$$

Proof. All that has to be done is to characterize the solutions of (6-8.11) in view of Lemma 6-8.2. However, (6-8.11) is a system of quasilinear first-order partial differential equations with the same principal part $v^i(x^j, \phi^\beta(x^j))\partial_i$. We saw in Section 2-5 that solutions can always be written in the form (6-8.13) for any system of functions $F^\alpha(g_1, \dots, g_{n+N-1})$ that satisfy the invertability condition (6-8.14), where the functions $g_1, \dots, g_{n+N-1}$ are a system of primitive integrals of (6-8.15). □

The reader should note that the restriction $N > 1$ is not essential; it only simplifies matters. For $N = 1$, some, but not all isovectors of the balance ideal are generated by their projection V_g onto $T(\mathscr{G})$. For those isovectors that are projectable, Theorem 6-8.1 applies directly. In the general case with $N = 1$, we have nonlinear rather than linear first order partial differential equations, as an examination of (6-4.7) and (6-4.8) clearly shows.

Theorem 6-8.1 shows that the structure of similarity solutions generated by an isovector field V of the balance ideal is determined to a large part by the system of primitive integrals

$$g_1(x^j, q^\beta), \dots, g_{n+N-1}(x^j, q^\beta)$$

of the linear partial differential equation $V_g\langle F\rangle = 0$.

Definition. The primitive integrals $g_1, \dots, g_{n+N-1}$ of $V_g\langle F\rangle = 0$ are the *similarity variables* of all similarity solutions that are generated by the isovector field V of the balance ideal.

The essential aspect of similarity solutions can now be made clear. The problem starts with the $n + N$ variables (x^i, q^α). Once a system of $n + N - 1$ similarity variables $\{g_1, \dots, g_{n+N-1}\}$ has been found, all quantities are expressible in terms of these new variables. *Introduction of similarity variables and similarity solutions reduces the intrinsic dimension of the problem by one*. This is particularly important when there are only two independent variables, for in that case the problem can be reduced to functions of only *one* new independent variable; that is, partial differential equations can be reduced to ordinary differential equations. The reader must clearly note that similarity solutions are very special in a very real sense—they satisfy additional constraint equations. Be this as it may, the simplification afforded by similarity constructs is often worth the price of the specialization.

Another and equally important property of similarity solutions and similarity variables can be seen from the following result.

Theorem 6-8.2. *Let*

$$\Phi | x^i = x^i, \qquad q^\alpha = \phi^\alpha(x^j), \qquad y_i^\alpha = \partial_i \phi^\alpha(x^j) \tag{6-8.16}$$

be a similarity solution that is generated by the isovector field V of the balance ideal, and let $\{\bar{\phi}^\alpha(\bar{x}^j; s)\}$ be the 1*-parameter family of functions that obtains from transport of Φ along the orbits of V so that $\{\bar{\phi}^\alpha(\bar{x}^j; s)\}$ represent the mappings $\Phi_V(s) = T_V(s) \circ \phi$. The similarity variables and the similarity solutions are invariant under the mapping $T_V(s)$,*

$$g_a(\bar{x}^j, \bar{q}^\beta) = g_a(x^j, q^\beta), \qquad a = 1, \ldots, n + N - 1, \tag{6-8.17}$$

$$\bar{\phi}^\alpha(\bar{x}^j; s) = \phi^\alpha(\bar{x}^j). \tag{6-8.18}$$

Proof. The functions $g_a(x^j, q^\beta)$ are primitive integrals of $V_g\langle F \rangle = 0$, where $V_g = v^i(x^j, q^\beta)\partial_i + \bar{v}^\alpha(x^j, q^\beta)\partial_\alpha$. Thus if $\{\bar{x}^i(s), \bar{q}^\alpha(s)\}$ are solutions of the orbital equations of V_g subject to the initial data $\bar{x}^i(0) = x^i$, $\bar{q}^\alpha(0) = q^\alpha$, we have $g_a(\bar{x}^j, \bar{q}^\beta) = g_a(x^j, q^\beta)$. It then follows directly from (6-8.13) that the equations that define the ϕ's in terms of the x's are also invariant under the substitution $x^i \mapsto \bar{x}^i$, $q^\alpha \mapsto \bar{q}^\alpha$; that is, the $\bar{\phi}$'s are the same functions of the $\bar{x}$'s as the ϕ's were functions of the x's. □

Similarity solutions generated by a $V \in TB(E_\alpha)$ may be thought of as *fixed points* of the mapping $T_V(s)$ of K to K. As such, we do not get new solutions by transport of similarity solutions generated by $V \in TB(E_\alpha)$ along the orbits of V. However, if U is any other isovector field of the balance ideal that does not belong to span(V), then transport of a similarity solution generated by V along the orbits of U will generate new solutions.

Some examples may prove useful here. The isovector fields of the balance ideal for linear diffusion in two spatial dimensions are given by (6-7.34) and (6-7.35). If we take $k_9 = 1$ and all other parameters zero, the generating vector field is

$$V_g = x^1x^3\partial_1 + x^2x^3\partial_2 + (x^3)^2\partial_3 - \left(x^3 + \frac{R}{4}(x^1)^2 + \frac{R}{4}(x^2)^2\right)\partial_q \tag{6-8.19}$$

and

$$\Phi | x^1 = x, \qquad x^2 = y, \qquad x^3 = t, \qquad q = T(x, y, t). \tag{6-8.20}$$

This is a case where $N = 1$ but the isovector field is projectable onto $T(\mathscr{G})$. Solutions of the orbital equations of V_g in the coordinates (x, y, t, q) subject to the initial data $\bar{x}(0) = x$, $\bar{y}(0) = y$, $\bar{t}(0) = t$, $\bar{q}(0) = q$ are

$$\bar{x}(s) = x(1 - st)^{-1}, \qquad \bar{y}(s) = y(1 - st)^{-1}, \tag{6-8.21}$$

$$\bar{t}(s) = t(1 - st)^{-1},$$

$$\bar{q} = q(1 - st)\exp\left(-\frac{Rs}{4}\frac{x^2 + y^2}{1 - st}\right).$$

Accordingly, the primitive integrals of $V_g\langle F(x, y, t, q)\rangle = 0$ are given by

$$g_1 = x/t, \qquad g_2 = y/t, \qquad g_3 = qt\exp\left(\frac{R}{4}\frac{x^2 + y^2}{t}\right), \tag{6-8.22}$$

and all similarity solutions satisfy

$$F\left(\frac{x}{t}, \frac{y}{t}, Tt\exp\left(\frac{R}{4}\frac{x^2 + y^2}{t}\right)\right) = k;$$

that is

$$T(x, y, t) = M\left(\frac{x}{t}, \frac{y}{t}\right)\frac{1}{t}\exp\left(-\frac{R}{4}\frac{x^2 + y^2}{t}\right) \tag{6-8.23}$$

for some function $M(u, v)$.

If $T(x, y, t)$ is any solution of the linear diffusion equation in two spatial dimensions, then transport along the orbits of any $V \in TB(E_1)$ embeds

$T(x, y, t)$ in the 1-parameter family of solutions. For V given by (6-8.19), we have

$$\bar{T}(\bar{x}, \bar{y}, \bar{t}; s) = T\left(\frac{\bar{x}}{1+s\bar{t}}, \frac{\bar{y}}{1+s\bar{t}}, \frac{\bar{t}}{1+s\bar{t}}\right)\frac{\exp\left(-\frac{Rs}{4}\frac{\bar{x}^2+\bar{y}^2}{1+s\bar{t}}\right)}{1+s\bar{t}}.$$

It is then a simple matter to substitute any similarity solution $T(x, y, t)$ given by (6-8.23) into this family to see that the fixed point property

$$\bar{T}(\bar{x}, \bar{y}, \bar{t}; s) = T(\bar{x}, \bar{y}, \bar{t})$$

is satisfied.

For a problem with $N = 2$, we take the graph space to have coordinates (x, t, ϕ, ψ) and an isovector field of the balance ideal to be generated by

(6-8.24) $$V_g = -x\partial_x + (1 - 2\phi)\partial_t + \phi\partial_\phi + x\partial_\psi.$$

The solution of the orbital equations of this V_g subject to the standard initial data is given by

(6-8.25) $$\bar{x} = xe^{-s}, \qquad \bar{t} = t + s - 2\phi(e^s - 1),$$

$$\bar{\phi} = \phi e^s, \qquad \bar{\psi} = \psi + x(1 - e^{-s}).$$

Accordingly, the primitive integrals of $V_g\langle F(x, t, \phi, \psi)\rangle = 0$ are

(6-8.26) $$g_1 = x\phi, \qquad g_2 = \psi + x, \qquad g_3 = t + \ln(x) + 2\phi.$$

All similarity solutions that are generated by V_g have the form $F^\alpha(g_1, g_2, g_3) = k^\alpha$; that is,

(6-8.27) $$\phi = g(w)x^{-1}, \qquad \psi = h(w) - x$$

where g and h are functions of the single variable

(6-8.28) $$w = t + \ln(x) + 2\phi.$$

Everything is thus determined in terms of functions of the single specific similarity variable $w = g_3$; that is, the problem is reduced from finding two functions ϕ and ψ of two variables (x, t) to the problem of finding two functions g and h of the *single* similarity variable w.

The reader familiar with similarity solutions of partial differential equations may be dismayed by seeing that the similarity variable in the above example depends upon the old independent variables and the dependent variable ϕ as well,

$$w = t + \ln(x) + 2\phi.$$

In this case, (6-8.27) only define the similarity solution implicitly. We have seen, however, that similarity solutions and similarity variables result in implicitly defined solutions as the rule rather than the exception. Explicit similarity variables and similarity solutions only result in those cases where the generating vector V_g of the isovector field of the balance ideal is particularly simple, such as in homogeneous scaling transformations from which the theory started. In this regard, it is also useful to note that the mapping of arbitrary solutions $\phi(x, t)$, $\psi(x, t)$ by transport along the orbits of the vector field V_g is given by

$$\bar{\phi}(\bar{x}, \bar{t}; s) = e^{s}\phi(\bar{x}e^{s}, \bar{t} - s + 2(1 - e^{-s})\bar{\phi}(\bar{x}, \bar{t}; s)), \tag{6-8.29}$$

$$\bar{\psi}(\bar{x}, \bar{t}; s) = \psi(\bar{x}e^{s}, \bar{t} - s + 2(1 - e^{-s})\bar{\phi}(\bar{x}, \bar{t}; s)) + (e^{s} - 1)\bar{x},$$

which follow immediately from (6-8.25). Particular note should be taken of the intrinsic implicit nature of this mapping of solutions onto solutions. A direct computation shows that we obtain the fixed point property

$$\bar{\phi}(\bar{x}, \bar{t}; s) = \phi(\bar{x}, \bar{t}), \qquad \bar{\psi}(\bar{x}, \bar{t}; s) = \psi(\bar{x}, \bar{t}) \tag{6-8.30}$$

whenver $\phi(x, t)$ and $\psi(x, t)$ are a similarity solution that satisfy (6-8.27) and (6-8.28).

6-9. INVERSE ISOVECTOR METHODS

There are interesting and often useful inverse problems that can be associated with isovectors. Suppose that we choose specific nontrivial functions $v^{i}(x^{j}, q^{\beta})$ and $\bar{v}^{\alpha}(x^{j}, q^{\beta})$ and from

$$V_g = v^{i}(x^{j}, q^{\beta})\partial_i + \bar{v}^{\beta}(x^{j}, q^{\beta})\partial_{\alpha}. \tag{6-9.1}$$

Then V_g generates an element $V \in TC(K)$ by

$$V = V_g + Z_i\langle V_g\langle q^{\alpha}\rangle - y_j^{\alpha} V_g\langle x^{j}\rangle\rangle \partial_{\alpha}^{i}. \tag{6-9.2}$$

An obvious basis for such a choice would be that the orbital equations of V_g can be integrated in closed form. The isovector problem can now be turned around by asking the inverse question: find all nontrivial balance systems E_α such that $\pounds_V E_\alpha \subset I\{C^\beta, dC^\beta, E_\beta\}$. The collection of all such balance systems will have the isovector V in common, and hence they will have the mapping $T_V(s)$ of solutions onto solutions in common. If V_g has been picked so that the orbital equations can be integrated in closed form, all of the balance systems will share the common mapping $T_V(s)$ that can be obtained in closed form. The *inverse isovector method* thus has strong computational advantages that can provide unexpected insights. Its use should not be dismissed too lightly even though it is a problem of an isovector field in search of a balance system.

The simplest way to understand what is going on is through examples. Let $\mu = dx \wedge dt$, so that $\mu_1 = dt$, $\mu_2 = -dx$, take $N = 1$ and

(6-9.3) $$V_g = f'(t)\,\partial_x + \partial_t.$$

We take kinematic space as a five-dimensional number space with coordinate functions (x, t, ϕ, y_1, y_2) and contact form

$$C^1 = d\phi - y_1\,dx - y_2\,dt.$$

It then follows that $V_g \lrcorner C^1 = -f'(t)y_1 - y_2$, and the similarities generated by V_g are obtained by solving

$$f'(t)\,\partial_x\phi + \partial_t\phi = 0;$$

that is,

(6-9.4) $$\phi = \rho(x - f(t)).$$

This serves as a motivation for the choice (6-9.3); all similarity solutions generated by V_g are constant on the family of curves $x = f(t) + k$ in the (x, t)-plane.

The inverse isovector method comes into play when we ask for all balance systems of the form

$$E_1 = dW^i \wedge \mu_i,$$

that is, laws of conservation. Hence we obtain the requirements

$$d\pounds_V\left(W^i\mu_i\right) = A\,dW^i \wedge \mu_i + B \wedge C^1 + R\,dC^1$$

where

$$V = f'(t)\,\partial_x + \partial_t - f''(t)\,y_1\partial_{y_2}.$$

Clearly a sufficient condition for satisfaction of these requirements is given by

$$£_V(W^1 dt - W^2 dx) = P(d\phi - y_1 dx - y_2 dt) + \Psi(\phi)\, d\phi.$$

Resolving this expression on the basis $(dx, dt, d\phi)$ and eliminating the quantity P yields

$$(6\text{-}9.5) \qquad V\langle W^2\rangle = -y_1\Psi(\phi), \qquad V\langle W^1\rangle = f''(t)W^2 + y_2\Psi(\phi).$$

The general solution of this system of quasilinear partial differential equations with the same principal part is given by

$$(6\text{-}9.6) \qquad W^1 = \Psi(\phi)ty_2 + f'(t)\alpha + \beta, \qquad W^2 = -\Psi(\phi)ty_1 + \alpha$$

where α and β are arbitrary functions of the variables

$$x - f(t),\ \phi,\ y_1,\ f'(t)y_1 + y_2.$$

Further, since the solutions of the orbital equations of V_g are given by

$$(6\text{-}9.7) \qquad \bar{x} = x + f(t+s) - f(t), \qquad \bar{t} = t + s, \qquad \bar{\phi} = \phi,$$

any solution $\phi(x, t)$ of the balance system $E_1 = d(W^i\mu_i)$ can be embedded in a 1-parameter family of solutions

$$(6\text{-}9.8) \qquad \bar{\phi}(\bar{x}, \bar{t}; s) = \phi(\bar{x} - f(\bar{t}) + f(\bar{t} - s), \bar{t} - s).$$

The vector field

$$(6\text{-}9.9) \qquad V_g = -x\partial_x + (1 - 2\phi)\partial_t + \phi\partial_\phi + x\partial_\psi$$

considered at the end of the last section is of interest since we know that it generates the implicit similarity

$$(6\text{-}9.10) \quad \phi = g(w)x^{-1}, \qquad \psi = h(w) - x, \qquad w = t + \ln(x) + 2\phi.$$

Since a balance system was not given for V_g, we have an obvious candidate for the inverse isovector method.

We first note that V_g generates the vector field

$$(6\text{-}9.11) \qquad V = V_g + 2y_1^1(1 + y_2^1)\frac{\partial}{\partial y_1^1} + y_2^1(1 + 2y_2^1)\frac{\partial}{\partial y_2^1}$$

$$+ (1 + y_1^2 + 2y_1^1 y_2^2)\frac{\partial}{\partial y_1^2} + 2y_2^1 y_2^2 \frac{\partial}{\partial y_2^2}$$

on kinematic space with coordinate functions $(x, t, \phi, \psi, y_1^1, y_2^1, y_1^2, y_2^2)$ and contact forms

$$C^1 = d\phi - y_1^1\,dx - y_2^1\,dt, \qquad C^2 = d\psi - y_1^2\,dx - y_2^2\,dt.$$

A sufficient condition for V to be an isovector field of the balance ideal is satisfaction of the conditions

(6-9.12) $$£_V E_\alpha \equiv \ \text{mod}\ \overline{C}, \qquad \alpha = 1, 2,$$

where

(6-9.13) $$E_\alpha = F_\alpha^0\,dx \wedge dt - d\left(F_\alpha^1\,dt - F_\alpha^2\,dx\right)$$

and the F's are functions on kinematic space. A detailed calculation then shows that these conditions will be satisfied if the functions $(F_\alpha^0, F_\alpha^1, F_\alpha^2)$ are solutions of

(6-9.14) $$V\langle F_\alpha^0\rangle = \left(1 + 2y_2^1\right)F_\alpha^0, \qquad V\langle F_\alpha^1\rangle = 2y_2^1 F_\alpha^1,$$

$$V\langle F_\alpha^2\rangle = -2y_1^1 F_\alpha^1 + F_\alpha^2.$$

Particular note should be taken that we need *only sufficient* conditions for V to be an isovector of the balance ideal, provided the conditions are not trivial. Necessity is another matter altogether that we do not need to address.

The method of characteristics applied to the quasilinear system (6-9.14) with the same principal part leads to the general solution

(6-9.15) $$F_\alpha^0 = \phi y_2^2 M_\alpha^0(\Sigma),$$

(6-9.16) $$F_\alpha^1 = y_2^2 M_\alpha^1(\Sigma),$$

(6-9.17) $$F_\alpha^2 = \phi M_\alpha^2(\Sigma) - \frac{y_1^1 y_2^2}{y_2^1} M_\alpha^1(\Sigma)$$

for any choice of the functions $(M_\alpha^0, M_\alpha^1, M_\alpha^2)$ of the list of arguments

(6-9.18) $$\Sigma = \left\{ x\phi, \psi + x, t + \ln(x) + 2\phi, \frac{y_1^1}{y_2^1},\right.$$

$$\left.\frac{\phi y_2^2}{y_2^1}, \frac{y_2^1}{\phi} + \frac{1}{\phi} - \frac{y_1^1 y_2^2}{\phi y_2^1}, \frac{y_2^1}{\phi\left(1 + 2y_2^1\right)}\right\}.$$

Any system of balance equations of the form (6-9.13), (6-9.15) through (6-9.17) admits V as an isovector field, and hence admits similarity solutions of the form

$$\phi = g(w)x^{-1}, \qquad \psi = h(w) - x, \tag{6-9.19}$$

in terms of the implicit similarity variable

$$w = t + 1\ln x + 2\phi. \tag{6-9.20}$$

When (6-9.19) and (6-9.20) are used to compute the various first partial derivatives, we obtain

$$y_1^1 = \phi_x = \frac{g' - g}{x(x - 2g')}, \qquad y_2^1 = \phi_t = \frac{g'}{x - 2g'}, \tag{6-9.21}$$

$$y_1^2 = \psi_x = \frac{h'(x - 2g)}{x(x - 2g')} - 1, \qquad y_2^2 = \psi_t = \frac{xh'}{x - 2g'}, \tag{6-9.22}$$

where the prime signifies differentiation with respect to the implicit similarity variable w. It is then a simple matter to see that the list Σ given by (6-9.18) becomes

$$\Sigma = \left\{ g, h, w, \frac{g' - g}{gg'}, \frac{gh'}{g'}, \frac{h'}{g'}, \frac{g'}{g} \right\} \tag{6-9.23}$$

for (6-9.19) and (6-9.20). Every entry in the list Σ thus becomes a function of only the single variable w!

The implicit similarity likewise leads to

$$F_\alpha^0 = \frac{gh'}{x - 2g'} M_\alpha^0(\Sigma), \tag{6-9.24}$$

$$F_\alpha^1 = \frac{xh'}{x - 2g'} M_\alpha^1(\Sigma),$$

$$F_\alpha^2 = \frac{g}{x} M_\alpha^2(\Sigma) - \frac{h'(g' - g)}{g'(x - 2g')} M_\alpha^1(\Sigma),$$

as follows directly from (6-9.15) through (6-9.17), (6-9.21) and (6-9.22). Accordingly, we have

$$F_\alpha^1\, dt - F_\alpha^2\, dx = \frac{h'}{g'} M_\alpha^1\, dg - \left(M_\alpha^2 + \frac{h'}{g'} M_\alpha^1 \right) \phi\, dx \tag{6-9.25}$$

where we have used the identity $x\,d\phi = dg - \phi\,dx$ (i.e., $g = x\phi$) several times in the simplification process. The balance forms thus become

$$(6\text{-}9.26)\qquad E_\alpha = \frac{gh'}{x-2g'} M_\alpha^0\,dx \wedge dt - d\left[\frac{h'}{g'} M_\alpha^1\,dg - \left(M_\alpha^2 + \frac{h'}{g'} M_\alpha^1\right)\phi\,dx\right].$$

However, the term

$$\frac{h'}{g'} M_\alpha^1\,dg(w) = h' M_\alpha^1\,dw$$

involves only functions of the single independent variable w and hence its exterior derivative vanishes identically. We thus have

$$(6\text{-}9.27)\qquad E_\alpha = \frac{gh'}{x-2g'} M_\alpha^0\,dx \wedge dt + d\left[\phi\left(M_\alpha^2 + \frac{h'}{g'} M_\alpha^1\right)\right] \wedge dx$$

$$= \frac{gh'}{x-2g'} M_\alpha^0\,dx \wedge dt + \frac{1}{x}\,d\left[g\left(M_\alpha^2 + \frac{h'}{g'} M_\alpha^1\right)\right] dx$$

since $g = x\phi$. Now, $w = t + \ln x + 2\phi$ yields

$$dx \wedge dw = (1 + 2\phi_t)\,dx \wedge dt = \frac{x}{x-2g'}\,dx \wedge dt,$$

so that

$$dx \wedge dt = \frac{x-2g'}{x}\,dx \wedge dw,$$

and (6-9.27) becomes

$$(6\text{-}9.28)\qquad E_\alpha = \frac{gh'}{x} M_\alpha^0\,dx \wedge dw + \frac{1}{x}\,d\left[g\left(M_\alpha^2 + \frac{h'}{g'} M_\alpha^1\right)\right] \wedge dx$$

$$= \frac{1}{x}\left\{gh' M_\alpha^0 - \frac{d}{dw}\left[g\left(M_\alpha^2 + \frac{h'}{g'} M_\alpha^1\right)\right]\right\} dx \wedge dw.$$

Since all quantities that appear in this expression are functions of the implicit similarity variable w only, *the implicit similarity* (6-9.19), (6-9.20) *reduces the system of balance equations to the system of ordinary differential equations*

$$\text{(6-9.29)} \qquad \frac{d}{dw}\left[g\left(M_\alpha^2 + \frac{h'}{g'}M_\alpha^1\right)\right] = gh'M_\alpha^0, \qquad \alpha = 1,2$$

for each choice of the M_α's as functions of the arguments

$$\Sigma = \left\{g(w), h(w), w, \frac{g'-g}{gg'}, \frac{gh'}{g'}, \frac{h'}{g'}, \frac{g'}{g}\right\}.$$

6-10. DIMENSION REDUCTION IN THE CALCULATION OF ISOVECTORS

There are certain practical matters associated with the computation of isovectors of a balance ideal that must be addressed. Theorem 6-4.1 tells us that any isovector V of the contact ideal ($V \in TC(K)$) necessarily has the form

$$\text{(6-10.1)} \qquad V = V_g + Z_i(V_g \lrcorner C^\alpha)\,\partial_\alpha^i,$$

where

$$\text{(6-10.2)} \qquad V_g = f^i(x^j, q^\beta)\,\partial_i + \bar{f}^\alpha(x^j, q^\beta)\,\partial_\alpha$$

is any vector field on graph space whenever $N > 1$. If we restrict attention to balance forms

$$\text{(6-10.3)} \qquad B_\alpha = W_\alpha(x^j, q^\beta, y_j^\beta)\mu - dW_\alpha^i(x^j, q^\beta, y_i^\beta) \wedge \mu_i,$$

Theorem 6-6.2 shows that any $V \in TC(K)$ is an isovector field of the balance ideal $I\{C^\alpha, dC^\alpha, B_\alpha, dB_\alpha\}$ if and only if

$$\text{(6-10.4)} \qquad £_V B_\alpha = A_\alpha^\beta B_\beta + F_{\alpha\beta} \wedge C^\beta + G_{\alpha\beta} \wedge dC^\beta,$$

for some $A_\alpha^\beta \in \Lambda^0(K)$, $F_{\alpha\beta} \in \Lambda^{n-1}(K)$, $G_{\alpha\beta} \in \Lambda^{n-2}(K)$.

An appreciation of the magnitude of the computations that are involved in securing satisfaction of the conditions (6-10.4) is most easily perceived by noting that $£_V B_\alpha$ is an n-form on the space K, where $\dim(K) = n + N + nN$. Thus, for $n = N = 4$, $\dim(\Lambda^n(K)) = \binom{24}{4} = 10{,}626$, and hence the arena for the computations indicated by (6-10.4) is a vector space of dimension 10,626. The sheer magnitude of such computations has thus relegated isovector methods to those of absolute last resort. There is, however, a strong dimension reduction argument that reduces the computations to manageable proportions.

The details of this dimension reduction argument are the subject of this section.

We start by noting that an expansion of the indicated exterior derivation in (6-10.3) gives

$$B_\alpha = W_\alpha \mu - \left(\partial_j W_\alpha^i\right) dx^j \wedge \mu_i - \left(\partial_\beta W_\alpha^i\right) dq^\beta \wedge \mu_i - \left(\partial_\beta^j W_\alpha^i\right) dy_j^\beta \wedge \mu_i .$$

Noting that $dx^i \wedge \mu_j = \delta_j^i \mu$, we see that

(6-10.5) $B_\alpha \subset \operatorname{span}\left\{\mu, dq^\beta \wedge \mu_i, dy_j^\beta \wedge \mu_i | 1 \le i \le n, 1 \le \beta \le N\right\}$

for all values of α.

Next we note that

(6-10.6) $$\pounds_V(W_\alpha \mu) = V\langle W_\alpha \rangle \mu + W_\alpha \, df^i \wedge \mu_i$$

when (6-10.1) and (6-10.2) are used. It the follows from $f^i = f^i(x^j, q^\beta)$ that $\pounds_V(W_\alpha \mu)$ may be written in the form

(6-10.7) $$\pounds_V(W_\alpha \mu) = P_\alpha \mu + Q_{\alpha\beta}^i \, dq^\beta \wedge \mu_i$$

where $P_\alpha = V\langle W_\alpha \rangle + W_\alpha \partial_j f^j$, $Q_{\alpha\beta}^i = W_\alpha \partial_\beta f^i$. In like manner, $\pounds_V(W_\alpha^i \mu_i)$ may be written in the form

(6-10.8) $$\pounds_V\left(W_\alpha^i \mu_i\right) = P_\alpha^i \mu_i + Q_{\alpha\beta}^{ij} \, dq^\beta \wedge \mu_{ji},$$

on noting that $\pounds_V \mu_i = df^j \wedge \mu_{ji}$ and that $dx^k \wedge \mu_{ji} = \delta_j^k \mu_i - \delta_i^k \mu_j$. Thus since $\pounds_V \, d(W_\alpha^i \mu_i) = d\pounds_V(W_\alpha^i \mu_i)$, (6-10.8) gives

(6-10.9) $$\pounds_V \, d\left(W_\alpha^i \mu_i\right) = dP_\alpha^i \wedge \mu_i + dQ_{\alpha\beta}^{ij} \wedge dq^\beta \wedge \mu_{ji} .$$

It thus follows directly from (6-10.3), (6-10.7) and an expansion of the right-hand side of (6-10.9) that

(6-10.10) $$\pounds_V B_\alpha = P_\alpha \mu + Q_{\alpha\beta}^i \, dq^\beta \wedge \mu_i + R_{\alpha\beta}^{ij} \, dy_j^\beta \wedge \mu_i + S_{\alpha\beta\gamma}^{ij} \, dq^\gamma \wedge dq^\beta \wedge \mu_{ji} + T_{\alpha\beta\gamma}^{ijk} \, dy_k^\gamma \wedge dq^\beta \wedge \mu_{ji} .$$

An inspection of (6-10.10) leads to a direct verification of the following result.

Lemma 6-10.1. *The Lie derivative of any balance n-form on K with respect to any $V \in TC(K)$ is contained in the subspace*

(6-10.11) $$\mathrm{BAL} = \operatorname{span}\left\{\mu, dq^\beta \wedge \mu_i, dy_j^\beta \wedge \mu_i, dq^\gamma \wedge dq^\beta \wedge \mu_{ji}, dy_k^\gamma \wedge dq^\beta \wedge \mu_{ij}\right\}.$$

It is clear from Lemma 6-10.1 that satisfaction of the conditions (6-10.4) may be achieved without loss of generality by restricting the members on the right-hand side of (6-10.4) to belong to the subspace BAL. Thus since $C^\alpha = dq^\alpha - y_i^\alpha\, dx^i$, $dC^\alpha = -dy_i^\alpha \wedge dx^i$, while the basis elements of BAL are either of degree zero or of degree one in dy_j^β, the terms $F_{\alpha\beta} \wedge C^\beta + G_{\alpha\beta} \wedge dC^\beta$ on the right-hand side of (6-10.4) may be replaced by

$$b_{\alpha i}^\beta C^\beta \wedge \mu_i + e_{\alpha\beta\gamma}^{ij} C^\gamma \wedge C^\beta \wedge \mu_{ji} + K_{\alpha\beta} \wedge dC^\beta,$$

where $b_{\alpha i}^\beta$, $e_{\alpha\beta\gamma}^{ij}$ belong to $\Lambda^0(K)$ while $K_{\alpha\beta} \in \Lambda^{n-2}(\mathcal{G})$ and $\mathcal{G}$ is graph space with the local coordinates (x^i, q^β). Further, (6-10.5) and (6-10.11) combine to show that $B_\alpha \in$ BAL for each value of α, so that we have the following result.

Lemma 6-10.2. *A vector field $V \in TC(K)$ is an isovector field of the balance ideal if and only if*

$$\text{(6-10.12)} \quad \pounds_V B_\alpha = a_\alpha^\beta B_\beta + b_{\alpha i}^\beta C^\beta \wedge \mu_i + e_{\alpha\beta\gamma}^{ij} C^\gamma \wedge C^\alpha \wedge \mu_{ji} + K_{\alpha\beta} \wedge dC^\beta,$$

for some a_α^β, $b_{\alpha i}^\beta$, $e_{\alpha\beta\gamma}^{ij}$ belonging to $\Lambda^0(K)$ and some $K_{\alpha\beta} \in \Lambda^{n-2}(\mathcal{G})$.

Although some reduction has been achieved, there is still an enormous amount of work remaining, for both the left-hand and the right-hand sides of (6-10.12) belong to the subspace BAL and the quantities a_α^β, b_α^β, $e_{\alpha\beta\gamma}^{ij}$ and $K_{\alpha\beta}$ have to be solved for and then eliminated. We therefore turn to a different avenue of approach.

The contact forms have the representation $C^\alpha = dq^\alpha - y_i^\alpha\, dx^i$, and hence $dq^\alpha \equiv y_i^\alpha\, dx^i \bmod C$. Thus if we replace dq^β wherever it occurs in (6-10.12) by $y_j^\beta\, dx^j$, the resulting equations will be congruent mod C to the original ones. This is most easily accomplished by introducing the projection operator

$$\text{(6-10.13)} \qquad \pi_q : dq^\beta \mapsto y_k^\beta\, dx^k,$$

in which case we have

$$\text{(6-10.14)} \qquad \pi_q C^\alpha = 0, \qquad 1 \le \alpha \le N.$$

Lemma 6-10.3. *The conditions* (6-10.12) *are equivalent to the conditions*

$$\text{(6-10.15)} \quad \pi_q(\pounds_V B_\alpha) = a_\alpha^\beta \pi_q(B_\beta) + k_{\alpha\beta}^{ij}\left(dy_i^\beta \wedge \mu_j - dy_j^\beta \wedge \mu_i\right)$$

and

$$\text{(6-10.16)} \qquad \pi_q(\text{BAL}) = \text{span}\left\{\mu, dy_j^\beta \wedge \mu_i\right\}.$$

Proof. We first establish (6-10.16). When (6-10.13) is used, (6-10.11) gives

$$\pi_q(\text{BAL}) = \text{span}\{\mu, y_k^\beta dx^k \wedge \mu_i, dy_j^\beta \wedge \mu_i, y_k^\gamma dx^k \wedge y_m^\beta dx^m \wedge \mu_{ji},$$

$$dy_k^\gamma \wedge y_m^\beta dx^m \wedge \mu_{ij}\}$$

$$= \text{span}\{\mu, dy_k^\beta \wedge \mu_j\}$$

because $dx^k \wedge \mu_j = \delta_j^k \mu$ and $dx^k \wedge \mu_{ij} = \delta_i^k \mu_j - \delta_j^k \mu_i$. We now apply π_q to both sides of (6-10.12) and use (6-10.14). The result is

$$\pi_q(\pounds_V B_\alpha) = a_\alpha^\beta \pi_q(B_\beta) + \pi_q(K_{\alpha\beta}) \wedge dC^\beta.$$

However, $K_{\alpha\beta} \in \Lambda^{n-2}(\mathscr{G})$ and hence $\pi_q(K_{\alpha\beta}) = -k_{\alpha\beta}^{ij}\mu_{ij}$. When this is used to evaluate $\pi_q(K_{\alpha\beta}) \wedge dC^\beta$, we have

$$\pi_q(K_{\alpha\beta}) \wedge dC^\beta = k_{\alpha\beta}^{ij}\mu_{ij} \wedge dy_k^\beta \wedge dx^k = k_{\alpha\beta}^{ij} dy_k^\beta \wedge dx^k \wedge \mu_{ij}$$

$$= k_{\alpha\beta}^{ij} dy_k^\beta \wedge (\delta_i^k \mu_j - \delta_j^i \mu_i)$$

$$= k_{\alpha\beta}^{ij}(dy_i^\beta \wedge \mu_j - dy_j^\beta \wedge \mu_i). \qquad \square$$

The last dimension reduction obtains by noting that

$$dC^\alpha \wedge \mu_{ij} = -dy_k^\alpha \wedge dx^k \wedge \mu_{ij} = dy_j^\alpha \wedge \mu_i - dy_i^\alpha \wedge \mu_j.$$

Thus if we define the projection operator π_y by

(6-10.7) $$\pi_y : dy_j^\beta \wedge \mu_i \mapsto \tfrac{1}{2}(dy_j^\beta \wedge \mu_i + dy_i^\beta \wedge \mu_j),$$

then

(6-10.18) $$\pi_y(dC^\alpha \wedge \mu_{ij}) = \pi_y(dy_j^\alpha \wedge \mu_i - dy_i^\alpha \wedge \mu_j) = 0.$$

Theorem 6-10.1. *If $V \in TC(K)$, then V is an isovector field of the balance ideal if and only if*

(6-10.19) $$\pi_y \pi_q(\pounds_V B_\alpha) = a_\alpha^\beta \pi_y \pi_q(B_\alpha), \qquad 1 \le \alpha \le N$$

are satisfied for some $a_\alpha^\beta \in \Lambda^0(K)$, *and these equations obtain over the subspace*

(6-10.20) $$\pi_y \pi_q(\text{BAL}) = \text{span}\left\{ \mu, \tfrac{1}{2}\left(dy_j^\beta \mu_i + dy_i^\beta \mu_j \right) \right\}$$

of $\Lambda^n(K)$ *of dimension* $1 + Nn(n+1)/2$.

Proof. It follows directly from Lemma 6-10.3 and (6-10.17) that (6-10.16) implies (6-10.20), while (6-10.15) implies (6-10.19). □

The results given by Theorem 6-10.1 constitute a significant reduction in the calculation of isovectors of the balance ideal. First, there are now only N^2 functions $\{a_\alpha^\beta\}$ to be solved for and then eliminated. Second, we have been able to progress from a problem over a subspace of dimension $\binom{n+N+Nn}{n}$ to a problem over a subspace of dimension $1 + Nn(n+1)/2$. Thus for the case where $n = N = 4$ we have a reduction in dimension from 10,626 down to 41. Computation of the isovector fields for all of the larger examples given in Section 6-7 were actually performed through use of this dimension reduction argument.

6-11. CONTACT FORMS OF HIGHER ORDER

The reader will have noticed that the theory given up to this point is applicable primarily to systems of second-order partial differential equations that are linear in the second derivatives:

$$A_{\alpha\beta}^{ij}\left(x^k, \phi^\gamma, \partial_k \phi^\gamma\right) \partial_i \partial_j \phi^\beta + B_\alpha\left(x^k, \phi^\gamma, \partial_i \phi^\gamma\right) = 0.$$

Fully general nonlinear problems,

$$F_\alpha\left(x^k, \phi^\beta, \partial_k \phi^\beta, \partial_k \partial_m \phi^\beta\right) = 0,$$

require direct access to the second-order derivatives through study of contact forms of second order,

$$C_j^\alpha = dy_j^\alpha - y_{ji}^\alpha\, dx^i.$$

Let M be an $n + N(1 + n + n^2)$ dimensional space with local coordinates $(x^i, q^\alpha, y_i^\alpha, y_{ij}^\alpha)$ and let

(6-11.1) $$\mu = dx^1 \wedge dx^2 \wedge \cdots \wedge dx^n$$

be the volume element of the space E_n of independent variables. As before, the contact forms are given by

(6-11.2) $$C^\alpha = dq^\alpha - y_i^\alpha\, dx^i, \qquad \alpha = 1, \ldots, N.$$

The new variables y_{ij}^{α} give rise to the *second-order contact* 1-*forms*

$$C_j^{\alpha} = dy_j^{\alpha} - y_{ji}^{\alpha}\, dx^i, \qquad \alpha = 1, \dots, N, \qquad j = 1, \dots, n, \tag{6-11.3}$$

and to the *second-order contact ideal*

$$C_2 = I\{C^1, \dots, C^N; C_1^1, \dots, C_n^N\}. \tag{6-11.4}$$

A map $\Phi: E_n \to M$ is said to be *regular* if and only if $\Phi^*\mu \neq 0$ and Φ^* annihilates the second order contact ideal. The collection of all regular maps is denoted by $R_2(M)$;

$$R_2(M) = \{\Phi: E_n \to M | \Phi^*\mu \neq 0, \Phi^*C_2 = 0\}. \tag{6-11.5}$$

Thus if Φ is regular, we also have

$$\Phi^*\, dC_2 = 0, \tag{6-11.6}$$

while (6-11.3) gives

$$\Phi^* y_{ij}^{\alpha} = \partial_j \Phi^* y_i^{\alpha} = \partial_j \partial_i \Phi^* q^{\alpha}. \tag{6-11.7}$$

Thus the second-order contact forms serve to identify y_{ij}^{α} with the second-order partial derivatives of q's with respect to the x's on any regular section of M. Further, (6-11.2) and (6-11.3) give

$$\begin{aligned} dC^{\alpha} &= -dy_i^{\alpha} \wedge dx^i = -\left(C_i^{\alpha} + y_{ij}^{\alpha}\, dx^j\right) \wedge dx^i \\ &= \tfrac{1}{2}\left(y_{ji}^{\alpha} - y_{ij}^{\alpha}\right) dx^j \wedge dx^i \bmod C_2. \end{aligned} \tag{6-11.8}$$

Thus since Φ^* annihilates dC^{α} if Φ is regular, we have

$$\Phi^* y_{ji}^{\alpha} = \Phi^* y_{ij}^{\alpha}. \tag{6-11.9}$$

Let V be a vector field on M; then V has a representation

$$V = v^i \partial_i + \bar{v}^{\alpha} \partial_{\alpha} + v_i^{\alpha} \partial_{\alpha}^i + v_{ij}^{\alpha} \partial_{\alpha}^{ij}, \tag{6-11.10}$$

where we have introduced the notation

$$\partial_{\alpha}^{ij} := \partial / \partial y_{ij}^{\alpha}. \tag{6-11.11}$$

Definition. A vector field V is said to be an *isovector field* of the second-order contact ideal C_2 if and only if

$$\pounds_V C_2 \subset C_2. \tag{6-11.12}$$

The collection of all isovector fields of the second-order contact ideal is denoted by $TC_2(M)$:

$$TC_2(M) = \{V \in T(M) | \pounds_V C_2 \subset C_2\}. \tag{6-11.13}$$

Arguments similar to those used in Section 6-4 give the following immediate results.

Theorem 6-11.1. *$TC_2(M)$ is a Lie subalgebra of $T(M)$ over $\mathbb{R}$, and $V \in TC_2(M)$ if and only if there exist elements $\{A^{\alpha}_{\beta}, B^{\alpha j}_{\beta}, K^{\alpha}_{i\beta}, L^{\alpha j}_{i\beta}\}$ of $\Lambda^0(M)$ such that*

$$\pounds_V C^{\alpha} = A^{\alpha}_{\beta} C^{\beta} + B^{\alpha j}_{\beta} C^{\beta}_j, \tag{6-11.14}$$

$$\pounds_V C^{\alpha}_i = K^{\alpha}_{i\beta} C^{\beta} + L^{\alpha j}_{i\beta} C^{\beta}_j. \tag{6-11.15}$$

Theorem 6-11.2. *If $V \in TC_2(M)$ and $\Phi \in R_2(M)$, then V serves to embed Φ in a 1-parameter family*

$$\Phi_V(s) = T_V(s) \circ \Phi \tag{6-11.16}$$

of regular maps for all s in a sufficiently small neighborhood of $s = 0$, where $T_V(s)$ is the flow of the vector field V in M.

It is clear from these results that all of the arguments given in previous sections of this chapter can be carried through simply by replacing the contact ideal C by the second-order contact ideal C_2. Accordingly, once the general solution of the conditions (6-11.14) and (6-11.15) has been obtained, isovector methods can be applied directly to general nonlinear second-order partial differential equations of the type $F(x^k, \phi^{\beta}, \partial_k \phi^{\beta}, \partial_k \partial_m \phi^{\beta}) = 0$.

We begin by noting that $\pounds_V C^{\alpha} = d\bar{v}^{\alpha} - v^{\alpha}_i \, dx^i - y^{\alpha}_i \, dv^i$. When this is substituted into the left-hand side of (6-11.14) and (6-11.2) through (6-11.3) are used, a resolution on the basis $(dx^k, dq^{\beta}, dy^{\beta}_k, dy^{\beta}_{km})$ of $\Lambda^1(M)$ gives

$$\partial_k \bar{v}^{\alpha} - v^{\alpha}_i - y^{\alpha}_i \partial_k v^i = -A^{\alpha}_{\beta} y^{\beta}_k - B^{\alpha l}_{\beta} y^{\beta}_{lk}, \tag{6-11.17}$$

$$\partial_{\beta} \bar{v}^{\alpha} - y^{\alpha}_i \partial_{\beta} v^i = A^{\alpha}_{\beta}, \tag{6-11.18}$$

$$\partial^k_{\beta} \bar{v}^{\alpha} - y^{\alpha}_i \partial^k_{\beta} v^i = B^{\alpha k}_{\beta}, \tag{6-11.19}$$

$$\partial^{km}_{\beta} \bar{v}^{\alpha} - y^{\alpha}_i \partial^{km}_{\beta} v^i = 0. \tag{6-11.20}$$

Equations (6-11.18) and (6-11.19) serve to determine A^{α}_{β} and $B^{\alpha k}_{\beta}$, respectively. When these are substituted into (6-11.17), introduction of the linear operators

$$M_i = Z_i + y^{\beta}_{mi}\partial^m_{\beta}, \qquad Z_i = \partial_i + y^{\beta}_i\partial_{\beta} \tag{6-11.21}$$

serves to determine the quantities v^{α}_k by

$$v^{\alpha}_k = M_k\langle \bar{v}^{\alpha}\rangle - y^{\alpha}_i M_k\langle v^i\rangle = M_k\langle \bar{v}^{\alpha} - y^{\alpha}_i v^i\rangle + v^i y^{\alpha}_{ik}, \tag{6-11.22}$$

and we are left with the conditions

$$\partial^{km}_{\beta}\left(\bar{v}^{\alpha} - y^{\alpha}_i v^i\right) = \partial^{km}_{\beta}(V \lrcorner C^{\alpha}) = 0. \tag{6-11.23}$$

The exact same procedure starting with (6-11.16) gives

$$\partial_k v^{\alpha}_j - v^{\alpha}_{jk} - y^{\alpha}_{ji}\partial_k v^i = -K^{\alpha}_{j\beta}y^{\beta}_k - L^{\alpha m}_{j\beta}y^{\beta}_{mk}, \tag{6-11.24}$$

$$\partial_{\beta}v^{\alpha}_j - y^{\alpha}_{ji}\partial_{\beta}v^i = K^{\alpha}_{j\beta}, \tag{6-11.25}$$

$$\partial^k_{\beta}v^{\alpha}_j - y^{\alpha}_{ji}\partial^k_{\beta}v^i = L^{\alpha k}_{j\beta}, \tag{6-11.26}$$

$$\partial^{km}_{\beta}v^{\alpha}_j - y^{\alpha}_{ji}\partial^{km}_{\beta}v^i = 0. \tag{6-11.27}$$

Here, (6-11.25) and (6-11.26) serve to determine $K^{\alpha}_{j\beta}$ and $L^{\alpha k}_{j\beta}$, respectively. When these determinations are substituted into (6-11.24), we have

$$v^{\alpha}_{jk} = M_k\langle v^{\alpha}_j - y^{\alpha}_{ji}v^i\rangle = M_k\langle V \lrcorner C^{\alpha}_j\rangle, \tag{6-11.28}$$

while (6-11.27) yields the conditions

$$\partial^{km}_{\beta}v^{\alpha}_j = y^{\alpha}_{ji}\partial^{km}_{\beta}v^i. \tag{6-11.29}$$

A direct integration of the conditions (6-11.23) leads to

$$\bar{v}^{\alpha} = y^{\alpha}_i v^i + F^{\alpha}\left(x^j, q^{\beta}, y^{\beta}_k\right), \tag{6-11.30}$$

and hence (6-11.22) gives

$$v^{\alpha}_k = M_k\langle F^{\alpha}\rangle + v^i y^{\alpha}_{ik}. \tag{6-11.31}$$

We now substitute (6-11.31) into (6-11.29). This gives the conditions

$$\partial^k_{\beta}F^{\alpha}\left(x^j, q^{\gamma}, y^{\gamma}_j\right) + \delta^{\alpha}_{\beta}v^k = 0. \tag{6-11.32}$$

It thus follows that

$$F^{\alpha} = f^{\alpha}\left(x^j, q^{\beta}\right) - y^{\alpha}_i f^i\left(x^j, q^{\beta}\right), \tag{6-11.33}$$

$$v^i = f^i\left(x^j, q^{\beta}\right). \tag{6-11.34}$$

Thus (6-11.30) yields

(6-11.35) $$\bar{v}^\alpha = f^\alpha(x^j, q^\beta),$$

while (6-11.31) and (6-11.28) give

(6-11.36) $$v_k^\alpha = Z_k\langle v^\alpha - y_i^\alpha v^i\rangle = Z_k\langle f^\alpha - y_i^\alpha f^i\rangle = Z_k\langle V \lrcorner C^\alpha\rangle$$

(6-11.37) $$\begin{aligned} v_{jk}^\alpha &= M_k\langle Z_j\langle V \lrcorner C^\alpha\rangle - y_{ji}^\alpha f^i\rangle \\ &= Z_k\langle Z_j\langle V \lrcorner C^\alpha\rangle - y_{ji}^\alpha f^i\rangle + y_{lk}^\gamma \partial_\gamma^l\langle Z_j\langle V \lrcorner C^\alpha\rangle\rangle. \end{aligned}$$

Theorem 6-11.3. *A vector field*

$$V = v^i\partial_i + \bar{v}^\alpha\partial_\alpha + v_i^\alpha\partial_\alpha^i + v_{ij}^\alpha\partial_\alpha^{ij}$$

is an isovector field of the second contact ideal C_2 if and only if

(6-11.38) $$v^i = f^i(x^j, q^\beta),$$

(6-11.39) $$\bar{v}^\alpha = f^\alpha(x^j, q^\beta),$$

(6-11.40) $$v_i^\alpha = Z_i\langle f^\alpha - y_j^\alpha f^j\rangle,$$

(6-11.41) $$\begin{aligned} v_{ij}^\alpha = Z_j\langle Z_i\langle f^\alpha - y_m^\alpha f^m\rangle - y_{ik}^\alpha f^k\rangle \\ + y_{mj}^\gamma \partial_\gamma^m\langle Z_i f^\alpha - y_m^\alpha f^m\rangle, \end{aligned}$$

where $\{Z_k\}$ are the linear differential operators

(6-11.42) $$Z_k = \partial_k + y_k^\beta\partial_\beta.$$

There are some immediate and useful results that obtain directly from Theorem 6-11.3 and its proof. A direct substitution of the evaluations obtained for the A's, B's, K's and L's into (6-11.14) and (6-11.15), together with (6-11.38) through (6-11.41), lead to the following results.

Corollary 6-11.1. *If $V \in TC_2(M)$, then*

(6-11.43) $$\pounds_V C^\alpha = \partial_\beta(\bar{v}^\alpha - y_i^\alpha v^i)C^\beta,$$

(6-11.44) $$\pounds_V C_j^\alpha = \partial_\beta(v_j^\alpha - y_{ji}^\alpha v^i)C^\beta + \partial_\beta^k v_j^\alpha C_k^\beta.$$

and hence any $V \in TC_2(M)$ maps the first-order contact ideal into itself. □

Next we note that a vector field

$$V_g = f^i(x^j, q^\beta)\,\partial_i + f^\alpha(x^j, q^\beta)\,\partial_\beta \tag{6-11.45}$$

on graph space $\mathscr{G}$ with local coordinates (x^j, q^β) serves to determine a vector field $V \in TC_2(M)$ by

$$\begin{aligned} V = {} & V_g + Z_i\langle V_g \lrcorner C^\alpha\rangle\,\partial_\alpha^i \\ & + \{Z_j\langle Z_i\langle V_g \lrcorner C^\alpha\rangle - y_{ik}^\alpha V_g \lrcorner\, dx^k\rangle + y_{mj}^\gamma \partial_\gamma^m\langle Z_i\langle V_g \lrcorner C^\alpha\rangle\rangle\}\,\partial_\alpha^{ij} \end{aligned} \tag{6-11.46}$$

because $V_g \lrcorner C^\alpha = f^\alpha - y_j^\alpha f^j$, $V_g \lrcorner\, dx^k = f^k$. Thus each $V \in TC_2(M)$ is generated by a $V_g \in T(\mathscr{G})$ by (6-11.46). In fact, isovector fields of the second-order contact ideal are second extensions of infinitesimal point transformations of $\mathscr{G}$ that are generated by V_g. This conclusion is also directly evident from Theorem 6-11.3, which shows that the deformations of the x's and the q's can depend only on the x's and the q's but not on the y's. Elements of $TC_2(M)$ and elements of $TC(K)$ thus differ only in the order of the extension, but not in substance.

Applications to nonlinear second-order PDE can differ markedly if we use the second-order contact structure. To see this, we first note that the second-order PDE can now be written as a system of scalar equations

$$\Phi^* W_\alpha(x^j, q^\beta, y_j^\beta, y_{km}^\beta) = 0, \qquad 1 \le \alpha \le N, \tag{6-11.47}$$

where $\Phi: E_n \to M$ is a regular map. As such, we now obtain the closed differential ideal

$$Q = I\{C^\alpha, dC^\alpha, C_j^\alpha, dC_j^\alpha, dW_\alpha\} \tag{6-11.48}$$

that is annihilated by any regular solving map; that is, $\Phi^* C^\alpha = 0$, $\Phi^* C_j^\alpha = 0$, $\Phi^*\, dW_\alpha = d\Phi^* W_\alpha = 0$ implies and is implied by $\Phi^* Q = 0$. Isovectors of the ideal Q then transport solutions into solutions, while the conditions that V be such an isovector field read

$$\pounds_V C^\alpha = A_\beta^\alpha C^\beta + B_\beta^{\alpha j} C_j^\beta + E^{\alpha\beta}\, dW_\beta, \tag{6-11.49}$$

$$\pounds_V C_j^\alpha = \bar{A}_{j\beta}^\alpha C^\beta + \bar{B}_{j\beta}^{\alpha k} C_k^\beta + \bar{E}_j^{\alpha\beta}\, dW_\beta, \tag{6-11.50}$$

$$\pounds_V\, dW_\alpha = A_{\alpha\beta} C^\beta + B_{\alpha\beta}^k C_k^\beta + E_\alpha^\beta\, dW_\beta. \tag{6-11.51}$$

Therefore, in view of the new terms $E^{\alpha\beta}\, dW_\beta$ and $\bar{E}_j^{\alpha\beta}\, dW_\beta$ on the right-hand sides of (6-11.50) and (6-11.51), isovector fields of the ideal Q are not necessarily contained in $TC_2(M)$.

CHAPTER SEVEN

CALCULUS OF VARIATIONS

7-1. FORMULATION OF THE PROBLEM

One of the oldest and most useful mathematical structures is the calculus of variations. It is the foundation upon which Lagrangian and Hamiltonian mechanics of particles and rigid bodies is constructed. In fact, it has now come to pass that a fundamental physical theory is not considered well constructed unless the governing field equations are derived from a variational principle. Some of the reasons for this state of affairs are that a variationally derived theory has a certain internal consistency and that physical symmetries of the system can be directly related to conservation laws that are satisfied by any and all solutions. Last, but not least is the fact that variational formulations lead to a complete theory of compatible boundary and initial data.

Let E_n be an n-dimensional number space with a coordinate cover (x^i), and let B_n be an n-dimensional, arcwise connected point set of E_n. The notation ∂B_n will be used for the boundary of B_n. We assume that ∂B_n is smooth with the possible exception of a finite number of edges and vertices.

Suppose that a physical system is described by a finite number N of state functions $\{\phi^1(x^j), \phi^2(x^j), \ldots, \phi^N(x^j)\} = \{\phi^\alpha(x^j)\}$ and that the system has a smooth *Lagrangian function*

$$L\big(x^j, \phi^\alpha(x^j), \partial_i\phi^\alpha(x^j)\big)$$

that is a known function of the independent variables (x^j), the dependent variables $\{\phi^\alpha(x^j)\}$, and the first derivatives $\{\partial_i\phi^\alpha(x^j)\}$. In simple cases, the

Lagrangian function is given by the difference of the kinetic energy and the potential energy per unit volume of B_n. The *action* of the system in the state represented by $\{\phi^\alpha(x^j)\}$ has the evaluation

$$A[\phi^\alpha] = \int_{B_n} L\big(x^j, \phi^\alpha(x^j), \partial_i\phi^\alpha(x^j)\big)\, dx^1\, dx^2 \cdots dx^n. \tag{7-1.1}$$

The action is thus a map from the space of state functions $\{\phi^\alpha(x^j)\}$ into the real line $\mathbb{R}$.

Most analyses of problems in the calculus of variations spend a significant amount of time in the study of the properties of the space of state functions; i.e., functional analysis and the accompanying theory of function spaces. The approach taken here is couched in geometric language and the emphasis is shifted to the study of collections of maps between spaces with appropriate structure.

A little reflection on the structure of the action (7-1.1), in the light of the material covered in the last chapter, shows that there is a simple and direct geometric reformulation. Let E_n be as before and consider an $(n + N + nN)$-dimensional kinematic space K with the coordinate cover $(x^i, q^\alpha, y_i^\alpha)$ and contact 1-forms

$$C^\alpha = dq^\alpha - y_i^\alpha\, dx^i. \tag{7-1.2}$$

We then have the contact ideal

$$C = I\{C^1, C^2, \dots, C^N\} \tag{7-1.3}$$

and its closure

$$\bar{C} = I\{C^1, \dots, C^N, dC^1, \dots, dC^N\}. \tag{7-1.4}$$

The collection of all regular maps from B_n into K is defined by

$$R(B_n) = \{\Phi : B_n \to K \,|\, \Phi^*\bar{C} = 0, \Phi^*\mu \neq 0\}, \tag{7-1.5}$$

where

$$\mu = dx^1 \wedge dx^2 \wedge \cdots \wedge dx^n \tag{7-1.6}$$

is the volume element of E_n. Accordingly, any $\Phi \in R(B_n)$ has a representation

$$\Phi | x^i = x^i, \qquad q^\alpha = \phi^\alpha(x^j), \qquad y_i^\alpha = \partial_i\phi^\alpha(x^j); \tag{7-1.7}$$

i.e., Φ^* annihilates the contact forms and hence the y's become the derivatives of the q's with respect to the x's on the range of Φ.

If L is the Lagrangian function for the system under study, an immediate computation shows that

$$\Phi^* L(x^j, q^\alpha, y_i^\alpha) = L(x^j, \phi^\alpha(x^j), \partial_i \phi^\alpha(x^j)),$$

and hence L may be considered as a scalar-valued function with domain K; i.e., $L \in \Lambda^0(K)$. A direct transcription of (7-1.1) then leads to the following reformulation.

Definition. The *action* associated with any $\Phi \in R(B_n)$ and any $L \in \Lambda^0(K)$ has the evaluation

(7-1.8) $$A[\Phi] = \int_{B_n} \Phi^*(L\mu).$$

Let Ψ be any other regular map of B_n into K,

$$\Psi | x^i = x^i, \qquad q^\alpha = \psi^\alpha(x^j), \qquad y_i^\alpha = \partial_i \psi^\alpha(x^j).$$

The action associated with Ψ, for given $L \in \Lambda^0(K)$ has the evaluation

$$A[\Psi] = \int_{B_n} \Psi^*(L\mu).$$

Accordingly, since $A[\Phi]$ and $A[\Psi]$ belong to $\mathbb{R}$, their values can be compared:

(7-1.9) $$A[\Psi] - A[\Phi] = \int_{B_n} \{\Psi^*(L\mu) - \Phi^*(L\mu)\}.$$

There are two facts that must be noted here. First, we have changed Φ to Ψ by changing the functions $\phi^\alpha(x^j)$ to $\psi^\alpha(x^j)$, but the (x^i) *have not been changed.* Second, if we can embed Φ and Ψ in a 1-parameter family of *regular maps* with parameter s such that $\Psi \equiv \Phi$ at $s = 0$, then a division of (7-1.9) by s followed by an evaluation of the limit as s tends to zero provides the basis for construction of a well defined differentiation process.

The work of the previous chapter shows that there is an immediate candidate for the embedding process: transport along the orbits of isovector fields of the contact ideal. Indeed, there is no real choice here, for transport along orbits of isovector fields of the contact ideal is the most general transport process that preserves membership in the collection $R(B_n)$ of regular maps; $\Phi_V(s) \in R(B_n)$ for all $\Phi \in R(B_n)$ and all $V \in TC(K)$, while $\Phi_U(s) \notin R(B_n)$ if $s \neq 0$ for any $\Phi \in R(B_n)$ and any $U \notin TC(K)$.

Theorem 6-4.1 shows that any isovector field V of the contact ideal is generated by a vector field

$$V_g = f^i(x^j, q^\beta)\partial_i + \bar{f}^\alpha(x^j, q^\beta)\partial_\alpha \tag{7-1.10}$$

on graph space through the relation

$$V = V_g + Z_i\langle \bar{f}^\alpha - y_j^\alpha f^j \rangle \partial_\alpha^i, \tag{7-1.11}$$

where

$$Z_i \coloneqq \partial_i + y_i^\beta \partial_\beta. \tag{7-1.12}$$

Accordingly, transport along the orbits of an arbitrary isovector field of the contact ideal will change the x's as well as the q's. It is therefore necessary that we restrict attention to those elements of $TC(K)$ that do not change the independent variables.

Definition. An isovector field U of the contact ideal is said to be *vertical* if and only if $U\langle x^i \rangle = 0$, $i = 1, \ldots, n$. The collection of all vertical isovector fields is denoted by

$$TV = \{ U \in TC(K) \mid U\langle x^i \rangle = 0,\ i = 1, \ldots, n \}. \tag{7-1.13}$$

Lemma 7-1.1. *A vector field $U \in T(K)$ is vertical if and only if*

$$U = \bar{f}^\alpha(x^j, q^\beta)\partial_\alpha + Z_i\langle \bar{f}^\alpha \rangle \partial_\alpha^i \tag{7-1.14}$$

for some choice of the functions $\{\bar{f}^\alpha(x^j, q^\beta)\}$.

Proof. By definition $U \in TV(K)$ if and only if $U \in TC(K)$ and $U\langle x^i \rangle = 0$, $i = 1, \ldots, n$. Since all $U \in TC(K)$ have the form given by (7-1.10) and (7-1.11), we have

$$U\langle x^i \rangle = U \lrcorner dx^i = f^i(x^j, q^\beta),$$

from which the result follows. □

Vertical isovector fields allow us to introduce the concept of the variation of a regular map in a natural way. The orbital equations of any $U \in TV(K)$ are

$$\frac{d\bar{x}^i}{ds} = 0, \qquad \frac{d\bar{q}^\alpha}{ds} = \bar{f}^\alpha(\bar{x}^j, \bar{q}^\beta), \qquad \frac{d\bar{y}_i^\alpha}{ds} = \bar{Z}_i\langle \bar{f}^\alpha(\bar{x}^j, \bar{q}^\beta) \rangle$$

subject to the initial data

$$\bar{x}^i(0) = x^i, \qquad \bar{q}^\alpha(0) = q^\alpha, \qquad \bar{y}_i^\alpha(0) = y_i^\alpha.$$

The map $T_U(s)$ of K to K is thus given by

$$T_U(s)|\bar{x}^i = x^i, \qquad \bar{q}^\alpha = q^\alpha + \bar{f}^\alpha(x^j, q^\beta)s + o(s),$$

$$\bar{y}_i^\alpha = y_i^\alpha + Z_i\langle \bar{f}^\alpha(x^j, q^\beta)\rangle s + o(s),$$

and hence composition with any $\Phi \in R(K)$ gives $\Phi_U(s) = T_U(s) \circ \Phi$ with

$$\Phi_U(s)|\bar{x}^i = x^i, \qquad \bar{\phi}^\alpha = \phi^\alpha + \bar{f}^\alpha(x^j, \phi^\beta)s + o(s),$$

$$\bar{y}_i^\alpha = \partial_i\phi^\alpha + \left(\partial_i + (\partial_i\phi^\beta)\partial_\beta\right)\langle \bar{f}^\alpha(x^j, \phi^\gamma)\rangle s + o(s).$$

The *variation* of any $\Phi \in R(K)$ (restriction to first-order terms in the parameter s) is thus defined by

$$\delta_U\phi^\alpha = \Phi^*(U \lrcorner dq^\alpha), \qquad \delta_U\,\partial_i\phi^\alpha = \Phi^*(U \lrcorner dy_i^\alpha). \tag{7-1.15}$$

Accordingly, for any $U \in TV(K)$ that is generated by

$$U_g = \bar{f}^\alpha(x^j, q^\beta)\,\partial_\alpha,$$

we have

$$\delta_U\phi^\alpha = \bar{f}^\alpha(x^j, \phi^\beta) \tag{7-1.16}$$

and

$$\delta_U\,\partial_i\phi^\alpha = \left(\partial_i + (\partial_i\phi^\beta)\partial_\beta\right)\langle \bar{f}^\alpha(x^j, \phi^\beta)\rangle = D_i\bar{f}^\alpha(x^j, \phi^\beta). \tag{7-1.17}$$

It is then an easy matter to see that (7-1.16) and (7-1.17) imply

$$\delta_U\,\partial_i\phi^\alpha = D_i\delta_U\phi^\alpha; \tag{7-1.18}$$

that is, *partial differentiation and variation are commuting operations*. This variation process clearly allows us to assign any smooth increments to the functions $\phi^\alpha(x^j)$, for

$$\bar{\phi}^\alpha = \phi^\alpha(x^j) + \bar{f}^\alpha(x^j, \phi^\beta(x^j))s + o(s)$$

$$= \phi^\alpha(x^j) + \delta_U\phi^\alpha s + o(s),$$

and hence $\delta_U \phi^\alpha$ can be assigned arbitrary function evaluations even when the arguments $\phi^\beta(x^j)$ of the $\bar{f}$'s are ignored.

Now that we have $U \in TV(K)$, any $\Phi \in R(K)$ can be embedded in the 1-parameter family of regular maps $\Phi_U(s)$. The values of $A[\Phi_U(s)]$ give a smooth function of the parameter s that can be compared with the value of $A[\Phi_U(0)] = A[\Phi]$. The following definition is then immediate.

Definition. The action functional $A[\Phi]$ is *stationary* in value if and only if $\Phi \in R(K)$ satisfies

$$\delta_U A[\Phi] = \lim_{s \to 0} \left(\frac{A[\Phi_U(s)] - A[\Phi]}{s} \right) = 0 \tag{7-1.19}$$

for all $U \in TV(K)$ such that $\Phi^*(U \lrcorner dq^\alpha)|_{\partial B_n} = \delta_U \phi^\alpha|_{\partial B_n} = 0$.

The restriction to variations that vanish on the boundary is classical. It arises from the fact that the ϕ's are to assume given values on the boundary and hence the neighboring functions, $\phi^\alpha + s\delta_U \phi^\alpha$, should assume the same values on the boundary.

7-2. FINITE VARIATIONS, STATIONARITY, AND THE EULER-LAGRANGE EQUATIONS

Let

$$\Phi : B_n \to K | x^i = x^i, \qquad q^\alpha = \phi^\alpha(x^j), \qquad y_i^\alpha = \partial_i \phi^\alpha(x^j) \tag{7-2.1}$$

be a regular map and let $L(x^j, q^\beta, y_j^\beta)$ be a given element of $\Lambda^0(K)$ (the Lagrangian function). The action associated with the state $\{\phi^\alpha(x^j)\}$ was shown to be equivalent to the map

$$A : R(B_n) \to \mathbb{R} \tag{7-2.2}$$

with the evaluation

$$A[\Phi] = \int_{B_n} \Phi^*(L\mu). \tag{7-2.3}$$

We have also seen that any vertical vector field $U \in TV(K)$ can be used to embed Φ in a 1-parameter family of maps $\Phi_U(s)$ with the same independent variables (x^i). On the other hand, we know from previous chapters that

$$\Phi_U(s) = T_U(s) \circ \Phi, \tag{7-2.4}$$

$$\Phi_U(s)^* = \Phi^* \circ T_U(s)^* = \Phi^* \exp(s\pounds_U), \tag{7-2.5}$$

and hence

$$A[\Phi_U(s)] = \int_{B_n} \Phi^* \exp(s\pounds_U)(L\mu). \tag{7-2.6}$$

The important quantities are therefore n-forms on K and their Lie derivatives with respect to vertical isovector fields of the contact ideal.

There are several technical lemmas that prove to be very useful to our purposes. In view of later requirements, we temporarily relax the condition that the isovectors of the contact ideal be vertical. As before, we use $\mu = dx^1 \wedge dx^2 \wedge \cdots \wedge dx^n$ for the volume element of E_n and the basis $\{\mu_i, i = 1, \ldots, n\}$ for $(n-1)$-forms. Here $\mu_i = \partial_i \lrcorner \mu$, so that

$$d\mu_i = 0, \qquad dx^i \wedge \mu_j = \delta^i_j \mu. \tag{7-2.7}$$

Further let $\mu_{ji} = \partial_j \lrcorner \mu_i$, so that

$$\mu_{ji} = -\mu_{ij}, \qquad d\mu_{ij} = 0, \qquad dx^k \wedge \mu_{ij} = \delta^k_i \mu_j - \delta^k_j \mu_i \tag{7-2.8}$$

(see Section 3-5).

Lemma 7-2.1. *The n-form*

$$J = C^\alpha \wedge (\partial^i_\alpha L)\mu_i = (\partial^i_\alpha L)(dq^\alpha \wedge \mu_i - y^\alpha_i \mu) \tag{7-2.9}$$

belongs to the closed contact ideal, $\bar{C}$. Hence

$$\Phi^* J = 0, \tag{7-2.10}$$

$$\Phi^* \pounds_V J = 0 \tag{7-2.11}$$

for all $\Phi \in R(B_n)$ and all $V \in TC(K)$. Further,

$$V \lrcorner (L\mu + J) = \{(L\delta^i_j - y^\alpha_j \partial^i_\alpha L) f^j + \bar{f}^\alpha \partial^i_\alpha L\} \mu_i \bmod \bar{C}, \tag{7-2.12}$$

where $V_g = f^i(x^i, q^\beta)\partial_i + \bar{f}^\alpha(x^i, q^\beta)\partial_\alpha$, and hence $V \lrcorner (L\mu + J)$ does not depend on the components of $V \in TC(K)$ in the y-directions.

Proof. That $J \in \bar{C}$ follows directly from (7-2.9) since each term of J has a contact form as a factor. Accordingly, since any $\Phi \in R(K)$ annihilates $\bar{C}$, we have $\Phi^* J = 0$. If V is any isovector field of the contact ideal, $\pounds_V J = 0 \bmod \bar{C}$ because $J \in \bar{C}$ and $\bar{C}$ is closed under transport by elements of $TC(K)$. Accordingly, $\Phi^* \pounds_V J = 0$. A direct computation shows that

$$\begin{aligned} V \lrcorner (L\mu + J) &= L(V \lrcorner \mu) + (\partial^i_\alpha L)\{(V \lrcorner C^\alpha)\mu_i - C^\alpha \wedge (V \lrcorner \mu_i)\} \\ &= L(V \lrcorner \mu) + (\partial^i_\alpha L)(V \lrcorner C^\alpha)\mu_i \bmod \bar{C}. \end{aligned}$$

The relation (7-2.12) then follows upon noting that

$$V \lrcorner \mu = f^i \mu_i, \qquad V \lrcorner C^\alpha = \bar{f}^\alpha - y_i^\alpha f^i$$

for V generated by $V_g = f^i \partial_i + \bar{f}^\alpha \partial_\alpha$. An inspection of (7-2.12) and the above calculations show that $V \lrcorner (L\mu + J) = V_g \lrcorner (L\mu + J)$ and hence $V \lrcorner (L\mu + J)$ does not depend on the y-components, $Z_i \langle \bar{f}^\alpha - y_j^\alpha f^j \rangle$, of V. □

The fact that $V \lrcorner (L\mu + J)$ does not depend on the y-components of V will have particular importance later on. For the time being, it is sufficient to note that the n-form J gives rise to an equivalent evaluation of the action that will significantly simplify our calculations.

Lemma 7-2.2. *The action $A[\Phi]$ has the evaluation*

(7-2.13) $$A[\Phi] = \int_{B_n} \Phi^*(L\mu + J)$$

for any regular map $\Phi: B_n \to K$.

Proof. Lemma 7-2.1 shows that $\Phi^* J = 0$ for any $\Phi \in R(B_n)$, and hence $\Phi^*(L\mu + J) = \Phi^*(L\mu)$. The result then follows directly from the definition of the action of the map Φ, (7-2.3). □

Lemma 7-2.3. *Let $\Phi \in R(B_n)$ and $V \in TC(K)$. The 1-parameter family of maps*

(7-2.14) $$\Phi_V(s) = T_V(s) \circ \Phi$$

belongs to $R(B_n)$ for all values of s such that $\exp(s£_V)\mu \neq 0$ *and yields*

(7-2.15) $$A[\Phi_V(s)] = \int_{B_n} \Phi^* \exp(s£_V)(L\mu + J).$$

Proof. The 1-parameter family of maps $\Phi_V(s)$ belongs to $R(B_n)$ for all values of s for which $\exp(s£_V)\mu \neq 0$ because V is an isovector field of the contact ideal and $\Phi \in R(B_n)$. The final result then follows upon noting that $\Phi_V(s)^* = \Phi^* \circ T_V(s)^* = \Phi^* \exp(s£_V)$. □

Lemma 7-2.3 provides a natural basis for the definitions of the finite variation and of the infinitesimal variation of the action of any regular map $\Phi: B_n \to K$.

Definition. The *finite variation* of the action of any regular map $\Phi: B_n \to K$ that is generated by a $V \in TC(K)$ is given by

$$\Delta_V(s)A[\Phi] = A[\Phi_V(s)] - A[\Phi]. \tag{7-2.16}$$

Definition. The *infinitesimal variation* of the action of any regular map $\Phi: B_n \to K$ that is generated by a $V \in TC(K)$ is given by

$$\delta_V A[\Phi] = \lim_{s \to 0}\left(s^{-1}\Delta_V(s)A[\Phi]\right). \tag{7-2.17}$$

Lemma 7-2.4. *The finite and the infinitesimal variations of the action of any regular map $\Phi: B_n \to K$ that are generated by a $V \in TC(K)$ have the evaluations*

$$\Delta_V(s)A[\Phi] = \int_{B_n} \Phi^*(\exp(s£_V) - 1)(L\mu + J), \tag{7-2.18}$$

$$\delta_V A[\Phi] = \int_{B_n} \Phi^* £_V(L\mu + J). \tag{7-2.19}$$

Proof. The results are immediate consequences of the definitions of the finite and infinitesimal variations of the action of a regular map and formula (7-2.15) of Lemma 7-2.3. □

These results have replaced $L\mu$ by the new n-form $L\mu + J$ and the analysis has been shifted from vertical vector fields to general isovector fields of the contact ideal. The replacing of $TV(K)$ by $TC(K)$ will have important ramifications in later sections. For the present, it is the transition $L\mu \mapsto L\mu + J$ that is central. This is made evident by the following results on noting that $£_V(L\mu + J) = V \lrcorner d(L\mu + J) + d\{V \lrcorner (L\mu + J)\}$.

Lemma 7-2.5. *The Cartan $(n+1)$-form*

$$F = d(L\mu + J) \tag{7-2.20}$$

has the evaluation

$$F = C^\alpha \wedge E_\alpha, \tag{7-2.21}$$

where

$$E_\alpha = (\partial_\alpha L)\mu - d(\partial_\alpha^i L) \wedge \mu_i \tag{7-2.22}$$

are the Euler-Lagrange n-forms. Thus F belongs to the closed contact ideal and

$$V \lrcorner F \equiv (V \lrcorner C^\alpha) E_\alpha \bmod \bar{C} \tag{7-2.23}$$

for any $V \in TC(K)$.

Proof. It follows directly from (7-2.9) that

$$L\mu + J = L\mu + (\partial_\alpha^i L) C^\alpha \wedge \mu_i,$$

and hence

$$d(L\mu + J) = (\partial_\alpha L)\, dq^\alpha \wedge \mu + (\partial_\alpha^i L)\, dy_i^\alpha \wedge \mu + d(\partial_\alpha^i L) \wedge C^\alpha \wedge \mu_i + (\partial_\alpha^i L)\, dC^\alpha \wedge \mu_i. \tag{7-2.24}$$

However, $dq^\alpha \wedge \mu = (C^\alpha + y_j^\alpha\, dx^j) \wedge \mu = C^\alpha \wedge \mu$, and $dC^\alpha \wedge \mu_i = -dy_j^\alpha \wedge dx^j \wedge \mu_i = -dy_j^\alpha \wedge \delta_i^j \mu = -dy_i^\alpha \wedge \mu$. When these are put back into (7-2.24), there is a direct cancellation and

$$F = d(L\mu + J) = (\partial_\alpha L) C^\alpha \wedge \mu + d(\partial_\alpha^i L) \wedge C^\alpha \wedge \mu_i$$

$$= C^\alpha \wedge \{(\partial_\alpha L)\mu - d(\partial_\alpha^i L) \wedge \mu_i\}.$$

It is then a simple matter to see that (7-2.21) implies (7-2.23) for all $V \in TC(K)$. □

Lemma 7-2.6. *The infinitesimal variation of the action of any regular map* $\Phi: B_n \to K$ *that is generated by an isovector field* V *of the contact ideal has the evaluation*

$$\delta_V A[\Phi] = \int_{B_n} \Phi^*(V \lrcorner C^\alpha) \Phi^* E_\alpha + \int_{\partial B_n} \Phi^*\{V \lrcorner (L\mu + J)\}. \tag{7-2.25}$$

Proof. Lemma 7-2.4 gave the evaluation

$$\delta_V A[\Phi] = \int_{B_n} \Phi^* £_V (L\mu + J). \tag{7-2.26}$$

However,

$$£_V(L\mu + J) = V \lrcorner d(L\mu + J) + d\{V \lrcorner (L\mu + J)\} \tag{7-2.27}$$

$$= V \lrcorner F + d\{V \lrcorner (L\mu + J)\}$$

$$= (V \lrcorner C^\alpha) E_\alpha + d\{V \lrcorner (L\mu + J)\} \bmod \overline{C},$$

where the last equality follows form (7-2.23). Since $\Phi \in R(B_n)$ annihilates anything in the closed contact ideal, allowing Φ^* to act on both sides of (7-2.27), we see that (7-2.26) gives

$$\delta_V A[\Phi] = \int_{B_n} \Phi^*(V \lrcorner C^\alpha) \Phi^* E_\alpha + \int_{B_n} \Phi^* d\{V \lrcorner (L\mu + J)\}. \tag{7-2.28}$$

The result then follows by an application of Stokes' theorem to the last term on the right-hand side of (7-2.28). □

All of the ground work has now been laid in order to characterize regular maps that render the action stationary.

Definition. A regular map $\Phi : B_n \to K$ is said to *stationarize* the action $A[\cdot]$ if and only if

$$\delta_U A[\Phi] = 0 \tag{7-2.29}$$

for all vertical vector fields U such that

$$\Phi^* (U \lrcorner C^\alpha)|_{\partial B_n} = 0. \tag{7-2.30}$$

We note that (7-2.30) is the same thing as $\delta_U \phi^\alpha|_{\partial B_n} = 0$ in the notation of the previous section.

Theorem 7-2.1. *A regular map*

$$\Phi | x^i = x^i, \qquad q^\alpha = \phi^\alpha(x^i), \qquad y_i^\alpha = \partial_i \phi^\alpha(x^i)$$

stationarizes the action $A[\cdot]$ if and only if Φ is such that the Euler-Lagrange equations

$$\Phi^* E_\alpha = 0, \qquad \alpha = 1, \ldots, N \tag{7-2.31}$$

are satisfied at all interior points of B_n.

Proof. Since vertical vector fields constitute a subspace of $TC(K)$ for which $U\langle x^i \rangle = 0$, Lemma 7-2.6 gives

$$\delta_U A[\Phi] = \int_{B_n} \Phi^*(U \lrcorner C^\alpha) \Phi^* E_\alpha + \int_{\partial B_n} \Phi^* \{ U \lrcorner (L\mu + J) \}. \tag{7-2.32}$$

However, any $U \in TV(K)$ gives $U \lrcorner C^\alpha = U \lrcorner dq^\alpha$ and hence (7-1.15) gives $\Phi^*(U \lrcorner C^\alpha) = \delta_U \phi^\alpha$. Next we note that any $V \in TC(K)$ is generated by $V_g = f^i(x^j, q^\beta)\partial_i + \bar{f}^\alpha(x^j, q^\beta)\partial_\alpha$ while any $U \in TV(K)$ is generated by $U_g = \bar{f}^\alpha(x^j, q^\beta)\partial_\alpha$. Thus any result for $V \in TC(K)$ can be cut down to a result for $U \in TV(K)$ by setting $f^i(x^j, q^\beta) = 0$. When this is done in (7-2.12), we have

$$\begin{aligned} U \lrcorner (L\mu + J) &= \bar{f}^\alpha(x^j, q^\beta)(\partial_\alpha^i L)\mu_i \bmod C \\ &= (U \lrcorner C^\alpha)(\partial_\alpha^i L)\mu_i \bmod C. \end{aligned}$$

Accordingly, when the boundary conditions (7-2.30) are used, we have

$$\Phi^* \{U \lrcorner (L\mu + J)\}|_{\partial B_n} = \Phi^*(U \lrcorner C^\alpha)\big|_{\partial B_n} \Phi^*(\partial^i_\alpha L)\big|_{\partial B_n} \mu_i = 0$$

and (7-2.32) reduces to

$$\delta_U A[\Phi] = \int_{B_n} \delta_U \phi^\alpha \, \Phi^* E_\alpha. \tag{7-2.33}$$

By definition Φ stationarizes $A[\cdot]$ if and only if $\delta_U A[\Phi] = 0$ for all $U \in TV(K)$ that satisfy the boundary conditions (7-2.30), and this will be the case if and only if

$$\int_{B_n} \delta_U \phi^\alpha \, \Phi^* E_\alpha = 0 \tag{7-2.34}$$

holds for *all* smooth functions $\{\delta_U \phi^\alpha\}$ that vanish on the boundary of B_n. However, the well known fundamental lemma of the calculus of variations tells us that this can be the case if and only if

$$\Phi^* E_\alpha = 0 \tag{7-2.35}$$

are satisfied at all interior points of B_n. □

The proof of Theorem 7-2.1 shows that the only place where a function space argument is needed is in the proof of the fundamental lemma of the calculus of variations. This lemma is clearly a deep and important one, if for no other reason than the fact that it reduces the global statement,

$$\int_{B_n} \delta_U \phi^\alpha \, \Phi^* E_\alpha = 0 \text{ for all } \delta_U \phi^\alpha,$$

to the pointwise local statement, $\Phi^* E_\alpha = 0$ at all interior points of B_n. The results obtained here are thus no better and no worse than those obtained in standard treatments of the calculus of variations; rather, they are simply organized along different lines.

The Euler-Lagrange equations (7-2.31) are systems of partial differential equations for the determination of the functions $\{\phi^\alpha(x^j)\}$ that specify a stationarizing map Φ by (7-2.1).

Lemma 7-2.7. *The Euler-Lagrange equations have the specific realization*

$$\text{(7-2.36)} \qquad \frac{\partial L(x^j, \phi^\beta, \partial_j \phi^\beta)}{\partial \phi^\alpha} = D_i \left(\frac{\partial L(x^j, \phi^\beta, \partial_j \phi^\beta)}{\partial(\partial_i \phi^\alpha)} \right), \qquad \alpha = 1, \ldots, N.$$

Proof. It follows directly from (7-2.22) that

$$\begin{aligned} \Phi^* E_\alpha &= \Phi^*(\partial_\alpha L)\mu - d\Phi^*(\partial_\alpha^i L) \wedge \mu_i \\ &= \Phi^*(\partial_\alpha L)\mu - D_j \Phi^*(\partial_\alpha^i L)\, dx^j \wedge \mu_i \\ &= \{\Phi^*(\partial_\alpha L) - D_i \Phi^*(\partial_\alpha^i L)\}\mu \end{aligned}$$

because $dx^j \wedge \mu_i = \delta_j^i \mu$. Thus since μ is a basis for $\Lambda^n(E_n)$, $\Phi^* E_\alpha = 0$ if and only if (2-7.36) hold. □

The following result is recorded here for later purposes.

Theorem 7-2.2. *If the action $A[\cdot]$ is rendered stationary by a regular map Φ, then the Euler-Lagrange equations*

$$\text{(7-2.37)} \qquad \Phi^* E_\alpha = 0$$

hold at all interior points of B_n and the infinitesimal variation of $A[\Phi]$ that is generated by an isovector field V of the contact ideal has the evaluation

$$\text{(7-2.38)} \qquad \delta_V A[\Phi] = \int_{\partial B_n} \Phi^*\{V \lrcorner (L\mu + J)\}.$$

Proof. The first part of the result, namely $\Phi^* E_\alpha = 0$ at all interior points of B_n, follows directly from Theorem 7-2.1. When this result is put into (7-2.25) of Lemma 7-2.6, only the boundary integral survives, and this is just (7-2.38). □

7-3 PROPERTIES OF EULER-LAGRANGE FORMS AND STATIONARIZING MAPS

A number of important and useful properties of the Euler-Lagrange n-forms and stationarizing maps are collected together in this section.

An inspection of (7-2.22) shows that the Euler-Lagrange n-forms are uniquely determined by the Lagrangian function $L \in \Lambda^0(K)$. In order to take full note of this fact, we introduce the amended notation

$$E_\alpha(L) = (\partial_\alpha L)\mu - d(\partial^i_\alpha L) \wedge \mu_i. \tag{7-3.1}$$

Lemma 7-3.1. *The Euler-Lagrange n-forms are linear mappings over* $\mathbb{R}$ *from* $\Lambda^0(K)$ *to* $\Lambda^n(K)$,

$$E_\alpha(aL_1 + bL_2) = aE_\alpha(L_1) + bE_\alpha(L_2), \tag{7-3.2}$$

and

$$\begin{aligned} E_\alpha(L_1 L_2) = {} & L_1 E_\alpha(L_2) + L_2 E_\alpha(L_1) \\ & - \{(\partial^i_\alpha L_1)\, dL_2 + (\partial^i_\alpha L_2)\, dL_1\} \wedge \mu_i. \end{aligned} \tag{7-3.3}$$

Proof. (7-3.2) follows directly from (7-3.1) for any $a, b \in \mathbb{R}$, and hence $E_\alpha(\cdot)$ is a linear map with domain $\Lambda^0(K)$ considered as a vector space over $\mathbb{R}$, and range contained in $\Lambda^n(K)$. In like manner (7-3.3) also follows directly from (7-3.1). This latter result shows that $E_\alpha(\cdot)$ *does not act as a derivation or as an antiderivation* on the algebra $\Lambda^0(K)$. □

Lemma 7-3.2. *The Euler-Lagrange n-forms are balance n-forms.*

Proof. Balance n-forms were defined in Section 6-5 to be any system of n-forms

$$B_\alpha = W_\alpha \mu - dW^i_\alpha \wedge \mu_i \tag{7-3.4}$$

for some collection of scalar-valued quantities $(W_\alpha, W^i_\alpha) \in \Lambda^0(K)$. The result then follows upon making the identifications

$$W_\alpha = \partial_\alpha L, \qquad W^i_\alpha = \partial^i_\alpha L. \tag{7-3.5}$$ □

The converse problem is a much more interesting problem with a long history; namely, the inverse problem of the calculus of variations.

Definition. A given balance system

$$B_\alpha = W_\alpha \mu - d(W^i_\alpha) \wedge \mu_i \tag{7-3.6}$$

on K is said to *admit a variational principle* if and only if there exists a Lagrangian function $L \in \Lambda^0(K)$ such that $B_\alpha = E_\alpha(L)$; that is,

$$W_\alpha = \partial_\alpha L, \qquad W^i_\alpha = \partial^i_\alpha L. \tag{7-3.7}$$

Lemma 7-3.3. *Let* $\{B_\alpha\}$ *be a given balance system of the form* (7-3.6) *and define the* 1*-form* W *by*

$$W = W_\alpha \, dq^\alpha + W_\alpha^i \, dy_i^\alpha . \tag{7-3.8}$$

The balance system $\{B_\alpha\}$ *admits a variational principle if and only if the* $(n+1)$*-form* $W \wedge \mu$ *is closed.*

Proof. If B_α admits a variational principle, then there exists a Lagrangian function $L \in \Lambda^0(K)$ for which (7-3.7) hold. In this case, we have

$$W \wedge \mu = \{(\partial_\alpha L)\, dq^\alpha + (\partial_\alpha^i L)\, dy_i^\alpha\} \wedge \mu = dL \wedge \mu = d(L\mu).$$

$W \wedge \mu$ is thus exact and hence closed. conversely, in order that $W \wedge \mu$ be closed we must have $d(W \wedge \mu) = 0$. Since K is an $(n+N+nN)$-dimensional number space, it is starshaped with respect to any of its points as center and hence the Poincaré lemma holds on K. We may thus conclude that there exists an n-form ρ on K such that $W \wedge \mu = d\rho$. However, μ is a simple n-form and hence $W \wedge \mu$ is a simple $(n+1)$-form on K. Thus, $W \wedge \mu = d\rho$ can hold if and only if $d\rho$ is a simple $(n+1)$-form on K. Since μ is also closed, we infer the existence of a zero form L such that $W \wedge \mu = d(L\mu) = dL \wedge \mu$. Resolution of these $(n+1)$-forms on the basis $\{dq^\alpha \wedge \mu, dy_i^\alpha \wedge \mu\}$ then gives (7-3.7). □

It is clear from this result that not every balance system on K admits a variational principle. For example, the balance system

$$B_1 = d\left(q^1 \, dx^1 + y_1^1 \, dx^2\right)$$

for the linear diffusion equation in one spatial dimension does not admit a variational principle on K with coordinates $(x^1, x^2, q^1, y_1^1, y_2^1)$ and contact form $C^1 = dq^1 - y_1^1 \, dx^1 - y_2^1 \, dx^2$. There is nothing else that can be done if $W \wedge \mu$ is not closed and we are unwilling to enlarge the problem a little. On the other hand, the extensive literature on the inverse problem of the calculus of variations shows quite clearly that an enlargement of the problem often meets with surprising success.

Theorem 7-3.1. *Any balance system*

$$B_\alpha = W_\alpha \mu - dW_\alpha^i \wedge \mu_i \tag{7-3.9}$$

on kinematic space K *can be extended to a balance system on a space* $\hat{K} = K \times E_{N+nN}$ *with coordinate functions* $(x^i, q^\alpha, y_i^\alpha, \hat{q}^\alpha, \hat{y}_i^\alpha)$ *and additional exterior forms*

$$\hat{C}^\alpha = d\hat{q}^\alpha - \hat{y}_i^\alpha \, dx^i, \tag{7-3.10}$$

$$\hat{B}_\alpha = \hat{W}_\alpha \mu - d\hat{W}_\alpha^i \wedge \mu_i \tag{7-3.11}$$

in such a fashion that $\{B_\alpha, \hat{B}_\alpha\}$ *admits a variational principle. This extension is achieved by the choice*

$$(7\text{-}3.12)\quad \hat{W}_\alpha = \partial_\alpha\big(W_\beta \hat{q}^\beta + W_\beta^j \hat{y}_j^\beta\big), \qquad \hat{W}_\alpha^i = \partial_\alpha^i\big(W_\beta \hat{q}^\beta + W_\beta^j \hat{y}_j^\beta\big)$$

that gives rise to the Lagrangian function

$$(7\text{-}3.13)\qquad L = W_\beta \hat{q}^\beta + W_\beta^j \hat{y}_j^\beta,$$

and the adjoint balance system $\{\hat{B}_\alpha\}$ *is linear in the adjoint variables* $(\hat{q}^\alpha, \hat{y}_i^\alpha)$.

Proof. A direct calculation based on (7-3.13) shows that

$$(7\text{-}3.14)\quad W_\alpha \equiv \hat{\partial}_\alpha L, \qquad W_\alpha^i \equiv \hat{\partial}_\alpha^i L, \qquad \hat{W}_\alpha = \partial_\alpha L, \qquad \hat{W}_\alpha^i = \partial_\alpha^i L,$$

where we have used the notation $\hat{\partial}_\alpha = \partial/\partial\hat{q}^\alpha$, $\hat{\partial}_\alpha^i = \partial/\partial\hat{y}_i^\alpha$, and hence (7-3.12) are established. In view of the pairings that are established by the identities (7-3.14), the 1-form W for the extended system is given by

$$(7\text{-}3.15)\qquad W = W_\alpha d\hat{q}^\alpha + W_\alpha^i d\hat{y}_i^\alpha + \hat{W}_\alpha dq^\alpha + \hat{W}_\alpha^i dy_i^\alpha$$

$$= dL - \partial_i L\, dx^i.$$

It thus follows that $W \wedge \mu = d(L\mu)$, and hence the system admits a variational principle by Lemma 7-3.3. Finally, when (7-3.12) are substituted into (7-3.11), it is readily seen that the adjoint balance system $\{\hat{B}_\alpha\}$ is linear in the adjoint variables $(\hat{q}^\alpha, \hat{y}_i^\alpha)$. □

What makes the theorem work is the pairing that is chosen in the representation of the 1-form W (see (7-3.15)), and the linearity of the Lagrangian in the adjoint variables $(\hat{q}^\alpha, \hat{y}_i^\alpha)$. This embedding of a system of nonlinear partial differential equations in a variational system by introduction of the appropriate collection of "adjoint" variables is a classical technique that appears to go back to the early works of Bateman. It is also the underpinning of much of the modern work on variational inequalities even though this may not be apparent on reading some of the more popular expositions.

The linearity of the adjoint balance equations in the adjoint variables is a very useful aspect of Theorem 7-3.1 that often leads to significant computational efficiencies. The balance equation of a nonlinear diffusion process in one spatial dimension is given by

$$B_1 = -d\big(Q(q^1)\mu_2 - y_1^1\mu_1\big)$$

for $\mu = dx^1 \wedge dx^2$. It is then an easy matter to see that (7-3.13) gives

$$L = Q(q^1)\hat{y}_2^1 - y_1^1\hat{y}_1^1,$$

and hence the adjoint balance system is

$$\hat{B}_1 = \frac{dQ(q^1)}{dq^1}\hat{y}_2^1\mu + d(\hat{y}_1^1\mu_1).$$

Thus if $\Phi: B_2 \to K|x^1 = x,\ x^2 = t,\ q^1 = \phi(x,t),\ \hat{q}^1 = \hat{\phi}(x,t)$, we have the specific balance equations

$$\Phi^*B_1 = \left(-\frac{dQ(\phi)}{d\phi}\partial_t\phi + \partial_x\partial_x\phi\right)dx \wedge dt,$$

$$\Phi^*\hat{B}_1 = \left(\frac{dQ(\phi)}{d\phi}\partial_t\hat{\phi} + \partial_x\partial_x\hat{\phi}\right)dx \wedge dt.$$

The important thing to note here is that diffusion processes do have variational principles; it is just that we must also include the additional adjoint variables in order to obtain them.

We now turn to the properties of regular maps that stationarize a given action. A particularly important role is played here by what are known as the components of the momentum-energy complex, as we shall see in a moment.

Definition. The components of the *momentum-energy complex* of a given Lagrangian function $L \in \Lambda^0(K)$ are given by

(7-3.16) $$T_j^i(L) = y_j^\alpha\partial_\alpha^i L - L\delta_j^i.$$

Lemma 7-3.4. *The components of the momentum-energy complex are linear in the Lagrangian over* $\mathbb{R}$,

(7-3.17) $$T_j^i(aL_1 + bL_2) = aT_j^i(L_1) + bT_j^i(L_2),$$

and we have

(7-3.18) $$T_j^i(L_1L_2) = L_1T_j^i(L_2) + L_2T_j^i(L_1) + L_1L_2\delta_j^i.$$

Proof. Both (7-3.17) and (7-3.18) follow directly from (7-3.16) by direct calculation. □

Theorem 7-3.2. *Any regular map* $\Phi: B_n \to K$ *satisfies the identities*

(7-3.19) $$-\Phi^*\{y_i^\alpha E_\alpha(L) + dT_i^j(L) \wedge \mu_j\} = \Phi^*\{\partial_i L\mu\}.$$

Thus if

$$(7\text{-}3.20)\qquad \Phi|x^i = x^i, \qquad q^\alpha = \phi^\alpha(x^j), \qquad y_i^\alpha = \partial_i\phi^\alpha(x^j)$$

stationarizes the action $A[\Phi] = \int_{B_n}\Phi^* L\mu$, *then the functions* $\{\phi^\alpha(x^j)\}$ *satisfy*

$$(7\text{-}3.21)\qquad D_j\{\Phi^* T_i^j(L)\} = -\Phi^*(\partial_i L).$$

Proof. We start with the relation

$$(7\text{-}3.22)\qquad £_V(L\mu + J) = (V\lrcorner C^\alpha)E_\alpha(L) + d\{V\lrcorner(L\mu + J)\}$$

that is valid for any $V \in TC(K)$ (see (7-2.27)). However, Lemma 7-2.1 has shown that $£_V J = 0 \bmod \bar{C}$, and hence (7-3.22) becomes

$$(7\text{-}3.23)\qquad £_V L\mu = (V\lrcorner C^\alpha)E_\alpha(L) + d\{V\lrcorner(L\mu + J)\} \bmod \bar{C}.$$

We now specialize $V \in TC(K)$ by $V = \partial_i$. This gives

$$£_{\partial_i} L\mu = \partial_i L\mu \qquad \text{and} \qquad V\lrcorner C^\alpha = -y_i^\alpha,$$

while (7-2.12) gives

$$\partial_i\lrcorner(L\mu + J) = \left(L\delta_i^j - y_i^\alpha\partial_\alpha^j L\right)\mu_j \bmod \bar{C} = -T_i^j(L)\mu_j \bmod \bar{C}.$$

When these evaluations are put into (7-3.23), we have

$$(7\text{-}3.24)\qquad (\partial_i L)\mu = -\left\{y_i^\alpha E_\alpha(L) + dT_i^j(L)\wedge\mu_j\right\} \bmod \bar{C},$$

from which the results follow. □

It is important to note here that, no matter what the values of n and N are, any solution of the N balance equations $\Phi^* E_\alpha(L) = 0$ will always satisfy n laws of balance

$$D_j\{\Phi^* T_i^j(L)\} = -\Phi^*(\partial_i L), \qquad i = 1, \ldots, n.$$

In actual practice, the Lagrangian function usually does not depend explicitly on the independent variables (x^i), in which case $\Phi^*(\partial_i L) = 0$. In these circumstances, any solution of the N balance equations $\Phi^* E_\alpha(L) = 0$ will always satisfy the n laws of conservation

$$D_j\{\Phi^* T_i^j(L)\} = 0, \qquad i = 1, \ldots, n.$$

It is for this reason that the $T_j^i(L)$ are referred to as components of the momentum-energy complex, for $D_j\{\Phi^* T_i^j(L)\} = 0$ on a base space E_4 of space-time look like laws of conservation of momentum and energy. Although the interpretation is not universal, it is essential to realize that such conservation laws obtain nevertheless.

7-4. NOETHERIAN VECTOR FIELDS AND THEIR ASSOCIATED CURRENTS

The existence of systems of quantities that are either balanced or conserved for any solution of the Euler-Lagrange equations is more fundamental than might be supposed from Theorem 7-3.2. The underlying concept was first pointed out in a seminal paper by E. Noether wherein deformation invariance of the action of a system was related to a system of conservation laws that are necessarily enforced by satisfaction of the Euler-Lagrange equations. Now, deformations are simply the changes that result from transport along the orbits of vector fields in kinematic space. We also know that deformations will preserve the correlation between the y's and the derivatives of the q's with respect to the x's if and only if the vector fields are isovector fields of the contact ideal. Accordingly, it is natural—and indeed necessary—to restrict attention to transport along orbits of isovector fields of the contact ideal. On the other hand, there is nothing that precludes deformations of both the independent and the dependent variables; indeed, the underlying idea of Noether's work arose through the consideration of deformation processes that are induced by an underlying group of transformations of the independent variables. General isovector fields of the contact ideal are thus required, in contrast to questions of stationarity that use only vertical vector fields. It is primarily for this reason that much of the work in previous sections dealt with general $V \in TC(K)$ rather than just $U \in TV(K)$.

Definition. A vector field $V \in TC(K)$ is a *Noetherian vector field* for a given action $A[\Phi]$ (for a given Lagrangian L) if and only if

$$\text{(7-4.1)} \qquad \pounds_V(L\mu + J) = 0.$$

The collection of all Noetherian vector fields for a given L is denoted by $N(L, K)$.

The reader should carefully note that Noetherian vector fields are specific isovector fields of the contact ideal that leave the fundamental n-form $L\mu + J$ invariant under transport. This is in sharp contrast with the process of stationarization where we demanded, in effect, that $\pounds_U(L\mu + J) = 0 \bmod \overline{C}$ hold for *all* vertical vector fields U.

Definition. A vector field $V \in TC(K)$ is a *Noetherian vector field of the first kind* for a given Lagrangian function L if and only if

$$\text{(7-4.2)} \qquad \pounds_V(L\mu + J) = 0 \bmod \bar{C}.$$

The collection of all Noetherian vector fields of the first kind for a given L is denoted by $N_1(L, K)$.

Definition. A vector field $V \in TC(K)$ is a *Noetherian vector field of the second kind* for a given L if and only if

$$\text{(7-4.3)} \qquad \pounds_V(L\mu + J) = d\beta(V) \bmod \bar{C}$$

for some $(n-1)$-form $\beta(V)$. The collection of all Noetherian vector fields of the second kind for a given L is denoted by $N_2(L, K)$.

We first establish the Lie algebra properties of Noetherian vector fields of all kinds.

Theorem 7-4.1. *$N(L, K)$, $N_1(L, K)$ and $N_2(L, K)$ form Lie subalgebras of the Lie algebra $TC(K)$ over $\mathbb{R}$ that satisfy the inclusion relations*

$$\text{(7-4.4)} \qquad N(L, K) \subset N_1(L, K) \subset N_2(L, K) \subset TC(K),$$

and the laws of composition for the elements $\beta(V)$ are given by $d\beta(aV_1 + bV_2) = d\{a\beta(V_1) + b\beta(V_2)\} \bmod \bar{C}$ and

$$\text{(7-4.5)} \qquad d\beta([V_1, V_2]) = d\{\pounds_{V_1}\beta(V_2) - \pounds_{V_2}\beta(V_1)\} \bmod \bar{C}.$$

Proof. The inclusion relations (7-4.4) are immediate consequences of the definitions of $N(L, K)$, $N_1(L, K)$, $N_2(L, K)$. Further, (7-4.1) through (7-4.3) show that it is sufficient to establish that $N_2(L, K)$ forms a Lie subalgebra of $TC(K)$ over $\mathbb{R}$. To this end let $\pounds_{V_1}(L\mu + J) = d\beta(V_1) \bmod \bar{C}$ and $\pounds_{V_2}(L\mu + J) = d\beta(V_2) \bmod \bar{C}$. Thus, for all $a, b \in \mathbb{R}$,

$$\begin{aligned}
\pounds_{(aV_1 + bV_2)}(L\mu + J) &= (a\pounds_{V_1} + b\pounds_{V_2})(L\mu + J) \\
&= a\,d\beta(V_1) + b\,d\beta(V_2) \bmod \bar{C} \\
&= d(a\beta(V_1) + b\beta(V_2)) \bmod \bar{C}.
\end{aligned}$$

$N_2(L, K)$ is thus a vector space over $\mathbb{R}$ and $d\beta(aV_1 + bV_2) = d\{a\beta(V_1) + b\beta(V_2)\} \bmod \bar{C}$. It is a simple matter to show that $N_2(L, K)$ is not a vector space over $\Lambda^0(K)$ because $\pounds_{fV}\omega = f\pounds_V\omega + df \wedge (V \lrcorner \omega)$. Finally, since $\pounds_V\bar{C} \subset \bar{C}$

for any $V \in TC(K)$ and $£_V$ commutes with d, we have

$$£_{[V_1, V_2]}(L\mu + J) = (£_{V_1}£_{V_2} - £_{V_2}£_{V_1})(L\mu + J)$$

$$= £_{V_1}\, d\beta(V_2) - £_{V_2}\, d\beta(V_1) \bmod \bar{C}$$

$$= d\big(£_{V_1}\beta(V_2) - £_{V_2}\beta(V_1)\big) \bmod \bar{C}.$$

Thus $[V_1, V_2]$ belongs to $N_2(L, K)$ for any $V_1, V_2 \in N_2(L, K)$, and the right-hand side of the above equality gives (7-4.5). □

The inclusion relations given in the last theorem show that $N_1(L, K)$ and $N_2(L, K)$ are nontrivial generalizations of $N(L, K)$ in general. As yet, however, there is no relation between any of the Lie algebras of Noetherian vector fields and invariance properties of the action. We now proceed to make this aspect of the problem clear. There is a natural vehicle for the analysis of this question, namely the total variation

$$\Delta_V(s)A[\Phi] = A[\Phi_V(s)] - A[\Phi],$$

that is

$$A[\Phi_V(s)] = A[\Phi] + \Delta_V(s)A[\Phi].$$

Accordingly, an explicit evaluation of $\Delta_V(s)A[\Phi]$, for all s in some neighborhood of $s = 0$, will lead to an explicit evaluation of the deformation that is induced in the action by transport along the orbits of a given $V \in TC(K)$.

Theorem 7-4.2. *If $V \in N(L, K)$, then the finite variation, $\Delta_V(s)A[\Phi]$, vanishes for all $\Phi \in R(B_n)$ and we have the global invariance*

(7-4.6) $$A[\Phi_V(s)] = A[\Phi]$$

for all regular maps Φ. Further, any regular map Φ that satisfies the Euler-Lagrange equations also satisfies the identities

(7-4.7) $$d\Phi^*\{V \lrcorner (L\mu + J)\} = 0.$$

Proof. If $V \in N(L, K)$, then $£_V(L\mu + J) = 0$ by (7-4.1). It then follows directly from the evaluation (7-2.18) of Lemma 7-2.4 that $\Delta_V(s)A[\Phi] = 0$ for all $\Phi \in R(K)$. The global invariance, (7-4.6), is thus established for *all* regular maps Φ. We now restrict attention to those regular maps that satisfy the

Euler-Lagrange equations, $\Phi^* E_\alpha(L) = 0$. For any $V \in N(L, K)$, (7-4.1) and (7-2.27) give

$$0 = \pounds_V(L\mu + J) = (V \lrcorner C^\alpha) E_\alpha(L) + d\{V \lrcorner (L\mu + J)\} \bmod \bar{C}$$

and hence $d\Phi^*\{V \lrcorner (L\mu + J)\} = 0$ for all $\Phi \in R(K)$ such that $\Phi^* E_\alpha(L) = 0$. □

The first thing that must be noted is that Theorem 7-4.2 gives a very strong result: a Noetherian vector field gives rise to the invariance $A[\Phi_V(s)] = A[\Phi]$ for *every* regular map Φ, not just those that satisfy the Euler-Lagrange equations. Stated another way, the 1-parameter family of maps $\Phi_V(s)$, for any $V \in N(L, K)$, are "level curves" of the map $A[\cdot]$ from $R(K)$ into $\mathbb{R}$. However, Theorem 7-4.1 shows that $N(L, K)$ forms a Lie algebra over $\mathbb{R}$, and hence $\exp(N(L, K))$ forms a Lie group. This shows that the *dimension of the level surfaces of* $A[\cdot]$ *in* $R(K)$ *is at least as large as the dimension of the Lie algebra* $N(L, K)$. The practical utility of Theorem 7-4.2 lies in the association of invariance of the action with the conservation laws (7-4.7) that are satisfied by any solution of the Euler-Lagrange equations, however. In order to see what is actually involved, let $V \in N(L, K)$ be generated by

$$V_g = f^i(x^j, q^\beta)\,\partial_i + \bar{f}^\alpha(x^j, q^\beta)\,\partial_\alpha.$$

(Remember that any $V \in N(L, K)$ also belongs to $TC(K)$.) The evaluation given by (7-2.12) gives

$$V \lrcorner (L\mu + J) = \left\{\left(L\delta_j^i - y_j^\alpha \partial_\alpha L\right) f^j + \bar{f}^\alpha \partial_\alpha^i L\right\} \mu_i \bmod \bar{C}.$$

When the definition of the momentum-energy complex (7-3.16) is used, it is seen that

$$V \lrcorner (L\mu + J) = \left\{\bar{f}^\alpha \partial_\alpha^i L - T_j^i(L) f^j\right\} \mu_i \bmod \bar{C}, \tag{7-4.8}$$

and hence (7-4.7) gives the conservation laws

$$D_i \Phi^*\left(\bar{f}^\alpha \partial_\alpha^i L - T_j^i(L) f^j\right) = 0. \tag{7-4.9}$$

Definition. The $(n - 1)$-form

$$\mathscr{J}(V, L) = \left\{\bar{f}^\alpha \partial_\alpha^i L - f^j T_j^i(L)\right\} \mu_i \tag{7-4.10}$$

is the *current* associated with any $V \in TC(K)$ and $L \in \Lambda^0(K)$.

The following lemma is an immediate consequence of Theorem 7-4.2 and Stokes' theorem.

Lemma 7-4.1. *If Φ is any regular map that satisfies the Euler-Lagrange equations and V is any Noetherian vector field for L, then the current associated with V and L satisfies*

$$d\Phi^*\mathcal{J}(V, L) = 0 \tag{7-4.11}$$

and we have the integral conservation laws

$$\int_{\partial B_n} \Phi^*\mathcal{J}(V, L) = 0. \tag{7-4.12}$$

Theorem 7-4.3. *If $V \in N_1(L, K)$, then the finite variation, $\Delta_V(s)A[\Phi]$, vanishes for all $\Phi \in R(B_n)$ and we have the global invariance*

$$A[\Phi_V(s)] = A[\Phi] \tag{7-4.13}$$

for all regular maps Φ. Further, any regular map Φ that satisfies the Euler-Lagrange equations also satisfies the identities

$$d\Phi^*\mathcal{J}(V, L) = 0 \tag{7-4.14}$$

and the integral conservation laws

$$\int_{\partial B_n} \Phi^*\mathcal{J}(V, L) = 0. \tag{7-4.15}$$

Proof. The only difference between $N(L, K)$ and $N_1(L, K)$ is that we have $\pounds_V(L\mu + J) = 0$ in the first case and $\pounds_V(L\mu + J) = 0 \bmod \bar{C}$ in the second. An inspection of the proof of Theorem 7-4.2 shows, however, that quantities in the ideal $\bar{C}$ make no difference. The proof is therefore the same as that for Theorem 7-4.2. □

Theorems 7-4.2 and 7-4.3 appear to be similar. This is not the case, however. A simple calculation shows that

$$\pounds_V(L\mu) = (V \lrcorner dL + L\partial_i f^i)\mu + L\partial_\alpha f^i dq^\alpha \wedge \mu_i.$$

When dq^α is eliminated between this equation and $C^\alpha = dq^\alpha - y_i^\alpha dx^i$, we obtain

$$\pounds_V(L\mu) = (V \lrcorner dL + LZ_i f^i)\mu + C^\alpha \wedge (L\partial_\alpha f^i)\mu_i.$$

Thus $V \in N(L, K)$ makes the statement

$$\{V \lrcorner dL + LZ_i f\}\mu + C^\alpha \wedge (L\partial_\alpha f^i)\mu_i = 0,$$

because $0 = \pounds_V(L\mu + J)$, while $V \in N_1(L, K)$ makes the specific statement

$$\{V \lrcorner dL + LZ_i f^i\}\mu = 0 \bmod \bar{C}$$

because $0 = \pounds_V(L\mu + J) \bmod \bar{C} = \pounds_V(L\mu) \bmod \bar{C}$ and $\pounds_V J = 0 \bmod \bar{C}$.

Theorem 7-4.4. *If $V \in N_2(L, K)$, then the finite variation, $\Delta_V(s)A[\Phi]$, vanishes to within a surface integral for any $\Phi \in R(K)$,*

$$A[\Phi_V(s)] = A[\Phi] + \int_{\partial B_n} \Phi^* \int_0^s \exp(p£_V)\beta(V)\, dp. \tag{7-4.16}$$

Further, any regular map Φ that satisfies the Euler-Lagrange equations also satisfies the identities

$$d\Phi^*\{\mathscr{J}(V, L) - \beta(V)\} = 0 \tag{7-4.17}$$

and the integral conservation laws

$$\int_{\partial B_n} \Phi^*\{\mathscr{J}(V, L) - \beta(V)\} = 0. \tag{7-4.18}$$

Proof. If $V \in N_2(L, K)$, then $£_V(L\mu + J) = d\beta(V) \bmod \bar{C}$ by (7-4.3). We note that

$$(\exp(s£_V) - 1)\omega = \int_0^s \frac{d}{dp}\exp(p£_V)\omega\, dp = \int_0^s \exp(p£_V)£_V\omega\, dp$$

is true for any $\omega \in \Lambda(K)$, and hence

$$\begin{aligned}(\exp(s£_V) - 1)(L\mu + J) &= \int_0^s \exp(p£_V)£_V(L\mu + J)\, dp \\ &= \int_0^s \exp(p£_V)\, d\beta(V)\, dp.\end{aligned}$$

However, d and $£_V$ commute and hence d and $\exp(p£_V)$ commute; that is,

$$(\exp(s£_V) - 1)(L\mu + J) = d\int_0^s \exp(p£_V)\beta(V)\, dp$$

for any $V \in N_2(L, K)$. It then follows directly from (7-2.18) that

$$\begin{aligned}\Delta_V(s)A[\Phi] &= \int_{B_n} \Phi^* d\int_0^s \exp(p£_V)\beta(V)\, dp \\ &= \int_{\partial B_n} \Phi^* \int_0^s \exp(p£_V)\beta(V)\, dp.\end{aligned}$$

Similarly, (7-4.3) and (7-2.27) give

$$\begin{aligned}d\beta(V) = £_V(L\mu + J) &= (V \lrcorner C^\alpha)E_\alpha(L) + d\{V \lrcorner (L\mu + J)\} \bmod \bar{C} \\ &= (V \lrcorner C^\alpha)E_\alpha(L) + d\mathscr{J}(V, L) \bmod \bar{C}.\end{aligned}$$

Hence, applying Φ^* to both sides of this equality for any $\Phi \in R(K)$ such that $\Phi^* E_\alpha(L) = 0$ establishes (7-4.17). (4-7.18) then follows by Stokes' theorem. □

7-5. BOUNDARY CONDITIONS AND THE NULL CLASS

Up to this point, we have had very little to say about boundary conditions for variational problems. Problems in the calculus of variations are particularly pretty in this respect for there is a complete accompanying theory of natural boundary data.

Let $L(x^i, q^\beta, y_i^\beta)$ be a given element of $\Lambda^0(K)$ and let $\Phi: B_n \to K$ be a regular map. We then have the action

$$A[\Phi] = \int_{B_n} \Phi^*(L\mu) = \int_{B_n} \Phi^*(L\mu + J). \tag{7-5.1}$$

If Φ satisfies the Euler-Lagrange equations,

$$\Phi^* E_\alpha(L) = 0, \qquad E_\alpha(L) = (\partial_\alpha L)\mu - d(\partial_\alpha^i) \wedge \mu_i, \tag{7-5.2}$$

Theorem 7-2.2 shows that

$$\delta_V A[\Phi] = \int_{\partial B_n} \Phi^*\{V \lrcorner (L\mu + J)\} \tag{7-5.3}$$

for any $V \in TC(K)$. We now specialize to vertical vector fields $U \in TV(K) \subset TV(K)$ so that U is generated by

$$U_g = \bar{f}^\alpha(x^j, q^\beta)\,\partial_\alpha, \tag{7-5.4}$$

(i.e., $f^i(x^j, q^\beta) = V\langle x^i \rangle = 0$). When (7-2.12) is used with $f^i = 0$, (7-5.3) gives the explicit evaluation

$$\delta_U A[\Phi] = \int_{\partial B_n} \Phi^*(\bar{f}^\alpha \partial_\alpha^i L\mu_i). \tag{7-5.5}$$

On the other hand, we have based our definition of stationarization on the process δ_U for vertical vector fields such that U_g vanishes on the boundary, ∂B_n, of B_n. It is thus clear that we can generalize the concept of stationarity significantly.

Definition. A regular map Φ is said to *weakly stationarize* the action $A[\cdot]$ if and only if

$$\delta_U A[\Phi] = 0 \tag{7-5.6}$$

for all vertical vector fields U such that the boundary data for U and Φ give

$$\int_{\partial B_n} \Phi^*(\bar{f}^\alpha \partial^i_\alpha L\mu_i) = 0. \tag{7-5.7}$$

Lemma 7-5.1. *If Φ stationarizes $A[\cdot]$, then Φ weakly stationarizes $A[\cdot]$ and hence Φ satisfies the Euler-Lagrange equations*

$$\Phi^* E_\alpha(L) = 0 \tag{7-5.8}$$

at all interior points of B_n.

Proof. By definition Φ stationarizes $A[\cdot]$ if and only if (7-5.6) holds for all U such that U_g vanishes on the boundary, $\bar{f}^\alpha(x^j, q^\beta)|_{\partial B_n} = 0$. Under these conditions, (7-5.7) holds and hence Φ weakly stationarizes $A[\cdot]$. □

An examination of (7-5.7) shows that this condition can be satisfied in ways other than just $\bar{f}^\alpha|_{\partial B_n} = 0$. Thus, *all stationarizing maps constitute a subset of all weakly stationarizing maps*. We thus proceed with the study of conditions under which (7-5.7) holds.

Let $\partial_1 B_n$ and $\partial_2 B_n$ be two subsets of ∂B_n such that

$$\partial B_n = \partial_1 B_n \cup \partial_2 B_n, \qquad \partial_1 B_n \cap \partial_2 B_n = \varnothing. \tag{7-5.9}$$

Definition. The subset $\partial_1 B_n$ is the support of the *natural Dirichlet data* if and only if

$$\phi^\alpha(x^j)|_{\partial_1 B_n} = \text{assigned functions on } \partial_1 B_n, \tag{7-5.10}$$

$$\Phi^* \bar{f}^\alpha(x^j, q^\beta)|_{\partial_1 B_n} = 0. \tag{7-5.11}$$

Definition. The subset $\partial_2 B_n$ is the support of the *natural Neumann data* if and only if

$$\Phi^*(\partial^i_\alpha L\mu_i)|_{\partial_2 B_n} = 0, \tag{7-5.12}$$

$$\bar{f}^\alpha(x^j, q^\beta)|_{\partial_2 B_n} \text{ unassigned.} \tag{7-5.13}$$

Theorem 7-5.1. *A regular map Φ weakly stationarizes $A[\cdot]$ if*
1. *Φ satisfies the Euler-Lagrange equations at all interior points of B_n, and*
2. *∂B_n admits the decomposition (7-5.9) such that $\partial_1 B_n$ is the support of the natural Dirichlet data and $\partial_2 B_n$ is the support of the natural Neumann data.*

Proof. If Φ satisfies the Euler-Lagrange equations at all interior points of B, then (7-5.5) gives

$$\delta_U A[\Phi] = \int_{\partial B_n} \Phi^*(\bar{f}^\alpha \partial_\alpha^i L\mu_i).$$

Accordingly, satisfaction of natural Dirichlet data on $\partial_1 B_n$ and natural Neumann data on $\partial_2 B_n$ leads to the simultaneous satisfaction of (7-5.6) and (7-5.7). □

The new thing that has emerged here is the possibility of assigning natural Neumann data

$$\Phi^*(\partial_\alpha^i L\mu_i)|_{\partial_2 B_n} = 0. \tag{7-5.14}$$

Such data is obviously quite different from Dirichlet data if for no other reason than the fact that $\Phi^*\partial_\alpha^i L|_{\partial_2 B_n}$ will involve boundary values of the y's (of $\partial_i\phi^\alpha(x^j)$). We note, however, that the natural Neumann data (7-5.14) is homogeneous. This restriction is overcome by a crafty reformulation of the problem.

Suppose that we can replace L by $L + \eta$ without changing the Euler-Lagrange equations; that is,

$$\Phi^*E_\alpha(L) = \Phi^*E_\alpha(L + \eta) = \Phi^*E_\alpha(L) + \Phi^*E_\alpha(\eta).$$

This would demand that η be such that

$$\Phi^*E_\alpha(\eta) = 0$$

for all $\Phi \in R(K)$. In this event, the collection of all regular maps that satisfy the Euler-Lagrange equations is left invariant under such changes. On the other hand, the natural Neumann data becomes

$$\Phi^*(\partial_\alpha^i L\mu_i + \partial_\alpha^i \eta\mu_i)_{\partial_2 B_n} = 0 \tag{7-5.15}$$

and we have the possibility of satisfying inhomogeneous Neumann data. It is thus important that we study what is known as the null class of Lagrangian functions.

Definition. The *null class* of Lagrangian functions is

$$\mathcal{N}(K) = \{\eta \in \Lambda^0(K) | \Phi^*E_\alpha(\eta) = 0 \quad \text{for all } \Phi \in R(K)\}. \tag{7-5.16}$$

Null class Lagrangians η give rise to a null class n-form $\eta\mu$ for which the associated action

$$a[\Phi] = \int_{B_n} \Phi^*(\eta\mu)$$

is rendered stationary by every regular map. The n-form $J = C^\alpha \wedge (\partial^i_\alpha L)\mu_i$ is clearly a null class n-forms since $J \in \bar{C}$ and hence any regular map annihilates J. In fact, it is convenient to consider all n-forms that belong to $\bar{C}$ to be *trivial* null class n-forms.

Theorem 7-5.2. *A scalar $\eta \in \Lambda^0(K)$ is a null Lagrangian if*

$$\eta\mu = d\rho \bmod \bar{C} \tag{7-5.17}$$

for some $\rho \in \Lambda^{n-1}(\mathcal{G})$, where $\mathcal{G}$ is graph space with the coordinate cover (x^i, q^α).

Proof. Let U be any vertical vector field such that U_g vanishes on ∂B_n. Since $\rho \in \Lambda^{n-1}(\mathcal{G})$, we have

$$\pounds_U \rho = \pounds_{RU_g} \rho$$

because U_g is the projection of U onto $T(G)$. It is then a simple matter to see that

$$\delta_U \int_{B_n} \Phi^* \eta\mu = \delta_U \int_{B_n} \Phi^* d\rho = \delta_U \int_{\partial B_n} \Phi^* \rho$$

$$= \int_{\partial B_n} \Phi^* \pounds_U \rho = \int_{\partial B_n} \Phi^* \pounds_{U_g} \rho = \int_{\partial B_n} \Phi^*(U_g \lrcorner d\rho),$$

and hence $\delta_U \int_{B_n} \Phi^* \eta\mu = 0$ for all $\Phi \in R(K)$ because U_g vanishes on ∂B_n. Thus $\Phi^* E_\alpha(\eta) = 0$ for all $\Phi \in R(K)$. □

What makes the theorem work is the fact that ρ is an $(n-1)$-form on graph space so that there are no dy's present. This in turn implies that there are no variations of derivatives on the boundary when Stokes' theorem is used. The reader should carefully note, however, that a null class Lagrangian η can indeed depend on the y's. In order to make this as clear as possible, the case $n = 4$ that most often arises in practice will be analyzed. In this case ρ is a 3-form on graph space with the local coordinates (x^i, q^α) and hence

$$\begin{aligned} \rho = {} & P^i(x,q)\mu_i + \tfrac{1}{2} P^{ij}_\alpha(x,q)\, dq^\alpha \wedge \mu_{ij} \\ & + \tfrac{1}{2} P_{i\alpha\beta}(x,q)\, dq^\alpha \wedge dq^\beta \wedge dx^i \\ & + \tfrac{1}{6} P_{\alpha\beta\gamma}(x,q)\, dq^\alpha \wedge dq^\beta \wedge dq^\gamma \end{aligned} \tag{7-5.18}$$

with

$$P^{ij}_\alpha = -P^{ji}_\alpha, \qquad P_{i\alpha\beta} = -P_{i\beta\alpha} \tag{7-5.19}$$

and $P_{\alpha\beta\gamma}$ completely skew-symmetric in (α, β, γ). Now simply note that

$dq^\alpha = y_m^\alpha \, dx^m \bmod C$ and $df(x^j, q^\beta) = Z_m\langle f(x^j, q^\beta)\rangle dx^m \bmod C$. The third and fourth terms on the right-hand side of (7-5.18) involve protracted calculations, so we confine attention to the first two terms. In this simplified case, we have

$$\hat{\rho} = P^i \mu_i + \tfrac{1}{2} P_\alpha^{ij} \, dq^\alpha \wedge \mu_{ij}$$

and hence

$$\begin{aligned} d\hat{\rho} &= dP^i \wedge \mu_i + \tfrac{1}{2}\, dP_\alpha^{ij} \wedge dq^\alpha \wedge \mu_{ij} \\ &= dx^m \wedge \left\{ Z_m\langle P^i \rangle \mu_i + \tfrac{1}{2} Z_m\langle P_\alpha^{ij} \rangle y_k^\alpha \, dx^k \wedge \mu_{ij} \right\} \bmod \bar{C} \\ &= \left\{ Z_i\langle P^i \rangle + \tfrac{1}{2}\left(y_i^\alpha Z_j - y_j^\alpha Z_i \right)\langle P_\alpha^{ij} \rangle \right\} \mu \bmod \bar{C}. \end{aligned}$$

When this is put into (7-5.17) it then follows that

$$\hat{\eta} = Z_i\langle P^i \rangle + \tfrac{1}{2}\left(y_i^\alpha Z_j - y_j^\alpha Z_i \right)\langle P_\alpha^{ij} \rangle \tag{7-5.20}$$

is a null Lagrangian. Notice the explicit y-dependence as well as that from $Z_i = \partial_i + y_i^\beta \partial_\beta$. The corresponding terms that come from the third and fourth terms in (7-5.18) are left to the reader. A little reflection will show, however, that *any null Lagrangian can have only polynomial dependence on the y's with the degrees of the polynomials no greater than* $\min(n, N)$.

Although we have obtained the null class $\mathcal{N}(K)$ in order to obtain access to inhomogeneous Neumann data, the null class figures prominently in almost all aspects of the calculus of variations. First of all, suppose that $L_1 \in \Lambda^0(K)$ is a Lagrangian function that yields a given system of partial differential equations as the Euler-Lagrange equations (7-2.36). If $\eta \in \mathcal{N}(K)$, then $L_2 = L_1 + \eta$ is another Lagrangian function that yields the given system of partial differential equations as Euler-Lagrange equations; simply observe that $\Phi^* E_\alpha(\eta) = 0$ for *all* regular maps $\Phi: B_n \to K$. In fact, the relation

$$L_2\mu \equiv L_1\mu + \eta\mu \bmod \bar{C}, \qquad \eta \in \mathcal{N}(K) \tag{7-5.21}$$

is symmetric, reflexive and transitive and thus defines an equivalence relation on $\Lambda^n(K)$. Let $\mathscr{L}(L_1)$ denote the equivalence class that contains $L_1\mu$, $\mathscr{L}(L_1)$ being referred to as the *Lagrangian class* of L_1. It thus follows that any $L \in \mathscr{L}(L_1)$ is a Lagrangian function that yields the given system of partial differential equations as Euler-Lagrange equations. An immediate consequence of this observation is the following important result: *a Lagrangian that yields a given system of partial differential equations as Euler-Lagrange equations is determined only to within membership in a Lagrangian class.* This, in turn, shows that solution to inverse problems in the calculus of variations are never unique;

when a solution exists, it is a Lagrangian class rather than a unique Lagrangian function. However, the reader should carefully note, that different elements of a Lagrangian class are distinguished by their corresponding natural Neumann data. For example, if $L_2 = L_1 + \eta$ for some $\eta \in \mathscr{N}(K)$, then the natural Neumann data for L_1 is $\Phi^*(\partial^i_\alpha L_1 \mu_i)|_{\partial_2 B_n} = 0$ while that for L_2 is $\Phi^*(\partial^i_\alpha L_1 \mu_i + \partial^i_\alpha \eta \mu_i)|_{\partial_2 B_n} = 0$.

The action associated with the Lagrangian L_1 is given by

$$A_1[\Phi] = \int_{B_n} \Phi^* L_1 \mu. \tag{7-5.22}$$

If $L_2 \in \mathscr{L}(L_1)$, then $L_2 \mu \equiv L_1 \mu + \eta \mu \bmod \bar{C}$ for some $\eta \in \mathscr{N}(K)$ and hence

$$\begin{aligned} A_2[\Phi] &= \int_{B_n} \Phi^* L_2 \mu = \int_{B_n} \Phi^*(L_1 + \eta)\mu \\ &= A_1[\Phi] + \int_{B_n} \Phi^* \eta \mu. \end{aligned} \tag{7-5.23}$$

Since $A_2[\Phi]$ and $A_1[\Phi]$ give the same regular stationarizing maps for the same natural Dirichlet data, the actions $A_1[\Phi]$ and $A_2[\Phi]$ are said to be *equivalent* (equivalent integrals). As it turns out, the entire theory of multiple integral problems in the calculus of variations (both stationarity and minimality) can be obtained by use of the theory of equivalent integrals. This approach is primarily due to the pioneering work of Carathéodory and is carefully summarized and extended in the book by Rund (see the list of References for Further Study).

Let $L \in \Lambda^0(K)$ be a Lagrangian function whose Euler-Lagrange equations are to be solved. The results of Section 7-4 show that the Noetherian vector fields $N(L, K)$, $N_1(L, K)$ and $N_2(L, K)$ form Lie subalgebras of $TC(K)$ and each independent element of each of these Lie algebras gives rise to a current that is conserved by every solution of the Euler-Lagrange equations. Since this is probably the most used and fundamental aspect of variational problems, the following question naturally arises: what happens when L is replaced by an equivalent Lagrangian function $L_1 \in \mathscr{L}(L)$?

Theorem 7-5.3. *Let L_1 be any Lagrangian function that belongs to the Lagrangian class $\mathscr{L}(L)$ of a given Lagrangian function L. The Lie algebras of Noetherian vector fields $N(L_1, K)$, $N_1(L_1, K)$, $N_2(L_1, K)$ satisfy the Lie algebra inclusions*

$$N(L_1, K) \subset N_2(L, K), \tag{7-5.24}$$

$$N_1(L_1, K) \subset N_2(L, K), \tag{7-5.25}$$

$$N_2(L_1, K) = N_2(L, K). \tag{7-5.26}$$

Proof. By definition, $L_1 \in \mathscr{L}(L)$ if and only if $L_1 = L + \eta$ and $\eta\mu = d\rho \bmod \bar{C}$, $\rho \in \Lambda^{n-1}(\mathscr{G})$ by Theorem 7-5.2. We therefore have

$$L_1\mu = L\mu + d\rho \bmod \bar{C}. \tag{7-5.27}$$

A vector field $V \in TC(K)$ belongs to $N(L, K)$ if and only if $£_V(L\mu + J) = 0$. Let J_1 denote the n-form J that is constructed from L_1 and introduce the notation $\alpha \overset{C}{=} \beta$ if and only if $\alpha \equiv \beta \bmod \bar{C}$. We then have $J \overset{C}{=} 0$ and $J_1 \overset{C}{=} 0$ by Lemma 7-2.1. If $V \in N(L, K)$, we have (remember $£_V\bar{C} \subset \bar{C}$ for any $V \in TC(K)$)

$$\begin{aligned} £_V(L_1\mu + J_1) &\overset{C}{=} £_V(L_1\mu) \overset{C}{=} £_V(L\mu + d\rho) \\ &\overset{C}{=} £_V(L\mu + J + d\rho) \overset{C}{=} £_V(L\mu + J) + d£_V\rho. \end{aligned} \tag{7-5.28}$$

Thus, for $\beta_1(V) = £_V\rho$, we see that $V \in N(L, K)$ implies

$$£_V(L_1\mu + J_1) = d\beta_1(V) \bmod \bar{C}, \qquad \beta_1(V) = £_V\rho; \tag{7-5.29}$$

that is, $V \in N_2(L_1, K)$. This establishes (7-5.24). By definition, $V \in N_1(L, K)$ if and only if $£_V(L\mu + J) = 0 \bmod \bar{C}$. When this is used in (7-5.28), we see that $V \in N_1(L, K)$ implies

$$£_V(L_1\mu + J_1) = d\beta_1(V) \bmod \bar{C}, \qquad \beta_1(V) = £_V\rho, \tag{7-5.30}$$

and (7-5.25) is established. Finally, $V \in N_2(L, K)$ if and only if $£_V(L\mu + J) \overset{C}{=} d\beta(V)$, and hence (7-5.28) gives

$$\begin{aligned} £_V(L_1\mu + J_1) &= d(\beta(V) + \beta_1(V)) \bmod \bar{C}, \\ \beta_1(V) &= £_V\rho. \end{aligned} \tag{7-5.31}$$

Thus $N_2(L_1, K) = N_2(L, K)$. □

We have seen that $N(L, K) \subset N_1(L, K) \subset N_2(L, K)$, and hence $N_2(L, K)$ is the maximal algebra of these Lie algebras. Further, Theorem 7-5.3 shows that the Lie algebra $N_2(L, K)$ is universal over $\mathscr{L}(L)$ in the sense that $N(\mathscr{L}(L), K) \subset N_2(L, K)$, $N_1(\mathscr{L}(L), K) \subset N_2(L, K)$ and $N_2(\mathscr{L}(L), K) = N_2(L, K)$. This universality also extends to the conserved currents, for (7-5.31) and the discussion given in Section 7-4 show that

$$d\Phi^*\{V \lrcorner (L_1\mu + J_1) - \beta(V) - \beta_1(V)\} = 0 \tag{7-5.32}$$

for any solution map Φ and hence

$$\int_{\partial B_n} \Phi^*\{V \lrcorner (L_1\mu + J_1) - \beta(V) - \beta_1(V)\} = 0. \tag{7-5.33}$$

This is a very useful result, for once $N_2(L, K)$ is known we see that everything else is then determined because $\beta_1(V) = \pounds_V\rho$. In fact, $L_1\mu = L\mu + d\rho \bmod \bar{C}$ serves to determine L_1 in terms of L and ρ, while L and L_1 differ in the variational sense only in that they have different Neumann data. The additional term $\Phi^*\beta_1(V)$ in (7-5.32) and (7-5.33) may thus be viewed as the current that is generated by the Neumann data. In the context of mechanics, this says that inhomogeneous Neumann data (boundary tractions) give rise to nontrivial compensating current distributions, a fact that is often overlooked in the use of variational methods.

An alternative view emerges if we set $\beta(V) = 0$ for any $V \in N(L, K)$ or for any $V \in N_1(L, K)$. The above discussion then shows that the equivalence replacement

(7-5.34) $$L\mu \mapsto L_1\mu = L\mu + d\rho \bmod \bar{C}$$

induces the replacement

(7-5.35) $$\beta(V) \mapsto \beta(V) + \pounds_V\rho \bmod \bar{C}.$$

Thus although $N_2(\mathscr{L}(L), K) = N_2(L, K)$, so that $N_2(L, K)$ is a subalgebra of $TC(K)$ that is invariant under the equivalence replacement (7-5.34), the associated $(n-1)$-forms $\beta(N_2(L, K))$ undergo the replacement (7-5.35). It thus follows from (7-5.30) that $N_1(L_1, K)$ is the Lie subalgebra of $N_2(L_1, K) = N_2(L, K)$ for which

(7-5.36) $$\pounds_V\rho = d\sigma - \beta(V) \bmod \bar{C}$$

for some $\sigma \in \Lambda^{n-2}(\mathscr{G})$. If $V \in (N_2(L_1, K) = N_2(L, K))$, then there is an $\alpha(V) \in \bar{C}$ such that

$$\pounds_V(L_1\mu + J_1) = d(\beta(V) + \pounds_V\rho) + \alpha(V).$$

Thus $N(L_1, K)$ is the Lie subalgebra of $N_2(L, K)$ for which

(7-5.37) $$d(\beta(V) + \pounds_V\rho) + \alpha(V) = 0.$$

This latter condition is very restrictive; $N(L_1, K)$ can be empty if ρ and $\beta(N_2(L, K))$ are sufficiently complicated.

7-6. PROBLEMS WITH BOUNDARY INTEGRALS AND DIFFERENTIAL CONSTRAINTS

As before, let K be $(n + N + nN)$-dimensional kinematic space with local coordinates $(x^i, q^\alpha, y_i^\alpha)$ and contact 1-forms

(7-6.1) $$C^\alpha = dq^\alpha - y_i^\alpha\, dx^i.$$

The collection of all regular maps of $B \subset E_n$ into K is denoted by

$$R(B) = \{\Phi: B \to K | \Phi^*\mu \neq 0, \quad \Phi^* C^\alpha = 0, \quad 1 \leq \alpha \leq N\}. \tag{7-6.2}$$

Contrary to what has been said previously, we now assume that functions L and $\{L^i | 1 \leq i \leq n\}$, belonging to $\Lambda^0(K)$, are given. These serve to define the *action with boundary terms* by

$$A[\Phi] = \int_B \Phi^*(L\mu) + \int_{\partial B} \Phi^*(L^i \mu_i) \tag{7-6.3}$$

for any regular map Φ. The new aspect here is the explicit inclusion of the boundary terms $\Phi^*(L^i\mu_i)$. The arguments given in previous sections can now be repeated almost word for word provided care is exercised to account for the boundary terms in (7-6.3). We therefore consider an added complication that often arises in practice; namely, we adjoin a system of differential constraints.

Let $\{\omega_a | 1 \leq a \leq r\}$ be a given system of r independent exterior differential forms on K (i.e., $\omega_a \in \Lambda(K)$, $1 \leq a \leq r\}$. These serve to define a system of *differential constraints* upon satisfaction of the conditions

$$\Phi^*\omega_a = 0, \quad 1 \leq a \leq r, \quad \Phi \in R(B). \tag{7-6.4}$$

For example, suppose that we have the given system

$$\omega_a = w^{ij}_{a\alpha}\, dy^\alpha_i \wedge \mu_j + w_a \mu$$

of n-forms on K. Since any $\Phi \in R(B)$ annihilates the contact ideal $\overline{C}$,

$$\Phi^*\omega_a = \left\{ w^{ij}_{a\alpha}\left(x^k, \phi^\beta, \partial_k\phi^\beta\right) \partial_i \partial_j \phi^\alpha + w_a\left(x^k, \phi^\beta, \partial_k \phi^\beta\right)\right\}\mu,$$

and hence satisfaction of the constraints means that the ϕ^α's must satisfy the system of second-order partial differential equations

$$w^{ij}_{a\alpha}\left(x^k, \phi^\beta, \partial_k\phi^\beta\right) \partial_i \partial_j \phi^\alpha + w^a\left(x^k, \phi^\beta, \partial_k \phi^\beta\right) = 0.$$

We started these considerations with the class of regular maps $R(B)$, and hence all such maps satisfy the constraints $\Phi^*\mu \neq 0$, $\Phi^*C^\alpha = 0$, $1 \leq \alpha \leq N$. Thus the given constraints $\Phi^*\omega_a = 0$ will be satisfied only for the class of *constraint maps*

$$R_c(B) = \{\Phi \in R(B) | \Phi^*\omega_a = 0, \quad 1 \leq a \leq r\}. \tag{7-6.5}$$

It is explicitly assumed that the class $R_c(B)$ of constraint maps is nonvacuous; that is, the system of constraints $\Phi^*C^\alpha = 0$, $\Phi^*\mu \neq 0$, $\Phi^*\omega_a = 0$ is consistent.

If Φ is any constraint map, the action $A[\Phi]$ is well defined and takes its values in $\mathbb{R}$. Thus as Φ ranges over $R_c(B)$, $A[\Phi]$ ranges over some subset of $\mathbb{R}$. The variational problem that we wish to solve may now be stated as follows. *Find all critical points of the action map* $A[\cdot]: R_c(B) \to \mathbb{R}$. It should be carefully noted that we require the critical points of the action map $A[\cdot]$ as a map from $R_c(B)$ into $\mathbb{R}$, rather than as a map from $R(B)$ into $\mathbb{R}$. Thus the action map is to be scrutinized only on the set $R_c(B)$ of maps that satisfy the constraints $\Phi^*\omega_a = 0$. This shows that we must assume that the constraints have already been solved and that the action map is evaluated only on such solutions. It is therefore essential that any element of $R_c(B)$ belong to at least a 1-parameter family of elements of $R_c(B)$. If this were not the case, the constraints would be satisfied only for a disconnected discrete set of regular maps and questions of stationarity (criticality) of the action map become mute. These observations show that it is essential that a characterization of the connectivity of the set $R_c(B)$ of constraint maps be obtained.

Since each $\Phi \in R_c(B)$ has B as domain and an n-dimensional subset of K as range, the connectivity of $R_c(B)$ becomes accessible by study of smooth deformations of K that carry elements of $R_c(B)$ into elements of $R_c(B)$. Now, smooth deformations of K may be thought of as arising from the action of a continuous family of automorphisms of K that contains the identity map, and these in turn may be realized in terms of transport along orbits of vector fields on K. We have seen, however, that problems in the calculus of variations require vertical vector fields in order that the independent variables are not deformed by the flow of the vector fields. We therefore confine attention to the collection of all *vertical* vector fields

(7-6.6) $$TV(K) = \{V \in T(K) \mid V = \bar{v}^\alpha \partial_\alpha + v_i^\alpha \partial_\alpha^i\}.$$

Any $\Phi \in R_c(B)$ is such that Φ^* annihilates each C^α and each ω_a. Thus since Φ^* commutes with exterior differentiation, Φ^* also annihilates each dC^α and each $d\omega_a$. Accordingly, Φ^* annihilates the *constraint ideal*

(7-6.7) $$\Omega = I\{C^\alpha, \omega_a, dC^\alpha, d\omega_a \mid 1 \le \alpha \le N, \quad 1 \le a \le r\}$$

of the exterior algebra $\Lambda(K)$ for any $\Phi \in R_c(B)$. Conversely, the ideal Ω is the largest ideal of $\Lambda(K)$ with generators of degree not exceeding n that is annihilated by $R_c(B)^*$.

The only essential difference between these ideas and those used previously is that the contact ideal $\bar{C}$ has been replaced by the constraint ideal Ω. We therefore consider the *vertical* isovectors of the constraint ideal

(7-6.8) $$T_c(K) = \{V \in TV(K) \mid \pounds_V \Omega \subset \Omega\}.$$

Arguments identical to those given in a number of places in the last two chapters accomplish the desired results.

Theorem 7-6.1. *If $\Phi \in R_c(B)$, then $\Phi_V(s) = T_V(s) \circ \Phi$ belongs to $R_c(B)$ for any $V \in T_c(K)$ and all s in a neighborhood of $s = 0$, $\Phi_V(0) = \Phi$, and $T_c(K)$ is the largest Lie subalgebra of $TV(K)$ for which $\Phi_V(s) \in R_c(K)$ for all s in a neighborhood of $s = 0$.*

This result shows that the connectivity of $R_c(B)$ is exactly that generated by the flows $\{T_V(s)\}$ for all vertical isovector fields V of the constraint ideal. Thus if the constraints are to remain satisfied, *the only changes that are permitted in any $\Phi \in R_c(B)$ are those that are generated by composition with $T_V(s)$ for some $V \in T_c(K)$.*

Let us first note that $\Phi_V(s)^* = \Phi^*\exp(s£_V)$ by (7-2.5). It is then a simple matter to use (7-2.16), (7-2.17) and (7-6.3) to obtain the following result.

Theorem 7-6.2. *A map $\Phi : B \to K$ renders the action integral* (7-6.3) *stationary in the presence of constraints if and only if*

(7-6.9) $$\Phi^*C^\alpha = 0, \qquad \Phi^*\omega_a = 0, \qquad \Phi^*\mu \neq 0$$

and

(7-6.10) $$\int_B \Phi^*£_V(L\mu) + \int_{\partial B} \Phi^*£_V(L^i\mu_i) = 0$$

for all $V \in T_c(K)$. Thus a map $\Phi \in R_c(B)$ renders $A[\Phi]$ stationary in the presence of constraints if and only if

(7-6.11) $$\int_B \Phi^*(\bar{v}^\alpha E_\alpha(L)) + \int_{\partial B} \Phi^*\{\bar{v}^\alpha(\partial^i_\alpha L + \partial_\alpha L^i) + v^\alpha_j \partial^j_\alpha L^i\}\mu_i = 0$$

is satisfied for all $V = \bar{v}^\alpha\partial_\alpha + v^\alpha_j\partial^j_\alpha \in T_c(K)$.

It is essential to note that the functions $\bar{v}^\alpha$ are not necessarily independent here, in contrast with past considerations, for V must belong to $T_c(K)$ and the presence of the constraint forms ω_a may demand specific relations between the $\bar{v}^\alpha$ and the v^α_j as well. This is easily seen when we write out the conditions that $V \in T_c(K)$:

(7-6.12) $$£_V C^\alpha = A^\alpha_\beta C^\beta + B^{\alpha a}\omega_a + K^{\alpha a}\, d\omega_a,$$

(7-6.13) $$£_V \omega_a = A_{a\beta} \wedge C^\beta + R^b_a \wedge \omega_b + S^b_a \wedge d\omega_b,$$

where the various coefficient forms are taken so that each term in any one of these equations is of the same degree. Thus in particular, we may use the fundamental lemma of the calculus of variations only with those "parts" of the collection $\{\bar{v}^\alpha | 1 \leq \alpha \leq N\}$ that are left arbitrary upon satisfaction of the conditions (7-6.12), (7-6.13). In this respect, it is worth noting that the absence of constraints $\{\omega_a\}$ means that $T_c(K)$ coincides with all vertical isovector fields of the contact ideal, $\bar{C}$, in which case $\bar{v}^\alpha = f^\alpha(x^j, q^\beta)$ where the functions f^α are arbitrary. In this case (7-6.11) can be used with the fundamental lemma of the calculus of variations to reproduce our previous results in the case where the action integral contains boundary integrals; i.e., $\Phi^* E_\alpha(L) = 0$ at all interior points of B while the data must satisfy

$$\int_{\partial B} \Phi^*\{ f^\alpha(\partial^i_\alpha L + \partial_\alpha L^i) + Z_j\langle f^\alpha\rangle \partial^j_\alpha L^i\}\mu_i = 0.$$

The simplest way of seeing what is involved is to look at specific problems. A wide class of such problems arises naturally in what is now termed *control theory*, so we will draw from this discipline. All that is required here is to take the realization of $\Phi: B \to K$ so that $\Phi^* q^\alpha = \phi^\alpha(x^j)$ for $1 \le \alpha \le m < N$, $\Phi^*(q^{m+\beta}) = u^\beta(x^j)$ for $1 \le \beta \le N - m$, where $\{\phi^\alpha(x^j)\}$ are the state variables and $\{u^\beta(x^j)\}$ are the control variables. The constraints $\{\omega_a | 1 \le a \le r\}$ then describe the evolution of the state in the presence of the control while $A[\Phi]$ becomes the penalty functional for the control process. The admissible set of control laws for the process is obtained by solving the conditions of stationarity of $A[\Phi]$ in the presence of the given constraints.

We consider problems with two independent variables ($n = 2$), one state variable ϕ, and one control variable u ($N = 2$ with $\Phi^* q^1 = \phi$, $\Phi^* q^2 = u$). In such cases, it is simplest to use the variables (ϕ, u) rather than (q^1, q^2). Thus (y^1_1, y^1_2) represent the first derivatives of ϕ when Φ^* acts, while (y^2_1, y^2_2) represent the first derivatives of the control variable u.

A reasonable control problem is that for which the constraint, $\Phi^*(\omega_1) = 0$, is

$$\frac{\partial^2 \phi}{\partial x^2} = \frac{\partial u}{\partial t}\frac{\partial \phi}{\partial x} - \frac{\partial u}{\partial x}\frac{\partial \phi}{\partial t} \tag{7-6.14}$$

that is,

$$\omega_1 = du \wedge d\phi - dt \wedge dy^1_1. \tag{7-6.15}$$

A specific physical realization of (7-6.14) obtains when ϕ undergoes one-dimensional convective diffusion in a compressible fluid medium with density ρ, one-dimensional fluid velocity v_x, and

$$\rho = -\partial u/\partial x, \qquad \rho v_x = \partial u/\partial t$$

so that the continuity equation $\partial_t \rho + \partial_x(\rho v_x) = 0$ is satisfied. The control u is thus realizable through appropriate realizations of the one-dimensional compressible fluid flow.

The first thing we note is that the constraint form ω_1 is a 2-form. Accordingly, $V = \bar{v}^\alpha \partial_\alpha + v_i^\alpha \partial_\alpha^i$ must be such that $v_i^\alpha = (\partial_i + y_i^\beta \partial_\beta) v^\alpha(x, t, \phi, u)$ in order to secure satisfaction of $\pounds_V C^\alpha \subset \Omega$. Here, $C^1 = d\phi - y_1^1\, dx - y_2^1\, dt$, $C^2 = du - y_1^2\, dx - y_2^2\, dt$ and $\mu = dx \wedge dt$, $\mu_1 = dt$, $\mu_2 = -dx$. Straightforward but lengthy calculation shows that $\pounds_V \omega_1 \subset \Omega$ if and only if

$$V_q = (a\phi + b)\partial_\phi + f(\phi)\partial_u, \tag{7-6.16}$$

where (a, b) are arbitrary constants and $f(\phi)$ is an arbitrary function of the variable ϕ.

It is now simply a matter of substituting (7-6.16) into (7-6.11). We assume that $\{L^i\}$ are such that $\partial_\alpha^j L^i = 0$, in which case we obtain

$$\int_B \phi \Phi^* E_\phi + \int_{\partial B} \phi \Phi^* \left(\partial_1^i L + \partial_\phi L^i \right) \mu = 0 \tag{7-6.17}$$

from the parameter a,

$$\int_B \Phi^* E_\phi + \int_{\partial B} \Phi^* \left(\partial_1^i L + \partial_\phi L^i \right) \mu_i = 0 \tag{7-6.18}$$

from the parameter b, and

$$\int_B f(\phi) \Phi^* E_u + \int_{\partial B} f(\phi) \Phi^* \left(\partial_2^i L + \partial_u L^i \right) \mu_i = 0 \tag{7-6.19}$$

for all smooth functions $f(\phi)$. Thus if we set $f(\phi) = \Sigma c_n \phi^n$, (7-6.19) gives

$$\int_B \phi^n \Phi^* E_u + \int_{\partial B} \phi^n \Phi^* \left(\partial_2^i L + \partial_u L^i \right) \mu_i = 0, \qquad n = 0, 1, 2 \dots . \tag{7-6.20}$$

Here L and L^i are the functions that serve to determine the penalty functional $A[\Phi] = \int_B \Phi^*(L\mu) + \int_{\partial B} \Phi^*(L^i \mu_i)$ and (E_ϕ, E_u) are the Euler-Lagrange 2-forms for the pair (ϕ, u), respectively. *The problem of stationarizing $A[\Phi]$ in the presence of the constraint* (7-6.15) *is thus solved by finding all pairs* $(\phi(x, t), u(x, t))$ *that satisfy the constraint* (7-6.14) *and the integral conditions* (7-6.17) *through* (7-6.20). It is interesting to note in this regard that (7-6.20) will be identically satisfied if (L, L^i) do not depend explicitly on the control variable, as is often the case. In this event, we would then have only the constraint (7-6.14) and the two integral conditions (7-6.17) and (7-6.18).

The reason why such direct results obtain is that we have been able to characterize all continuous deformations of K that preserve satisfaction of the constraint equation (7-6.14), even though explicit solution of the constraint equation has not been obtained. Comparison for purposes of stationarization of $A[\Phi]$ then occurs only in this class of deformations, for only this class guarantees that the constraints will be satisfied for $\Phi_V(s)$ if they are satisfied for $\Phi_V(0) = \Phi$. Accordingly, since (7-6.16) shows that all such deformations are generated by $\phi\partial_\phi$, ∂_ϕ, and $f(\phi)\partial_u$ for all smooth functions $f(\phi)$, integral conditions rather than field equations result; there are no arbitrary functions of (x, t) for which the fundamental lemma of the calculus of variations may be used.

For a second example consider the situation in which the constraint is $\omega_1 = d(uy_1^1\, dt - uy_2^1\, dx) - dx \wedge dt$, that is,

$$\frac{\partial}{\partial x}\left(u\frac{\partial \phi}{\partial x}\right) + \frac{\partial}{\partial t}\left(u\frac{\partial \phi}{\partial t}\right) = 1, \tag{7-6.21}$$

subject to the Dirichlet boundary data

$$\phi|_{\partial B} = 0. \tag{7-6.22}$$

Here ϕ is the state variable, u is the control variable, and the notation is the same as that in the previous example. Disregarding for the moment the boundary data (7-6.22), a lengthy but straightforward calculation shows that $V_q = R(x, t, \phi, u)\partial_\phi + S(x, t, \phi, u)\partial_u$ generates an element of $T_c(K)$ if and only if

$$\partial_u R = \partial_x R = \partial_t R = 0, \qquad \partial_\phi S + u\partial_\phi\partial_\phi R = 0,$$

$$u\partial_u S - S = 0, \qquad \partial_x S + 2u\partial_\phi\partial_x R = 0,$$

$$\partial_t S + 2u\partial_\phi\partial_t R = 0, \qquad u\partial_\phi R + u^2(\partial_x\partial_x + \partial_t\partial_t)R + S = 0.$$

Thus all elements of $T_c(K)$ are generated by

$$V_q = f(\phi)\partial_\phi - u\frac{df(\phi)}{d\phi}\partial_u \tag{7-6.23}$$

where $f(\phi)$ is an arbitrary smooth (C^∞) function of its indicated argument. Indeed, (7-6.23) defines an infinite dimensional Lie algebra for

$$[X(f), X(g)] = X\left(f\frac{dg}{d\phi} - g\frac{df}{d\phi}\right),$$

where $X(f)$ is the continuum of operators

$$X(f) = f\partial_\phi - u\frac{df}{d\phi}\partial_u.$$

The deformations generated by (7-6.23) are all deformations of K that preserve satisfaction of the differential constraint (7-6.21). Accordingly, if we require

$$f(0) = 0, \tag{7-6.24}$$

then all such deformations will also preserve satisfaction of the Dirichlet boundary data (7-6.22). Thus (7-6.23) and (7-6.24) define all deformations that preserve the constraints.

The stationarity conditions now follow directly from (7-6.11) for any $A[\phi]$ for which $\partial^i_\alpha L^j = 0$:

$$0 = \int_B \left\{ f(\phi)\Phi^* E_\phi - u\frac{df(\phi)}{d\phi}\Phi^* E_u \right\} + \int_{\partial B} \left\{ f(\phi)\Phi^*\left(\partial^i_1 L + \partial_\phi L^i\right) - u\frac{df(\phi)}{d\phi}\Phi^*\left(\partial^i_2 L + \partial_u L^i\right)\right\}\mu_i. \tag{7-6.25}$$

If we formally write $f(\phi) = \Sigma_1^\infty C_n \phi^n$, in view of (7-6.24) then (7-6.25) and the boundary conditions (7-6.24) give

$$0 = \int_B \{\phi\Phi^* E_\phi - u\Phi^* E_u\} - \int_{\partial B} u\Phi^*\left(\partial^i_1 L + \partial_\phi L^i\right)\mu_i \tag{7-6.26}$$

for C_1 and

$$0 = \int_B \{\phi^n \Phi^* E_\phi - nu\phi^{n-1}\Phi^* E_u\}, \qquad n = 2, 3, \ldots \tag{7-6.27}$$

for $C_2, C_3, \ldots$. Again, we obtain an infinite system of integral conditions rather than Euler-Lagrange field equations with Lagrange multipliers.

In addition to providing access to the stationarity conditions, knowledge of $T_c(K)$ provides a significant amount of information about the nature of solutions of the constraint equations. If we take $f(\phi) = \phi^{n+1}$ for n a positive integer, an isovector field of the constraints (7-6.21), (7-6.22) is generated by

$$V_q = \phi^{n+1}\partial_\phi - (n+1)u\phi^n\partial_u.$$

The flow of this isovector field is obtained by solving $d\Phi(s)/ds = \Phi(s)^{n+1}$, $dU(s)/ds = -(n+1)U(s)\Phi(s)^n$ subject to the initial data $\Phi(0) = \phi$, $U(0) = u$;

$$\Phi = \phi(1 - ns\phi^n)^{-1/n}, \qquad U = u|1 - ns\phi^n|^{(n+1)/n}. \tag{7-6.28}$$

Thus if $(\phi(x,t), u(x,t))$ is a solution of (7-6.21) subject to the Dirichlet data (7-6.22), then (7-6.28) defines a 1-parameter family of solutions $(\phi(x,t;s), U(x,t;s))$. The pathology inherent in solutions of the constraint is in clear evidence from the first of (7-6.28), although the second shows that $U(x,t;s)$ remains bounded for bounded $(\phi(x,t), u(x,t))$.

REFERENCES FOR FURTHER STUDY

Chapter 6

Bluman, G. W. and J. D. Cole, *Similarity Methods for Differential Equations*. Springer-Verlag, Berlin, 1974.

Delph, T. J., Isovector Fields and Self-similar Solutions for Power Law Creep. *Int. J. Engng. Sci.* **21** (1983), 1061–1067.

Edelen, D. G. B., *Isovector Methods for Equations of Balance*. Sijthoff and Noordhoff, Alphen aan den Rijn, The Netherlands, 1980.

Edelen, D. G. B., Isovector Fields for Problems in the Mechanics of Solids and Fluids. *Int. J. Engng. Sci.* **20** (1982), 803–815.

Edelen, D. G. B., Implicit Similarities and Inverse Isovector Methods. *Arch. Rational Mech. Anal.* **82** (1983), 181–189.

Estabrook, F. B., Differential Geometry as a Tool for Applied Mathematics. In A. Dold and B. Eckmann, eds., *Geometric Approaches to Differential Equations*, Lecture Notes in Mathematics No. 810. Springer-Verlag, Berlin, 1980.

Harrison, B. K. and F. B. Estabrook, Geometric Approach to Invariance Groups and Solutions of Partial Differential Equations. *J. Math. Phys.* **12** (1971), 653–666.

Hermann, R., *The Geometry of Non-Linear Differential Equations, Bäcklund Transformations, and Solitons*. Math. Sci. Press, Brookline, Maine, 1976.

Ibragimov, N. H. and R. L. Anderson, Lie-Bäcklund Tangent Transformations. *J. Math. Anal. Applications* **59** (1977), 145–162.

Miura, N., ed. *Bäcklund Transformations, the Inverse Scattering Method, Solitons and Their Applications*. Lecture Notes in Mathematics No. 515. Springer-Verlag, Berlin, 1976.

Ovsiannikov, L. V., *Group Analysis of Differential Equations*. Academic Press, New York, 1982.

Tabor, M. and Y. M. Treve, eds., *Mathematical Methods in Hydrodynamics and Integrability of Dynamical Systems*. AIP Conference Proceedings No. 88, American Institute of Physics, New York, 1982.

Chapter 7

Anderson, I. M. and T. Duchamp, On the Existence of Global Variational Principles. *Am. J. Math.* **102** (1980), 781–868.

Carathéodory, C., *Calculus of Variations and Partial Differential Equations of the First Order, Part II: Calculus of Variations*, R. B. Dean Translation. Holden-Day, San Francisco, 1967.

Courant, R. and D. Hilbert, *Methods of Mathematical Physics*, Vol. I. Interscience, New York, 1953.

Dedecker, P., On the Generalization of Symplectic Geometry to Multiple Integrals in the Calculus of Variations. In *Differential Geometric Methods in Mathematical Physics*, Lecture Notes in Mathematics Vol. 570. Springer-Verlag, Berlin, 1977.

Edelen, D. G. B., *Isovector Methods for Equations of Balance*. Sijthoff and Noordhoff, Alphen aan den Rijn, The Netherlands, 1980.

Edelen, D. G. B., Aspects of Variational Arguments in the Theory of Elasticity; Fact or Folklore. *Int. J. Solid Structures* **17** (1981), 729–740.

Gelfand, I. M. and S. V. Fomin, *Calculus of Variations*. Prentice-Hall, Englewood Cliffs, New Jersey, 1963.

Goldschmidt, H. and S. Sternburg, The Hamiltonian-Cartan Formalism in the Calculus of Variations. *Ann. Inst. Fourier* **23** (1973), 203–269.

Hermann, R., *Differential Geometry and the Calculus of Variations*. Interdisciplinary Mathematics, Vol. XVII. Math. Sci. Press, Brookline, Maine, 1977.

Lovelock, D. and H. Rund, *Tensors, Differential Forms, and Variational Principles*. Wiley-Interscience, New York, 1975.

Morse, P. M. and H. Feshback, *Methods of Theoretical Physics*, Part I. McGraw-Hill, New York, 1953.

Rund, H., *The Hamilton-Jacobi Theory in the Calculus of Variations*. D. Van Nostrand Co., London, 1966.

PART THREE

APPLICATIONS TO PHYSICS

CHAPTER EIGHT

MODERN THERMODYNAMICS

8-1. FORMULATION OF THE PROBLEM

Classical thermodynamics of homogeneous systems, as it is usually presented, involves simple manipulations of exterior forms of degree one that came about as consequences of a number of physically motivated statements that are anything but simple. These statements usually involve quantities such as empirical temperature, internal energy, free energy, work, forces, entropy, chemical potentials, heat, quasistatic processes, etc., that are conjoined by the first and second laws of thermodynamics (read thermostatics). It would thus seem useful to take full advantage of our knowledge of the exterior calculus to simplify and reduce the number of physically complicated statements that proliferate thermodynamics. This has already been done in large part by the fundamental simplification achieved by Carathéodory through implementation of the notion of inaccessibility for adiabatic processes.

Let S denote a material system that occupies a finite region of Euclidean three-dimensional space. The homogeneous *mechanical substate* of the system S at any given time is assumed to be characterized by finite values of finitely many mechanical substate variables $x^1, x^2, \ldots, x^n$. For example, if part of S consists of a cylindrical container with a moveable piston whose interior is filled with a fluid or gas, we would identify x^1 with the volume of the fluid or gas. If another part of S consists of a fixed volume that is filled with a mixture of chemicals in a solvent, we would identify $x^2, x^3, \ldots,$ with the concentrations of the distinct chemical compounds of the mixture. The important thing here is that the mechanical substate of S can be characterized by the values of an n-dimensional mechanical substate column matrix $\mathbf{x}$ that locates a point in the n-dimensional number space E_n. For convenience, let $\mathbf{x}_0$ denote a specific n-dimensional column matrix that quantifies a *reference mechanical substate*.

The assumption that the mechanical substate of S can be characterized by a finite number of values at any given time involves a high degree of idealization that is "tucked in" by the innocuous sounding qualifier "homogeneous." What this means is that a liquid or gaseous part of the system has exactly the same properties at every point of that part, and that a deformable solid part has the same deformation at every point of that part. Properties of any given part of the system S cannot vary from point to point in that part; each is *homogeneous* and can thus be quantified by the values of the salient mechanical properties at any one point of that part.

We have purposefully referred to $\mathbf{x}$ as a mechanical *substate*, for thermodynamic studies recognize another state variable that characterizes thermal properties, as the name implies. For the purposes of this discussion, we adopt the concept of a homogeneous empirical temperature θ of the system S. An empirical temperature is one that would be measured by a thermometer or similar instrument in the laboratory. Again, it is taken as a homogeneous property of the system S; that is, every point of the system has the same empirical temperature at a given time. The value of θ is to be reckoned on some fixed, but arbitrary scale of empirical temperature. The thermodynamic state of the system S is thus quantified by the values of the $n + 1$ state variables $(\mathbf{x}, \theta)$ at any given time. Accordingly, the state space of the system S is an $(n + 1)$-dimensional number space E_{n+1}.

We suppose that the system S is acted upon by a system of forces that can be represented by n functions $F_1(\mathbf{x}, \theta), F_2(\mathbf{x}, \theta), \ldots, F_n(\mathbf{x}, \theta)$ of the thermodynamic state $(\mathbf{x}, \theta)$. Here $F_1(\mathbf{x}, \theta)$ is the effective force that acts on the system in the "x^1-direction", etc. In particular, we assume that the functions $\{F_i(\mathbf{x}, \theta)\}$ are continuous and differentiable mapping from $D \times [a, b] \subset E_{n+1}$ into E_n, where D is a starshaped region of E_n with center $\mathbf{x}_0$. Thus if $\mathbf{x}$ is in D then the set of points $\lambda\mathbf{x} + (1 - \lambda)\mathbf{x}_0$ is in D for all λ such that $0 \leq \lambda \leq 1$ (see Section 5-2). The domain of the functions $\{F_i(\mathbf{x}, \theta)\}$ is thus an $(n + 1)$-dimensional cylinder $D \times [a, b]$ with the starshaped base D. It is further assumed that a choice of positive *versus* negative action of the forces has been made in such a way that the infinitesimal work, $F_i(\mathbf{x}, \theta)\, dx^i$, that occurs as a consequence of the mechanical displacements dx^i is positive when the forces do work on the system (is negative when the system S performs work on the agencies that give rise to the forces).

Our purpose in the next few sections is to reduce all calculations of thermodynamically significant quantities to calculations that can be performed from knowledge of the forces that act on the thermodynamic system.

8-2. THE WORK FUNCTION FOR THERMOSTATICALLY CONSERVATIVE FORCES

The *work* 1-form of a thermodynamic system S with state coordinates $(\mathbf{x}, \theta)$ is defined by

$$W = F_i(\mathbf{x}, \theta)\, dx^i, \tag{8-2.1}$$

and hence W belongs to $\Lambda^1(E_{n+1})$. A direct calculation based on (8-2.1) shows that

$$dW = \tfrac{1}{2}(\partial_j F_i - \partial_i F_j)\, dx^j \wedge dx^i + \partial_\theta F_i\, d\theta \wedge dx^i, \tag{8-2.2}$$

where we have used the notation $\partial_\theta \coloneqq \partial/\partial\theta$.

There is a clear physical distinction between the state variables $\mathbf{x}$ on the one hand and the empirical temperature θ on the other. We accordingly refer to an exterior form on E_{n+1} that is restricted to the hyperplane $\theta = \text{constant}$ as being *thermostatically restricted*. If $\omega \in \Lambda^k(E_{n+1})$, the thermostatic restriction of ω is denoted by $\omega|_\theta$. Clearly, a thermostatic restriction can be realized by the action of the pull back τ^* of a map $\tau: E_n \to E_{n+1}|\mathbf{x} = \mathbf{x},\ \theta = \text{constant}$, if desired. In particular, (8-2.2) gives

$$(dW)|_\theta = \tfrac{1}{2}(\partial_j F_i - \partial_j F_i)\, dx^j \wedge dx^i. \tag{8-2.3}$$

For the time being, we confine our attention to thermodynamic systems whose forces are *mechanically conservative*; that is,

$$\partial_i F_j(\mathbf{x}, \theta) = \partial_j F_i(\mathbf{x}, \theta) \tag{8-2.4}$$

for all $(\mathbf{x}, \theta)$ in $D \times [a, b]$. An inspection of (8-2.3) shows that this restriction can also be expressed by

$$(dW)|_\theta = 0. \tag{8-2.5}$$

A mechanically conservative thermodynamic system is one for which the work 1*-form* W *is a closed* 1*-form on any hypersurface* $\theta = $ *constant.*

If $\mathbf{x}$ is any point in D, then each of the points

$$\mathbf{x}(\lambda) = \lambda\mathbf{x} + (1 - \lambda)\mathbf{x}_0, \qquad 0 \le \lambda \le 1 \tag{8-2.6}$$

belongs to D. Thus each of the functions

$$F_i(\mathbf{x}(\lambda), \theta) = F_i(\lambda\mathbf{x} + (1 - \lambda)\mathbf{x}_0, \theta)$$

is well defined for all $\lambda \in [0,1]$, all $\mathbf{x} \in D$, and all $\theta \in [a,b]$. The quantity

$$\Psi(\mathbf{x},\theta) = \psi_0(\theta) + \int_0^1 (x^i - x_0^i) F_i(\mathbf{x}(\lambda),\theta)\, d\lambda \tag{8-2.7}$$

is well defined on $D \times [a,b]$ and is continuously differentiable in $(\mathbf{x},\theta)$ if $\psi_0(\theta)$ is. Now (8-2.7) shows that $\Psi(\mathbf{x},\theta)$ is equal to within the function $\psi_0(\theta)$, to the mechanical work that is done by $\{F_i(\mathbf{x},\theta)\}$ on the system S in going from the mechanical substate $\mathbf{x}_0$ to the mechanical substate $\mathbf{x}$ by means of the linear process (8-2.6) that takes place at constant empirical temperature θ. In fact, if we interpret the symbol λ in (8-2.7) as proportional to time, then $\Psi(\mathbf{x},\theta)$ is the work that obtains at constant empirical temperature for the mechanical history $\mathbf{x}(\lambda)$, $0 \le \lambda \le 1$, that is linear in time; in other words, a quasistatic history leaves the macroscopic mechanical kinetic energy unchanged. It is therefore consistent to refer to the function $\Psi(\mathbf{x},\theta)$ as the *thermostatic work function* of the system S. It then follows that $\psi_0(\theta)$ is the value of the thermostatic work function in the reference mechanical state $\mathbf{x}_0$ (i.e., $\Psi(\mathbf{x}_0,\theta) = \psi_0(\theta)$). Hence $\psi_0(\theta)$ is not determined by the forces $F_i(\mathbf{x},\theta)$ and so may be assigned freely. As we shall see, the ability to assign $\psi_0(\theta)$ is important because $\partial_\theta \Psi(\mathbf{x}_0,\theta) = \partial_\theta \psi_0(\theta)$.

Comparison of (8-2.7) with the definition of the linear homotopy operator given in Chapter 5 shows that $\Psi(\mathbf{x},\theta)$ would be HW if there were no temperature variable, θ, present; in which case we would have $W = dHW = d\Psi$. The temperature variable is present, however, and the work 1-form is only closed on hypersurfaces $\theta =$ constant. This fact gives rise to the arguments $F_i(\mathbf{x}(\lambda),\theta)$ in (8-2.7) (i.e., θ rather than the 1-parameter family $\theta(\lambda)$), and hence $\Psi(\mathbf{x},\theta)$ does not arise by application of the linear homotopy operator to the 1-form W on E_{n+1}. It is therefore simpler to proceed directly rather than try to use homotopy operators since these operators do not commute with hypersurface restriction. Elementary calculations based on (8-2.7) show that

$$F_i(\mathbf{x},\theta) = \partial_i \Psi(\mathbf{x},\theta), \tag{8-2.8}$$

from which it follows that

$$d\Psi(\mathbf{x},\theta) = F_i(\mathbf{x},\theta)\, dx^i + \partial_\theta \Psi(\mathbf{x},\theta)\, d\theta. \tag{8-2.9}$$

Let W denote the work 1-form of the system S with mechanically conservative forces $\{F_i(\mathbf{x},\theta)\}$.

$$W = F_i(\mathbf{x},\theta)\, dx^i. \tag{8-2.10}$$

When (8-2.9) is used, we obtain

$$W = d\Psi - \partial_\theta \Psi\, d\theta. \tag{8-2.11}$$

We may thus write

$$W = d\Psi(\mathbf{x}, \theta) + r(\mathbf{x}, \theta)\, d\theta \tag{8-2.12}$$

where the function $r(\mathbf{x}, \theta)$ is defined by

$$r(\mathbf{x}, \theta) = -\partial_\theta \Psi(\mathbf{x}, \theta). \tag{8-2.13}$$

Exterior differentiation of (8-2.12) yields

$$dW = dr \wedge d\theta, \tag{8-2.14}$$

while (8-2.13) and (8-2.14) combine to give

$$W \wedge dW = d\Psi \wedge dr \wedge d\theta = d(\Psi dr \wedge d\theta), \tag{8-2.15}$$

$$dW \wedge dW = d(W \wedge dW) = 0. \tag{8-2.16}$$

These results provide all of the information required in order to compute the Darboux class and the rank of the work 1-form W (see Section 4-4); (8-2.16) shows that the Darboux class is ≤ 3.

The work 1-form W has Darboux class 2 and rank 2 if and only if

$$d\Psi \wedge dr \wedge d\theta = 0, \tag{8-2.17}$$

in which case we have

$$\Psi(\mathbf{x}, \theta) = \phi(r(\mathbf{x}, \theta), \theta). \tag{8-2.18}$$

The work 1-form W has Darboux class 3 and rank 2 if and only if

$$d\Psi \wedge dr \wedge d\theta \neq 0, \tag{8-2.19}$$

in which case $\Psi(\mathbf{x}, \theta)$, $r(\mathbf{x}, \theta)$ and θ are functionally independent. We exclude the possibility that W has Darboux class 1 and rank 0, for in that case $dW = dr \wedge d\theta = 0$ and the forces $F_i(\mathbf{x}, \theta)$ would be independent of the empirical temperature θ.

If R_2 is any compact, two-dimensional surface in $D \times [a, b]$ with a smooth one-dimensional boundary ∂R_2, then Stokes' theorem and (8-2.14) show that

$$\int_{\partial R_2} W = \int_{R_2} dW = \int_{R_2} dr \wedge d\theta, \tag{8-2.20}$$

where we have taken the orientation of $D \times [a, b]$ to be that induced by $dx^1 \wedge dx^2 \wedge \cdots \wedge dx^n \wedge d\theta$. Thus if we introduce $r(\mathbf{x}, \theta)$ as a new coordinate function in place of one of the x's, (8-2.20) shows that the work done on the system S in any cycle $\partial R_2 \subset D \times [a, b]$, $\int_{\partial R_2} W$, is equal to the projection of the area spanned by the cycle ∂R_2 on the two-dimensional coordinate plane spanned by $(dr, d\theta)$ with the orientation that is induced by $dx_1 \wedge \cdots \wedge dx_n \wedge d\theta$. However, in view of (8-2.8) and (8-2.13) we also have $dr \wedge d\theta = -\partial_\theta F_i \, dx^i \wedge d\theta$, and hence

(8-2.21)

$$\int_{\partial R_2} W = \int_{R_2} dr \wedge d\theta = -\int_{R_2} \partial_\theta F_i \, dx^i \wedge d\theta$$

$$= -\int_{R_2} \partial_\theta F_i(\mathbf{x}(u_1, u_2), \theta(u_1, u_2)) \left(\frac{\partial x^i}{\partial u_1} \frac{\partial \theta}{\partial u_2} - \frac{\partial x^i}{\partial u_2} \frac{\partial \theta}{\partial u_1} \right) du_1 \wedge du_2$$

where $\mathbf{x} = \mathbf{x}(u_1, u_2)$, $\theta = \theta(u_1, u_2)$ are the parametric equations for R_2. These results give a complete characterization of work properties for cyclic processes of the system S. They also show the drastic simplification that obtains by introduction of the quantity $r(\mathbf{x}, \theta)$; simply compare (8-2.20) with (8-2.21).

8-3. INTERNAL ENERGY, HEAT ADDITION, AND IRREVERSIBILITY

Having confined our attention to quasistatic processes—processes that occur in such a fashion that there is no change in the macroscopic kinetic energy of the system S—the law of conservation of energy (first law) states that there exists an *internal energy* scalar function E and a *heat addition* 1-form Q such that

(8-3.1)
$$dE = Q + W, \qquad dQ \not\equiv 0$$

holds for every quasistatic process.

We have taken the point of view that we will determine all quantities in terms of expressions involving the applied mechanical forces. Such a representation for W has already been obtained in terms of the quantities $\Psi(\mathbf{x}, \theta)$, $r(\mathbf{x}, \theta)$. The above equation is thus seen to relate an arbitrary undetermined 1-form Q and the unknown exact 1-form dE to the known 1-form W. Clearly, additional information must be provided if specific expressions are to be computed for E and Q. However, the fact that the 1-form dE is exact provides us with a certain amount of information which we proceed to obtain.

If (8-2.12) is substituted into (8-3.1), we obtain

(8-3.2)
$$Q = d(E - \Psi) - r\,d\theta.$$

Exterior differentiation of (8-3.2) then yields

(8-3.3) $$dQ = -dr \wedge d\theta$$

(8-3.4) $$Q \wedge dQ = -d(E - \Psi) \wedge dr \wedge d\theta = -d((E - \Psi)\, dr \wedge d\theta)$$

(8-3.5) $$d(Q \wedge dQ) = dQ \wedge dQ = 0.$$

A comparison with (8-2.14) through (8-2.16) shows that

(8-3.6) $$dQ = -dW, \qquad Q \wedge dQ = W \wedge dW - dE \wedge dW,$$
$$dQ \wedge dQ = dW \wedge dW = 0,$$

and hence

(8-3.7) $$\int_{R_2} dQ = -\int_{R_2} dW = -\int_{\partial R_2} dr \wedge d\theta = \oint Q = -\oint W,$$

(8-3.8) $$0 = \int_{R_4} d(Q \wedge dQ) = \int_{\partial R_4} Q \wedge dQ = -\int_{\partial R_4} d(E - \Psi) \wedge dr \wedge d\theta$$
$$= \int_{\partial R_4} W \wedge dW - \int_{\partial R_4} dE \wedge dW.$$

Thus Q *has Darboux class* 2 *and rank* 2 *if and only if* $d(E - \Psi) \wedge dr \wedge d\theta = 0$, *and class* 3 *and rank* 2 *if and only if* $d(E - \Psi) \wedge dr \wedge d\theta \neq 0$. Q has the same rank as W but can be of different class.

Carathéodory's innovative approach to thermodynamics is well known to serious students of the subject. In effect Carathéodory's condition requires that the 1-form Q of heat addition be such that there are states $(\bar{\mathbf{x}}, \bar{\theta})$ in any neighborhood of a given state $(\mathbf{x}, \theta) \subset D \times [a, b]$ that cannot be reached by quasistatic processes that satisfy $Q = 0$. Now $Q = 0$ is a Pfaff equation, and the known results concerning such equations show that the inaccessibility condition of Carathéodory is that $Q = 0$ be a completely integrable Pfaffian; that is, Q must be such that

(8-3.9) $$Q \wedge dQ = 0.$$

Stated another way, *Carathéodory's inaccessibility condition* is satisfied if and only if Q has rank 2 *and* class 2 (see Theorem 4-4.3). Darboux's theorem and the condition that the class of Q is 2 give the existence of scalar valued functions $u(\mathbf{x}, \theta)$ and $v(\mathbf{x}, \theta)$ such that $Q = u(\mathbf{x}, \theta)\, dv(\mathbf{x}, \theta)$.

Although we will not pursue the theory directly in terms of the implied scalar functions $u(\mathbf{x}, \theta)$ and $v(\mathbf{x}, \theta)$, it is worth noting an equivalent statement that follows from (8-3.4)

$$0 = \int_{R_3} Q \wedge dQ = -\int_{R_3} d((E - \Psi)\, dr \wedge d\theta) = \int_{\partial R_3} (E - \Psi)\, dr \wedge d\theta,$$

where R_3 is any regular three-dimensional region in $D \times [a, b]$ with a smooth boundary ∂R_3. Thus although $dQ \neq 0$, in general, and hence $\int_{R_2} dQ = -\int_{R_2} dr \wedge d\theta = \int_{\partial R_2} Q \neq 0$, inaccessibility does imply that E must be related to Ψ in such a fashion that $\int_{\partial R_3} (E - \Psi)\, dr \wedge d\theta = 0$ for every two-dimensional closed surface ∂R_3 that is contained in $D \times [a, b]$.

We now proceed directly from (8-3.4) and (8-3.9). These equations combine to give $0 = d(E - \Psi) \wedge dr \wedge d\theta$, and hence we infer the existence of a scalar valued function $f(r, \theta)$ such that

(8-3.10) $$E(\mathbf{x}, \theta) - \Psi(\mathbf{x}, \theta) = f(r(\mathbf{x}, \theta), \theta).$$

Thus if $f(r, \theta)$ can be determined, we will have determined the unknown scalar valued function $E(\mathbf{x}, \theta)$ in terms of the known quantities $\Psi(\mathbf{x}, \theta)$ and $r(\mathbf{x}, \theta)$. Further, when (8-3.10) is substituted into (8-3.2), we obtain

(8-3.11) $$Q = df(r, \theta) - r\, d\theta.$$

The inaccessibility condition thus reduces the 1-form Q to a 1-form on the two-dimensional manifold with coordinate functions (r, θ). Inaccessibility is actually a strong dimension reduction condition for the 1-form of heat addition. As a last remark, we note that (8-3.9) and (8-3.10) yield

(8-3.12) $$0 = \int_{\partial R_3} f(r, \theta)\, dr \wedge d\theta$$

for any $\partial R_3 \subset D \times [a, b]$.

Most treatments of thermodynamics use a strong additivity condition for internal energy and free energy at this point in the argument in order to proceed further. It turns out, however, that this condition can be relaxed in a significant way.

Let S_1 and S_2 be two systems with mechanical substates $\mathbf{x}_1$ and $\mathbf{x}_2$, respectively, that are described in terms of the same empirical temperature. We denote the isothermal work functions for these two systems by $\Psi_1(\mathbf{x}_1, \theta)$, $\Psi_2(\mathbf{x}_2, \theta)$, and the applied forces that act on the two separate systems by $\mathbf{F}^1(\mathbf{x}_1, \theta)$, $\mathbf{F}^2(\mathbf{x}_2, \theta)$, respectively. When the systems S_1 and S_2 are brought into contact at given empirical temperature θ, we obtain the composite system S. We denote the isothermal work function for the composite system by $\Psi(\mathbf{x}_1, \mathbf{x}_2, \theta)$. Additional interaction forces come into play when the systems S_1 and S_2 are brought into contact. These forces are now assumed to be such that they yield the following relation between the isothermal work functions:

$$\Psi(\mathbf{x}_1, \mathbf{x}_2, \theta) = \Psi_1(\mathbf{x}_1, \theta) + \Psi_2(\mathbf{x}_2, \theta) + \phi(\mathbf{x}_1, \mathbf{x}_2). \tag{8-3.13}$$

Strict additivity would obtain when $\phi(\mathbf{x}_1, \mathbf{x}_2) = 0$, in which case the interaction forces would do no net work on the composite system. This is what usually is assumed in classical treatments. Our assumption amounts to the requirement that the work that is done on the composite system by the interaction forces is independent of the empirical temperature. The upshot of the condition (8-3.13) is that we have the relation

$$r = r_1 + r_2 \tag{8-3.14}$$

where $r = -\partial_\theta \Psi(\mathbf{x}_1, \mathbf{x}_2, \theta)$, $r_1 = -\partial_\theta \Psi_1(\mathbf{x}_1, \theta)$, $r_2 = -\partial_\theta \Psi_2(\mathbf{x}_2, \theta)$; that is, *the variable r is additive under compositions of systems.*

The composition properties of internal energy are obtained from the requirement that $E - \Psi$ be additive under system composition. A combination of this requirement with (8-3.13) gives

$$E(\mathbf{x}_1, \mathbf{x}_2, \theta) = E_1(\mathbf{x}_1, \theta) + E_2(\mathbf{x}_2, \theta) + \phi(\mathbf{x}_1, \mathbf{x}_2), \tag{8-3.15}$$

and hence (8-3.10) and (8-3.14) yield the relation

$$f(r_1, \theta) + f(r_2, \theta) = f(r, \theta) = f(r_1 + r_2, \theta). \tag{8-3.16}$$

The solution of the function equation (8-3.16) is given by

$$f(r, \theta) = g(\theta) r, \tag{8-3.17}$$

for some continuously differentiable function $g(\theta)$. A combination of (8-3.17) with our previous results now gives us

$$E(\mathbf{x}, \theta) = \Psi(\mathbf{x}, \theta) + g(\theta) r(\mathbf{x}, \theta), \tag{8-3.18}$$

$$Q = g(\theta)\, dr + \big(g(\theta)' - 1\big) r\, d\theta = gr\left(\frac{dr}{r} + \frac{g' - 1}{g}\, d\theta\right) \tag{8-3.19}$$

$$= g(\theta) r\, d\left\{\ln|r| + \ln|g(\theta)| - \int \frac{d\theta}{g(\theta)}\right\}$$

so that we have identified two functions u and v such that $Q = u\, dv$.

Up to this point, we have been working with an arbitrary but fixed empirical temperature. If we make a differentiable monotone transformation of the temperature, $T = T(\theta)$, $dT/d\theta > 0$, we can define the isothermal work function $\overline{\Psi}(\mathbf{x}, T)$ relative to the temperature T by

(8-3.20) $$\overline{\Psi}(\mathbf{x}, T) = \Psi(\mathbf{x}, \theta(T)).$$

Thus if we define $\bar{r}(\mathbf{x}, T)$ by

(8-3.21) $$\bar{r}(\mathbf{x}, T) = -\partial_T \overline{\Psi}(\mathbf{x}, T),$$

then (8-3.20), (8-3.21) and $r = -\partial_\theta \Psi(\mathbf{x}, \theta)$ give

(8-3.22) $$r = \bar{r}\, dT/d\theta.$$

With $\overline{E}(\mathbf{x}, T)$ defined in terms of $E(\mathbf{x}, \theta)$ in a manner analogous to (8-3.20), (8-3.18) and (8-3.19) yield

(8-3.23)

$$\overline{E}(\mathbf{x}, T) = \overline{\Psi}(\mathbf{x}, T) + g(\theta)\bar{r}\, dT/d\theta,$$

(8-3.24) $$Q = g(\theta)\frac{dT}{d\theta}\, d\bar{r} + g(\theta)\bar{r}\frac{dT}{d\theta}\, d\left(\ln\left|g\frac{dT}{d\theta}\right| - \int \frac{d\theta}{g(\theta)}\right).$$

We now choose $T(\theta)$ to be such that $g(\theta)\, dT/d\theta = T$. The above equations then reduce to $\overline{E} = \overline{\Psi} + T\bar{r}$, $Q = T\, d\bar{r}$. This observation and the requirement $dT/d\theta > 0$ establishes the following basic result.

There exists a nonnegative preferred thermodynamic temperature T that is determined by

(8-3.25) $$T(\theta) = T_0 \exp\left(\int \frac{d\theta}{g(\theta)}\right), \qquad g(\theta) > 0,$$

to within an arbitrary positive multiplicative constant T_0 such that

(8-3.26) $$Q = T\, d\bar{r}, \qquad \bar{r}(\mathbf{x}, T) = -\partial_T \overline{\Psi}(\mathbf{x}, T),$$

(8-3.27) $$\overline{E}(\mathbf{x}, T) = \overline{\Psi}(\mathbf{x}, T) + T\bar{r}(\mathbf{x}, T) = \overline{\Psi}(\mathbf{x}, t) - T\partial_T \overline{\Psi}(\mathbf{x}, T)$$

hold on $D \times [T(a), T(b)]$.

This result establishes the existence of a *thermodynamic temperature T*, an *entropy* function $\bar{r}(\mathbf{x}, T)$, an internal energy function $\overline{E}(x, T)$, and the latter two functions are determined solely from knowledge of the thermostatic work function $\overline{\Psi}(\mathbf{x}, T)$. We also note for future reference that

(8-3.28) $$\int_{\partial R_2} Q = \int_{R_2} dT \wedge d\bar{r} = -\int_{\partial R_2} W$$

(8-3.29) $$0 = \int_{\partial R_3} T\bar{r}\, dT \wedge d\bar{r} = \tfrac{1}{4}\int_{\partial R_3} d(T^2) \wedge d(\bar{r}^2),$$

while $\bar{E}(\mathbf{x}, T)$ is still defined by (8-3.27) even when the right-hand side of (8-3.27) does not define a proper Legendre transformation (i.e., when $\partial^2\bar{\Psi}/\partial T^2$ is allowed to vanish on $[T(a), T(b)]$). The relation $\int_{\partial R_2} Q = \int_{R_2} dT \wedge d\bar{r}$ shows why Carnot cycles figure so heavily in classical thermodynamics: Carnot cycles are those that project onto the $(T, \bar{r})$-plane as simple rectangles. We also note that (8-3.7) becomes

$$\bar{\Psi}(\mathbf{x}, T) = \psi_0(T) + \int_0^1 (x^i - x_0^i) \cdot \bar{F}_i(\lambda\mathbf{x} + (1-\lambda)\mathbf{x}_0, T)\, d\lambda, \tag{8-3.30}$$

and hence

$$\bar{r}(\mathbf{x}_0, T) = -\partial_T \psi_0(T), \tag{8-3.31}$$

$$\partial_T \bar{r}(\mathbf{x}_0, T) = -\partial_T^2 \psi_0(T). \tag{8-3.32}$$

The function $\psi_0(T)$ is thus essential in establishing several of the properties usually contained in classical thermodynamics; nonvanishing heat capacity in the reference state $\mathbf{x}_0$, properly defined Legendre transformation from T to $\bar{r}$ at $\mathbf{x} = \mathbf{x}_0$, etc.

As is well known, the first half of the second law of thermodynamics is actually contained in the Carathéodory inaccessibility condition $Q \wedge dQ = 0$. The second half of the second law deals with processes of the real world for which the balance of energy (first law) now contains a term that reflects the actual changes in the kinetic energy; it deals with real processes rather than quasistatic processes. (The classic notion of reversible processes has been replaced in this work by the notion of processes for which the balance of energy reduces to the first law $dE = Q + W$.) For real processes, the second half of the second law states that the heat addition 1-form Q^* for actual processes obeys the inequality

$$Q^* \leq T\, d\bar{r} = Q. \tag{8-3.33}$$

Hence we have

$$\int_{\partial R_2} Q^* \leq \int_{\partial R_2} Q = \int_{R_2} dT \wedge d\bar{r} \tag{8-3.34}$$

for any cycle ∂R_2 that is contained in $D \times [T(a), T(b)]$.

Although the area integral, $\int_{R_2} dT \wedge d\bar{r}$, is to be calculated for quasistatic processes, the integral does not know this; that is, the area spanned by the projection of the actual cycle onto the $(T, \bar{r})$-plane does not distinguish whether the boundary of this projected area results from quasistatic paths or actual paths of the system that constitute the history of the cycle (i.e., $\bar{r} = \bar{r}(\mathbf{x}, T)$ is determined uniquely by the cycle $\mathbf{x} = \mathbf{x}(\tau)$, $T = T(\tau)$). It thus follows that the quantity

(8-3.35) $$\mathscr{R}(\partial R_2) = \int_{\partial R_2} Q^* \bigg/ \int_{R_2} dT \wedge d\bar{r}$$

satisfies the inequality

(8-3.36) $$\mathscr{R}(\partial R_2) \leq 1$$

with equality holding only if the cycle ∂R_2 consists of quasistatic processes. It thus follows that the quantity $\mathscr{R}(\partial R_2)$ provides a measure of the irreversibility of cyclic processes. However, this measure involves the quantity $\int_{\partial R_2} Q^*$ so that it depends upon measurement of heat addition rather than expended or extracted work, and is thus contrary to the intention to express basic quantities in terms of the applied forces.

Our previous analysis has established the relation

$$\int_{\partial R_2} Q = -\int_{\partial R_2} W$$

so that $\int_{\partial R_2} Q$ can be replaced by the *extractable* work, $-\int_{\partial R_2} W$, when dealing with the quasistatic processes. If W^* denotes the 1-form of work done on the system in an *actual* process, the second half of the second law can also be stated in the form

(8-3.37) $$W^* \geq W.$$

We can then introduce the quantity $\mathscr{I}(\partial R_2)$ by

(8-3.38) $$\mathscr{I}(\partial R_2) = -\int_{\partial R_2} W^* \bigg/ \int_{R_2} dT \wedge d\bar{r}.$$

It thus follows from $\int_{\partial R_2} Q = \int_{R_2} dT \wedge d\bar{r} = -\int_{\partial R_2} W$ and (8-3.37) that

(8-3.39) $$\mathscr{I}(\partial R_2) \geq -\int_{\partial R_2} W \bigg/ \int_{R_2} dT \wedge d\bar{r} = 1,$$

with equality holding only for reversible cycles. Accordingly, $\mathscr{I}(\partial R_2)$ *provides a measure of the irreversibility of any actual cyclic process*, and this measure is uniquely determined by the work extracted from the body by the cyclic process and the area of the projection of the process on the $(T, \bar{r})$-plane. It is therefore convenient to refer to $\mathscr{I}(\partial R_2) - 1$ as the *irreversibility* of the cycle ∂R_2; a quasistatic cycle is thus assigned an irreversibility of zero.

8-4. HOMOGENEOUS SYSTEMS WITH INTERNAL DEGREES OF FREEDOM

As before, we use S to denote a material system that occupies a finite region of Euclidean three-dimensional space and assume that S is in a homogeneous mechanical state. In constrast with previous sections, however, the homogeneous mechanical substate of S is now assumed to be characterized by finite values of finitely many "external" mechanical variables $x^1, \ldots, x^n$ *and* by finite values of finitely many "internal" variables $\alpha_1, \ldots, \alpha_N$ that describe internal degrees of freedom of the system S. The mechanical substate of S is now characterized by the values of the n-dimensional mechanical substate column matrix $\mathbf{x}$ *and* by the N-dimensional internal substate column matrix $\boldsymbol{\alpha}$. We use $\mathbf{x}_0$ to designate a fixed reference external mechanical substate. We also adopt the notion of empirical temperature and an associated quantifying variable θ referred to some fixed but arbitrary empirical temperature scale. The state of S is thus described by the $1 + n + N$ state variables $(\mathbf{x}, \boldsymbol{\alpha}, \theta)$ that define the coordinates of a point in $(n + N + 1)$-dimensional number space E_{n+N+1}.

We suppose that the system S is acted upon by a system of external forces that can be represented by n-functions $F_i(\mathbf{x}, \boldsymbol{\alpha}, \theta)$ of the thermodynamic state $(\mathbf{x}, \boldsymbol{\alpha}, \theta)$. In particular, we assume that the forces are continuous and differentiable mappings from $D \times E_N \times [a, b]$ into E_n, where D is a starshaped region with respect to the external reference mechanical substate $\mathbf{x}_0$. It is further assumed that the signs of the forces are so chosen that the infinitesimal work $F_i(\mathbf{x}, \boldsymbol{\alpha}, \theta)\, dx^i$ that occurs as a consequence of the external mechanical displacement $\{dx^i\}$ is positive when these forces do work on the system S and is negative when the system S performs work on the agencies that give rise to the forces.

The forces $\{F_i(\mathbf{x}, \boldsymbol{\alpha}, \theta)\}$ are assumed to be conservative with respect to the external mechanical variables in the sense that $\partial_{x_i} F_j = \partial_{x_j} F_i$; that is,

$$d\big(F_i(\mathbf{x}, \boldsymbol{\alpha}, \theta)\, dx^i\big)\big|_{\boldsymbol{\alpha}, \theta} = 0. \tag{8-4.1}$$

Thus $F_i(\mathbf{x}, \boldsymbol{\alpha}, \theta)\, dx^i$ is an exact 1-form on the surface $\boldsymbol{\alpha} =$ constant vector, $\theta =$ constant. The distinction between the external mechanical variables $\mathbf{x}$ and the internal variables $\boldsymbol{\alpha}$ can now be made precise, for, like the variable θ, there are no external forces that are associated with the internal variables $\boldsymbol{\alpha}$ and hence there is no work contribution in an arbitrary displacement $d\boldsymbol{\alpha}$.

As before, our assumption that D is starshaped with respect to $\mathbf{x}_0$ allows us to define a thermostatic work function $\Psi(\mathbf{x}, \boldsymbol{\alpha}, \theta)$ on $D \times E_N \times [a, b]$ by

$$\Psi(\mathbf{x}, \boldsymbol{\alpha}, \theta) = \Psi_0(\boldsymbol{\alpha}, \theta) + \int_0^1 (\mathbf{x} - \mathbf{x}_0) \cdot \mathbf{F}(\lambda \mathbf{x} + (1 - \lambda)\mathbf{x}_0, \boldsymbol{\alpha}, \theta)\, d\lambda. \tag{8-4.2}$$

A simple calculation shows that $F_i(\mathbf{x}, \boldsymbol{\alpha}, \theta) = \partial_i \Psi(\mathbf{x}, \boldsymbol{\alpha}, \theta)$, and hence

$$d\Psi = F_i(\mathbf{x}, \boldsymbol{\alpha}, \theta)\, dx^i + \partial_{\alpha_a} \Psi\, d\alpha_a + \partial_\theta \Psi\, d\theta. \tag{8-4.3}$$

Let W denote the work 1-form of the system S. Since there is no work associated with the internal variables $\boldsymbol{\alpha}$, we have

$$W = F_i(\mathbf{x}, \boldsymbol{\alpha}, \theta)\, dx^i. \tag{8-4.4}$$

Use of (8-4.3) now gives us

$$W = d\Psi - \partial_{\alpha_a}\Psi\, d\alpha_a - \partial_\theta \Psi\, d\theta,$$

and hence we may write

$$W = d\Psi - \beta^a\, d\alpha_a + r\, d\theta \tag{8-4.5}$$

where we have introduced the functions $r(\mathbf{x}, \boldsymbol{\alpha}, \theta)$ and $\beta^a(\mathbf{x}, \boldsymbol{\alpha}, \theta)$ by

$$r(\mathbf{x}, \boldsymbol{\alpha}, \theta) = -\partial_\theta \Psi(\mathbf{x}, \boldsymbol{\alpha}, \theta), \tag{8-4.6}$$

$$\beta^a(\mathbf{x}, \boldsymbol{\alpha}, \theta) = \partial_{\alpha_a}\Psi(\mathbf{x}, \boldsymbol{\alpha}, \theta), \qquad a = 1, \ldots, N. \tag{8-4.7}$$

We note at this point that (8-4.5) yields

$$dW = dr \wedge d\theta - d\beta^a \wedge d\alpha_a, \tag{8-4.8}$$

while

$$W|_{\boldsymbol{\alpha}} = d\Psi|_{\boldsymbol{\alpha}} + r\, d\theta, \qquad dW|_{\boldsymbol{\alpha}} = dr|_{\boldsymbol{\alpha}} \wedge d\theta, \tag{8-4.9}$$

$$W|_{\boldsymbol{\alpha}} \wedge dW|_{\boldsymbol{\alpha}} = d\Psi|_{\boldsymbol{\alpha}} \wedge dr|_{\boldsymbol{\alpha}} \wedge d\theta,$$

$$dW|_{\boldsymbol{\alpha}} \wedge dW|_{\boldsymbol{\alpha}} = d(W|_{\boldsymbol{\alpha}} \wedge dW|_{\boldsymbol{\alpha}})|_{\boldsymbol{\alpha}} = 0.$$

It thus follows from (8-4.8) that W will have rank 4 if $dr \wedge d\theta \wedge d\beta^a \wedge d\alpha_a \neq 0$ for at least one a, $1 \le a \le N$, or if $d\beta^a \wedge d\alpha_a \wedge d\beta^b \wedge d\alpha_b \neq 0$ for at least one pair (a, b) in the range 1 through N. We explicitly assume throughout this section that the rank of W is greater than or equal to 4. If the rank of W is less than 4, the problem reduces to that dealt with in the previous section.

Under restriction to quasistatic processes, the first law states that there exists an *internal energy* scalar function E and a *heat addition* 1-form Q such that

$$dE = Q + W, \qquad dQ \not\equiv 0 \tag{8-4.10}$$

holds for every quasistatic process. As before, the function E and the 1-form Q must be determined by imposition of additional conditions. However, the fact that dE is an exact 1-form allows us to obtain certain additional relations.

If (8-4.5) is substituted into (8-4.10), we obtain

$$Q = d(E - \Psi) - r\,d\theta + \beta^a\,d\alpha_a. \tag{8-4.11}$$

Exterior differentiation of (8-4.11) then yields

$$dQ = -dr \wedge d\theta + d\beta^a \wedge d\alpha_a$$

and hence (8-4.8) shows that

$$dQ = -dW. \tag{8-4.12}$$

Thus *Q and W have the same rank*, but can have different classes. In particular, since we have assumed that W has rank ≥ 4, we see that Q must also have rank ≥ 4. We note for future reference that

$$Q|_{\boldsymbol{\alpha}} = d(E - \Psi)|_{\boldsymbol{\alpha}} - r\,d\theta, \qquad dQ|_{\boldsymbol{\alpha}} = -dr|_{\boldsymbol{\alpha}} \wedge d\theta, \tag{8-4.13}$$

$$Q|_{\boldsymbol{\alpha}} \wedge dQ|_{\boldsymbol{\alpha}} = -d(E - \Psi)|_{\boldsymbol{\alpha}} dr|_{\boldsymbol{\alpha}} \wedge d\theta = -d((E - \Psi)\,dr|_{\boldsymbol{\alpha}} \wedge d\theta)|_{\boldsymbol{\alpha}},$$

$$dQ|_{\boldsymbol{\alpha}} \wedge dQ|_{\boldsymbol{\alpha}} = d(Q|_{\boldsymbol{\alpha}} \wedge dQ|_{\boldsymbol{\alpha}})|_{\boldsymbol{\alpha}} = 0$$

and hence the restriction of our results to the surface $\boldsymbol{\alpha} =$ constant vector yields equations that agree with the corresponding equations established in the previous section. These equations show that $Q|_{\boldsymbol{\alpha}}$ has rank 2.

If we were to try to apply Carathéodory's inaccessibility condition, $Q \wedge dQ = 0$, to the system S with internal state variables, then exterior differentiation would give $d(Q \wedge dQ) = dQ \wedge dQ = 0$ and the rank of Q would have to be less than or equal to 2. This is clearly impossible since W has rank ≥ 4 and this implies that Q must have rank ≥ 4. Put another way, every point in $D \times E_N \times [a, b]$ is accessible from every other point by solutions of $Q = 0$ since Q has rank ≥ 4. On the other hand, the system of equations (8-4.13) shows that the restriction of Q to the surface $\boldsymbol{\alpha} =$ constant yields a 1-form for which it is possible to require complete integrability of the corresponding Pfaffian ($Q|_{\boldsymbol{\alpha}}$ has rank 2). Further, the restriction to the surface $\boldsymbol{\alpha} =$ constant reduces the system S to a new system whose behavior exactly mimics the behavior of the systems considered in Section 3. These considerations motivate the following restricted inaccessibility condition: *the 1-form Q is such that*

$$Q|_{\boldsymbol{\alpha}} \wedge dQ|_{\boldsymbol{\alpha}} = 0 \tag{8-4.14}$$

holds for every $\boldsymbol{\alpha} \in E_N$. Under this condition, we have inaccessibility in each surface $\boldsymbol{\alpha} =$ constant vector.

We now proceed exactly as before. The equations (8-4.13) show that the condition (8-4.14) is satisfied if and only if $0 = d(E - \Psi)|_{\boldsymbol{\alpha}} \wedge dr|_{\boldsymbol{\alpha}} \wedge d\theta$, from which we infer the existence of a scalar valued function $f(r, \theta, \boldsymbol{\alpha})$ such that

$$E(\mathbf{x},\boldsymbol{\alpha},\theta)-\Psi(\mathbf{x},\boldsymbol{\alpha},\theta)=f(r(\mathbf{x},\boldsymbol{\alpha},\theta),\theta,\boldsymbol{\alpha}). \tag{8-4.15}$$

Thus since $Q|_{\boldsymbol{\alpha}}=d(E-\Psi)|_{\boldsymbol{\alpha}}-r\,d\theta$ and $df|_{\boldsymbol{\alpha}}=\partial_r f\,dr+\partial_\theta f\,d\theta$, we obtain

$$Q|_{\boldsymbol{\alpha}}=df|_{\boldsymbol{\alpha}}-r\,d\theta=\partial_r f\,dr+(\partial_\theta f-r)\,d\theta. \tag{8-4.16}$$

Let S_1 and S_2 be two systems with mechanical substates $(\mathbf{x}_1,\boldsymbol{\alpha})$ and $(\mathbf{x}_2,\boldsymbol{\alpha})$, respectively (i.e., S_1 and S_2 have the same internal substate). We denote the isothermal work functions of these two separate systems by $\Psi_1(\mathbf{x}_1,\boldsymbol{\alpha},\theta)$ and $\Psi_2(\mathbf{x}_2,\boldsymbol{\alpha},\theta)$ and the internal energy functions by $E_1(\mathbf{x}_1,\boldsymbol{\alpha},\theta)$ and $E_2(\mathbf{x}_2,\boldsymbol{\alpha},\theta)$. Let the two systems S_1 and S_2 be composed to yield the composite system S and denote the resulting isothermal work function and the internal energy function by $\Psi(\mathbf{x}_1,\mathbf{x}_2,\boldsymbol{\alpha},\theta)$ and $E(\mathbf{x}_1,\mathbf{x}_2,\boldsymbol{\alpha},\theta)$, respectively. We assume that these quantities are related by

$$\Psi=\Psi_1+\Psi_2+\phi(\mathbf{x}_1,\mathbf{x}_2,\boldsymbol{\alpha}), \tag{8-4.17}$$

$$E=E_1+E_2+\phi(\mathbf{x}_1,\mathbf{x}_2,\boldsymbol{\alpha}); \tag{8-4.18}$$

that is, the nonadditive parts are temperature independent. This has the effect that

$$r=r_1+r_2 \tag{8-4.19}$$

where $r=-\partial_\theta\Psi$, $r_1=-\partial_\theta\Psi_1$, $r_2=-\partial_\theta\Psi_2$, so that *the variable r is additive under composition of systems with the same internal substate*.

We now substitute (8-4.17) through (8-4.19) into (8-4.15) to obtain

$$f(r_1,\theta,\boldsymbol{\alpha})+f(r_2,\theta,\boldsymbol{\alpha})=f(r,\theta,\boldsymbol{\alpha})=f(r_1+r_2,\theta,\boldsymbol{\alpha}). \tag{8-4.20}$$

The solution of this function equation is given by

$$f(r,\theta,\boldsymbol{\alpha})=g(\theta,\boldsymbol{\alpha})r \tag{8-4.21}$$

for some continuously differentiable function $g(\theta,\boldsymbol{\alpha})$. A combination of our previous results with (8-4.21) now gives us

$$E(\mathbf{x},\boldsymbol{\alpha},\theta)=\psi(\mathbf{x},\boldsymbol{\alpha},\theta)+g(\theta,\boldsymbol{\alpha})r(\mathbf{x},\boldsymbol{\alpha},\theta), \tag{8-4.22}$$

$$Q|_{\boldsymbol{\alpha}}=g(\theta,\boldsymbol{\alpha})r\left(\frac{dr|_{\boldsymbol{\alpha}}}{r}+\frac{\partial_\theta g-1}{g}\,d\theta\right), \tag{8-4.23}$$

where $r(\mathbf{x},\boldsymbol{\alpha},\theta)$ is given by

$$r(\mathbf{x},\boldsymbol{\alpha},\theta)=-\partial_\theta\Psi(\mathbf{x},\boldsymbol{\alpha},\theta). \tag{8-4.24}$$

We now proceed, as in Section 3, to consider transformations of the temperature scale. In distinction to Section 3, however, we now must allow the temperature transformations to depend on the internal variables $\boldsymbol{\alpha}$, $T = T(\theta, \boldsymbol{\alpha})$, $\partial_\theta T > 0$, and assume invertibility for each $\boldsymbol{\alpha}$ so that $\theta = \theta(T, \boldsymbol{\alpha})$. We then define the isothermal work function $\bar{\Psi}(\mathbf{x}, \boldsymbol{\alpha}, T)$ relative to the temperature T by

(8-4.25) $$\bar{\Psi}(\mathbf{x}, \boldsymbol{\alpha}, T) = \Psi(\mathbf{x}, \boldsymbol{\alpha}, \theta(T, \boldsymbol{\alpha})).$$

Thus if we define $\bar{r}(\mathbf{x}, \boldsymbol{\alpha}, T)$ by

(8-4.26) $$\bar{r}(\mathbf{x}, \boldsymbol{\alpha}, T) = -\partial_T \bar{\Psi}(\mathbf{x}, \boldsymbol{\alpha}, T),$$

then (8-4.25), (8-4.26) and $r = -\partial_\theta \Psi(\mathbf{x}, \boldsymbol{\alpha}, \theta)$ give

(8-4.27) $$r = \bar{r}\partial_\theta T.$$

With $\bar{E}(\mathbf{x}, \boldsymbol{\alpha}, T)$ defined in an analogous manner, we obtain

(8-4.28) $$\bar{E} - \bar{\Psi} = g(\theta, \boldsymbol{\alpha})\bar{r}\partial_\theta T,$$

(8-4.29) $$Q|_{\boldsymbol{\alpha}} = g\bar{r}\partial_\theta T\left(\frac{d\bar{r}|_{\boldsymbol{\alpha}}}{r} + \left(\frac{\partial_\theta^2 T}{\partial_\theta T} + \frac{\partial_\theta g - 1}{g}\right) d\theta\right).$$

If we choose $T(\theta, \boldsymbol{\alpha})$ to be such that $g(\theta, \boldsymbol{\alpha})\partial_\theta T = T$, then these equations reduce to $\bar{E} = \bar{\Psi} + T\bar{r}$, $Q|_{\boldsymbol{\alpha}} = T\,d\bar{r}$. This observation and the requirement $\partial_\theta T > 0$ establishes the following basic result. *There exists a nonnegative preferred thermodynamic temperature $T(\theta, \boldsymbol{\alpha})$ that is determined by*

(8-4.30) $$T(\theta, \boldsymbol{\alpha}) = T_0(\boldsymbol{\alpha})\exp\left(\int \frac{d\theta}{g(\theta, \boldsymbol{\alpha})}\right)$$

to within an arbitrary positive function $T_0(\boldsymbol{\alpha})$ such that

(8-4.31) $$Q|_{\boldsymbol{\alpha}} = T\,d\bar{r}|_{\boldsymbol{\alpha}}, \qquad \bar{r} = -\partial_T \bar{\Psi}(\mathbf{x}, \boldsymbol{\alpha}, T)$$

(8-4.32) $$\bar{E}(\mathbf{x}, \boldsymbol{\alpha}, T) = \bar{\Psi}(\mathbf{x}, \boldsymbol{\alpha}, T) + T\bar{r}(\mathbf{x}, \boldsymbol{\alpha}, T) = \bar{\Psi} - T\partial_T \bar{\Psi}$$

hold on $D \times E_N \times [T(a, \boldsymbol{\alpha}), T(b, \boldsymbol{\alpha})]$.

What now remains is to remove the restriction of Q to $\boldsymbol{\alpha}$ = constant vector. Now $Q|_{\boldsymbol{\alpha}} = d(E - \Psi)|_{\boldsymbol{\alpha}} - r\,d\theta = d(\bar{E} - \bar{\Psi})|_{\boldsymbol{\alpha}} - \bar{r}\partial_\theta T\,d\theta = d(\bar{E} - \bar{\Psi})|_{\boldsymbol{\alpha}} - \bar{r}\,dT|_{\boldsymbol{\alpha}} = T\,d\bar{r}|_{\boldsymbol{\alpha}}$, so that

(8-4.33) $$-\bar{r}\,dT|_{\boldsymbol{\alpha}} = T\,d\bar{r}|_{\boldsymbol{\alpha}} - d(\bar{E} - \bar{\Psi})|_{\boldsymbol{\alpha}}.$$

On the other hand, (8-4.11) gives

$$Q = d(E - \Psi) - r\,d\theta + \beta^a d\alpha_a = d(\bar{E} - \bar{\Psi}) - r\partial_\theta T\,d\theta + \beta^a d\alpha_a$$

$$= d(\bar{E} - \bar{\Psi}) - \bar{r}\,dT|_{\boldsymbol{\alpha}} + \beta^a d\alpha_a,$$

and hence use of (8-4.33) yields

$$\text{(8-4.34)} \qquad Q = d(\bar{E} - \bar{\Psi}) + T\,d\bar{r}|_{\boldsymbol{\alpha}} - d(\bar{E} - \bar{\Psi})|_{\boldsymbol{\alpha}} - \beta^a d\alpha_a$$

$$= d(T\bar{r}) - d(T\bar{r})|_{\boldsymbol{\alpha}} + T\,dr|_{\boldsymbol{\alpha}} + \beta^a d\alpha_a$$

$$= T\,d\bar{r} + \left(\beta^a + r\partial_{\alpha_a} T\right) d\alpha_a.$$

However, $\bar{\Psi}(\mathbf{x}, \boldsymbol{\alpha}, T(\theta, \boldsymbol{\alpha})) = \Psi(\mathbf{x}, \boldsymbol{\alpha}, \theta)$ implies that $\beta^a = \partial_{\alpha_a}\Psi = \partial_{\alpha_a}\bar{\Psi} + \partial_T\bar{\Psi}\partial_{\alpha_a}T = \bar{\beta}_a - \bar{r}\partial_{\alpha_a}T$, where $\bar{\beta}_a = \partial\bar{\Psi}(\mathbf{x}, \boldsymbol{\alpha}, T)/\partial\alpha_a$. When this is substituted into (8-4.34) we obtain

$$\text{(8-4.35)} \qquad Q = T\,d\bar{r} + \bar{\beta}^a d\alpha_a$$

with

$$\text{(8-4.36)} \qquad \bar{r} = -\partial_T\bar{\Psi}(\mathbf{x}, \boldsymbol{\alpha}, T), \qquad \bar{\beta}^a = \partial_{\alpha_a}\bar{\Psi}(\mathbf{x}, \boldsymbol{\alpha}, T).$$

Accordingly, *the* 1-*form* $\bar{\beta}^a d\alpha_a$ *may be viewed as the heat addition* 1-*form that results from the internal degrees of freedom when the entropy* $\bar{r}$ *is held constant*, while the 1-form $T\,d\bar{r}$ represents the classical 1-form of heat addition that results when the processes occur so as to satisfy the constraint $\boldsymbol{\alpha}$ = constant vector. In this same context, we note that (8-4.32) gives

$$\text{(8-4.37)} \qquad d\bar{E} = T\,d\bar{r} + \bar{F}_i\,dx^i + \bar{\beta}^a d\alpha_a.$$

Thermodynamic temperatures $T(\theta, \boldsymbol{\alpha})$ are not a usual occurrence in thermostatics, and hence it is useful to examine the conditions under which

$$\text{(8-4.38)} \qquad \partial_{\alpha_a} T(\theta, \boldsymbol{\alpha}) = 0, \qquad a = 1, \ldots, N.$$

It is easily seen from (8-4.30) that (8-4.38) can hold only if

$$\text{(8-4.39)} \qquad \partial_{\alpha_a} g(\theta, \boldsymbol{\alpha}) = 0, \qquad a = 1, \ldots, N.$$

Further, (8-4.37) gives

$$\text{(8-4.40)} \qquad T = \partial_{\bar{r}}\bar{E}, \qquad \bar{\beta}_a = \partial_{\alpha_a}\bar{E},$$

and hence satisfaction of (8-4.38) would imply

$$\partial_{\bar{r}}\bar{\beta}_a = 0, \qquad a = 1, \ldots, N, \tag{8-4.41}$$

$$\bar{E} = e_1(\mathbf{x}, \bar{r}) + e_2(\mathbf{x}, \boldsymbol{\alpha}). \tag{8-4.42}$$

Accordingly, we may say that there exists a strong thermodynamic temperature $T = T(\theta)$ if and only if (8-4.39) through (8-4.42) hold. In general, these are very stringent conditions that will be satisfied for few real materials. Thus in general, internal degrees of freedom for which W has rank ≥ 4 imply that the thermodynamic temperature is a function of the empirical temperature and the internal state vector $\boldsymbol{\alpha}$.

When actual rather than quasistatic processes are considered, a severe problem is encountered with a precise statement of the second half of the second law since we now have $Q = T\,d\bar{r} + \bar{\beta}^a\,d\alpha_a$. One can proceed by requiring

$$Q^* - \bar{\beta}^a\,d\alpha_a \leq Q - \bar{\beta}^a\,d\alpha_a, \tag{8-4.43}$$

but this begs the question in that it assumes that the $d\boldsymbol{\alpha}$ for quasistatic processes agrees with the $d\boldsymbol{\alpha}$ for actual processes, and this need not be the case. It would appear that the balance of energy equation and $\bar{E} = \bar{\Psi} + T\bar{r}$ should be used to obtain an equation of evolution of $\bar{r}$ and then require that the production of $\bar{r}$ be nonnegative, as is often done in modern irreversible thermodynamics (see Section 8-6).

8-5. HOMOGENEOUS SYSTEMS WITH NONCONSERVATIVE FORCES

We use S to denote a material system in three-dimensional Euclidean space that is in a homogeneous mechanical substate characterized by an n-dimensional mechanical substate $\mathbf{x}$ in the n-dimensional space E_n. As before, let $\mathbf{x}_0$ denote a reference mechanical substate of S and let θ denote the homogeneous empirical temperature of S in the state $(\mathbf{x}, \theta)$ with respect to some fixed but arbitrary scale of empirical temperature.

We suppose that the system S is acted upon by a system of forces that can be represented as functions of the thermodynamic state $(\mathbf{x}, \theta)$. In particular, we assume that $\{F_i(\mathbf{x}, \theta)\}$ is a continuous and differentiable mapping from $D \times [a, b]$ into E_n, where D is a starshaped region with respect to the reference mechanical substate $\mathbf{x}_0$. The signs of the components of the field $\{F_i(x, \theta)\}$ are assumed to be so chosen that the infinitesimal work $F_i(\mathbf{x}, \theta)\,dx^i$ that occurs as a consequence of the mechanical displacement $d\mathbf{x}$ is positive when the forces do work on the system S and is negative when the system S performs work on the agencies that give rise to the forces.

All of the above assumptions are exactly the same as those made in Section 8-2 up to the point where the forces are assumed to be mechanically conservative. We now explicitly assume the contrary; that is, the forces that act on the system S are mechanically nonconservative. By this we mean that the forces $\{F_i(\mathbf{x}, \theta)\}$ are such that

$$d\left(F_i(\mathbf{x}, \theta)\, dx^i\right)\big|_\theta \neq 0 \tag{8-5.1}$$

holds on some nonempty subset of D. Such systems of forces arise in many important physical problems; the best known example is the theory of Cauchy elastic solids in which the matrix of first derivatives of stress with respect to strain is not symmetric. It would therefore seem to be of interest to investigate to what extent a thermodynamics can be constructed for such systems. Clearly, there are problems in such a construction, for (8-5.1) says that the work done in any displacement is dependent upon the path followed by that displacement —work ceases to be a function of only the beginning and endpoints of a displacement. Accordingly, there is significant difficulty associated with the concept of an isothermal work function (Helmholtz free energy). We give two intrinsically different procedures for overcoming this difficulty in the following two parts of this section and leave it to the reader to select that which appears most natural.

A. Thermodynamics Based on the Exact Part of the Work 1-Form

It is well known (Chapter Five) that any 1-form has a unique exact part that is determined in terms of a homotopy integral. In effect, this construction is equivalent to selecting a system of preferred paths and then computing the work function for these preferred paths.

Since D is starshaped with respect to $\mathbf{x}_0$, each of the points

$$\mathbf{x}(\lambda) = \lambda\mathbf{x} + (1 - \lambda)\mathbf{x}_0, \qquad 0 \leq \lambda \leq 1 \tag{8-5.2}$$

belongs to D if $\mathbf{x}$ belongs to D. Thus $\{F_i(\mathbf{x}(\lambda), \theta)\}$ is well defined for all $\lambda \in [0, 1]$, all $\mathbf{x} \in D$, and all $\theta \in [a, b]$. If $\psi_0(\theta)$ is continuous and differentiable on $[a, b]$, then the quantity

$$\Psi(\mathbf{x}, \theta) = \psi_0(\theta) + \int_0^1 (x^i - x_0^i) F_i(\mathbf{x}(\lambda), \theta)\, d\lambda \tag{8-5.3}$$

is well defined on $D \times [a, b]$ and is continuously differentiable in $(\mathbf{x}, \theta)$. Now (8-5.3) shows that $\Psi(\mathbf{x}, \theta)$ is equal, to within the function $\psi_0(\theta)$, to the mechanical work that is done by $\{F_i(\mathbf{x}, \theta)\}$ on the system S in going from the mechanical substate $\mathbf{x}_0$ to the mechanical substate $\mathbf{x}$ by means of the generalized linear process (8-5.2) that takes place at constant empirical temperature θ. Thus an evaluation of $\Psi(\mathbf{x}, \theta)$ by means of linear processes (8-5.2) exactly mimics the procedure given in Section 8-2 for conservative mechanical systems. The difference now is that we obtain different evaluations of the work if we deviate from the set of linear processes given by (8-5.2). Thus an analysis based on (8-5.3) singles out the linear processes (8-5.2) as a preferred set of paths connecting $\mathbf{x}_0$ with any $\mathbf{x}$ in D.

This procedure is not as arbitrary as it might seem on first glance, for a simple calculation shows that

$$F_i(\mathbf{x},\theta) = \partial_i \Psi(\mathbf{x},\theta) + U_i(\mathbf{x},\theta) \tag{8-5.4}$$

where

$$U_i(\mathbf{x},\theta) = \int_0^1 (x^j - x_0^j)\left(\frac{\partial F_i(\mathbf{x}(\lambda),\theta)}{\partial x^j(\lambda)} - \frac{\partial F_j(\mathbf{x}(\lambda),\theta)}{\partial x^i(\lambda)}\right)\lambda\, d\lambda \tag{8-5.5}$$

and hence

$$(x^i - x_0^i)U_i(\mathbf{x},\theta) = 0, \qquad U_i(\mathbf{x}_0,\theta) = 0. \tag{8-5.6}$$

It is then a simple matter to show that the decomposition (8-5.4) is unique for $\{U_i(\mathbf{x},\theta)\}$ satisfying the conditions (8-5.6) and that $\Psi(\mathbf{x},\theta)$ is unique to within the additive function $\psi_0(\theta)$ (see Chapter Five). Thus in particular, we have the unique representation

$$F_i(\mathbf{x},\theta)\, dx^i = d\Psi(\mathbf{x},\theta)|_\theta + U_i(\mathbf{x},\theta)\, dx^i. \tag{8-5.7}$$

If we introduce the 1-forms W and U by

$$W = F_i(\mathbf{x},\theta)\, dx^i, \qquad U = U_i(\mathbf{x},\theta)\, dx^i, \tag{8-5.8}$$

we obtain the unique decomposition

$$W = d\Psi(\mathbf{x},\theta) + r(\mathbf{x},\theta)\, d\theta + U, \tag{8-5.9}$$

where the quantity $r(\mathbf{x},\theta)$ is defined by

$$r(\mathbf{x},\theta) = -\partial_\theta \Psi(\mathbf{x},\theta). \tag{8-5.10}$$

We note in passing that (8-5.9) implies

$$W|_\theta = d\Psi(\mathbf{x},\theta)|_\theta + U, \qquad d(W|_\theta)|_\theta = dU|_\theta, \tag{8-5.11}$$

$$dW = dr \wedge d\theta + dU. \tag{8-5.12}$$

Thus the 1-form U accounts for all of the mechanical nonconservative structure of W, and in particular, we have $dU \not\equiv 0$. It also follows from (8-5.12) that the rank of W is ≥ 4 since

$$dW \wedge dW = 2\,dr \wedge d\theta \wedge dU + dU \wedge dU$$

does not vanish in general.

We now proceed exactly as in Section 8-3 and introduce the internal energy $E(\mathbf{x}, \theta)$ and the 1-form Q of heat addition that are connected by the first law

(8-5.13) $$dE = Q + W.$$

When (8-5.9) is substituted into (8-5.13) and the resulting equation is solved for Q, we obtain

(8-5.14) $$Q = d(E - \Psi) - r\,d\theta - U.$$

This relation implies that

(8-5.15) $$dQ = -dr \wedge d\theta - dU = -dW,$$

and hence Q and W have the same rank ≥ 4. Further, we have

(8-5.16) $$\int_{\partial S_2} Q = \int_{S_2} dQ = -\int_{S_2} dr \wedge d\theta - \int_{S_2} dU = -\int_{S_2} dr \wedge d\theta - \int_{\partial S_2} U,$$

as an immediate consequence of (8-5.15) and Stokes' theorem.

We now come up against the same problem as encountered in Section 8-4 concerning inaccessibility, for Q has rank ≥ 4 in general. In this case, the problem is further compounded, for we no longer have the simple expedient of looking at what happens on the surfaces $\boldsymbol{\alpha}$ = constant vector. In fact, the 1-form U does not admit integral manifolds, in general, as shown by (8-5.6), and hence there is no simple way of characterizing inaccessibility by means of a restriction to a surface in $D \times [a, b]$. This unavoidable problem is the basic drawback of the procedure now under discussion for its resolution requires the following artificial procedure. We partition the 1-form Q given by (8-5.14) by means of

(8-5.17) $$Q = Q_1 + Q_2,$$

where

(8-5.18) $$Q_1 = d(E - \Psi) - r\,d\theta, \qquad Q_2 = -U,$$

and require that the 1-form Q_1 satisfy the condition of inaccessibility:

(8-5.19) $$Q_1 \wedge dQ_1 = 0.$$

This allows us to take over all of the results of Section 8-3 for the 1-form Q_1

since the first of (8-5.18) and (8-5.19) agree exactly with the corresponding equations of Section 8-3. The same assumptions concerning semiadditivity and transformations of temperature scales then give us the following results:

(8-5.20) $$\bar{E}(\mathbf{x}, T) = \bar{\Psi}(\mathbf{x}, T) + \bar{r}(\mathbf{x}, T)T, \qquad T = T(\theta)$$

(8-5.21) $$Q_1 = T\,d\bar{r}, \qquad \bar{r}(\mathbf{x}, T) = -\partial_T \bar{\Psi}(\mathbf{x}, T).$$

Thus we again establish the existence of a thermodynamic temperature T and an entropy function, but these are now based on the 1-form Q_1 rather than Q itself. Putting this back into (8-5.17) gives

(8-5.22) $$Q = T\,d\bar{r} - U;$$

accordingly, the mechanical nonconservative part of W represented by U, acts as an independent heat addition 1-form as well.

It is clear from this procedure that we have based the thermodynamic considerations on work and energy functions that derive from the unique exact part, $d\Psi|_\theta$, of $(F_i(\mathbf{x}, \theta)\,dx^i)|_\theta$ and hence on the preferred linear processes $\{\mathbf{x}(\lambda)\}$. This procedure has the merit that it preserves the intrinsic notion of an isothermal work function (8-5.3) that involves an experimentally determinable integral, but has the decided drawback of rendering the inaccessibility condition in a form that cannot be checked through experiment since it does not refer to inaccessibility on any surface in the domain of definition $D \times [a, b]$.

B. Darboux Decomposition and Testable Inaccessibility

The results of the previous part were obtained under the unique decomposition $F_i = \partial_i \Psi + U_i$, $U_i(\mathbf{x}_0, \theta) = 0$, $(x^i - x_0^i)U_i(\mathbf{x}, \theta) = 0$ that preserved the notion of an experimentally determinable isothermal work function (for preferred paths), but led to an inaccessibility condition with no geometric interpretation nor testable properties. This part is based upon the nonunique representation of a 1-form given by the Darboux theorem. In this instance, we lose the simple interpretation of Ψ as an isothermal work function but gain a geometrically interpretable inaccessibility condition that is also experimentally verifiable.

The Darboux theorem states (see Chapter Four) that any 1-form P of odd class $K = 2\rho + 1$ on a given domain admits the representation

(8-5.23) $$P = d\phi + \sum_{a=1}^{\rho} u^a\,dv_a$$

where the K functions (ϕ, u_a, v_a) are functionally independent at each regular point in the domain. Clearly, this representation is not unique, for $d(uv) =$

$u\,dv + v\,du$, and hence (8-5.23) can also be written as

$$P = d\left(\phi + \sum_{a=1}^{\rho} u^a v_a\right) - \sum_{a=1}^{\rho} v_a\,du^a$$

or any of the many other possible forms obtainable by use of the previously noted differential identity. We accordingly make use of the various ways of writing a 1-form so as to represent the class of $W|_\theta$'s to be considered in this part in the particular form

$$W|_\theta = d\Psi(\mathbf{x},\theta)|_\theta + \sum_{a=1}^{\rho} u^a(\mathbf{x},\theta)\,dv_a(\mathbf{x}). \tag{8-5.24}$$

Careful note should be taken of the fact that the ρ functions $\{v_a\}$ do not depend on the empirical temperature so that $dv_a|_\theta = dv_a$ and that the functions (Ψ, u^a, v_a) are not uniquely determined. However, we also note that

$$d(W|_\theta)|_\theta = \sum_{a=1}^{\rho} du^a|_\theta \wedge dv_a,$$

and hence the terms $u^a\,dv_a$ describe the aspects of the forces that are not mechanically conservative.

Since the functions $\{v_a(\mathbf{x})\}$ are functionally independent at regular points of D, we can complete them to a function basis $(\omega_\gamma(\mathbf{x}), v_a(\mathbf{x}))$ at regular points of D, where γ ranges from 1 through $n - \rho > 0$. This allows us to effect a transformation of coordinate patches at regular points of D so that $x^i = \mu^i(\omega_\gamma, v_a)$ defines a regular coordinate map at regular points of D. Under these circumstances (8-5.24) yields

$$W = d\Psi(\omega_\gamma, v_b, \theta) + r(\omega_\gamma, v_b, \theta)\,d\theta + \sum_{a=1}^{\rho} u^a(\omega_\gamma, v_b, \theta)\,dv_a, \tag{8-5.25}$$

where the function $r(\omega_\gamma, v_b, \theta)$ is defined by

$$r(\omega_\gamma, v_b, \theta) = -\partial_\theta \Psi(\omega_\gamma, v_b, \theta). \tag{8-5.26}$$

This yields

$$dW = dr \wedge d\theta + \sum_{a=1}^{\rho} du^a \wedge dv_a, \tag{8-5.27}$$

and we see that we are in exactly the case studied in Section 8-4 if we identify the variables $\{v_a\}$ with the internal variables and the variables $\{\omega_\gamma\}$ with the external variables. There is one difference, however, and that is that the coefficients $\{u^a(\omega_\gamma, v_b, \theta)\}$ that appear in the forms $u^a\, dv_a$ are no longer partial derivatives of Ψ with respect to the internal variables, as was the case in Section 8-4.

Introduction of the internal energy $E(\omega_\gamma, v_b, \theta)$ and the 1-form of heat addition Q by means of the first law $dE = Q + W$ gives us

$$Q = d(E - \Psi) - r\, d\theta - \sum_{a=1}^{\rho} u^a\, dv_a, \tag{8-5.28}$$

$$dQ = -dW, \tag{8-5.29}$$

$$Q|_{v_a} = d(E - \Psi)|_{v_a} - r\, d\theta, \tag{8-5.30}$$

$$d(Q|_{v_a})|_{v_a} = -dr|_{v_a} \wedge d\theta. \tag{8-5.31}$$

We can thus follow the procedure established in Section 8-4 and require inaccessibility on the surface $v_a = c_a$ (= constant) for all constants c_a. The same additional assumptions concerning semiadditivity and transformations of temperature scales as made in Section 8-4 accordingly give us the following results:

$$\bar{E}(\omega_\gamma, v_b, T) = \bar{\Psi} + \bar{r}(\omega_\gamma, v_b, T)T, \qquad T = T(\theta, v_b) \tag{8-5.32}$$

$$Q|_{v_a} = T\, d\bar{r}, \qquad \bar{r}(\omega_\gamma, v_b, T) = -\partial_T \bar{\Psi}(\omega_\gamma, v_b, T), \tag{8-5.33}$$

$$Q = T\, d\bar{r} + \sum_{a=1}^{\rho} \left(\bar{r}\partial_{v_a} T - \bar{u}^a\right) dv_a. \tag{8-5.34}$$

Again, we have the additional ρ modes of heat addition given by $(\bar{r}\partial_{v_a}T - \bar{u}^a)\, dv_a$, a $= 1, \ldots, \rho$, that now depend on the thermodynamic temperature *and* the entropy. This comes about because $\bar{u}^a \neq \partial_{v_a}\Psi$, in constrast to the systems considered in Section 8-4.

8-6. NONEQUILIBRIUM THERMODYNAMICS

Our considerations up to this point have been restricted to systems S with homogeneous states that undergo quasistatic (quasiequilibrium) processes. We now relax both of these constraints. To this end let $(z^1, z^2, z^3) = (z^m)$ be a Cartesian coordinate cover of three-dimensional space in which the system S finds itself at any time t. The external variables $\{x^i\}$ and the internal variables $\{\alpha_a\}$ of actual processes of S become functions of position and time ($x^i = x^i(z^m, t)$, $\alpha_a = \alpha_a(z^m, t)$). If we look at any point $P:(z^m)$ in E_3 at a given time t, we assume that we can find an equilibrium configuration of a *companion equilibrium system* S_{eq} whose mechanical state $(\mathbf{x}, \boldsymbol{\alpha})_{\text{eq}}$ agrees with $(\mathbf{x}(z^m, t), \boldsymbol{\alpha}(z^m, t))$. Under these conditions (8-4.37) holds,

$$dE = T\,d\bar{r} + \bar{F}_i\,dx^i + \bar{\beta}^a\,d\alpha_a, \tag{8-6.1}$$

where $\{\bar{F}_i(\mathbf{x}, \boldsymbol{\alpha}, T)\}$ are the forces in the accompanying equilibrium system and $\{\bar{\beta}^a(\mathbf{x}, \boldsymbol{\alpha}, T)\}$ are the internal intensive variables in the accompanying equilibrium process. We now solve (8-6.1) for the 1-form $T\,d\bar{r}$ to obtain

$$T\,d\bar{r} = d\bar{E} - \bar{F}_i\,dx^i - \bar{\beta}^a\,d\alpha_a. \tag{8-6.2}$$

Although (8-6.2) is just a rearrangement of (8-6.1) it has certain important consequences. First, it shows that we may consider the entropy $\bar{r}$ to be a function of the list of variables $(\bar{E}, x^i, \alpha_a)$; we have the fundamental equation of state

$$\bar{r} = \bar{r}(\bar{E}, x^i, \alpha_a).$$

In this case, quantities $(T, \bar{F}_i, \bar{\beta}^a)$ necessarily become functions of the same variables $(\bar{E}, x^i, \alpha_a)$. It then follows directly from a resolution of (8-6.2) on the basis $(d\bar{E}, dx^i, d\alpha_a)$ that

$$\frac{1}{T} = \frac{\partial \bar{r}}{\partial \bar{E}}, \qquad \bar{F}_i = -\frac{1}{T}\frac{\partial \bar{r}}{\partial x^i}, \qquad \bar{\beta}_a = \frac{1}{T}\frac{\partial \bar{r}}{\partial \alpha_a}. \tag{8-6.3}$$

The quantities $(T, \bar{F}_i, \bar{\beta}^a)$ may thus be considered as *known* functions of $(\bar{E}, x^i, \alpha_a)$.

The actual nonequilibrium processes that occur in real bodies necessarily differ from the accompanying equilibrium processes. In particular, the actual forces $\{F_i\}$ will be different from the forces $\{\bar{F}_i\}$ that are computed on the basis of an equilibrium process and the extensive internal quantities $\{\beta^a\}$ will also be different from the $\{\bar{\beta}^a\}$. Further, since the thermodynamic temperature can vary from one point to another in the system S, there will be accompanying exchanges of energy between different points. This latter effect is described by the components of energy flux $\{q^m(z^n, t)\}$ per unit time. Under these circumstances, the first law of thermodynamics (balance of energy) becomes

$$\frac{d\bar{E}}{dt} = -\frac{\partial q^m}{\partial z^m} + F_i\frac{dx^i}{dt} + \beta^a\frac{d\alpha_a}{dt}. \tag{8-6.4}$$

We now eliminate $d\bar{E}/dt$ between (8-6.4) and what obtains by restriction of

(8-6.1) to the actual dynamical process:

$$\frac{d\bar{E}}{dt} = T\frac{d\bar{r}}{dt} + \bar{F}_i\frac{dx^i}{dt} + \bar{\beta}^a\frac{d\alpha_a}{dt}.$$

The result is

$$T\frac{d\bar{r}}{dt} = -\frac{\partial q^m}{\partial z^m} + (F_i - \bar{F}_i)\frac{dx^i}{dt} + (\beta^a - \bar{\beta}^a)\frac{d\alpha_a}{dt}. \tag{8-6.5}$$

It is then a simple matter to divide both sides by T to obtain a law of balance for the entropy $\bar{r}$:

$$\frac{d\bar{r}}{dt} + \frac{\partial(q^m/T)}{\partial z^m} = \dot{\Theta}. \tag{8-6.6}$$

Here

$$\dot{\Theta} = q^m\frac{\partial(1/T)}{\partial z^m} + \frac{1}{T}(F_i - \bar{F}_i)\frac{dx^i}{dt} + \frac{1}{T}(\beta^a - \bar{\beta}^a)\frac{d\alpha_a}{dt} \tag{8-6.7}$$

is the *entropy production*. A process is said to be *thermodynamically admissible* if and only if the quantities $(q^m, F_i - \bar{F}_i, \beta^a - \bar{\beta}^a)$ are such as to satisfy the *dissipation inequality* (second law) $\dot{\Theta} \geq 0$; that is,

$$q^m\frac{\partial(1/T)}{\partial z^m} + \frac{1}{T}(F_i - \bar{F}_i)\frac{dx^i}{dt} + \frac{1}{T}(\beta^a - \bar{\beta}^a)\frac{d\alpha_a}{dt} \geq 0. \tag{8-6.8}$$

(Remember that $(\bar{F}_i, \bar{\beta}^a, T)$ are given by (8-6.3).)

8-7. REFORMULATION AS A WELL POSED MATHEMATICAL PROBLEM

The foregoing discussion shows that the state variables of the system are now $(\mathbf{x}, \boldsymbol{\alpha}, T)$, so we certainly must provide for the possibility that (q^m, F_i, β^a) depend on these variables. This is easily accomplished by introducing a $(1 + n + N)$-dimensional state space E_{1+n+N} with local coordinates $(x^i, \alpha_a, T) = (\boldsymbol{\omega})$. It is also clear from (8-6.8) that (q^m, F_i, β^a) must depend on the quantities $((\partial/\partial z^m)(1/T),\ dx^i/dt,\ d\alpha_a/dt)$. These latter quantities are accordingly used to define a coordinate cover of a space E_{3+n+N} of *thermodynamic forces*

$$(X^A) = \left(\frac{\partial}{\partial z^m}\left(\frac{1}{T}\right),\quad \frac{dx^i}{dt},\quad \frac{d\alpha_a}{dt}\right). \tag{8-7.1}$$

The conjugate *thermodynamic fluxes* are defined by

(8-7.2) $$(J_A) = \left(q^m, \quad \frac{F_i - \bar{F}_i}{T}, \quad \frac{\beta^a - \bar{\beta}^a}{T}\right).$$

Any possible system of constitutive relations for $(q^m, (F_i - \bar{F}_i)/T, (\beta^a - \bar{\beta}^a)/T)$, and hence for (q^m, F_i, β^a) can thus be written

(8-7.3) $$J_A = j_A(X^B, \omega).$$

It is then a simple matter to see that the dissipation inequality (8-6.8) assumes the following particularly simple form:

(8-7.4) $$X^A j_A(X^B; \omega) \geq 0.$$

Our problem is therefore that of determining all functions $\{j_A(X^B; \omega)\}$ *that satisfy the inequality* (8-7.4). We also note in passing that the entropy production has the evaluation

(8-7.5) $$\dot{\Theta}(X^B; \omega) = X^A j_A(X^B; \omega).$$

One immediate consequence of the requirement (8-7.4) is that *the thermodynamic fluxes vanish with the thermodynamic forces for all* ω,

(8-7.6) $$j_A(\mathbf{0}; \omega) = 0.$$

To see this, we note that (8-7.4) and (8-7.5) combine to give $\dot{\Theta}(X^B; \omega) = X^A j_A(X^B; \omega) \geq 0$, and hence $\dot{\Theta}(\mathbf{0}; \omega) = 0 = \min_{\mathbf{X}}(\dot{\Theta}(X^B; \omega))$. Thus we conclude that $(\partial\dot{\Theta}/\partial X^A)|_{\mathbf{X}=\mathbf{0}} = j_A(\mathbf{0}; \omega) = 0$.

Starting with the functions $j_A(X^B; \omega)$, we construct the 1-form $D \in \Lambda^1(E_{3+n+N})$ by

(8-7.7) $$D = j_A(X^B; \omega)\, dX^A,$$

where the entries of ω are considered frozen parameters that do not participate in the exterior calculus on E_{3+n+N}. Let H be the linear homotopy operator on E_{3+n+N} with center $(X_0^A) = \mathbf{0}$. The results of Chapter Five then show that there is then the unique decomposition

(8-7.8) $$D = d\Phi|_\omega + U$$

with

(8-7.9) $$\Phi(X^B; \omega) = H(D)(X^B; \omega)$$

a 0-form. The antiexact part of D is given by

$$U(X^B;\omega) = H(dD|_\omega)(X^B;\omega) = U_A(X^B;\omega)\,dX^A \tag{8-7.10}$$

with

$$X^A U_A(X^B;\omega) = 0, \qquad U_A(\mathbf{0};\omega) = 0 \tag{8-7.11}$$

since the center has coordinates $(X_0^A) = (\mathbf{0})$.

We now put (8-7.8) back into (8-7.5) to obtain

$$\dot{\Theta}(X^B;\omega) = X^A \partial_A \Phi(X^B;\omega) \tag{8-7.12}$$

since $X^A U_A(X^B;\omega) = 0$. Notice that the only choice for the center is $X_0^A = 0$, for otherwise we would have $(X^A - X_0^A)U_A(X^B;\omega) = 0$. We may thus consider the scalar function $\Phi(X^B;\omega)$ to be a *dissipation potential*, while the antiexact part U of D is the *nondissipative part* since it makes an identically zero contribution to the entropy production.

The above consideration show that the dissipation inequality $X^A j_A(X^B;\omega) \geq 0$ reduces to the scalar differential inequality

$$X^A \partial_A \Phi(X^B;\omega) \geq 0 \tag{8-7.13}$$

under the unique decomposition

$$j_A(X^B;\omega) = \partial_A \Phi(X^B;\omega) + U_A(X^B;\omega). \tag{8-7.14}$$

The general solution of this inequality is given by

$$\Phi(X^B;\omega) = \int_0^1 p(\lambda X^B;\omega)\frac{d\lambda}{\lambda}, \tag{8-7.15}$$

where $p(X^B;\omega)$ is an arbitrary nonnegative valued function of its indicated arguments such that $p(\lambda X^B;\omega) = o(\lambda)$.

The general solution of the dissipation inequality $X^A j_A(X^B;\omega) \geq 0$ *is given by*

$$j_A(X^B;\omega) = \partial_A \Phi(X^B;\omega) + U_A(X^B;\omega), \tag{8-7.16}$$

where $\Phi(X^B;\omega) = \int_0^1 \frac{1}{\lambda} p(\lambda X^B;\omega)\,d\lambda$, *for some* $p(\lambda X^B;\omega) = o(\lambda)$ *such that* $p(X^B;\omega) \geq 0$, *is the dissipation potential, and* $\{U_A(X^B;\omega)\}$ *are such that*

$$X^A U_A(X^B;\omega) = 0, \qquad U_A(\mathbf{0};\omega) = 0. \tag{8-7.17}$$

If we have $U_A(X^B;\omega)=0$, (8-7.16) translates into

$$q^m = \frac{\partial\Phi}{\partial\left(\partial_m\left(\frac{1}{T}\right)\right)}, \qquad \frac{1}{T}(F_i - \bar{F}_i) = \frac{\partial\Phi}{\partial\left(\frac{dx^i}{dt}\right)},$$

$$\frac{1}{T}(\beta^a - \bar{\beta}^a) = \frac{\partial\Phi}{\partial\left(\frac{d\alpha_a}{dt}\right)}$$

as follows directly from (8-7.1), and hence (q^m, F_i, β^a) are then uniquely determined once $\Phi(X^B;\omega)$ is given. However,

$$\dot{\Theta}(X^B;\omega) = X^A\partial_A\Phi(X^B;\omega)$$

and hence

$$\dot{\Theta}(\lambda X^B;\omega) = \lambda X^A\frac{\partial\Phi(\lambda X^B;\omega)}{\partial(\lambda X^A)} = p(\lambda X^B;\omega);$$

that is

(8-7.18) $$\Phi(X^B;\omega) = \int_0^1 \dot{\Theta}(\lambda X^B;\omega)\frac{d\lambda}{\lambda}.$$

The unique decomposition of a 1-form into exact and antiexact parts, for given center, has led to the proof of the existence of a dissipation potential $\Phi(X^B;\omega)$ for any nonequilibrium process and to the existence of a nondissipative part $\{U_A(X^B;\omega)\}$ of the thermodynamic fluxes. In particular, we have

(8-7.19) $$D = j_A(X^B;\omega)\,dX^A = d\Phi|_\omega + U,$$

and hence

(8-7.20) $$D - U = d\Phi|_\omega$$

is closed on $\omega =$ constant:

(8-7.21) $$d(D-U)|_\omega = 0.$$

These last relations are simply Onsager's reciprocity relations for nonlinear processes

(8-7.22) $$\frac{\partial}{\partial X^A}(j_B - U_B) = \frac{\partial}{\partial X^B}(j_A - U_A).$$

CHAPTER NINE

ELECTRODYNAMICS WITH ELECTRIC AND MAGNETIC CHARGES

9-1. THE FOUR-DIMENSIONAL BASE SPACE

The arena or base space for this discussion is a four-dimensional number space E_4 that is the Cartesian product of ordinary three-dimensional Euclidean space and the real line. It is sufficient for our purposes to work with a fixed, global coordinate cover (x^i) of E_4 with

$$x^1 = x, \qquad x^2 = y, \qquad x^3 = z, \qquad x^4 = t,$$

since the results for any other regular coordinate cover or covers can be obtained directly by the known mapping properties of exterior forms and vector fields. The standard orientation of E_4 is that for which the volume element μ has the representation

$$\mu = dx^1 \wedge dx^2 \wedge dx^3 \wedge dx^4 = dx \wedge dy \wedge dz \wedge dt. \tag{9-1.1}$$

Functions of all four x's will be denoted by $f(x)$, thus if $V = V(x)$ is a vector field on E_4, we write

$$V(x) = v^i(x)\,\partial_i.$$

The standard notation of the exterior calculus introduced in Part One will be used throughout.

A significant portion of the discussion will use the "top down" or conjugate bases for $\Lambda^3(E_4)$ and $\Lambda^2(E_4)$ that were introduced in Section 3-5. An explicit account of these bases is given as an aid to the reader. Since μ is the natural basis for $\Lambda^4(E_4)$, the conjugate base elements $\{\mu_i\}$ of $\Lambda^3(E_4)$ are defined by

(9-1.2) $$\mu_i = \partial_i \lrcorner \mu$$

so that

(9-1.3) $$\mu_1 = dx^2 \wedge dx^3 \wedge dx^4, \qquad \mu_2 = -dx^1 \wedge dx^3 \wedge dx^4,$$
$$\mu_3 = dx^1 \wedge dx^2 \wedge dx^4, \qquad \mu_4 = -dx^1 \wedge dx^2 \wedge dx^3.$$

The particular ordering of the differentials that occurs in (9-1.3) induces the transformation law

(9-1.4) $$\mu_i'(y) = \Delta(y,x)\frac{\partial x^j}{\partial y^i}\mu_j(x)$$

for a coordinate map $y^i = \phi^i(x^k)$ with $\Delta(y,x) = \det(\partial y^i/\partial x^j)$; that is, (9-1.4) follows directly from (9-1.2) and the known transformation properties of μ and $\{\partial_i\}$. The base elements $\{\mu_i\}$ exhibit the following properties

(9-1.5) $$d\mu_i = 0, \qquad dx^i \wedge \mu_j = \delta^i_j \mu.$$

Since $\{\mu_i\}$ is a basis for $\Lambda^3(E_4)$, any 3-form ω can be written uniquely as

(9-1.6) $$\omega = W^i \mu_i,$$

in which case (9-1.4) and $\omega(x) = W^i(x)\mu_i(x) = W^{j\prime}(y)\mu_j'(y) = \omega(y)$ induces the transformation law

(9-1.7) $$W^{i\prime}(y) = \Delta(y,x)^{-1}\frac{\partial y^i}{\partial x^j}W^j(x).$$

Similarly, if $\omega = W^i\mu_i$ is a *pseudo* 3-form (i.e., a 3-form with the transformation law $\omega(x) = \operatorname{sign}(\Delta(y,x))\omega(y)$ for a coordinate map $\Phi: (x^i) \to (y^j)$, we obtain

(9-1.8) $$W^{i\prime}(y) = \Delta(y,x)^{-1}\operatorname{sign}(\Delta(y,x))\frac{\partial y^i}{\partial x^j}W^j(x).$$

If $V = v^i\partial_i$ is any vector (*pseudovector*) field on E_4 (i.e., an element of $T(E_4)$), (9-1.2) shows that

(9-1.9) $$\mathscr{V} = V \lrcorner \mu = v^i \mu_i$$

is a corresponding 3-form (pseudo 3-form). The analysis of electromagnetic theory will demand use of the new quantities referred to as pseudovector fields and pseudo k-forms because of the characteristic behavior of magnetic fields under changes of orientation.

If ω is an arbitrary 3-form or pseudo 3-form, (9-1.5) shows that

(9-1.10) $$d\omega = d(W^i \mu_i) = (\partial_i W^i)\mu.$$

Thus if $V = v^i \partial_i$ is a vector or pseudovector field,

(9-1.11) $$d(V \lrcorner \mu) = d(v^i \mu_i) = (\partial_i v^i)\mu.$$

This shows that the vector (pseudovector) $V = v^i \partial_i$ is conserved if and only if the corresponding 3-form (pseudo 3-form) $V \lrcorner \mu$ is closed; that is, $d(V \lrcorner \mu) = 0$.

Starting with the basis μ_i, we construct a collection of 2-forms μ_{ji} by

(9-1.12) $$\mu_{ji} = \partial_j \lrcorner \mu_i.$$

The set $\{\mu_{ij} | i < j\}$ forms a basis for $\Lambda^2(E_4)$ and exhibits the following properties:

(9-1.13) $$d\mu_{ij} = 0, \qquad \mu_{ij} = -\mu_{ji},$$
$$dx^k \wedge \mu_{ij} = \delta_i^k \mu_j - \delta_j^k \mu_i.$$

Thus if $\omega \in \Lambda^2(E_4)$, we may write

(9-1.14) $$\omega = \tfrac{1}{2} W^{ij} \mu_{ij}, \qquad W^{ij} = -W^{ji}$$

and hence

(9-1.15) $$d\omega = (\partial_i W^{ij})\mu_j.$$

The explicit evaluation is given by

$$\mu_{21} = dx^3 \wedge dx^4, \qquad \mu_{31} = -dx^2 \wedge dx^4, \qquad \mu_{41} = dx^2 \wedge dx^3,$$
$$\mu_{12} = -dx^3 \wedge dx^4, \qquad \mu_{32} = dx^1 \wedge dx^4, \qquad \mu_{42} = -dx^1 \wedge dx^3,$$
$$\mu_{13} = dx^2 \wedge dx^4, \qquad \mu_{23} = -dx^1 \wedge dx^4, \qquad \mu_{43} = dx^1 \wedge dx^2,$$
$$\mu_{14} = -dx^2 \wedge dx^3, \qquad \mu_{24} = dx^1 \wedge dx^3, \qquad \mu_{34} = -dx^1 \wedge dx^2.$$

If ω is any 2-form on E_4, we may alternatively write

(9-1.16) $$\omega = \tfrac{1}{2} W^{ij} \mu_{ij} = \tfrac{1}{2} W_{ij}\, dx^i \wedge dx^j,$$
$$W^{ij} = -W^{ji}, \qquad W_{ij} = -W_{ji}.$$

The relations between W^{ij} and W_{ji} are then easily seen to be

(9-1.17)
$$W_{12} = -W^{34}, \qquad W_{13} = W^{24}, \qquad W_{14} = -W^{23},$$
$$W_{23} = -W^{14}, \qquad W_{24} = W^{13}, \qquad W_{34} = -W^{12}.$$

Thus if we consider the quantities W^{ij} and W_{ij} as entries of antisymmetric 4-by-4 matrices $((W^{ij}))$ and $((W_{ij}))$, respectively, (9-1.16) defines the mappings Γ and Γ^{-1} by

(9-1.18)
$$\Gamma((W^{\alpha\beta})) = \left(\begin{pmatrix} 0 & -W^{34} & W^{24} & -W^{23} \\ & 0 & -W^{14} & W^{13} \\ & & 0 & -W^{12} \\ & & & 0 \end{pmatrix}\right) = ((W_{\alpha\beta})) \equiv \mathbf{W}_*,$$

$$\Gamma^{-1}((W_{\alpha\beta})) = \left(\begin{pmatrix} 0 & -W_{34} & W_{24} & -W_{23} \\ & 0 & -W_{14} & W_{13} \\ & & 0 & -W_{12} \\ & & & 0 \end{pmatrix}\right) = ((W^{\alpha\beta})) \equiv \mathbf{W}^*;$$

that is $\Gamma\mathbf{W}^* = \mathbf{W}_*$, $\Gamma^{-1}\mathbf{W}_* = \mathbf{W}^*$.

The matrix representation of the components of a 2-form ω, by $\mathbf{W}^*$ for the antisymmetric matrix of coefficients relative to the basis μ_{ij}, and by $\mathbf{W}_*$ for the antisymmetric matrix of coefficients relative to the basis $dx^i \wedge dx^j$, is useful in another respect. If α and β are two 2-forms, we have

(9-1.19)
$$\begin{aligned} \alpha \wedge \beta &= \left(\tfrac{1}{2}A_{ij}\,dx^i \wedge dx^j\right) \wedge \left(\tfrac{1}{2}B^{ke}\mu_{ke}\right) \\ &= \tfrac{1}{4}A_{ij}B^{ke}\,dx^i \wedge dx^j \wedge \mu_{ke} \\ &= \tfrac{1}{4}A_{ij}B^{ke}\left(\delta^j_k\delta^i_e - \delta^j_e\delta^i_k\right)\mu \\ &= \tfrac{1}{4}\left(A_{ek} - A_{ke}\right)B^{ke}\mu = \tfrac{1}{2}A_{ek}B^{ke}\mu \\ &= \tfrac{1}{2}tr(\mathbf{A}_*\mathbf{B}^*)\mu. \end{aligned}$$

This result is instrumental in computing invariants formed from 2-forms, as we shall see in later sections; in fact, (9-1.19) simplifies the construction of the Lagrangian densities for the variational principles underlying electrodynamics and many other disciplines.

9-2. ELECTRODYNAMICS WITH FREE MAGNETIC CHARGE AND CURRENT

Maxwell's equations with macroscopic distributions of free electric and magnetic charges and currents are the starting point for this discussion. We include free magnetic charges and currents for two reasons. First, certain structural results of the theory can be missed without their inclusion. Second, analogs of magnetic charges and currents arise naturally in gauge theories that will be treated in the next chapter. The reader should bear with us in this, for all magnetic charges and currents can be set to zero to ascertain results that correspond to classic electromagnetic field theory.

Let q denote the density of free electric charge and let $\vec{J}$ denote the three-dimensional vector field of free electric current. The known transformation properties of Maxwell's equations under parity transformations $P(\vec{x} \mapsto -\vec{x})$ and time reversal $T(t \mapsto -t)$ demand that magnetic monopoles be represented in terms of pseudodensities and that three-dimensional magnetic current be represented by three-dimensional pseudovector fields of free magnetic current. This is because magnetic quantities reverse signs under changes of orientation while electrical quantities do not. Thus let g denote the pseudodensity of free magnetic charge and let $\vec{G}$ denote the three-dimensional pseudovector field of free magnetic current. If we use the standard coordinate cover $(x, y, z, t) = (x^i)$ of E_4 and the standard meanings for the three-dimensional electromagnetic field "vectors," $\vec{E}$ = electric field intensity, $\vec{H}$ = magnetic field intensity, $\vec{B}$ = magnetic induction, $\vec{D}$ = electric displacement in the M.K.S. system, then Maxwell's equations read

$$(9\text{-}2.1) \qquad \vec{\nabla} \times \vec{H} - \partial_t \vec{D} = \vec{J}, \qquad \vec{\nabla} \cdot \vec{D} = q,$$

$$(9\text{-}2.2) \qquad -\vec{\nabla} \times \vec{E} - \partial_t \vec{B} = \vec{G}, \qquad \vec{\nabla} \cdot \vec{B} = g.$$

The minus signs on the left-hand side of the first of (9.2.2) are included so that we secure conservation of free magnetic charge,

$$(9\text{-}2.3) \qquad \vec{\nabla} \cdot \vec{G} + \partial_t g = 0,$$

as well as conservation of free electric charge,

$$(9\text{-}2.4) \qquad \vec{\nabla} \cdot \vec{J} + \partial_t q = 0.$$

The classic treatment, as given above, obtains the conservation of free electric and magnetic charge from the field equations. For our purposes, it is easier, and perhaps preferable in a conceptual sense, to obtain the field equations from the statements of conservation of free electric and magnetic charge. To this end, we define the vector field $\mathscr{J}$ of free electric current on E_4 by

$$(9\text{-}2.5) \qquad \mathscr{J} = J^1\partial_1 + J^2\partial_2 + J^3\partial_3 + q\partial_4 = J^i\partial_i.$$

Similarly, define the pseudovector field $\mathscr{G}$ of free magnetic current on E_4 by

(9-2.6) $$\mathscr{G} = G^1\partial_1 + G^2\partial_2 + G^3\partial_3 + g\partial_4 = G^i\partial_i.$$

If we use (9-1.11), we see that

$$d(\mathscr{J} \lrcorner \mu) = (\partial_i J^i)\mu = (\vec{\nabla} \cdot \vec{J} + \partial_t q)\mu.$$

Accordingly, we conclude that *free electric charge and free magnetic charge are conserved if and only if the* 3-*form* $\mathscr{J} \lrcorner \mu$ *and the pseudo* 3-*form* $\mathscr{G} \lrcorner \mu$ *are closed.*

Since E_4 is globally starshaped with respect to any of its points, the Poincaré lemma holds on E_4: any closed form (pseudoform) is an exact form (pseudoform). Thus $d(\mathscr{J} \lrcorner \mu) = 0$, $d(\mathscr{G} \lrcorner \mu) = 0$ imply that a 2-form $\mathscr{H}$ and a pseudo 2-form $\mathscr{F}$ exist on E_4 such that

(9-2.7) $$\mathscr{J} \lrcorner \mu = d\mathscr{H}, \qquad \mathscr{G} \lrcorner \mu = d\mathscr{F}.$$

These deceptively simple relations between forms and pseudoforms are actually Maxwell's equations. In order to see this, we proceed as follows. We first use (9-1.9) to obtain

(9-2.8) $$\mathscr{J} \lrcorner \mu = J^i\mu_i, \qquad \mathscr{G} \lrcorner \mu = G^i\mu_i.$$

Similarly, we use the basis $\{\mu_{ij}\}$ to represent $\mathscr{H}$ and $\mathscr{F}$ by

(9-2.9) $$\mathscr{H} = \tfrac{1}{2}H^{ij}\mu_{ij}, \qquad \mathscr{F} = \tfrac{1}{2}F^{ij}\mu_{ij}.$$

Equations (9-2.7) through (9-2.9) then combine with (9-1.15) to yield the relations

(9-2.10) $$J^i = \partial_j H^{ji}, \qquad G^i = \partial_j F^{ji}.$$

It is now simply a matter of comparing the first of (9-2.10) with (9-2.1) and the second of (9-2.10) with (9-2.2) in order to replicate Maxwell's equations by the identifications

(9-2.11) $$((H^{ij})) = \left(\left(\begin{matrix} 0 & -H_z & H_y & D_x \\ & 0 & -H_x & D_y \\ & & 0 & D_z \\ & & & 0 \end{matrix}\right)\right) = \mathbf{H}^*,$$

(9-2.12) $$((F^{ij})) = \left(\left(\begin{matrix} 0 & E_z & -E_y & B_x \\ & 0 & E_x & B_y \\ & & 0 & B_z \\ & & & 0 \end{matrix}\right)\right) = \mathbf{F}^*.$$

We also note for later use that the mapping Γ given by (9-1.18) yields

$$(9\text{-}2.13)\qquad ((H_{ij})) = \left(\left(\begin{matrix} 0 & -D_z & D_y & H_x \\ & 0 & -D_x & H_y \\ & & 0 & H_z \\ & & & 0 \end{matrix}\right)\right) = \mathbf{H}_*,$$

$$(9\text{-}2.14)\qquad ((F_{ij})) = \left(\left(\begin{matrix} 0 & -B_z & B_y & -E_x \\ & 0 & -B_x & -E_y \\ & & 0 & -E_z \\ & & & 0 \end{matrix}\right)\right) = \mathbf{F}_*$$

with $\mathscr{H} = \frac{1}{2}H_{ij}\,dx^i \wedge dx^j$, $\mathscr{F} = \frac{1}{2}F_{ij}\,dx^i \wedge dx^j$.

A combination of the above results gives the following fully symmetric formulation of electrodynamics.

Maxwell's equations for the electromagnetic field can be given in an E_4 by

$$(9\text{-}2.15)\qquad \mathscr{J} \lrcorner \mu = d\mathscr{H}, \qquad \mathscr{G} \lrcorner \mu = d\mathscr{F},$$

and these equations are the direct consequences of the conservation of free electric charge,

$$(9\text{-}2.16)\qquad d(\mathscr{J} \lrcorner \mu) = 0$$

and of free magnetic charge

$$(9\text{-}2.17)\qquad d(\mathscr{G} \lrcorner \mu) = 0.$$

In the absence of free magnetic charges and currents, Maxwell's equations reduce to

$$(9\text{-}2.18)\qquad \mathscr{J} \lrcorner \mu = d\mathscr{H}, \qquad 0 = d\mathscr{F}.$$

9-3. GENERAL SOLUTIONS OF MAXWELL'S EQUATIONS

An abuse of language has grown up with respect to the meaning of a solution or a general solution of Maxwell's equations. What is commonly meant is a solution of Maxwell's equations (9-2.1) and (9-2.2) for given $\vec{J}$, q, $\vec{G}$, and g and *given constitutive relations* whereby the field vectors $\vec{B}$ and $\vec{D}$ are related to the field vectors $\vec{E}$ and $\vec{H}$. Now, mathematically, the system (9-2.1) and (9-2.2) constitutes a well defined system of first-order partial differential equations for the determination of $\vec{E}$, $\vec{H}$, $\vec{B}$, and $\vec{D}$ in terms of $\vec{J}$, q, $\vec{G}$, and g.

Viewed from this perspective, *we define a solution of Maxwell's equations to be any collection of field vectors* $\vec{E}$, $\vec{H}$, $\vec{B}$, *and* $\vec{D}$ *that satisfy* (9-2.1) *and* (9-2.2) *for given* $\vec{J}$, q, $\vec{G}$, *and* g; that is, a solution of Maxwell's equations without constitutive relations. At first glance, the prejudice of common exposure might lead the reader to believe that such considerations are utter nonsense, for whoever solves Maxwell's equations without some kind of constitutive relations (either relations or some equivalent statement)? This is indeed a prejudice, for general solutions of Maxwell's equations can be obtained quite easily, as we now proceed to show.

We saw in the last section that Maxwell's equations are given by

$$\mathscr{J} \lrcorner \mu = d\mathscr{H}, \qquad \mathscr{G} \lrcorner \mu = d\mathscr{F} \tag{9-3.1}$$

under the identifications (9-2.5), (9-2.6), (9-2.11), and (9-2.12). Since $\mathscr{J}$ and $\mathscr{G}$ are assumed given, we view the system (9-3.1) as a system of exterior differential equations for the determination of the 2-form $\mathscr{H}$ and the pseudo 2-form $\mathscr{F}$. Such systems of exterior differential equations are exactly the kind that can be solved readily through use of a linear homotopy operator.

The space E_4 is starshaped with respect to any of its points as center. It is therefore computationally advantageous to take the center of E_4 to be the origin of the coordinate cover (x^i); that is, we take $x_0^i = 0$. Let H be the linear homotopy operator with center $x_0^i = 0$ that was constructed in Chapter Five. The forms $\mathscr{H}$ and $\mathscr{F}$ admit the unique representation

$$\mathscr{H} = dH(\mathscr{H}) + H(d\mathscr{H}), \qquad \mathscr{F} = dH(\mathscr{F}) + H(d\mathscr{F}). \tag{9-3.2}$$

Thus if we set

$$\mathscr{K} = -d\alpha - H(\mathscr{H}), \qquad \mathscr{A} = -d\beta - H(\mathscr{F}) \tag{9-3.3}$$

and use (9-3.1) to evaluate $d\mathscr{H}$ and $d\mathscr{F}$, (9-3.2) gives

$$\mathscr{H} = -d\mathscr{K} + H(\mathscr{J} \lrcorner \mu), \qquad \mathscr{F} = -d\mathscr{A} + H(\mathscr{G} \lrcorner \mu). \tag{9-3.4}$$

Conversely, if $\mathscr{H}$ and $\mathscr{F}$ are given by (9-3.4), then

$$d\mathscr{H} = dH(\mathscr{J} \lrcorner \mu) = \mathscr{J} \lrcorner \mu - Hd(\mathscr{J} \lrcorner \mu) = \mathscr{J} \lrcorner \mu$$

$$d\mathscr{F} = dH(\mathscr{G} \lrcorner \mu) = \mathscr{G} \lrcorner \mu - Hd(\mathscr{G} \lrcorner \mu) = \mathscr{G} \lrcorner \mu$$

because $d(\mathscr{J} \lrcorner \mu) = 0$, $d(\mathscr{G} \lrcorner \mu) = 0$. Further, since α and β are arbitrary in (9-3.3) the 1-form $\mathscr{K}$ and the pseudo 1-form $\mathscr{A}$ may be chosen arbitrarily. We have thus established the following basic result.

The general solution of Maxwell's equations $\mathcal{J} \lrcorner \mu = d\mathcal{H}$, $\mathcal{G} \lrcorner \mu = d\mathcal{F}$ *is given by*

$$\mathcal{H} = -d\mathcal{K} + H(\mathcal{J} \lrcorner \mu), \tag{9-3.5}$$

$$\mathcal{F} = -d\mathcal{A} + H(\mathcal{G} \lrcorner \mu), \tag{9-3.6}$$

where $\mathcal{K}$ *is an arbitrary* 1-*form and* $\mathcal{A}$ *is an arbitrary pseudo* 1-*form*. (The minus sign in front of the term $d\mathcal{A}$ in (9-3.6) has been introduced in order to simplify comparison with classic results, as will become evident almost immediately.)

Before obtaining the implied representations of $\vec{E}$, $\vec{H}$, $\vec{B}$, and $\vec{D}$, some remarks would seem to be in order in view of the result that $\mathcal{A}$ is a pseudo 1-form. If we restrict the problem to be a classical one for which $\mathcal{G} = 0$ (i.e., no magnetic monopoles), then the second of (9-3.1) becomes $d\mathcal{F} = 0$ and (9-3.6) becomes $\mathcal{F} = -d\mathcal{A}$. However, the identical vanishing of $\mathcal{G}$ eliminates the requirement that $\mathcal{F}$ be a pseudo 2-form with the consequence that we may take $\mathcal{A}$ to be a 1-form rather than a pseudo 1-form. This is the classical result that $\vec{E}$ and $\vec{B}$ can be represented in terms of a "4-vector" potential. On the other hand, as soon as $\mathcal{G}$ is nonzero anywhere in M_4, its pseudovector density character becomes manifest. This, in turn, forces us to take $\mathcal{F}$ to be a pseudo 2-form, in which case the transformation properties of $\vec{E}$ and $\vec{B}$ exhibit the pseudostructure that is implied by the identification (9-2.12). We note, however, that the matrix representation of $\mathcal{F}$ in terms of $\mathbf{F}_*$, as given by (9-2.14), gives the correct behavior under parity transformations $P(\vec{x} \mapsto -\vec{x})$ (i.e., $\vec{E} \mapsto -\vec{E}$, $\vec{B} \mapsto \vec{B}$) and under time reversal $T(t \mapsto -t)$ (i.e., $\vec{E} \mapsto \vec{E}$, $\vec{B} \mapsto -\vec{B}$) provided $\mathcal{F}$ is a pseudo 2-form. It thus emerges that *electrodynamics with just one monopole demands that* $\vec{E}$ *and* $\vec{B}$ *transform in such a manner that* (9-2.12) *gives a pseudotensor density law of transformation for* $\mathbf{F}^*$, *and that* $\mathcal{A}$ *is a pseudo* 1-*form*. It follows from this that a study of electrodynamics without the inclusion of monopoles can lead to self consistent by spurious results.

9-4. THE RAY OPERATOR AND EXPLICIT REPRESENTATIONS

The general solution of Maxwell's equations in its abstract form

$$\mathcal{H} = -d\mathcal{K} + H(\mathcal{J} \lrcorner \mu), \tag{9-4.1}$$

$$\mathcal{F} = -d\mathcal{A} + H(\mathcal{G} \lrcorner \mu), \tag{9-4.2}$$

now needs to be "cut down" so as to obtain explicit representations for the three-dimensional field vectors $\vec{E}$, $\vec{H}$, $\vec{B}$, and $\vec{D}$. The more difficult part of this procedure entails the terms that involve the homotopy operator H, so we start with these.

We first represent $\mathscr{J} \lrcorner \mu$ and $\mathscr{G} \lrcorner \mu$ in terms of the basis $\{\mu_i\}$ by use of (9-1.9). This gives

(9-4.3) $$\mathscr{J} \lrcorner \mu = J^i \mu_i, \qquad \mathscr{G} \lrcorner \mu = G^i \mu_i.$$

The definition of the linear homotopy operator H then yields

(9-4.4) $$H(\mathscr{J} \lrcorner \mu) = \int_0^1 x^i \partial_i \lrcorner \big(J^j(\lambda x)\mu_j\big)\lambda^2 \, d\lambda$$
$$= x^i \left(\int_0^1 J^j(\lambda x)\lambda^2 \, d\lambda \right) \mu_{ij}.$$

Thus if we introduce the linear operator h on functions defined on E_4 by

(9-4.5) $$h\langle f \rangle(x) = \int_0^1 f(\lambda x)\lambda^2 \, d\lambda,$$

and use the fact that μ_{ij} is antisymmetric in the indices (i, j), (9-4.3) through (9-4.5) give the explicit evaluations

(9-4.6) $$2H(\mathscr{J} \lrcorner \mu) = \big(x^i h\langle J^j \rangle - x^j h\langle J^i \rangle\big)\mu_{ij},$$

(9-4.7) $$2H(\mathscr{G} \lrcorner \mu) = \big(x^i h\langle G^j \rangle - x^j h\langle G^i \rangle\big)\mu_{ij}.$$

It is obvious that the operator h will enter into the explicit evaluations of $\vec{E}$, $\vec{H}$, $\vec{B}$, and $\vec{D}$ in a very prominent way. We therefore take up a more detailed study of the geometric and physical implementations of the action of this operator on functions on E_4. The easiest way of doing this is to introduce four-dimensional spherical coordinates $\{R, \phi_1, \phi_2, \phi_3\}$ on E_4 so that the natural volume element becomes $R^3 \, dR \wedge d\Omega$, where $d\Omega$ is the element of differential solid angle on the unit 4-sphere. Although we still refrain from introducing a metric structure in E_4, R is related to our original coordinate cover $\{x, y, z, t\}$ by

$$R^2 = x^2 + y^2 + z^2 + t^2.$$

The four-dimensional spherical coordinate cover of E_4 may thus be viewed as defining a geometric (but not necessarily physical) measure of the *separation* between any point and the origin of the coordinate system $\{x, y, z, t\}$. Since $J(\lambda x) = J(\lambda x^1, \lambda x^2, \lambda x^3, \lambda x^4) = J(\lambda x, \lambda y, \lambda z, \lambda t)$, and the coordinates $\{x, y, z, t\}$ are held fixed during the integration process involved in (9-4.5), use of the spherical coordinate system gives the significant simplification $J(\lambda x) = \hat{J}(\lambda R, \phi_1, \phi_2, \phi_3)$. Substitution of this result into (9-4.5) gives us

(9-4.8) $$h\langle J \rangle(R, \phi_1, \phi_2, \phi_3,) = \int_0^1 \hat{J}(\lambda R, \phi_1, \phi_2, \phi_3)\lambda^2 \, d\lambda,$$

from which it follows that h performs a "radial" integration in E_4. Since $\{R, \phi_1, \phi_2, \phi_3\}$ are held constant during the integration process, the change of variables $p = \lambda R$ gives

$$h\langle J\rangle(R, \phi_1, \phi_2, \phi_3) = R^{-3}\int_0^R \hat{J}(p, \phi_1, \phi_2, \phi_3)\, p^2\, dp. \tag{9-4.9}$$

Now, suppose that $\hat{J}(R, \phi_1, \phi_2, \phi_3)$ vanishes for $R > R_0$. We then obtain

$$h\langle J\rangle(R, \phi_1, \phi_2, \phi_3) = R^{-3}\int_0^{R_0} \hat{J}(p, \phi_1, \phi_2, \phi_3)\, p^2\, dp;$$

that is, $h\langle J\rangle$ smears a nonzero J out radially in four-dimensions with a R^{-3} fall off outside the support of J. It thus seems acceptable, both geometrically and physically, to refer to the operator h as the *ray operator*, and, when need arises, to refer to $\{h\langle J^i\rangle\}$ as the ray generated by $\{J^i\}$.

The ray operator h and the rays generated by $\{J^i\}$ and $\{G^i\}$ are in many ways reminiscent of the "string" that Dirac introduced. This is easily seen for (9-4.9) shows that if J has a Dirac function singularity in more than one coordinate, then $h\langle J\rangle$ is singular along the entire radial ray from the singularity out to infinity.

Now that we have the representations (9-4.6) and (9-4.7), the rest is easy. The standard representation of 1-forms and pseudo 1-forms gives

$$2\,d\mathscr{K} = (\partial_i K_j - \partial_j K_i)\, dx^i \wedge dx^j, \qquad 2\,d\mathscr{A} = (\partial_i A_j - \partial_j A_i)\, dx^i \wedge dx^j.$$

Thus a substitution of (9-4.6), (9-4.7), and (9-4.9) into (9-4.1) and (9-4.2) yields

$$2\mathscr{H} = -(\partial_i K_j - \partial_j K_i)\, dx^i \wedge dx^j + (x^i h\langle J^j\rangle - x^j h\langle J^i\rangle)\mu_{ij}, \tag{9-4.10}$$

$$2\mathscr{F} = -(\partial_i A_j - \partial_j A_i)\, dx^i \wedge dx^j + (x^i h\langle G^j\rangle - x^j h\langle G^i\rangle)\mu_{ij}.$$

All that now remains is to use the mapping Γ to write the expressions in (9-4.10) in terms of the same basis $\{dx^i \wedge dx^j | i < j\}$ with $2\mathscr{H} = H_{ij}\, dx^i \wedge dx^j$, $2\mathscr{F} = \mathscr{F}_{ij}\, dx^i \wedge dx^j$ and the identifications (9-2.13) and (9-2.14). If we use the representations $\{A_i\} = \{\vec{A}, A_4\}$, $\{K_i\} = \{\vec{K}, K_4\}$ and $\{x^i\} = \{\vec{r}, t\}$, the following results are obtained:

$$\vec{H} = -\vec{\nabla} K_4 + \partial_t \vec{K} - \vec{r} \times h\langle \vec{J}\rangle, \tag{9-4.11}$$

$$\vec{D} = \vec{\nabla} \times \vec{K} + h\langle q\rangle \vec{r} - t h\langle \vec{J}\rangle,$$

$$\vec{E} = \vec{\nabla} A_4 - \partial_t \vec{A} + \vec{r} \times h\langle \vec{G}\rangle, \tag{9-4.12}$$

$$\vec{B} = \vec{\nabla} \times \vec{A} + h\langle g\rangle \vec{r} - t h\langle \vec{G}\rangle.$$

The relations (9-4.11) *and* (9-4.12) *constitute a solution of Maxwell's equations for any choice of the scalar potential* K_4, *the vector potential* $\vec{K}$, *the pseudoscalar potential* A_4, *and the pseudovector potential* $\vec{A}$.

Electrodynamics with free electric and magnetic charges provides the ideal situation in which the reader can grasp the full power and scope of the linear homotopy operator. The general explicit solution of Maxwell's equations given by (9-4.11) and (9-4.12) completely delineates the content of Maxwell's equations and renders their solution a closed subject. The reader must carefully note, however, that the meaning of a general solution used here is different from that commonly used by workers in electromagnetic theory: a solution for us means a solution of Maxwell's equations by themselves, while physicists usually mean a solution that satisfies both Maxwell's equations and the constitutive relations of the electromagnetic field. In fact, it is the constitutive relations rather than Maxwell's equations that give relations between the quantities $\vec{K}$, K_4, $\vec{A}$, A_4 and thus lead to many of the characteristic properties of the electromagnetic field. For instance, the general solution (9-4.11) and (9-4.12) does not imply wave propagation properties of the solutions. As we shall see later on, wave properties of solutions arise from the electromagnetic constitutive relations and the resulting differential relations between quantities $\vec{K}$, K_4, $\vec{A}$ and A_4.

9-5. PROPERTIES OF THE FIELD EQUATIONS AND THEIR SOLUTIONS

The first thing we note is that the general solution

$$\text{(9-5.1)} \qquad \mathscr{H} = -d\mathscr{K} + H(\mathscr{J} \lrcorner \mu), \qquad \mathscr{F} = -d\mathscr{A} + H(\mathscr{G} \lrcorner \mu)$$

of Maxwell's field equations

$$\text{(9-5.2)} \qquad \mathscr{J} \lrcorner \mu = d\mathscr{H}, \qquad \mathscr{G} \lrcorner \mu = d\mathscr{F}$$

involves both the 1-form $\mathscr{K}$ and the pseudo 1-form $\mathscr{A}$. Since $d^2\phi = 0$ for any scalar or pseudoscalar quantity, *the solutions* (9-5.1) *of Maxwell's field equations are invariant under the 2-fold gauge transformations*

$$\text{(9-5.3)} \qquad \mathscr{A} \mapsto \mathscr{A} + d\Phi, \qquad \mathscr{K} \mapsto \mathscr{K} + d\Psi$$

where Φ *is a pseudoscalar-valued function on* E_4 *and* Ψ *is a scalar-valued function on* E_4. There are two things that should be noted here. First, the general solution of Maxwell's equations are invariant under two-fold gauge transformations (9-5.3) rather than the customary one-fold gauge transformation $\mathscr{A} \mapsto \mathscr{A} + d\Phi$ that is customary in electromagnetic theory. The reason for this is that Maxwell's equations *without* constitutive relations give rise to the independent quantities $\mathscr{A}$ and $\mathscr{K}$, rather than to the single quantity $\mathscr{A}$, and each of these quantities can be subjected to an independent gauge transformation. Second, the occurrence of gauge transformations (i.e., transformations that map solutions onto solutions but leave the coordinate cover of the base space E_4 unchanged) in electrodynamics is the point of departure for a general theory of gauge fields. This is the subject of the next chapter.

Maxwell's equations together with the ether relations are commonly viewed as a theory that is invariant under the Lorentz group. Without the constitutive assumptions reflected in the ether relations, such an association is meaningless. This follows directly from (9-5.1) and (9-5.2) which express Maxwell's field equations and their solutions in generally covariant form. Accordingly, *Maxwell's field equations and their solutions are generally covariant and admit no preferred group of transformations on* E_4. It is thus the constitutive assumptions rather than the field equations and their solutions that single out certain preferred groups of transformations on E_4. A full discussion of this aspect of the theory will be given in the next section that deals with the constitutive theory and its implications.

In contrast with the above results, there is indeed a specific group of transformations that is naturally associated with Maxwell's equations and their solutions, although this group has nothing to do with groups of transformations on the coordinate manifold, E_4. As is well known, Maxwell's equations without magnetic currents admit the operation of electric charge conjugation. The extension of this operation to theories that include magnetic currents is usually accomplished by use of the known "exchange relations"

$$\begin{array}{cccc} q \mapsto g, & g \mapsto -q, & \vec{J} \mapsto \vec{G}, & \vec{G} \mapsto -\vec{J}, \\ K_4 \mapsto A_4, & A_4 \mapsto -K_4, & \vec{K} \mapsto \vec{A}, & \vec{A} \mapsto -\vec{K}, \\ \vec{E} \mapsto \vec{H}, & \vec{H} \mapsto -\vec{E}, & \vec{B} \mapsto -\vec{D}, & \vec{D} \mapsto \vec{B}. \end{array} \tag{9-5.4}$$

These exchange relations are the point of departure for the theory given here.

We start by noting that the field equations, (9-5.2), can be combined into a single 2-component matrix equation

$$\begin{Bmatrix} \mathscr{J} \\ \mathscr{G} \end{Bmatrix} \lrcorner \mu = d \begin{Bmatrix} \mathscr{H} \\ \mathscr{F} \end{Bmatrix}, \tag{9-5.5}$$

and their solutions can be similarly organized by writing

$$\begin{Bmatrix} \mathscr{H} \\ \mathscr{F} \end{Bmatrix} = -d \begin{Bmatrix} \mathscr{K} \\ \mathscr{A} \end{Bmatrix} + H\left(\begin{Bmatrix} \mathscr{J} \\ \mathscr{G} \end{Bmatrix} \lrcorner \mu \right). \tag{9-5.6}$$

Care must be exercised, however, for the first entries in these column matrix representations are ordinary forms while the second entries are pseudoforms.

This distinction among the entries of the column matrices is essential and constitutes the basis for our considerations. In order to account for the differences in transformation properties among the entries, we let n denote the unit scalar-valued function on E_4 and p denote the unit pseudoscalar-valued function on E_4. Let $PL(r_1, r_2)$ denote the *pseudolinear* (matrix) *group*; that is, the group of linear automorphisms of the vector space of $(r_1 + r_2)$-entried column matrices of constants whose first r_1 entries are scalars and whose remaining r_2 entries are pseudoscalars. Thus $PL(r_1, 0) = GL(r_1)$, $PL(0, r_2) = GL(r_2)$ where $GL(r)$ is the general linear matrix group (the matrix group of all nonsingular r-by-r matrices). A matrix $\mathbf{S}$ that represents an arbitrary element of $PL(r_1, r_2)$ has the natural block form $\mathbf{S} = \begin{pmatrix} \mathbf{N} & \mathbf{P} \\ \mathbf{P} & \mathbf{N} \end{pmatrix}$ ($\mathbf{N}$ = scalar-valued matrix, $\mathbf{P}$ = pseudoscalar-valued matrix). This representation follows immediately from the properties of matrix multiplication and the multiplication table $n_1 n_2 = n_3$, $np_1 = p_1 n = p_2$, $p_1 p_2 = n$. The group property of $PL(1,1)$ and the linearity and transformation properties of the relations (9-5.5) and (9-5.6) now lead to the following immediate result. *The group $PL(1,1)$ generates the group $\mathscr{D}$ of duality transformations by*

(9-5.7) $$\begin{Bmatrix} \mathscr{H} \\ \mathscr{F} \end{Bmatrix} \mapsto \mathbf{S} \begin{Bmatrix} \mathscr{H} \\ \mathscr{F} \end{Bmatrix}, \quad \begin{Bmatrix} \mathscr{K} \\ \mathscr{A} \end{Bmatrix} \mapsto \mathbf{S} \begin{Bmatrix} \mathscr{K} \\ \mathscr{A} \end{Bmatrix}, \quad \begin{Bmatrix} \mathscr{J} \\ \mathscr{G} \end{Bmatrix} \mapsto \mathbf{S} \begin{Bmatrix} \mathscr{J} \\ \mathscr{G} \end{Bmatrix}$$

with

$$\mathbf{S} = \begin{pmatrix} n_1 & p_1 \\ p_2 & n_2 \end{pmatrix} \in PL(1,1),$$

and the group $\mathscr{D}$ is a 4-parameter continuous group that maps the field equations and their solutions onto field equations and their solutions.

Let $PO(1,1)$ denote the pseudo-orthogonal subgroup of $PL(1,1)$ (the subgroup whose elements $\mathbf{S}$ satisfy $\mathbf{S}^T\mathbf{S} = \mathbf{I}$) and let $PO(1,1)^+$ denote the component of this subgroup that is continuously connected to the identity $\mathbf{I}$. In particular, if $S = \begin{pmatrix} 0 & 1 \\ -1 & 0 \end{pmatrix}$, one obtains the standard exchange relations (9-5.4). This is also evident from the vector form of the solutions that is given by (9-4.11) and (9-4.12).

The duality transformation that is generated by

$$\mathbf{S} = \begin{pmatrix} -1 & 0 \\ 0 & -1 \end{pmatrix} \in PO(1,1)^+$$

yields

(9-5.8) $$\begin{Bmatrix} \mathscr{H} \\ \mathscr{F} \end{Bmatrix} \mapsto \begin{Bmatrix} -\mathscr{H} \\ -\mathscr{F} \end{Bmatrix}, \quad \begin{Bmatrix} \mathscr{K} \\ \mathscr{A} \end{Bmatrix} \mapsto \begin{Bmatrix} -\mathscr{K} \\ -\mathscr{A} \end{Bmatrix}, \quad \begin{Bmatrix} \mathscr{J} \\ \mathscr{G} \end{Bmatrix} \mapsto \begin{Bmatrix} -\mathscr{J} \\ -\mathscr{G} \end{Bmatrix}.$$

This, however, is nothing more than the known charge conjugation properties of the field equations and their solutions, where, by charge conjugation, we mean both conjugation of electric charge and conjugation of magnetic charge. *Charge conjugation is a duality transformation that belongs to the subgroup* $PO(1,1)^+$. Further, the duality transformation that is generated by

$$\mathbf{S} = \begin{pmatrix} -1 & 0 \\ 0 & 1 \end{pmatrix} \in PO(1,1)^-$$

yields

$$\begin{Bmatrix} \mathscr{H} \\ \mathscr{F} \end{Bmatrix} \mapsto \begin{Bmatrix} -\mathscr{H} \\ \mathscr{F} \end{Bmatrix}, \quad \begin{Bmatrix} \mathscr{K} \\ \mathscr{A} \end{Bmatrix} \mapsto \begin{Bmatrix} -\mathscr{K} \\ \mathscr{A} \end{Bmatrix}, \quad \begin{Bmatrix} \mathscr{J} \\ \mathscr{G} \end{Bmatrix} \mapsto \begin{Bmatrix} -\mathscr{J} \\ \mathscr{G} \end{Bmatrix}$$

(i.e., conjugation of only electric charge), while

$$\mathbf{S} = \begin{pmatrix} 1 & 0 \\ 0 & -1 \end{pmatrix} \in PO(1,1)^-$$

generates

$$\begin{Bmatrix} \mathscr{H} \\ \mathscr{F} \end{Bmatrix} \mapsto \begin{Bmatrix} \mathscr{H} \\ -\mathscr{F} \end{Bmatrix}, \quad \begin{Bmatrix} \mathscr{K} \\ \mathscr{A} \end{Bmatrix} \mapsto \begin{Bmatrix} \mathscr{K} \\ -\mathscr{A} \end{Bmatrix}, \quad \begin{Bmatrix} \mathscr{J} \\ \mathscr{G} \end{Bmatrix} \mapsto \begin{Bmatrix} \mathscr{J} \\ -\mathscr{G} \end{Bmatrix}$$

(i.e., conjugation of only magnetic charge). *Conjugation of only electric charge and conjugation of only magnetic charge are duality transformations that are generated by elements of* $PO(1,1)^-$.

Since the duality group is a 4-parameter group that is generated by $PL(1,1)$, it is significantly larger than the class of duality transformations usually considered. This is already evident from the above results concerning separate conjugation of electric and magnetic charges. As further examples,

$$\mathbf{S} = \begin{pmatrix} b & 0 \\ 0 & b \end{pmatrix}, \qquad b \neq 0,$$

generates a duality transformation that multiplies sources, potentials and fields by the fixed constant b (i.e., doubling the sources and the potentials doubles the fields), while

$$\mathbf{S} = \begin{pmatrix} a & 0 \\ 0 & b \end{pmatrix}, \qquad ab \neq 0,$$

generates a duality transformation that increases the electric sources by the factor a and increases the magnetic sources by the factor b. Another example is

$$\mathbf{S} = p\begin{pmatrix} 0 & 1 \\ 1 & 0 \end{pmatrix},$$

which generates the exchange relations

$$\vec{D} \rightleftarrows \vec{B}, \qquad \vec{H} \rightleftarrows -\vec{E}, \qquad \mathscr{K} \rightleftarrows \mathscr{A}, \qquad \mathscr{J} \rightleftarrows \mathscr{G}$$

to within the appropriate multiple of the pseudoscalar unit p. This last set of transformations is clearly inconsistent with any constitutive theory for which $\vec{D}$ is a linear function of $\vec{E}$, and $\vec{H}$ is a linear function of $\vec{B}$. Now, our formulation of Maxwell's equations and their solutions has not involved constitutive assumptions. Accordingly, the duality group obtained without constitutive assumptions should be significantly richer than that which obtains under imposition of specific constitutive assumptions. In fact, the duality group $\mathscr{D}$ is a universal group that contains the restricted duality group that is consistent with any given system of constitutive relations. This restriction process will be examined in the next section for the case of the classic ether relations, and provides additional insight into the implications and structure of the full duality group.

We note, in particular, that the full implication of general duality transformations is far from understood, although the following results provide some additional understanding. The duality transformation that is generated by

$$\mathbf{S} = \begin{pmatrix} n_1 & p_1 \\ p_2 & n_2 \end{pmatrix}$$

is given in component form by

$$\text{(9-5.9)} \qquad \vec{H} \mapsto n_1\vec{H} - p_1\vec{E}, \qquad \vec{D} \mapsto n_1\vec{D} + p_1\vec{B},$$

$$\vec{E} \mapsto -p_2\vec{H} + n_2\vec{E}, \qquad \vec{B} \mapsto p_2\vec{D} + n_2\vec{B};$$

$$\text{(9-5.10)} \qquad K_\alpha \mapsto n_1K_\alpha + p_1A_\alpha, \qquad A_\alpha \mapsto p_2K_\alpha + n_2A_\alpha;$$

$$\text{(9-5.11)} \qquad J^\alpha \mapsto n_1J^\alpha + p_1G^\alpha, \qquad G^\alpha \mapsto p_2J^\alpha + n_2A_\alpha.$$

The principal difference between the solutions given here and those usually reported is that $\mathscr{H}$ and $\mathscr{F}$ are obtained in a completely similar format, both with respect to the occurrence of the potential 1-forms $\mathscr{K}$ and $\mathscr{A}$ and with respect to the sources $\mathscr{J}$ and $\mathscr{G}$. This identical structural similarity comes about because of the properties of the homotopy operator, H, and its implied ray operator, h. In this respect, it is useful to write the solution in the form

$$\text{(9-5.12)} \qquad \{\mathscr{H}, \mathscr{F}\}^T = \{\mathscr{H}_f, \mathscr{F}_f\}^T + \{\mathscr{H}_s, \mathscr{F}_s\}^T$$

where $\{\ ,\ \}$ signifies a row matrix and T denotes the transpose. Here

$$\text{(9-5.13)} \qquad \{\mathscr{H}_f, \mathscr{F}_f\}^T = -d\{\mathscr{K}, \mathscr{A}\}^T$$

is the *source free part* of the solution and

$$\{\mathscr{H}_s, \mathscr{F}_s\}^T = H\{\mathscr{J} \lrcorner \mu, \mathscr{G} \lrcorner \mu\}^T \tag{9-5.14}$$

is the *source part* of the solution. The occurrence of the operator H in the source part of the solution for both $\mathscr{H}$ and $\mathscr{F}$ is of particular significance. We have already seen that H gives rise to the ray operator h and this operator has the effect that it takes any function with compact support and smears it out radially to infinity in four-dimensions with an R^{-3} decay, i.e., it introduces intrinsically nonlocal effects. The source part (9-5.14) of the solution of Maxwell's equations thus contains the rays to infinity that are generated from the free electric current as well as those generated by the free magnetic current. This is amply shown by the vector component solutions (9-4.11) and (9-4.12). The theory and solutions presented here thus are completely symmetric in that they yield both electric and magnetic rays to infinity (strings), rather than just magnetic rays to infinity. Why indeed, in view of the exchange relations (9-5.4) of Maxwell's equations, should a theory give only rays to infinity from magnetic sources, as has been the case up to the present? Of greater importance, however, is that the theory developed here gives a clear separation between the potentials $\mathscr{K}$ and $\mathscr{A}$ (the source free parts of the solution) and the rays to infinity (the source part of the solution). It is clear that $\mathscr{K}$ and $\mathscr{A}$ can have very nice behavior even though the rays to infinity that are generated by $\mathscr{J}$ and $\mathscr{G}$ can be singular if $\mathscr{J}$ and $\mathscr{G}$ are singular. In fact, it follows directly from (9-5.1) and $Hd + dH =$ identity, $H(H(\cdot)) \equiv 0$, that

$$\mathscr{K} = -H(\mathscr{H}) + d\Psi, \qquad \mathscr{A} = -H(\mathscr{F}) + d\Phi, \tag{9-5.15}$$

where Ψ is an arbitrary scalar-valued function on E_4 and Φ is an arbitrary pseudoscalar-valued function on E_4. Thus unless Ψ and Φ are singular, the forms $\mathscr{K}$ and $\mathscr{A}$ are smoother than the fields $\mathscr{H}$ and $\mathscr{F}$, respectively. We thus see that the potentials can remain well behaved even if the rays to infinity become singular, and the rays to infinity arise from the electric current as well as from the magnetic current.

9-6. CONSTITUTIVE RELATIONS AND WAVE PROPERTIES

A characteristic feature of this formulation of electrodynamics is that it leads to somewhat cumbersome constitutive relations. Let ε_0 and μ_0 denote the free space permittivity and permeability, respectively, so that $\varepsilon_0\mu_0 = c^{-2}$ and c is the speed of light in vacuum. The standard representation of a homogeneous, polarizable and magnetizable medium is given by the constitutive relations

$$\vec{D} = \varepsilon_0(\vec{E} + \vec{P}), \qquad \vec{H} = \mu_0^{-1}(\vec{B} + \vec{M}), \tag{9-6.1}$$

where $\vec{P}$ and $\vec{M}$ are the free-space normalized polarization and magnetization vectors, respectively. An inspection of (9-2.11) and (9-2.14) shows that the relations (9-6.1) are equivalent to the matrix relations

$$\mathscr{H}^* = \varepsilon_0 c^2 \mathbf{a}^T(\mathbf{F}_* + \mathbf{M}_*)\mathbf{a}, \tag{9-6.2}$$

where the superior T denotes the transpose operation,

$$\mathbf{a} = \operatorname{diag}(\pm 1, \pm 1, \pm 1, \mp c^{-2}), \tag{9-6.3}$$

and $\mathbf{M}_*$ is the matrix of coefficients of the polarization-magnetization form $\mathscr{M} = M_{ij}\, dx^i \wedge dx^j$, with

$$2\mathbf{M}_* = 2((M_{ij})) = \left\|\begin{pmatrix} 0 & -M_z & M_y & -P_x \\ & 0 & -M_x & -P_y \\ & & 0 & -P_z \\ & & & 0 \end{pmatrix}\right\|. \tag{9-6.4}$$

Now $\{F_{ij}\}$ is a covariant pseudotensor, and hence $\{M_{ij}\}$ must likewise be a covariant pseudotensor in order that the sum $F_{ij} + M_{ij}$ be invariantly defined; i.e., $\mathscr{M}$ is a pseudo 2-form. On the other hand, H^{ij} is a contravariant tensor density. Accordingly, (9-6.1) and (9-6.2) can hold only if ε_0 is a pseudoscalar density, in which case $\varepsilon_0 \mu_0 = c^{-2}$ implies that μ_0 is an antipseudoscalar density. We thus obtain the transformation laws

$$'\varepsilon_0 = \Delta(y, x)^{-1} \operatorname{sign} \Delta(y, x)\varepsilon_0, \qquad '\mu_0 = \Delta(y, x) \operatorname{sign} \Delta(y, x)\mu_0 \tag{9-6.5}$$

for ε_0 and μ_0 under coordinate transformations $y^i = f^i(x^j)$.

The intrinsic cumbersomeness of these constitutive relations is clear, for (9-6.2) relates the coefficient matrices of 2-forms, where the quantities $\mathbf{H}^*$ determine $\mathscr{H}$ relative to the basis $\{\mu_{ij} | i < j\}$ while F_* determines $\mathscr{F}$ relative to the basis $\{dx^i \wedge dx^j | i < j\}$. This is, to some extent, the price we pay for not introducing a metric, for (9-6.2) and (9-6.3) would become very simple on noting that the entries of the matrix $\mathbf{a}$ are precisely those of the inverse of the standard metric tensor of special relativity.

Having noted the general form of the constitutive relations, we restrict attention from now on to the case of the *vacuum ether relations*, in view of their fundamental importance. We therefore have the relations

$$\mathbf{H}^* = \varepsilon_0 c^2 \mathbf{a}^T \mathbf{F}_* \mathbf{a}. \tag{9-6.6}$$

The *material symmetry group* of these constitutive relations consists of the subgroup of $GL(4)$ whose matrix representation $\mathbf{L}$ yields an identical satisfaction of

$$'\mathbf{H}^* = {}'\varepsilon_0 c^2 \mathbf{a}^T\, 'F_* \mathbf{a} \tag{9-6.7}$$

as a consequence of (9-6.6) and $y^\alpha = L^\alpha_\beta x^\beta$. Since ε_0 transforms according to (9-6.5) and $'\mathbf{H}^* = \Delta(y, x)\mathbf{L}\mathbf{H}^*\mathbf{L}^T$, $\mathbf{F}_* = \operatorname{sign}\Delta(y, x)\mathbf{L}^{T^{-1}}\mathbf{F}_*\mathbf{L}^{-1}$, (9-6.7) holds if and only if

$$\mathbf{L}\mathbf{a}\mathbf{L}^T = \mathbf{a}. \tag{9-6.8}$$

It thus follows, in view of (9-6.3), that *the material symmetry group of the vacuum ether relations is the Lorentz group*. This is, of course, well known. It is noted here solely to point up the fact that this group has nothing to do with Maxwell's equations per se, since these equations are obtained and solved without the constitutive relations (9-6.6). However, since Maxwell's equations are generally covariant, the invariance group of Maxwell's equations *and* the ether relations taken together is the Lorentz group. We also note that the Lorentz group has no *a priori* relation to a structure on E_4 unless one makes the additional assumption that the material symmetry group of the ether relations is also the isometry group of the metric structure. This additional assumption is, however, exactly what distinguishes special relativity from classical mechanics.

The obvious thing that remains is to obtain the implications of the vacuum ether relations. Since the ether relations read $\vec{D} = \varepsilon_0\vec{E}$, $\vec{B} = \mu_0\vec{H}$, a substitution of the solutions given by (9-4.11) and (9-4.12) into these relations yields the conditions

$$\vec{\nabla} \times \vec{K} + h\langle g\rangle\vec{r} - th\langle\vec{J}\rangle = \varepsilon_0\left(\vec{\nabla}A_4 - \partial_t\vec{A} + \vec{r} \times h\langle\vec{G}\rangle\right), \tag{9-6.9}$$

$$\vec{\nabla} \times \vec{A} + h\langle q\rangle\vec{r} - th\langle\vec{G}\rangle = -\mu_0\left(\vec{\nabla}K_4 - \partial_t\vec{K} + \vec{r} \times h\langle\vec{J}\rangle\right). \tag{9-6.10}$$

The ether relations thus establish a system of first-order differential relations between $\vec{K}$, K_4, $\vec{A}$, and A_4. If we take the divergence and curl of the relations (9-6.9) and (9-6.10) and use the Lorentz gauge conditions

$$\partial_i\left(a^{ij}A_j\right) = 0, \qquad \partial_i\left(a^{ij}K_j\right) = 0 \tag{9-6.11}$$

we see that each component of $\mathscr{A}$ and each component of $\mathscr{K}$ satisfy an inhomogeneous wave equation; for example,

$$\nabla^2 A_4 - c^{-2}\partial_t^2 A_4 = \varepsilon_0^{-1}q - \vec{\nabla}\cdot\left(\vec{r} \times h\langle\vec{G}\rangle\right),$$

$$\nabla^2 K_4 - c^{-2}\partial_t^2 K_4 = -\mu_0^{-1}g - \vec{\nabla}\cdot\left(\vec{r} \times h\langle\vec{J}\rangle\right),$$

with similar relations for $\vec{A}$ and $\vec{K}$. Of course, in obtaining these relations, we have made use of the fact that $d(\mathscr{J} \lrcorner \mu) = 0$ and $dH + Hd =$ identity imply that the ray operator on both J^i and G^i satisfies the relations

$$J^i = 3h\langle J^i\rangle + x^k\frac{\partial h\langle J^i\rangle}{\partial x^k} - x^i\frac{\partial h\langle J^k\rangle}{\partial x^k}. \tag{9-6.12}$$

This is a lengthy and tedious procedure and is not overly germane to the discussion, for we already have general solutions of Maxwell's equations. What is really needed is the relations between $\mathscr{A}$ and $\mathscr{K}$ that is induced by the vacuum ether relations. The question of the relation between $\mathscr{A}$ and $\mathscr{K}$ is relatively easy to settle. We already have the general relations $\mathscr{H} = -d\mathscr{K} + H(\mathscr{J} \lrcorner \mu)$. Thus if we use the constitutive relations (9-6.6), we obtain the relations

(9-6.13) $$-d\mathscr{K} + H(\mathscr{J} \lrcorner \mu) = \varepsilon_0 c^2 a^{ik} F_{kl} a^{lj} \mu_{ij}$$

and these may be viewed as a system of first-order partial differential equations for the determination of the coefficients of $\mathscr{K}$. As such, their integrability conditions, namely $d^2\mathscr{K} = 0$ must be satisfied. Taking the exterior derivative of (9-6.13), and noting that $dH + Hd =$ identity, $d(\mathscr{J} \lrcorner \mu) = 0$ imply $dH(\mathscr{J} \lrcorner \mu) = \mathscr{J} \lrcorner \mu = J^i \mu_i$, we obtain the relations

(9-6.14) $$J^i = 2\varepsilon_0 c^2 \partial_j \big(a^{jk} F_{kl} a^{li} \big).$$

Since the entries of $\{F^{ij}\}$ depend on the first derivatives of the entries of $\{A^i\}$, (9-6.14) are seen to be a system of second order partial differential equations for the determination of $\{A^i\}$ which we will obtain explicitly in the next paragraph. Assume now that the A's have been determined so as to secure satisfaction of (9-6.14), where these equations may be viewed as the residue relations of Maxwell's equations by the ether relations. We can then determine $\mathscr{K}$ by direct integration using the linear homotopy operator H. Thus since $HHU = 0$ for all $U \in \Lambda(E_4)$ (9-6.13) yields

(9-6.15) $$\mathscr{K} = -d\Psi - \varepsilon_0 c^2 H\big(a^{ik} F_{kl} a^{lj} \mu_{ij} \big).$$

The upshot of all of this is that everything is determined provided $\mathscr{A}$ is chosen so as to secure satisfaction of (9-6.14). We thus turn to the problem of evaluating the right-hand side of (9-6.14). Now,

$$\mathscr{F} = F_{ij}\, dx^i \wedge dx^j = -d\mathscr{A} + H(\mathscr{G} \lrcorner \mu),$$

and hence

(9-6.16) $$2F_{ij} = -(\partial_i A_j - \partial_j A_i) + L_{ijkl} x^k h\langle G^l \rangle,$$

where $L_{ijkl} x^k h\langle G^l \rangle$ is what obtains from writing

$$H(\mathscr{G} \lrcorner \mu) = x^k h\langle G^l \rangle \mu_{kl}$$

in terms of the basis $\{dx^i \wedge dx^j | i < j\}$ (note that L_{ijkl} is antisymmetric in the pair (i, j) and in the pair (k, l)). A substitution of (9-6.16) into the right-hand side of (9-6.14) yields the relations

$$\text{(9-6.17)} \qquad \varepsilon_0^{-1}c^{-2}J^j = -a^{ik}\partial_i\partial_k\left(a^{jm}A_m\right) + a^{jl}\partial_l\partial_m\left(a^{mi}A_i\right)$$
$$+\partial_i\left(a^{ik}L_{klmn}a^{lj}x^m h\langle G^n\rangle\right).$$

Satisfaction of the Lorentz gauge condition then yields the following determining inhomogeneous wave equations for the functions A_i:

$$\text{(9-6.18)} \qquad a^{ij}\partial_i\partial_j(A_k) = -\varepsilon_0^{-1}c^{-2}J^m a_{mk} + \partial_i\left(a^{ij}L_{jklm}x^l h\langle G^m\rangle\right).$$

We note that (9-6.18) reduces to the classic results whenever there are no magnetic currents ($\mathcal{G} = 0$).

It now remains to obtain the restriction of the general duality group that results from imposition of the constitutive relations $\vec{D} = \varepsilon_0\vec{E}$, $\vec{B} = \mu_0\vec{H}$. A substitution of these relations into the general duality transformation (9-5.9) generated by $\mathbf{S} = \begin{pmatrix} n_1 & p_1 \\ p_2 & n_2 \end{pmatrix}$ yields the relations

$$\text{(9-6.19)} \qquad \vec{H} \mapsto n_1\vec{H} - p_1\vec{E}, \qquad \vec{D} = \varepsilon_0\vec{E} \mapsto n_1\varepsilon_0\vec{E} + p_1\mu_0\vec{H},$$
$$\vec{E} \mapsto -p_2\vec{H} + n_2\vec{E}, \qquad \vec{B} = \mu_0\vec{H} \mapsto p_2\varepsilon_0\vec{E} + n_2\mu_0\vec{H}.$$

Consistency of these relations demands that

$$\text{(9-6.20)} \qquad n_2 = n_1, \qquad p_2 = -\gamma p_1, \qquad \gamma = \mu_0/\varepsilon_0.$$

Now, (9-6.5) shows that $'(\mu_0/\varepsilon_0) = \Delta(y,x)^2\mu_0/\varepsilon_0$, while (9-6.7) shows that the material symmetry group of the vacuum constitutive relations has $\Delta(y,x)^2 = 1$. Accordingly γ is a natural number for all transformations of E_4 that preserve the vacuum constitutive relations. It is also clear that matrices of the form $\begin{pmatrix} n & p \\ -\gamma p & n \end{pmatrix}$ with $n^2 + \gamma p^2 \neq 0$ form a group $PV(1,1)$ under matrix multiplication (the pseudovacuum group) and this group is continuously connected to the identity matrix. *The vacuum duality subgroup, $\mathcal{D}V$, that is obtained by the restriction of consistency with the vacuum ether relations, is a 2-parameter group that is continuously connected to the identity. It is generated by all matrices* $\mathbf{S}$ *of the form* $\begin{pmatrix} n & p \\ -\gamma p & n \end{pmatrix}$ *with* $n^2 + \gamma p^2 > 0$, $\gamma = \mu_0/\varepsilon_0$.

The structure of the vacuum duality group, $\mathcal{D}V$, of vacuum electrodynamics is easily obtained through a reparametrization. Let η denote the unit pseudoscalar, and let Θ_γ^+ be the subgroup of $PL(1,1)$ whose elements satisfy

$$\text{(9-6.21)} \qquad \boldsymbol{\theta}_\gamma \mathrm{Diag}(\gamma,1)\boldsymbol{\theta}_\gamma = \mathrm{Diag}(\gamma,1), \qquad \det(\boldsymbol{\theta}_\gamma) = 1.$$

Θ_γ^+ is thus a 1-parameter subgroup of $PV(1,1)$ whose elements have the form

(9-6.22)
$$\boldsymbol{\theta}_\gamma = \begin{pmatrix} \cos\eta\theta & \gamma^{-1/2}\sin\eta\theta \\ -\gamma^{1/2}\sin\eta\theta & \cos\eta\theta \end{pmatrix}.$$

Any matrix $\mathbf{S} = \begin{pmatrix} n & p \\ -\gamma p & n \end{pmatrix}$ with $n^2 + \gamma p^2 > 0$ can be written as

(9-6.23)
$$\mathbf{S} = e^{\xi}\boldsymbol{\theta}_\gamma$$

with

(9-6.24)
$$\xi = \tfrac{1}{2}\ln(n^2 + \gamma p^2), \qquad \tan\eta\theta = \gamma^{1/2}p/n.$$

Thus $\mathscr{D}V$ is generated by the 2-parameter commutative product group whose 1-parameter factor groups consist of the homothetic group, $\{\mathrm{Diag}(e^{\xi}, e^{\xi})\}$, and the group Θ_γ^+. We make specific note that the reduction of the full duality group by the vacuum ether relations excludes a number of significant duality transformations. Among these are separate conjugation of electric charge and of magnetic charge, and separate homogeneous scalings of the electric and magnetic charge densities. It is thus the vacuum ether relations, not Maxwell's equations or their general solutions, that preclude these operations.

9-7. VARIATIONAL FORMULATION OF THE FIELD EQUATIONS

The construction of a variational principle starts with the formation of four-dimensional volume invariants (4-forms) from the basis field quantities. Since $\mathscr{H}$, $\mathscr{K}$ and $\mathscr{J}\lrcorner\mu$ are forms while $\mathscr{F}$, $\mathscr{A}$ and $\mathscr{G}\lrcorner\mu$ are pseudo-forms, we let η denote a constant pseudoscalar unit. The following list can then be assembled,

(9-7.1)
$$\mathscr{I}_1 = \mathscr{F}\wedge\mathscr{F} = 2\,\mathrm{tr}(\mathbf{F}_*\mathbf{F}^*)\mu = 2\vec{B}\cdot\vec{E}\mu,$$

$$\mathscr{I}_2 = \eta\mathscr{F}\wedge\mathscr{H} = 2\eta\,\mathrm{tr}(\mathbf{F}_*\mathbf{H}^*)\mu = \frac{\eta}{2}(\vec{E}\cdot\vec{D} - \vec{B}\cdot\vec{H})\mu,$$

$$\mathscr{I}_3 = \mathscr{H}\wedge\mathscr{H} = 2\,\mathrm{tr}(\mathbf{H}_*\mathbf{H}^*)\mu = -2\vec{H}\cdot\vec{D}\mu,$$

$$\mathscr{I}_4 = \eta\mathscr{A}\wedge(\mathscr{J}\lrcorner\mu) = \eta(\mathscr{J}\lrcorner\mathscr{A})\mu = \eta J^iA_i\mu,$$

$$\mathscr{I}_5 = \mathscr{A}\wedge(\mathscr{G}\lrcorner\mu) = (\mathscr{G}\lrcorner\mathscr{A})\mu = G^iA_i\mu,$$

$$\mathscr{I}_6 = \mathscr{K}\wedge(\mathscr{J}\lrcorner\mu) = (\mathscr{J}\lrcorner\mathscr{K})\mu = J^iK_i\mu,$$

$$\mathscr{I}_7 = \eta\mathscr{K}\wedge(\mathscr{G}\lrcorner\mu) = \eta(\mathscr{G}\lrcorner\mathscr{K})\mu = \eta G^iK_i\mu.$$

This list clearly splits into two sublists; the first consisting of those invariants that involve the pseudoscalar unit, η, and the second consisting of those invariants that do not involve this pseudoscalar unit. There is a further distinction, for the invariants in the first sublist are bilinear in the naturally associated triplets $(\mathscr{F}, \mathscr{G}, \mathscr{A})$ and $(\mathscr{H}, \mathscr{J}, \mathscr{K})$, while the invariants in the second sublist are quadratic in the entries of each of the triplets separately.

Since a bilinear structure is significantly simpler to deal with than a quadratic structure, we confine our attention to the first sublist. Now, the usual variational procedure that leads to Maxwell's equations in the presence of the ether relations is one in which the potential 1-form $\mathscr{A}$ is varied. Accordingly, we express $\mathscr{H}$ and $\mathscr{F}$ in terms of $\mathscr{K}$, $\mathscr{J}$, $\mathscr{A}$, and $\mathscr{G}$ by the relations

(9-7.2) $$\{\mathscr{H}, \mathscr{F}\}^T = -d\{\mathscr{K}, \mathscr{A}\}^T + H(\{\mathscr{J}, \mathscr{G}\}^T \lrcorner \mu).$$

This yields the following sublist

(9-7.3) $$\mathscr{I}_2 = \eta(-d\mathscr{A} + \mathscr{H}(\mathscr{G} \lrcorner \mu) \wedge (-d\mathscr{K} + H(\mathscr{J} \lrcorner \mu))$$

$$\mathscr{I}_4 = \eta \mathscr{A} \wedge (\mathscr{J} \lrcorner \mu), \qquad \mathscr{I}_7 = \eta \mathscr{K} \wedge (\mathscr{G} \lrcorner \mu)$$

that involves only $\mathscr{K}$, $\mathscr{J}$, $\mathscr{A}$, and $\mathscr{G}$. We thus consider the action functional

(9-7.4) $$S_1(B_4) = \int_{B_4} \eta\{(-d\mathscr{A} + H(\mathscr{G} \lrcorner \mu) \wedge (-d\mathscr{K} + H(\mathscr{J} \lrcorner \mu)) + \mathscr{A} \wedge (\mathscr{J} \lrcorner \mu) + \mathscr{K} \wedge (\mathscr{G} \lrcorner \mu)\}$$

where B_4 is the closure of any arcwise connected, simply connected open set of E_4.

A particular property of the action functional $S_1(B_4)$ should be noted before we proceed in computation of the variations of $S_1(B_4)$. Since $dH + Hd$ = identity, we have

$$\mathscr{J} \lrcorner \mu = dH(\mathscr{J} \lrcorner \mu) + Hd(\mathscr{J} \lrcorner \mu), \qquad \mathscr{G} \lrcorner \mu = dH(\mathscr{G} \lrcorner \mu) + Hd(\mathscr{G} \lrcorner \mu).$$

If we substitute these relations into (9-7.4) and note that

$$-d\mathscr{A} \wedge H(\mathscr{J} \lrcorner \mu) + \mathscr{A} \wedge dH(\mathscr{J} \lrcorner \mu) = -d[\mathscr{A} \wedge H(\mathscr{J} \lrcorner \mu)],$$

Stokes' theorem gives us

(9-7.5) $$S_1(B_4) = \int_{B_4} \eta\{d\mathscr{A} \wedge d\mathscr{K} + H(\mathscr{J} \lrcorner \mu) \wedge H(\mathscr{G} \lrcorner \mu) + \mathscr{A} \wedge Hd(\mathscr{J} \lrcorner \mu) + \mathscr{K} \wedge Hd(\mathscr{G} \lrcorner \mu)\} + \int_{\partial B_4} \eta\{\mathscr{A} \wedge H(\mathscr{J} \lrcorner \mu) + \mathscr{K} \wedge H(\mathscr{G} \lrcorner \mu)\},$$

where ∂B_4 denotes the three-dimensional boundary of the four-dimensional region B_4.

We can now proceed with the calculation of the variations that are induced in $S_1(B_4)$ by variations $\delta\mathscr{A} = \delta A_i \, dx^i$ of the pseudo 1-form $\mathscr{A}$ and by variations $\delta\mathscr{K} = \delta K_i \, dx^i$ of the 1-form $\mathscr{K}$. A straightforward calculation based on (9-7.4) or on (9-7.5) yields

(9-7.6)

$$\delta S_1(B_4) = \int_{B_4} \eta\{\delta\mathscr{A} \wedge Hd(\mathscr{J} \lrcorner \mu) + \delta\mathscr{K} \wedge Hd(\mathscr{G} \lrcorner \mu)\}$$

$$+ \int_{\partial B_4} \eta\{\delta\mathscr{A} \wedge [d\mathscr{K} - H(\mathscr{J} \lrcorner \mu)] + \delta\mathscr{K} \wedge [d\mathscr{A} - H(\mathscr{G} \lrcorner \mu)]\}.$$

Thus $\delta S_1(B_4) = 0$ for all variations $\delta\mathscr{A}$ and $\delta\mathscr{K}$ that vanish on ∂B_4 if and only if

$$\text{(9-7.7)} \qquad Hd(\mathscr{J} \lrcorner \mu) = 0, \qquad Hd(\mathscr{G} \lrcorner \mu) = 0.$$

However, $dH \, d\mathscr{U} \equiv d\mathscr{U}$ for any $\mathscr{U}$, and we see that (9-7.7) are satisfied if and only if

$$\text{(9-7.8)} \qquad d(\mathscr{J} \lrcorner \mu) = 0, \qquad d(\mathscr{G} \lrcorner \mu) = 0;$$

that is, if and only if electric and magnetic charges are conserved. *The action $S_1(B_4)$ is stationary with respect to all variations $\delta\mathscr{A}$ and $\delta\mathscr{K}$ that vanish on ∂B_4 if and only if electric and magnetic charges are conserved.* We thus have a variational principle that recovers the starting point of the theory.

The variational principle just obtained is not altogether satisfactory. Although it leads to the basic statements of conservation of electric and magnetic charges, it does not give the full system of field equations or their solutions. This could have been anticipated, however, for we used the first integrals of the field equations, (9-7.2), in the construction of the variational principle. The ideal situation would be one in which the variational principle would yield both the field equations and their first integrals, as well as expressions for the forces that act on the electric and magnetic charge distributions. We now proceed to this larger task with the variational principle given above as a guide.

We consider the action functional

$$\text{(9-7.9)} \quad S_2(B_4) = \int_{B_4} \eta\{-d\mathscr{A} \wedge \mathscr{H} - d\mathscr{K} \wedge \mathscr{F} + \mathscr{A} \wedge (\mathscr{J} \lrcorner \mu) + \mathscr{K} \wedge (\mathscr{G} \lrcorner \mu)$$

$$+ H(\mathscr{G} \lrcorner \mu) \wedge \mathscr{H} + H(\mathscr{J} \lrcorner \mu) \wedge \mathscr{F} - \mathscr{H} \wedge \mathscr{F}\}.$$

This functional has the property that it reduces to the functional $S_1(B_4)$

whenever

$$\{\mathscr{H},\mathscr{F}\}^T = -d\{\mathscr{K},\mathscr{A}\}^T + H(\{\mathscr{J},\mathscr{G}\}^T \lrcorner \mu)$$

are satisfied throughout B_4. The distinction between $S_1(B_4)$ and $S_2(B_4)$ is evident, for in the first, only $\mathscr{A}$, $\mathscr{G}$, $\mathscr{K}$, and $\mathscr{J}$ occurred with variations of $\mathscr{A}$ and $\mathscr{K}$ only, while $S_2(B_4)$ involves all of the field quantities $\mathscr{F},\mathscr{A},\mathscr{G},\mathscr{H},\mathscr{K},\mathscr{J}$, and we may consider independent variations of $\mathscr{A}$, $\mathscr{F}$, $\mathscr{K}$, and $\mathscr{H}$. A straightforward computation and Stokes' theorem give us the following consequences of the variations generated by $\delta\mathscr{A},\delta\mathscr{K},\delta\mathscr{H},\delta\mathscr{F}$

$$\text{(9-7.10)}\quad \delta S_2(B_4) = \int_{B_4} \eta\{\delta\mathscr{A}\wedge[-d\mathscr{H}+\mathscr{J}\lrcorner\mu] + \delta\mathscr{K}\wedge[-d\mathscr{F}+\mathscr{G}\lrcorner\mu]$$

$$+\delta\mathscr{H}\wedge[H(\mathscr{G}\lrcorner\mu) - d\mathscr{A} - \mathscr{F}]$$

$$+\delta\mathscr{F}\wedge[H(\mathscr{G}\lrcorner\mu) - d\mathscr{K} - \mathscr{H}]\}$$

$$-\int_{\partial B_4} \eta\{\delta\mathscr{A}\wedge\mathscr{H} + \delta\mathscr{K}\wedge\mathscr{F}\}.$$

The action $S_2(B_4)$ is stationary with respect to all variations $\delta\mathscr{A},\delta\mathscr{K},\delta\mathscr{H},\delta\mathscr{F}$, such that $\delta\mathscr{A}$ and $\delta\mathscr{K}$ vanish on ∂B_4, if and only if

(9-7.11) $$d\mathscr{H} = \mathscr{J}\lrcorner\mu,$$

(9-7.12) $$d\mathscr{F} = \mathscr{G}\lrcorner\mu,$$

(9-7.13) $$\mathscr{F} = -d\mathscr{A} + H(\mathscr{G}\lrcorner\mu),$$

(9-7.14) $$\mathscr{H} = -d\mathscr{K} + H(\mathscr{J}\lrcorner\mu)$$

hold throughout the interior of B_4. Since a substitution of (9-7.13) and (9-7.14) into (9-7.11) and (9-7.12) yields the relations $d(\mathscr{J}\lrcorner\mu) = 0$, $d(\mathscr{G}\lrcorner\mu) = 0$, the variational principle $\delta S_2(B_4) = 0$ subsumes the whole theory presented up to this point. Further, if $S_2(B_4)$ is rendered stationary relative to the arbitrary variations $\delta\mathscr{F}$ and $\delta\mathscr{H}$, we obtain

$$\{\mathscr{H},\mathscr{F}\}^T = -d\{\mathscr{K},\mathscr{A}\}^T + H(\{\mathscr{J},\mathscr{G}\}^T \lrcorner \mu)$$

throughout B_4. Under these circumstances, we have seen that $S_2(B_4)$ reduces to $S_1(B_4)$. We thus have

(9-7.15) $$(S_2(B_4) | \delta_{\mathscr{H}} S_2 = \delta_{\mathscr{F}} S_2 = 0) = S_1(B_4),$$

which ties the two variational principles together very nicely. We further note that we do not require $\delta\mathcal{H}$ and $\delta\mathcal{F}$ to vanish on ∂B_4; the variations $\delta_{\mathcal{H}}S_2$ and $\delta_{\mathcal{F}}S_2$ that occur in (9-7.15) are unconstrained variations.

We take specific note of the fact that the variational principles given above are not invariant under duality transformations. This is not altogether unexpected, for duality transformations map Maxwell's equations and their solutions onto Maxwell's equations and their solutions, but are unrelated to deformation processes on the underlying manifold E_4. Accordingly, Noetherian theorems are not applicable and there is no underlying reason to expect invariance of the variational principles under duality transformations.

9-8. FIELD INDUCED MOMENTUM AND FORCE DENSITIES

The one question still outstanding is that of the evaluation of the field induced momentum and forces that are associated with the densities of electric and magnetic currents (i.e., the forces that act on charged particles in the presence of the electromagnetic field). Such an evaluation obviously necessitates a system of constitutive relations whereby the basic mechanical variables of velocity are related to certain of the electromagnetic source quantities. For the purposes of this discussion, we assume that

(9-8.1) $$J^i = qv^i(x^j), \qquad G^i = gu^i(x^j),$$

where $\mathscr{V} = v^i\partial_i$ is the four-dimensional velocity field of the electric charge distribution and $\mathscr{U} = u^i\partial_i$ is the four-dimensional velocity field of the magnetic charge distribution. It is, of course, assumed that $\mathscr{V}$ vanishes outside the support of $q(x^i)$, that $\mathscr{U}$ vanishes outside the support of $g(x^i)$, and that $v^4(x^i) = 1$, $u^4(x^i) = 1$ on the supports of q and g, respectively.

We now substitute the relations (9-8.1) into the action functional $S_2(B_4)$. This gives

(9-8.2)

$$S_2(B_4) = \int_{B_4} \eta\{-d\mathscr{A}\wedge\mathcal{H} - d\mathscr{K}\wedge\mathcal{F} + q\mathscr{A}\wedge(\mathscr{V}\lrcorner\mu) + g\mathscr{K}\wedge(\mathscr{U}\lrcorner\mu)$$
$$+ H(g\mathscr{U}\lrcorner\mu)\wedge\mathcal{H} + H(q\mathscr{V}\lrcorner\mu)\wedge\mathcal{F} - \mathcal{H}\wedge\mathcal{F}\},$$

and hence the part $S_{2m}(B_4)$ of the action $S_2(B_4)$ that depends on the mechanical variables $\mathscr{V}$ and $\mathscr{U}$ is given by

(9-8.3) $$S_{2m}(B_4) = \int_{B_4} \eta\{q\mathscr{A}\wedge(\mathscr{V}\lrcorner\mu) + H(q\mathscr{V}\lrcorner\mu)\wedge\mathcal{F}$$
$$+ g\mathscr{K}\wedge(\mathscr{U}\lrcorner\mu) + H(g\mathscr{U}\lrcorner\mu)\wedge\mathcal{H}\}.$$

In order to bring the velocity variables outside of the operator H, we use the adjoint operator H^+ that is defined in Section 5-19. Since this operator satisfies the identity

$$\int_{R_4} \{\alpha \wedge H(\beta) - H^+(\alpha) \wedge \beta\} = 0,$$

it follows that

$$(9\text{-}8.4)\quad \int_{B_4} H(q\mathscr{V} \lrcorner \mu) \wedge \mathscr{F} = \int_{B_4} \mathscr{F} \wedge H(q\mathscr{V} \lrcorner \mu) = \int_{B_4} H^+(\mathscr{F}) \wedge (q\mathscr{V} \lrcorner \mu),$$

and hence (9-8.3) is equivalent to

$$(9\text{-}8.5)\quad S_{2m}(B_4) = \int_{R_4} \eta\{q(\mathscr{A} + H^+(\mathscr{F})) \wedge (\mathscr{V} \lrcorner \mu)$$

$$+ g(\mathscr{K} + H^+(\mathscr{H})) \wedge (\mathscr{U} \lrcorner \mu)\}$$

$$= \int_{B_4} \eta\{q\mathscr{V} \lrcorner (\mathscr{A} + H^+(\mathscr{F})) + g\mathscr{U} \lrcorner (\mathscr{K} + H^+(\mathscr{H}))\}\mu.$$

If we use $\{\mathscr{H}, \mathscr{F}\}^T = -d\{\mathscr{K}, \mathscr{A}\}^T + H\{\mathscr{J} \lrcorner \mu, \mathscr{G} \lrcorner \mu\}^T$ and $H^+H \equiv 0$, (9-8.5) can then be written directly in terms of $\mathscr{A}$, $\mathscr{V}$, $\mathscr{K}$, and $\mathscr{U}$ as

$$(9\text{-}8.6)\quad S_{2m}(B_4) = \int_{B_4} \{q\mathscr{V} \lrcorner (\mathscr{A} - H^+(d\mathscr{A})) + g\mathscr{U} \lrcorner (\mathscr{K} - \mathscr{H}^+(d\mathscr{K}))\}\mu.$$

Our task is now that of constructing the variation of $S_{2m}(B_4)$ that is induced by variations in the orbits of the electric and magnetic charges. Since we need to vary the orbits, it is necessary to go over a Lagrangian coordinate description. For this purpose, we assume that B_4 is a four-dimensional region that is contained between the two hyperplanes $t = t_0$ and $t = t_1$. We may then describe the orbits of the electric charges by putting

$$(9\text{-}8.7)\qquad x^i = \Phi^i(X^1, X^2, X^3, t), \qquad \Phi^4(X^a, t) = t$$

on the support of q, where $\{X^a\}$ are the spatial coordinates of the electrically charged "particle" at $t = t_0$. Thus $\{\Phi^i(X^a, t)\}$ are the space-time coordinates of the particle at time t that crossed the hyperplane $t = t_0$ at the point with spatial coordinates $\{X^a\}$. The orbits of the magnetic charges is similarly described by putting

$$(9\text{-}8.8)\qquad x^i = \Psi^i(Y^a, t), \qquad \Psi^4(Y^a, t) = t.$$

The velocity fields of the "particles" are thus given by

$$(9\text{-}8.9) \qquad v^i = d\Phi^i/dt, \qquad u^i = d\Psi^i/dt.$$

Here, d/dt means the derivative with respect to t with $\{X^a\}$ or $\{Y^a\}$ held fixed; that is, the derivative following the particle.

For simplicity, let us deal with the term $\int_{B_4} \eta q(\mathscr{V} \lrcorner \mathscr{A})\mu$ in (9-8.6) for all of the other terms will follow the same pattern of argument. Since q vanishes outside its support, we have

$$(9\text{-}8.10) \qquad \int_{B_4} \eta q(\mathscr{V} \lrcorner \mathscr{A})\mu = \int_{\text{supt}(q)} \eta q(\mathscr{V} \lrcorner \mathscr{A})\mu.$$

Now, $\{X^a, t\}$ constitute a regular coordinate cover of supt(q) because the velocity field $\mathscr{V}$ on supt(q) is autonomous; that is, the functions $\Phi^i(X^a, t)$ are obtained by solving the autonomous system

$$(9\text{-}8.11) \qquad d\Phi^i/dt = v^i(\Phi^j), \qquad \Phi^a(t_0) = X^a, \qquad \Phi^4(t_0) = t_0.$$

We can thus refer the evaluation of the integral on the right-hand side of (9-8.10) to the coordinate cover $\{X^a, t\}$ and obtain

$$(9\text{-}8.12) \qquad \int_{B_4} \eta q(\mathscr{V} \lrcorner \mathscr{A})\mu$$

$$= \int_{\text{supt}(q)} \eta q(\Phi^j) \frac{\partial(\Phi^1, \Phi^2, \Phi^3, \Phi^4)}{\partial(X^1, X^2, X^3, t)} \frac{d\Phi^i}{dt} A_i(\Phi^k)\, dX^1 dX^2 dX^3\, dt.$$

We saw in the previous section that stationarity of $S_2(B_4)$ with respect to $\mathscr{H}$, $\mathscr{K}$, $\mathscr{F}$, and $\mathscr{A}$ demands conservation of electric and magnetic charges. Accordingly, we must demand that any variations in the orbits of the electric and magnetic charges be such that they preserve the electric and magnetic charge densities. This is accomplished by the requirements that

$$(9\text{-}8.13) \qquad q(\Phi^j) \frac{\partial(\Phi^1, \Phi^2, \Phi^3, \Phi^4)}{\partial(X^1, X^2, X^3, t)} = q_0(X^1, X^2, X^3),$$

$$g(\Psi^j) \frac{\partial(\Psi^1, \Psi^2, \Psi^3, \Psi^4)}{\partial(Y^1, Y^2, Y^3, t)} = g_0(Y^1, Y^2, Y^3),$$

for these equations are nothing more than the Lagrangian, as opposed to the Eulerian, statements of charge conservation. Clearly, $q_0(X^a)$ is the electric charge distribution on

$$\text{supt}(q) \cap (t = t_0)$$

and $g_0(X^a)$ is the magnetic charge distribution on

$$\text{supt}(g)\cap(t=t_0).$$

Thus since

$$\text{supt}(q)=(\text{supt}(q)\cap(t=t_0))\times[t_0,t_1]$$

in the coordinate cover $\{X^a,t\}$, (9-8.12) becomes

$$\int_{R_4}\eta q(\mathscr{V}\lrcorner\mathscr{A})\mu=\int_{t_0}^{t_1}\eta\int_{\mathscr{S}}q_0(X^a)\frac{d\Phi^i}{dt}A_i(\Phi^j)\,dX^1dX^2dX^3dt \tag{9-8.14}$$

where $\mathscr{S}=\text{supt}(q)\cap(t=t_0)$. For simplicity, we set

$$\text{supt}(q)\cap(t=t_0)=Q_3,\qquad \text{supt}(g)\cap(t=t_0)=G_3.$$

We can now write (9-8.5) in the equivalent form

(9-8.15)

$$S_{2m}(B_4)=\int_{t_0}^{t_1}\eta\left\{\int_{Q_3}q_0(X^a)\frac{d\Phi^i}{dt}(A_i-H^+(d\mathscr{A})_i)(\Phi^j)\,dX^1dX^2dX^3\right.$$
$$\left.+\int_{G_3}g_0(Y^a)\frac{d\Psi^i}{dt}(K_i-H^+(d\mathscr{K})_i)(\Psi^j)\,dY^1dY^2dY^3\right\}dt.$$

This shows that

$$\mathscr{L}_q=\eta q_0(X^a)\frac{d\Psi^i}{dt}(A_i-H^+(d\mathscr{A})_i)(\Psi^j) \tag{9-8.16}$$

is the Lagrangian per unit volume of Q_3 of the orbits of the electric charge and that

$$\mathscr{L}_g=\eta g_0(Y^a)\frac{d\Psi^i}{dt}(K_i-H^+(d\mathscr{K})_i)(\Psi^i) \tag{9-8.17}$$

is the Lagrangian per unit volume of G_3 of the orbits of the magnetic charge. The standard definition of the field induced momentum per unit volume [i.e., $p^i=\partial\mathscr{L}/\partial(d\Phi^i/dt)$] gives us the following immediate results:

$$P_{qi}=\eta q_0(X^a)(A_i-H^+(d\mathscr{A})_i)(\Phi^j), \tag{9-8.18}$$

$$P_{gi}=\eta g_0(Y^a)(K_i-H^+(d\mathscr{K})_i)(\Psi^j). \tag{9-8.19}$$

Let $\mathscr{L}_{mq}$ denote the Lagrangian per unit volume of Q_3 that results from purely mechanical properties of the electric charge and let $\mathscr{L}_{mg}$ denote the Lagrangian per unit volume of G_3 that results from purely mechanical properties of the magnetic charge. If we perform variations in the orbits of the electric and magnetic charges that vanish on the hypersurfaces $t = t_0$ and $t = t_1$, the vanishing of the induced variations in $\int_{t_0}^{t_1}(\mathscr{L}_{mq} + \mathscr{L}_q)\,dt$ and $\int_{t_0}^{t_1}(\mathscr{L}_{mg} + \mathscr{L}_g)\,dt$ give the Euler-Lagrange equations

(9-8.20) $$\frac{d}{dt}\left(\frac{\partial \mathscr{L}_{mq}}{\partial\, d\Phi^i/dt}\right) - \frac{\partial \mathscr{L}_{mq}}{\partial \Phi^i} = \bar{f}_{qi} = \frac{\partial \mathscr{L}_q}{\partial \Phi^i} - \frac{d}{dt}\left(\frac{\partial \mathscr{L}_q}{\partial\, d\Phi^i/dt}\right),$$

$$\frac{d}{dt}\left(\frac{\partial \mathscr{L}_{mg}}{\partial\, d\Psi^i/dt}\right) - \frac{\partial \mathscr{L}_{mg}}{\partial \Psi^i} = \bar{f}_{gi} = \frac{\partial \mathscr{L}_q}{\partial \Psi^i} - \frac{d}{dt}\left(\frac{\partial \mathscr{L}_g}{\partial\, d\Psi^i/dt}\right),$$

where $\bar{f}_{qi}$ are the components of the force per unit volume of Q_3 that acts on the electric charge and $\bar{f}_{gi}$ are the components of the force per unit volume of G_3 that acts on the magnetic charge. These forces are given, however, relative to the natural basis induced by the coordinate covers $\{X^a, t\}$ and $\{Y^a, t\}$ respectively. It thus follows from (9-8.16), (9-8.17) and (9-8.20) that

(9-8.21) $$\bar{f}_{qi} = \eta q \frac{d\Phi^j}{dt}\{\partial_i[A_j - H^+(d\mathscr{A})_j] - \partial_j[A_i - H^+(d\mathscr{A})_i]\},$$

$$\bar{f}_{gi} = g_0 \frac{d\Psi^j}{dt}\{\partial_i[K_j - H^+(d\mathscr{K})_j] - \partial_j[K_i - H^+(d\mathscr{K})_i]\}$$

in the coordinate covers $\{X^a, t\}$ and $\{Y^a, t\}$, respectively. Since the variations in Φ^i and Ψ^i vanish on the hypersurfaces $t = t_0$ and $t = t_1$, we have

$$\delta\int_{t_0}^{t_1}\mathscr{L}_q\,dt = \int_{t_0}^{t_1}\left\{\frac{\partial \mathscr{L}_q}{\partial \Phi^i} - \frac{d}{dt}\frac{\partial \mathscr{L}_q}{\partial(d\Phi^i/dt)}\right\}\delta\Phi^i\,dt = \int_{t_0}^{t_1}\bar{f}_{qi}\,\delta\Phi^i\,dt$$

from the definition of the variational process and (9-8.20). A combination of (9-8.6) with (9-8.15) through (9-8.17) thus yields

(9-8.22) $$\bar{\delta}S_{2m}(B_4) = \int_{t_0}^{t_1}\eta\int_{Q_3} q_0\left\{\frac{d\Phi^j}{dt}(\partial_i[A_j - H^+(d\mathscr{A})_j]\right.$$

$$\left. - \partial_j[A_i - H^+(d\mathscr{A})_i]\,\delta\Phi^i\right\} dX^1 dX^2 dX^3\,dt$$

$$+ \int_{t_0}^{t_1}\eta\int_{G_3} g_0\left\{\frac{d\Psi^j}{dt}(\partial_i[K_j - H^+(d\mathscr{K})_j]\right.$$

$$\left. - \partial_j[K_i - H^+(d\mathscr{K})_i])\,\delta\Psi^i\right\} dY^1 dY^2 dY^3\,dt,$$

where $\bar{\delta}$ denotes variation of the orbits. It is now a simple matter to reverse the whole process in order to recover an integral over B_4 with respect to the coordinate cover $\{x^i\}$. When (9-8.13) is used, and we note that

$$\frac{d\Phi^j}{dt}\{\partial_i A_j - \partial_j A_i\}\, dx^i = -\mathscr{V} \lrcorner d\mathscr{A},$$

(9-8.22) gives

$$(9\text{-}8.23)\quad \bar{\delta} S_{2m}(B_4) = \int_{B_4} \{ f_q \wedge (\delta\Phi^i \lrcorner \mu) + f_g \wedge (\delta\Psi^i \lrcorner \mu)\}$$

$$= -\int_{B_4} \eta\{[q\mathscr{V} \lrcorner d(\mathscr{A} - H^+(d\mathscr{A}))] \wedge (\delta\Phi^i \lrcorner \mu)$$

$$+ [g\mathscr{U} \lrcorner d(\mathscr{K} - H^+(d\mathscr{K}))] \wedge (\delta\Psi^i \lrcorner \mu)\}$$

where $f_q = \bar{f}_{qi}\, dx^i$ is the force density 1-form that acts on the electric charge distribution and $f_g = \bar{f}_{gi}\, dx^i$ is the force density 1-form that acts on the magnetic charge distribution. We thus obtain the following final results: *the electromagnetic force density* 1-*forms that act on the electric and magnetic charge distributions are given by*

$$(9\text{-}8.24)\qquad f_q = -\eta q \mathscr{V} \lrcorner d(\mathscr{A} - H^+(d\mathscr{A})),$$

$$(9\text{-}8.25)\qquad f_g = -\eta g \mathscr{U} \lrcorner d(\mathscr{K} - H^+(d\mathscr{K})).$$

The first thing we note is that these force densities satisfy the standard requirements of any four-dimensional formulation of force distributions:

$$(9\text{-}8.26)\qquad \mathscr{V} \lrcorner f_q = 0, \qquad \mathscr{U} \lrcorner f_g = 0.$$

Further, since f_q and f_g are linear in the velocity variables $\mathscr{V}$ and $\mathscr{U}$, respectively, they behave properly under transformations $t = T(s)$ of the dynamical parameter. This same fact is reflected by (9-8.15); that is, $S_{2m}(B_4)$ is invariant under $t = T(s)$. Thus if a Lorentz metric is introduced in E_4, the forces transform properly under Lorentz transformations.

Second, if there are no magnetic charges, that is, $\mathscr{G} = 0$, we have $\mathscr{F} = -d\mathscr{A}$. In this event, (9-2.14) shows that the three-dimensional vector part of f_q is $-\eta q \mathscr{V} \lrcorner d\mathscr{A} = \eta q(\vec{E} + \vec{v} \times \vec{B})$. Since that is exactly the Lorentz force to within the factor η, we conclude that the numerical value of η is given by

$$(9\text{-}8.27)\qquad \eta = 1.$$

Third, the results given by (9-8.24) and (9-8.25) are exact rather than approximate consequences of the variational principle that yields the field equations and their solutions. On the other hand, the Lorentz force $q(\vec{E} + \vec{v} \times \vec{B})$ is correct only in the approximation wherein radiation forces are neglected. Since the term $-q\mathscr{V} \lrcorner d\mathscr{A}$ leads to the Lorentz force law, the remaining term $q\mathscr{V} \lrcorner dH^{+}(d\mathscr{A})$ can be tentatively identified with the radiation forces. *The expressions* (9-8.24) *and* (9-8.25) *give exact expressions for the radiation forces that act on the electric and magnetic current distributions as a consequence of the electromagnetic radiation generated by the motions of the charges that comprise these currents*:

(9-8.28) $\quad \mathscr{f}_q = \mathscr{f}_{qL} + \mathscr{f}_{qR},$

(9-8.29) $\quad \mathscr{f}_{qL} = -\eta q\mathscr{V} \lrcorner d\mathscr{A} =$ Lorentz force on electric charges,

(9-8.30) $\quad \mathscr{f}_{qR} = \eta q\mathscr{V} \lrcorner dH^{+}(d\mathscr{A}) =$ radiation force on electric charges;

(9-8.31) $\quad \mathscr{f}_g = \mathscr{f}_{gL} + \mathscr{f}_{gR},$

(9-8.32) $\quad \mathscr{f}_{gL} = -\eta g\mathscr{U} \lrcorner d\mathscr{K} =$ Lorentz force on magnetic charges,

(9-8.33) $\quad \mathscr{f}_{gR} = \eta g\mathscr{U} \lrcorner dH^{+}(d\mathscr{K}) =$ radiation force on magnetic charges.

Fourth, (9-8.28) through (9-8.33) show that the forces depend on the fields only through their "source free" parts

(9-8.34) $$\mathscr{F}_{\not} = -d\mathscr{A}, \qquad \mathscr{H}_{\not} = -d\mathscr{K}$$

of the fields

(9-8.35) $$\mathscr{F} = -d\mathscr{A} + H(\mathscr{G} \lrcorner \mu), \qquad \mathscr{H} = -d\mathscr{K} + H(\mathscr{J} \lrcorner \mu).$$

Accordingly, the forces $\mathscr{f}_q$ and $\mathscr{f}_g$ do not depend on the unphysical "rays to infinity" that are generated by the action of H on $\mathscr{G} \lrcorner \mu$ and $\mathscr{J} \lrcorner \mu$. In fact, (9-8.28) through (9-8.35) yield

(9-8.36) $$\mathscr{f}_{qL} = q\mathscr{V} \lrcorner \mathscr{F}_{\not}, \qquad \mathscr{f}_{qR} = -\eta q\mathscr{V} \lrcorner dH^{+}\mathscr{F}_{\not},$$

(9-8.37) $$\mathscr{f}_{gL} = g\mathscr{U} \lrcorner \mathscr{H}_{\not}, \qquad \mathscr{f}_{gR} = -\eta g\mathscr{U} \lrcorner dH^{+}\mathscr{H}_{\not}.$$

The electric current may thus pass through the rays to infinity that are generated by either $\mathscr{G} \lrcorner \mu$ or $\mathscr{J} \lrcorner \mu$, and similarly for the magnetic current. We note in particular, however, that $\mathscr{f}_{qL} = \eta q\mathscr{V} \lrcorner \mathscr{F}$ only if $H(\mathscr{G} \lrcorner \mu) = 0$ and $\mathscr{f}_{gL} = \eta q\mathscr{U} \lrcorner \mathscr{H}$ only if $H(\mathscr{J} \lrcorner \mu) = 0$, and hence $\vec{\mathscr{f}}_{qL} = q(\vec{E} + \vec{v} \times \vec{B})$ only when $H(\mathscr{G} \lrcorner \mu) = 0$; that is

$$\mathscr{f}_{qL} = q\mathscr{V} \lrcorner \mathscr{F}$$

only when the electric current does not intersect the rays to infinity generated by $\mathscr{G} \lrcorner \mu$.

Fifth, (9-8.28) through (9-8.35) can be written in the equivalent form

(9-8.38)
$$\mathscr{f}_q = -\eta q \mathscr{V} \lrcorner d(\mathscr{A} + \mathscr{A}_R).$$
$$\mathscr{f}_g = -\eta g \mathscr{U} \lrcorner d(\mathscr{K} + \mathscr{K}_R),$$

where

(9-8.39)
$$\mathscr{A}_R = -H^+(d\mathscr{A}) = H^+\mathscr{F}_f$$

can be viewed as the *radiation pseudopotential* 1-form of the electric current, and

(9-8.40)
$$\mathscr{K}_R = -H^+(d\mathscr{K}) = H^+\mathscr{H}_f$$

can be viewed as the *radiation potential* 1-form of the magnetic current. The smoothness of the forces $\mathscr{f}_q$ and $\mathscr{f}_g$ are thus determined exclusively by the smoothness of $\mathscr{A}$, $\mathscr{A}_R$, and $\mathscr{K}$, $\mathscr{K}_R$, respectively. In order to see just what this entails, we use (9-2.14) and (9-8.39) to evaluate $\mathscr{A}_R$ in the case $\mathscr{G} = 0$. This yields

(9-8.41)
$$\mathscr{A}_R = \{\vec{r} \times h^-(\vec{B}) + th^-(\vec{E})\} \cdot d\vec{x} - (\vec{r} \cdot h^-(\vec{E}))\, dt,$$

where

(9-8.42)
$$h^-(\Phi) = -\int_1^\infty \Phi(\lambda\vec{r}, \lambda t)\lambda\, d\lambda$$

is the "antiray operator" that is induced by the operator H^+. If we use the vacuum ether relations $\vec{E} = \varepsilon_0^{-1}\vec{D}$, $\vec{B} = \mu_0\vec{H}$, (9-8.41) becomes

(9-8.43)
$$\mathscr{A}_R = \{\mu_0\vec{r} \times h^-(\vec{H}) + \varepsilon_0^{-1}th^-(\vec{D})\} \cdot d\vec{x} - \varepsilon_0^{-1}(\vec{r} \cdot h^-(\vec{D}))\, dt,$$

which is linear in the field vectors $\vec{H}$ and $\vec{D}$. The radiation force is then given by

$$\begin{aligned}\mathscr{f}_{qR} &= -\eta q \mathscr{V} \lrcorner d\mathscr{A}_R \\ &= \eta q\{\partial_t(\mu_0\vec{r} \times h^-(\vec{H}) + \varepsilon_0^{-1}th^-(\vec{D})) + \varepsilon_0^{-1}\vec{\nabla}(\vec{r} \cdot h^{-1}(\vec{D})) \\ &\qquad - \vec{v} \times \vec{\nabla} \times (\mu_0\vec{r} \times h^-(\vec{H}) + \varepsilon_0^{-1}th^-(\vec{D}))\} \cdot d\vec{x} \\ &\quad - \eta q\{\varepsilon_0^{-1}\vec{v} \cdot \vec{\nabla}(r \cdot h^-(\vec{D})) \\ &\qquad + \vec{v} \cdot \partial_t(\mu_0\vec{r} \times h^-(\vec{H}) + \varepsilon_0^{-1}th^-(\vec{D}))\}\, dt\end{aligned}$$

in terms of $h^-(\vec{H})$, $h^-(\vec{D})$ and their derivatives.

CHAPTER TEN

GAUGE THEORIES

10-1. THE ORIGIN OF GAUGE THEORIES

The theory of gauge fields that has grown up since the mid-1960s is one of the clearest examples of the intrinsic utility of the exterior calculus. Above and beyond this, gauge theories are now considered to be a fundamental aspect of nature that underlies all of physical reality. It is therefore appropriate to conclude these discussions with an account of the theory of gauge fields.

The underlying space is taken to be an n-dimensional number space E_n. In practice, n will have the value 4 and E_4 will be identified with space time. For the purposes of this discussion, we assume that E_n is referred to a fixed coordinate cover $(x^i,\ i = 1, \ldots, n)$. Other coordinate covers can be introduced later by use of the known mapping properties of vector fields and exterior differential forms if the need should arise.

The underlying space E_n is assumed to be populated by a known system of fields $\Psi(x^j)$. For purposes of discussion, we assume that the known system of fields may be organized as the entries of a column matrix: $\Psi(x^j) = \{\Psi^1(x^j), \ldots, \Psi^N(x^j)\}^T$, where the superior T denotes the transpose operation. This assumption is not essential to the theory, being primarily a notational convenience. The whole discussion can be put through with an arbitrary collection of N fields $\{\Psi^A(x^j),\ A = 1, \ldots, N\}$, but we would then lose the convenience of matrix notation.

Essential to the theory is the assumption that the fields $\Psi(x^j)$ satisfy a system of field equations that derive from a variational principle. We therefore have a Lagrangian function $L_0(x^j, \Psi(x^j), \partial_i\Psi(x^j))$ for the system and the results established in Chapter Seven. In particular, invariance of the associated action, $\int L_0 \mu$, under a 1-parameter family of transformations (the system admits a nontrivial Noetherian vector field), implies conservation of an associated current $(n-1)$-form (see Section 7-4).

Let G_0 be a connected r-parameter group of nonsingular N-by-N matrices that acts on the fields $\mathbf{\Psi}(x^j)$ from the left,

$$(10\text{-}1.1) \qquad {}'\mathbf{\Psi}(x^j) = \mathbf{A}\mathbf{\Psi}(x^j), \qquad \partial_i \mathbf{A} = \mathbf{0}, \qquad \mathbf{A} \in G_0.$$

Written out, this means

$$'\Psi^B(x^j) = A^B_C \Psi^C(x^j), \qquad B = 1, \ldots, N.$$

The action of any $\mathbf{A} \in G_0$ on any $\mathbf{\Psi}(x^j)$ is thus *homogeneous*; it acts on $\mathbf{\Psi}(x^j)$ at each point $P:(x^j) \in E_n$ in exactly the same way because the entries of the matrix $\mathbf{A}$ are constants over all of E_n. It thus follows that

$$(10\text{-}1.2) \qquad \partial_i {}'\mathbf{\Psi}(x^j) = \mathbf{A}\,\partial_i \mathbf{\Psi}(x^j), \qquad \mathbf{A} \in G_0;$$

partial differentiation and the action of G_0 commute. The action of G_0 thus knows nothing about the underlying space E_n on which the Ψ-fields live and the underlying space E_n is not affected in any way by the action of G_0 on the field variables.

Association of the group G_0 with the fields $\mathbf{\Psi}(x^j)$ comes about through the fact that the group G_0 is an internal symmetry group (*gauge group*) of the theory. By this we mean that the Lagrangian function L_0 is invariant under the action of G_0:

$$(10\text{-}1.3) \qquad L_0\left(x^j, \mathbf{A}\mathbf{\Psi}, \mathbf{A}\,\partial_i \mathbf{\Psi}\right) = L_0\left(x^j, \mathbf{\Psi}, \partial_i \mathbf{\Psi}\right)$$

for all $\mathbf{A} \in G_0$. Since the action of G_0 does not change the coordinate cover of E_n, each 1-parameter family of elements of G_0 that contains the identity matrix $\mathbf{I}$ gives a 1-parameter family of transformations that leaves the associated action invariant (the action of $\mathbf{A}$ on $\mathbf{\Psi}$ generates a vertical isovector field of the contact ideal that is a Noetherian vector field). Accordingly, there is an associated current $(n-1)$-form that is conserved. We thus see that there are r linearly independent current $(n-1)$-forms that are conserved since G_0 is an r-parameter matrix Lie group. The r independent current $(n-1)$-forms obtained in this way give direct statements about the physics of the fields $\mathbf{\Psi}(x^j)$ that are an essential part of modern field theory.

It has now become somewhat fashionable to take the gauge group G_0 as fundamental, in which case the Ψ-fields are relegated to the subordinate position of simply being elements of a representation space for the action of the group G_0. In fact, the Ψ-fields are actually ignored altogether in the study of what are called "free" gauge fields. To some extent, this is unfortunate, for we can quickly lose sight of the fact that the physics begins with the $\mathbf{\Psi}(x^j)$

fields as the primary quantities. Indeed, the group G_0 under consideration is "discovered," so to speak, by noting the invariance of the Lagrangian L_0 (see (10-1.3)) and the associated family of r linearly independent current $(n-1)$-forms that are necessarily conserved. Granted, once the group G_0 is in place, a great deal follows, but the uninitiated often are left with a certain amount of anxiety as to where G_0 comes from in the first place if the $\Psi(x^j)$-fields are dismissed. In this regard, it is just not enough to say that the Ψ-fields constitute a representation space for G_0. The $\Psi(x^i)$ fields will therefore be retained in a primary role in these discussions.

There are certain properties of the matrix Lie group G_0 that we will need in what follows. Since G_0 is a connected r-parameter matrix Lie group, it is generated by exponentiation of its matrix Lie algebra. Let $\{\boldsymbol{\gamma}_\alpha, \alpha = 1, \ldots, r\}$ be a basis for the matrix Lie algebra of G_0 with the Lie product $[\mathbf{A}, \mathbf{B}] = \mathbf{AB} - \mathbf{BA}$. Since each $\boldsymbol{\gamma}_\alpha$ is an N-by-N matrix, we may always write

$$\boldsymbol{\gamma}_\alpha \boldsymbol{\gamma}_\beta = a^\rho_{\alpha\beta} \boldsymbol{\gamma}_\rho + \mathbf{S}_{\alpha\beta}, \qquad \mathbf{S}_{\alpha\beta} = \mathbf{S}_{\beta\alpha}. \tag{10-1.4}$$

Consequently, in view of the symmetry of $\mathbf{S}_{\alpha\beta}$ in the indices α, β,

$$[\boldsymbol{\gamma}_\alpha, \boldsymbol{\gamma}_\beta] = \boldsymbol{\gamma}_\alpha \boldsymbol{\gamma}_\beta - \boldsymbol{\gamma}_\beta \boldsymbol{\gamma}_\alpha = \left(a^\rho_{\alpha\beta} - a^\rho_{\beta\alpha}\right) \boldsymbol{\gamma}_\rho,$$

and hence

$$[\boldsymbol{\gamma}_\alpha, \boldsymbol{\gamma}_\beta] = C^\rho_{\alpha\beta} \boldsymbol{\gamma}_\rho, \qquad C^\rho_{\alpha\beta} = a^\rho_{\alpha\beta} - a^\rho_{\beta\alpha}, \tag{10-1.5}$$

where $C^\rho_{\alpha\beta}$ are the constants of structure of G_0. The Cartan-Killing form of G_0 is then defined by

$$C_{\alpha\beta} = C^\rho_{\alpha\eta} C^\eta_{\beta\rho} \qquad (= C_{\beta\alpha}). \tag{10-1.6}$$

We note for later reference that each of the quantities $\boldsymbol{\gamma}_\alpha, a^\rho_{\alpha\beta}, \mathbf{S}_{\alpha\beta}, C^\rho_{\alpha\beta}, C_{\alpha\beta}$ are constants or matrices of constants on E_n.

Gauge fields and gauge theories come about as a simple and direct way of reconciling the following seemingly contradictory requirements:

1. Allow the group G_0 to act differently at different points of E_n.
2. Preserve conservation of r independent current $(n-1)$-forms.

10-2. THE MINIMAL REPLACEMENT CONSTRUCT

Suppose the transformations of the group G_0 are now allowed to depend upon position in E_n. We denote this new group by G and elements of G by $\mathbf{A}(x^j)$. Under these circumstances, (10-1.1) is replaced by

$$'\boldsymbol{\Psi}(x^j) = \mathbf{A}(x^j)\boldsymbol{\Psi}(x^j). \tag{10-2.1}$$

The inhomogeneity of the action of an element of G gives

$$\partial_i' \Psi(x^j) = \partial_i(\mathbf{A}\Psi) = \mathbf{A}\,\partial_i\Psi + (\partial_i \mathbf{A})\Psi \tag{10-2.2}$$

instead of the simple commutation law (10-1.2). Thus $\mathbf{A}$ no longer factors from the left and the invariance of the Lagrangian L_0 is lost:

$$L_0(x^j, '\Psi, \partial_i'\Psi) = L_0(x^j, \mathbf{A}\Psi, \mathbf{A}\,\partial_i\Psi + (\partial_i\mathbf{A})\Psi)$$

$$\neq L_0(x^j, \Psi, \partial_i\Psi).$$

Essential changes are required if we are to preserve conservation of r independent current $(n-1)$-forms, for the existence of such quantities arises directly from invariance properties of the Lagrangian function (of the action). Here it is necessary to note that replacing G_0 by G does not change the coordinate cover in any way, so that invariance of L_0 and invariance of the action are the same thing.

It is clear from the outset that preservation of the invariance of the Lagrangian L_0 under the inhomogeneous action of G requires something other than just the simple replacement of $\partial_i\Psi$ by $\partial_i'\Psi$. What is needed is an operator D_i and its image $'D_i$ under the action of G which are such that

$$'D_i'\Psi(x^j) = \mathbf{A}(x^j)\,D_i\Psi(x^j). \tag{10-2.3}$$

A direct comparison with (10-2.2) shows that the operator D_i has to account for the troublesome term $(\partial_i\mathbf{A})\Psi$ and to be compatible with the action of the group G. Let us therefore introduce a collection $\{\Gamma_i(x^j)\}$ of new fields (compensating gauge fields) that take their values in the Lie algebra of G_0 and transform under the action of G by

$$'\Gamma_i = \mathbf{A}\Gamma_i\mathbf{A}^{-1} - (\partial_i\mathbf{A})\mathbf{A}^{-1}. \tag{10-2.4}$$

We then have

$$\partial_i\mathbf{A} = \mathbf{A}\Gamma_i - {'\Gamma_i}\mathbf{A} \tag{10-2.5}$$

and hence (10-2.5) can be used to eliminate all terms involving $\partial_i\mathbf{A}$. When (10-2.5) is substituted into (10-2.2), we obtain

$$\partial_i'\Psi = \mathbf{A}\,\partial_i\Psi + (\mathbf{A}\Gamma_i - {'\Gamma_i}\mathbf{A})\Psi$$

$$= \mathbf{A}(\partial_i\Psi + \Gamma_i\Psi) - {'\Gamma_i}'\Psi$$

when (10-2.1) is used. Accordingly, we define the operators D_i and their images

$'D_i$ by

$$D_i\Psi = \partial_i\Psi + \Gamma_i\Psi, \tag{10-2.6}$$

$$'D_i'\Psi = \partial_i'\Psi + {'\Gamma_i'}\Psi, \tag{10-2.7}$$

so that the conditions (10-2.3) are then satisfied as a consequence of (10-2.4). The operators D_i are called *gauge covariant derivatives* and the $\{\Gamma_i\}$ are the corresponding *gauge connection* matrices of 1-forms.

The upshot of this is that a replacement of $\partial_i\Psi$ by $D_i\Psi$ as arguments of the Lagrangian function L_0 restores the invariance of L_0,

$$L_0(x^j, {'\Psi}, {'D_i'}\Psi) = L_0(x^j, \mathbf{A}\Psi, \mathbf{A}D_i\Psi) = L_0(x^j, \Psi, D_i\Psi)$$

by (10-1.3). This gives rise to the now famous *minimal replacement* of Yang and Mills:

$$\mathscr{M}_\Gamma : (x^j, \Psi, \partial_i\Psi) \mapsto (x^j, \Psi, D_i\Psi) \tag{10-2.8}$$

for any choice of the Γ_i that take their values in the Lie algebra of G_0 and transform under the action of G by (10-2.4). Thus, in particular, we have

$$\mathscr{M}_\Gamma L_0(x^j, \Psi, \partial_i\Psi) = L_0(x^j, \Psi, D_i\Psi) \tag{10-2.9}$$

and $\mathscr{M}_\Gamma L_0$ is invariant under the action of G. *The existence of r conserved current $(n-1)$-forms is thus preserved for the Lagrangian $\mathscr{M}_\Gamma L_0$.*

Replacing the original Lagrangian L_0 by the new Lagrangian $\mathscr{M}_\Gamma L_0$ is not as drastic as it may seem on first reading. In point of fact, since $\mathscr{M}_\Gamma$ is defined for any choice of $\{\Gamma_i\}$, we could make the choice $\Gamma_i = \mathbf{0}$, in which case (10-2.4) gives

$$'\Gamma_i = -(\partial_i\mathbf{A})\mathbf{A}^{-1}.$$

We then have $D_i\Psi = \partial_i\Psi$ and

$$'D_i'\Psi = \partial_i({'\Psi}) - (\partial_i\mathbf{A})\mathbf{A}^{-1}{'\Psi} = \mathbf{A}\partial_i\Psi.$$

Restriction of $\mathbf{A}$ to G_0 then shows that $\mathscr{M}_\Gamma$ reduces to the identity mapping. Accordingly, when we write $L_0(x^j, \Psi, \partial_i\Psi)$ we have no way of knowing that this is not really

$$\mathscr{M}_\Gamma L_0 = L_0(x^j, \Psi, D_i\Psi)$$

for any appropriate choice of $\{\Gamma_i\}$ consistent with the restriction of G to G_0. Put another way, if $L_0(x^j, \Psi, \partial_i\Psi)$ is determined from experiments under the assumption of restriction to G_0, it is not inconsistent to say that $\mathcal{M}_\Gamma L_0$ is what is actually there but undetectable because of the restriction.

10-3. MINIMAL COUPLING

The connection matrices $\{\Gamma_i\}$ take their values in the Lie algebra of G_0. Therefore, each of the Γ_i's can be expressed in terms of the basis $\{\gamma_\alpha, \alpha = 1, \ldots, r\}$ for the Lie algebra of G_0 by

$$\Gamma_i = W_i^\alpha(x^j)\gamma_\alpha. \tag{10-3.1}$$

Here the $W_i^\alpha(x^j)$ are the *Yang-Mills potential functions* associated with the inhomogeneous action of the gauge group G. The minimal replacement (10-2.8), which is necessary in order to preserve the invariance of the Lagrangian function under the inhomogeneous action of the gauge group G, gives rise to the new fields $W_i^\alpha(x^j)$ that are *coupled* to the original fields $\Psi(x^j)$ by

$$D_i\Psi = \partial_i\Psi + W_i^\alpha\gamma_\alpha\Psi. \tag{10-3.2}$$

Additional field equations are thus required for the determination of the new fields $\{W_i^\alpha(x^j)\}$.

We proceed by replacing the original Lagrangian, now $\mathcal{M}_\Gamma L_0$, by the new Lagrangian

$$L = \mathcal{M}_\Gamma L_0 + s_1 L_1, \tag{10-3.3}$$

where the new term $L_1 = L_1(x^j, W_i^\alpha(x^j), \partial_k W_i^\alpha(x^j))$ depends only on the fields $\{W_i^\alpha\}$ and their first derivatives. We must also require that L_1 *be invariant under the action of* G, for otherwise L would not be invariant under G and all of our work would be for naught. Here, s_1 is a coupling constant that *minimally couples* the W_i^α fields to the Ψ fields; that is, L_1 does not depend on the Ψ fields so the coupling arises only from terms in $\mathcal{M}_\Gamma L_0$ through (10-3.2). The Lagrangian L_1 is referred to as the Yang-Mills free field Lagrangian (it is assumed that $\mathcal{M}_\Gamma L_0$ vanishes with vanishing Ψ). Variation of the total action with respect to the $\{W_i^\alpha\}$ fields then gives the new field equations for the determination of the Yang-Mills fields, while variation with respect to the Ψ fields gives the field equations for the Ψ fields that are minimally coupled to the gauge fields $\{W_i^\alpha\}$ (see Section 10-6).

10-4. GAUGE CONNECTION 1-FORMS AND THE GAUGE COVARIANT EXTERIOR DERIVATIVE

Significant simplifications result if we collect together certain terms that appear in the previous discussion. To see this, we note that multiplication of (10-3.2) by dx^i and summing on i give

$$D_i\Psi\, dx^i = d\Psi + W_i^\alpha\, dx^i \gamma_\alpha \Psi = d\Psi + \Gamma_i\, dx^i \Psi.$$

We therefore introduce a system of Lie algebra-valued *gauge connection* 1-*forms* $\mathbf{\Gamma}$ by

$$\mathbf{\Gamma} = \mathbf{\Gamma}_i\, dx^i = W^\alpha \gamma_\alpha, \qquad W^\alpha = W_i^\alpha\, dx^i. \tag{10-4.1}$$

The 1-forms $\{W^\alpha\} \in \Lambda^1(E_n)$ are referred to as *Yang-Mills potential* 1-*forms*. The *gauge covariant exterior derivative* of the $\mathbf{\Psi}$ fields is then defined by

$$D\mathbf{\Psi} = d\mathbf{\Psi} + \mathbf{\Gamma}\mathbf{\Psi} = d\mathbf{\Psi} + W^\alpha \gamma_\alpha \mathbf{\Psi}. \tag{10-4.2}$$

with

$$'(D\mathbf{\Psi}) = d'\mathbf{\Psi} + '\mathbf{\Gamma}'\mathbf{\Psi}. \tag{10-4.3}$$

It thus follows directly from (10-2.3) that the gauge covariant exterior derivative of $\mathbf{\Psi}$ transforms under the action of G by

$$'(D\mathbf{\Psi}) = \mathbf{A} D\mathbf{\Psi}. \tag{10-4.4}$$

The transformation law (10-4.4) is the reason for referring to $D\mathbf{\Psi}$ as the gauge covariant exterior derivative of $\mathbf{\Psi}$, for the statement $D\mathbf{\Psi} = \mathbf{0}$ implies $'(D\mathbf{\Psi}) = \mathbf{0}$ for every $\mathbf{A} \in G$: *if a gauge covariant quantity vanishes in any one gauge, it vanishes for all choices of gauge* (all choices of $\mathbf{A} \in G$).

The covariance noted above is not a happenstance, for (10-2.4) and (10-4.1) imply that $\mathbf{\Gamma}$ has the transformation law

$$'\mathbf{\Gamma} = \mathbf{A}\mathbf{\Gamma}\mathbf{A}^{-1} - (d\mathbf{A})\mathbf{A}^{-1} \tag{10-4.5}$$

for any $\mathbf{A} \in G$. We therefore have

$$d\mathbf{A} = \mathbf{A}\mathbf{\Gamma} - '\mathbf{\Gamma}\mathbf{A} \tag{10-4.6}$$

and hence any troublesome terms such as $d\mathbf{A}$ or $d\mathbf{A}^{-1}$ can be eliminated in favor of the connection matrix $\mathbf{\Gamma}$ and its image under $\mathbf{A} \in G$. The elimination of $d\mathbf{A}^{-1}$ follows from the fact that $\mathbf{A}^{-1}\mathbf{A} = \mathbf{I}$ and hence

$$d\mathbf{A}^{-1} = -\mathbf{A}^{-1}(d\mathbf{A})\mathbf{A}^{-1}. \tag{10-4.7}$$

It may seem on first reading that reference to $\mathbf{\Gamma}$ as a gauge connection is somewhat arbitrary. This is not the case as is easily shown by use of the results established in Section 5-11. Let $D\mathbf{\Psi}$ be identified with $\mathbf{\Omega}$ and construct the differential system of degree one and class N. We then have

$$(10\text{-}4.8)\qquad d\boldsymbol{\Omega} = -\boldsymbol{\Gamma} \wedge \boldsymbol{\Omega} + \boldsymbol{\Sigma}, \qquad d\boldsymbol{\Sigma} = -\boldsymbol{\Gamma} \wedge \boldsymbol{\Sigma} + \boldsymbol{\Theta} \wedge \boldsymbol{\Omega},$$

$$d\boldsymbol{\Gamma} = -\boldsymbol{\Gamma} \wedge \boldsymbol{\Gamma} + \boldsymbol{\Theta}, \qquad d\boldsymbol{\Theta} = -\boldsymbol{\Gamma} \wedge \boldsymbol{\Theta} + \boldsymbol{\Theta} \wedge \boldsymbol{\Gamma},$$

where $\boldsymbol{\Gamma}$, $\boldsymbol{\Sigma}$, $\boldsymbol{\Theta}$ are the associated connection, torsion, and curvature forms, respectively. Now $\boldsymbol{\Omega} = D\boldsymbol{\Psi}$ and (10-4.4) show that $\boldsymbol{\Omega}$ has the transformation law

$$(10\text{-}4.9)\qquad '\boldsymbol{\Omega} = \mathbf{A}\boldsymbol{\Omega}$$

for any $\mathbf{A} \in G$. We thus have $d\,'\boldsymbol{\Omega} = (d\mathbf{A}) \wedge \boldsymbol{\Omega} + \mathbf{A}\,d\boldsymbol{\Omega} = (\mathbf{A}\boldsymbol{\Gamma} - {}'\boldsymbol{\Gamma}\mathbf{A}) \wedge \boldsymbol{\Omega} + \mathbf{A}\,d\boldsymbol{\Omega} = \mathbf{A}(d\boldsymbol{\Omega} + \boldsymbol{\Gamma} \wedge \boldsymbol{\Omega}) - {}'\boldsymbol{\Gamma} \wedge {}'\boldsymbol{\Omega}$; that is,

$$d\,'\boldsymbol{\Omega} + {}'\boldsymbol{\Gamma} \wedge {}'\boldsymbol{\Omega} = \mathbf{A}(d\boldsymbol{\Omega} + \boldsymbol{\Gamma} \wedge \boldsymbol{\Omega}).$$

Accordingly, if we define the covariant exterior derivative of $\boldsymbol{\Omega}$ by

$$(10\text{-}4.10)\qquad D\boldsymbol{\Omega} = d\boldsymbol{\Omega} + \boldsymbol{\Gamma} \wedge \boldsymbol{\Omega},$$

we have the transformation law

$$(10\text{-}4.11)\qquad '(D\boldsymbol{\Omega}) = \mathbf{A}\,D\boldsymbol{\Omega}.$$

Further, the first of (10-4.8) can be written as

$$(10\text{-}4.12)\qquad D\boldsymbol{\Omega} = \boldsymbol{\Sigma}$$

which shows that the associated torsion has the transformation law

$$(10\text{-}4.13)\qquad '\boldsymbol{\Sigma} = \mathbf{A}\boldsymbol{\Sigma}.$$

If exactly the same argument is repeated, starting with (10-4.13), we see that the covariant exterior derivative of $\boldsymbol{\Sigma}$ can be defined by

$$(10\text{-}4.14)\qquad D\boldsymbol{\Sigma} = d\boldsymbol{\Sigma} + \boldsymbol{\Gamma} \wedge \boldsymbol{\Sigma}$$

and the second of (10-4.8) becomes

$$(10\text{-}4.15)\qquad D\boldsymbol{\Sigma} = \boldsymbol{\Theta} \wedge \boldsymbol{\Omega}.$$

Thus since $\boldsymbol{\Omega}$ and $\boldsymbol{\Sigma}$ have the transformation laws (10-4.9) and (10-4.13), while (10-4.13) and (10-4.14) imply

$$(10\text{-}4.16)\qquad '(D\boldsymbol{\Sigma}) = \mathbf{A}\,D\boldsymbol{\Sigma},$$

the curvature 2-forms have the transformation law

$$'\mathbf{\Theta} = \mathbf{A}\mathbf{\Theta}\mathbf{A}^{-1}. \tag{10-4.17}$$

The important thing here is that the first and second of (10-4.8) assume the *gauge covariant form*

$$D\mathbf{\Omega} = \mathbf{\Sigma}, \qquad D\mathbf{\Sigma} = \mathbf{\Theta} \wedge \mathbf{\Omega}. \tag{10-4.18}$$

Hence satisfaction of these equations in any one gauge implies their satisfaction for all choices of gauge (for all $\mathbf{A} \in G$).

The differential system (10-4.8) was constructed by starting with the identification $\mathbf{\Omega} = D\mathbf{\Psi}$. A straightforward calculation shows that $D\mathbf{\Omega} = D\,D\mathbf{\Psi} = \mathbf{\Theta}\mathbf{\Psi}$, and hence (10-4.12) gives the explicit evaluation

$$\mathbf{\Sigma} = \mathbf{\Theta}\mathbf{\Psi}.$$

The associated matrix of torsion 2-forms is thus realized by the matrix of curvature 2-forms, $\mathbf{\Theta}$, acting as a linear operator on the $\mathbf{\Psi}$-fields. This serves to point up the important fact that the associated matrix of curvature 2-forms have direct interpretations in terms of the $\mathbf{\Psi}$-fields. In fact, $D\,D\mathbf{\Psi} = \mathbf{\Theta}\mathbf{\Psi}$ is a concise and useful summary of the relations between $\mathbf{\Psi}$, $\mathbf{\Gamma}$ and $\mathbf{\Theta}$. Contrary to the exterior derivative, DD is not identically zero, a situation that will be taken up presently since it is fundamental.

There is a further convenience that derives from construction of the differential system from $D\mathbf{\Psi}$, namely the explicit evaluation of the associated curvature 2-forms

$$\mathbf{\Theta} = d\mathbf{\Gamma} + \mathbf{\Gamma} \wedge \mathbf{\Gamma}. \tag{10-4.19}$$

Since $\mathbf{\Gamma}$ has the representation $\mathbf{\Gamma} = W^\alpha \boldsymbol{\gamma}_\alpha$ and $d\boldsymbol{\gamma}_\alpha = 0$, (10-4.19) give

$$\mathbf{\Theta} = dW^\alpha \boldsymbol{\gamma}_\alpha + W^\alpha \boldsymbol{\gamma}_\alpha \wedge W^\beta \boldsymbol{\gamma}_\beta = dW^\alpha \boldsymbol{\gamma}_\alpha + W^\alpha \wedge W^\beta \boldsymbol{\gamma}_\alpha \boldsymbol{\gamma}_\beta$$

$$= dW^\alpha \boldsymbol{\gamma}_\alpha + W^\alpha \wedge W^\beta \tfrac{1}{2}(\boldsymbol{\gamma}_\alpha \boldsymbol{\gamma}_\beta - \boldsymbol{\gamma}_\beta \boldsymbol{\gamma}_\alpha)$$

since $W^\alpha \wedge W^\beta = -W^\beta \wedge W^\alpha$. However, (10-1.5) gives

$$(\boldsymbol{\gamma}_\alpha \boldsymbol{\gamma}_\beta - \boldsymbol{\gamma}_\beta \boldsymbol{\gamma}_\alpha) = C^\rho_{\alpha\beta} \boldsymbol{\gamma}_\rho,$$

where the C's are the constants of structure of G, and we obtain

$$\mathbf{\Theta} = F^\rho \boldsymbol{\gamma}_\rho, \qquad F^\rho = dW^\rho + \tfrac{1}{2} C^\rho_{\alpha\beta} W^\alpha \wedge W^\beta. \tag{10-4.20}$$

Since the F^ρ are 2-forms, we may write

$$F^\rho = \tfrac{1}{2} F^\rho_{ij}\, dx^i \wedge dx^j, \tag{10-4.21}$$

in which case (10-4.21) gives

$$F^{\rho}_{ij} = \partial_i W^{\rho}_j - \partial_j W^{\rho}_i + C^{\rho}_{\alpha\beta} W^{\alpha}_i W^{\beta}_j. \tag{10-4.22}$$

The quantities $\{F^{\rho}_{ij}\}$ are referred to as the components of the *Yang-Mills field tensors* (for each of the r values of the gauge group index ρ). It is useful to note for later reference that $F^{\rho} = dW^{\rho}$ for $C^{\rho}_{\alpha\beta} = 0$, in which case each F^{α} looks like the field 2-form of an electromagnetic field with potential 1-form W^{ρ} (see Chapter Nine).

The process of covariant exterior differentiation is not confined solely to forms that derive from the fields $\boldsymbol{\Psi}$. All that is required is that the exterior forms transform under the action of G is a nice way.

Let $\boldsymbol{\eta}$ be a (column) matrix of k-forms that transforms under the action of G by $'\boldsymbol{\eta} = \mathbf{A}\boldsymbol{\eta}$. Exterior differentiation gives $d\,'\boldsymbol{\eta} = (d\mathbf{A}) \wedge \boldsymbol{\eta} + \mathbf{A}\, d\boldsymbol{\eta}$, in which case (10-4.6) can be used to eliminate $d\mathbf{A}$. The result is

$$d\,'\boldsymbol{\eta} + {}'\boldsymbol{\Gamma} \wedge {}'\boldsymbol{\eta} = \mathbf{A}(d\boldsymbol{\eta} + \boldsymbol{\Gamma} \wedge \boldsymbol{\eta}).$$

The covariant exterior derivative of $\boldsymbol{\eta}$ can thus be defined by $D\boldsymbol{\eta} = d\boldsymbol{\eta} + \boldsymbol{\Gamma} \wedge \boldsymbol{\eta}$ and has the transformation law $'(D\boldsymbol{\eta}) = \mathbf{A}\, D\boldsymbol{\eta}$ for any $\mathbf{A} \in G$. Now, the transformation law $'(D\boldsymbol{\eta}) = \mathbf{A}\, D\boldsymbol{\eta}$ allows us to do the same thing over again. In this case, something quite different emerges, for

$$\begin{aligned} D\,D\boldsymbol{\eta} &= d(d\boldsymbol{\eta} + \boldsymbol{\Gamma} \wedge \boldsymbol{\eta}) + \boldsymbol{\Gamma} \wedge (d\boldsymbol{\eta} + \boldsymbol{\Gamma} \wedge \boldsymbol{\eta}) \\ &= d\boldsymbol{\Gamma} \wedge \boldsymbol{\eta} - \boldsymbol{\Gamma} \wedge d\boldsymbol{\eta} + \boldsymbol{\Gamma} \wedge d\boldsymbol{\eta} + \boldsymbol{\Gamma} \wedge \boldsymbol{\Gamma} \wedge \boldsymbol{\eta} \\ &= (d\boldsymbol{\Gamma} + \boldsymbol{\Gamma} \wedge \boldsymbol{\Gamma}) \wedge \boldsymbol{\eta} = \boldsymbol{\Theta} \wedge \boldsymbol{\eta} \end{aligned}$$

when (10-4.19) is used. Thus, in contrast with the ordinary exterior derivative, DD is not zero, but rather has the evaluation $\boldsymbol{\Theta} \wedge \boldsymbol{\eta}$ in terms of $\boldsymbol{\eta}$ and the curvature 2-forms. We therefore have

$$\boldsymbol{\eta} \in \Lambda^k_{N,1}(E_n), \qquad {}'\boldsymbol{\eta} = \mathbf{A}\boldsymbol{\eta},$$

$$D\boldsymbol{\eta} = d\boldsymbol{\eta} + \boldsymbol{\Gamma} \wedge \boldsymbol{\eta}, \qquad {}'(D\boldsymbol{\eta}) = \mathbf{A}\, D\boldsymbol{\eta}, \tag{10-4.23}$$

$$D\,D\boldsymbol{\eta} = \boldsymbol{\Theta} \wedge \boldsymbol{\eta}. \tag{10-4.24}$$

If $\boldsymbol{\rho}$ is a (row) matrix of k-forms that transforms under the action of G by $'\boldsymbol{\rho} = \boldsymbol{\rho}\mathbf{A}^{-1}$, exterior differentiation gives

$$d\,'\boldsymbol{\rho} = (d\boldsymbol{\rho})\mathbf{A}^{-1} + (-1)^k \boldsymbol{\rho} \wedge d\mathbf{A}^{-1}.$$

When (10-4.7) is used to eliminate $d\mathbf{A}^{-1}$, we see that

$$d\,'\boldsymbol{\rho} - (-1)^k\, {}'\boldsymbol{\rho} \wedge {}'\boldsymbol{\Gamma} = \left(d\boldsymbol{\rho} - (-1)^k \boldsymbol{\rho} \wedge \boldsymbol{\Gamma}\right)\mathbf{A}^{-1}.$$

Hence the covariant exterior derivative is defined by $D\rho = d\rho - (-1)^k \rho \wedge \Gamma$ with the transformation law $'(D\rho) = (D\rho)\mathbf{A}^{-1}$. A straightforward calculation then gives

$$D\,D\rho = d\big(d\rho - (-1)^k \rho \wedge \Gamma\big) - (-1)^{k+1}\big(d\rho - (-1)^k \rho \wedge \Gamma\big) \wedge \Gamma$$

$$= -\rho \wedge (d\Gamma + \Gamma \wedge \Gamma) = -\rho \wedge \Theta.$$

We therefore have

$$\rho \in \Lambda^k_{1,N}(E_n), \qquad '\rho = \rho\mathbf{A}^{-1}$$

(10-4.25) $$D\rho = d\rho - (-1)^k \rho \wedge \Gamma, \qquad '(D\rho) = (D\rho)\mathbf{A}^{-1},$$

(10-4.26) $$D\,D\rho = -\rho \wedge \Theta.$$

Identical arguments lead to the following results:

$$\omega \in \Lambda^k_{N,N}(E_n), \qquad '\omega = \mathbf{A}\omega\mathbf{A}^{-1},$$

(10-4.27) $$D\omega = d\omega + \Gamma \wedge \omega - (-1)^k \omega \wedge \Gamma, \qquad '(D\omega) = \mathbf{A}(D\omega)\mathbf{A}^{-1},$$

(10-4.28) $$D\,D\omega = \Theta \wedge \omega - \omega \wedge \Theta.$$

In this regard, it is of interest to note that the curvature equations, $d\Theta = -\Gamma \wedge \Theta + \Theta \wedge \Gamma$, become the gauge covariant equations

(10-4.29) $$D\Theta = \mathbf{0}$$

since $\Theta \in \Lambda^2_{N,N}(E_n)$ and $'\Theta = \mathbf{A}\Theta\mathbf{A}^{-1}$. The gauge covariant equations $D\Theta = \mathbf{0}$ are usually referred to as the *first half of the Yang-Mills field equations* although they are identically satisfied as a consequence of the relations $\Theta = d\Gamma + \Gamma \wedge \Gamma$ for any $\Gamma = W^\alpha \gamma_\alpha$ (for any $W^\alpha \in \Lambda^1(E_n)$). Since Θ has the representation $\Theta = F^\alpha \gamma_\alpha$, by (10-4.20) and $\Gamma = W^\alpha \gamma_\alpha$, we have

(10-4.30)
$$\begin{aligned} D\Theta &= d\Theta + \Gamma \wedge \Theta - \Theta \wedge \Gamma \\ &= (dF^\alpha)\gamma_\alpha + W^\alpha \gamma_\alpha \wedge F^\beta \gamma_\beta - F^\alpha \gamma_\alpha \wedge W^\beta \gamma_\beta \\ &= (dF^\alpha)\gamma_\alpha + W^\alpha(\gamma_\alpha\gamma_\beta - \gamma_\beta\gamma_\alpha) \wedge F^\beta \\ &= (dF^\alpha)\gamma_\alpha + W^\alpha C^\rho_{\alpha\beta} \wedge F^\beta \gamma_\rho \\ &= \big(dF^\rho + C^\rho_{\alpha\beta} W^\alpha \wedge F^\beta\big)\gamma_\rho. \end{aligned}$$

Accordingly, since $\{\gamma_\rho, \rho = 1, \ldots, r\}$ is a basis for the Lie algebra of G_0, the first half of the Yang-Mills field equations $D\boldsymbol{\Theta} = \mathbf{0}$ have the explicit component form

(10-4.31) $$dF^\rho + C^\rho_{\alpha\beta} W^\alpha \wedge F^\beta = 0.$$

10-5. GAUGE COVARIANT DIFFERENTIATION OF GROUP ASSOCIATED QUANTITIES

The considerations of this section are motivated by the result stated in (10-4.30):

(10-5.1) $$D\boldsymbol{\Theta} = \left(dF^\alpha + C^\alpha_{\rho\beta} W^\rho \wedge F^\beta\right)\gamma_\alpha$$

for

(10-5.2) $$\boldsymbol{\Theta} = F^\alpha \gamma_\alpha, \qquad {}'\boldsymbol{\Theta} = \mathbf{A}\boldsymbol{\Theta}\mathbf{A}^{-1},$$

and $D\boldsymbol{\Theta} = d\boldsymbol{\Theta} + \boldsymbol{\Gamma} \wedge \boldsymbol{\Theta} - \boldsymbol{\Theta} \wedge \boldsymbol{\Gamma}$. Here $\boldsymbol{\Theta}$ is an N-by-N matrix of 2-forms that acts on the $\boldsymbol{\Psi}$ fields from the left to determine the column matrix of associated torsion 2-forms $\boldsymbol{\Sigma} = \boldsymbol{\Theta}\boldsymbol{\Psi}$. Thus $\boldsymbol{\Theta}$ is also a 2-form valued linear operator on the space of the $\boldsymbol{\Psi}$ fields and the $\boldsymbol{\Psi}$ fields live in a column matrix space that is a representation space for the action of the gauge group G, ${}'\boldsymbol{\Psi} = \mathbf{A}\boldsymbol{\Psi}$, $\mathbf{A} \in G$. On the other hand, $\boldsymbol{\Theta} = F^\alpha \gamma_\alpha$ may be viewed as a realization of $\boldsymbol{\Theta}$ in terms of 2-forms F^α that live in the group space of the gauge group and the bridging quantities between the two spaces are the γ_α. It would therefore be useful to obtain a gauge covariant exterior differentiation process that acts on group space quantities such as F^α in (10-5.2) to give (10-5.1); i.e., $DF^\alpha = dF^\alpha + C^\alpha_{\rho\beta} W^\rho \wedge F^\beta$. As it turns out, the results obtained through this study will prove to be essential in establishing the general gauge covariance of the field equations that serve to determine the compensating fields W^α.

Let $\mathbf{T}$ be a matrix of k-forms that takes values in the matrix Lie algebra of G_0 and transforms under G by

(10-5.3) $$'\mathbf{T} = \mathbf{A}\mathbf{T}\mathbf{A}^{-1}, \qquad \mathbf{A} \in G.$$

There is then a well defined gauge covariant exterior derivative

(10-5.4) $$D\mathbf{T} = d\mathbf{T} + \boldsymbol{\Gamma} \wedge \mathbf{T} - (-1)^k \mathbf{T} \wedge \boldsymbol{\Gamma}$$

with the transformation law

(10-5.5) $$'(D\mathbf{T}) = \mathbf{A}(D\mathbf{T})\mathbf{A}^{-1}.$$

Since $\mathbf{T}$ takes its values in the matrix Lie algebra of G_0, we have

$$\mathbf{T} = T^{\alpha}\gamma_{\alpha}. \tag{10-5.6}$$

When this is substituted into (10-5.3), it follows that

$$'\mathbf{T} = \mathbf{A}(T^{\alpha}\gamma_{\alpha})\mathbf{A}^{-1} = T^{\alpha}\mathbf{A}\gamma_{\alpha}\mathbf{A}^{-1} \tag{10-5.7}$$

under the action of $\mathbf{A} \in G$. Now, we may interpret

$$\bar{\gamma}_{\alpha} = \mathbf{A}\gamma_{\alpha}\mathbf{A}^{-1} \tag{10-5.8}$$

as a change of basis for the matrix Lie algebra of G_0 that is induced by the action of $\mathbf{A} \in G$ ($\mathbf{A}\gamma_{\alpha}\mathbf{A}^{-1}$ is the adjoint representation of G as a group of automorphisms of G_0). The fact that each $\bar{\gamma}_{\alpha}$ also belongs to the matrix Lie algebra of G_0 shows that $\bar{\gamma}_{\alpha}$ can be expressed in terms of the original basis $\{\gamma_{\beta}, \beta = 1, \ldots, r\}$. Accordingly, there exist functions $a^{\alpha}_{\beta}(x^j)$ such that

$$\bar{\gamma}_{\alpha} = \mathbf{A}\gamma_{\alpha}\mathbf{A}^{-1} = a^{\beta}_{\alpha}\gamma_{\beta}, \qquad \det\left(a^{\beta}_{\alpha}\right) \neq 0. \tag{10-5.9}$$

In fact, the regular map $\mathbf{A} \to ((a^{\beta}_{\alpha}))$ that is defined by (10-5.9) gives the adjoint representation of G on its matrix Lie algebra. When this result is put back into (10-5.7), we obtain

$$'\mathbf{T} = T^{\alpha}a^{\beta}_{\alpha}\gamma_{\beta} = {'T^{\beta}}\gamma_{\beta} \tag{10-5.10}$$

where the latter equality obtains because $'\mathbf{T}$ also takes its values in the matrix Lie algebra of G_0. Thus, since $\{\gamma_{\beta}, \beta = 1, \ldots, r\}$ is a basis for the Lie algebra, (10-5.10) gives the transformation law

$$'T^{\beta} = a^{\beta}_{\alpha}T^{\alpha} \tag{10-5.11}$$

of the group space quantities $\{T^{\alpha}\}$ that is induced by the map

$$\mathbf{A} \in G \overset{(10\text{-}5.9)}{\to} \left(\left(a^{\alpha}_{\beta}\right)\right).$$

We now proceed to exterior differentiation. When (10-5.6) and $\Gamma = W^{\alpha}\gamma_{\alpha}$ are substituted into (10-5.4), straightforward calculation gives

$$\begin{aligned} D\mathbf{T} &= d\mathbf{T}^{\alpha}\gamma_{\alpha} + W^{\alpha}\gamma_{\alpha} \wedge T^{\beta}\gamma_{\beta} - (-1)^k T^{\beta}\gamma_{\beta} \wedge W^{\alpha}\gamma_{\alpha} \\ &= dT^{\alpha}\gamma_{\alpha} + W^{\alpha} \wedge T^{\beta}(\gamma_{\alpha}\gamma_{\beta} - \gamma_{\beta}\gamma_{\alpha}) \\ &= \left(dT^{\rho} + C^{\rho}_{\alpha\beta}W^{\alpha} \wedge T^{\beta}\right)\gamma_{\rho}. \end{aligned}$$

Let the 1-forms Γ^{α}_{β} be defined by

$$\Gamma^{\alpha}_{\beta} = C^{\alpha}_{\rho\beta} W^{\rho}. \tag{10-5.12}$$

The above result can then be written as

$$D\mathbf{T} = \left(dT^{\alpha} + \Gamma^{\alpha}_{\beta} \wedge T^{\beta}\right)\boldsymbol{\gamma}_{\alpha}. \tag{10-5.13}$$

Since the left-hand side of (10-5.13) is a gauge covariant exterior derivative, we refer to the quantities Γ^{α}_{β} as the *connection* 1-*forms on group space* that give rise to the gauge covariant derivative

$$DT^{\alpha} = dT^{\alpha} + \Gamma^{\alpha}_{\beta} \wedge T^{\beta}, \qquad D\mathbf{T} = (DT^{\alpha})\boldsymbol{\gamma}_{\alpha}. \tag{10-5.14}$$

Clearly, what now needs to be done is to obtain the law of transformation for the connection 1-forms on group space. Let $'\Gamma^{\alpha}_{\beta}$ denote the connection 1-forms that obtain under the action that is induced by $\mathbf{A} \in G$. We may then write

$$'(DT^{\alpha}) = d\,'T^{\alpha} + '\Gamma^{\alpha}_{\beta} \wedge 'T^{\beta}, \qquad '(D\mathbf{T}) = '(DT^{\alpha})\boldsymbol{\gamma}_{\alpha}. \tag{10-5.15}$$

However, (10-5.5) gives $'(D\mathbf{T}) = \mathbf{A}(D\mathbf{T})\mathbf{A}^{-1}$, and hence

$$\begin{aligned} '(DT^{\alpha})\boldsymbol{\gamma}_{\alpha} &= \mathbf{A}(DT^{\beta})\boldsymbol{\gamma}_{\beta}\mathbf{A}^{-1} = (DT^{\beta})\mathbf{A}\boldsymbol{\gamma}_{\beta}\mathbf{A}^{-1} \\ &= (DT^{\beta})a^{\alpha}_{\beta}\boldsymbol{\gamma}_{\alpha}. \end{aligned}$$

This gives the transformation law

$$'(DT^{\alpha}) = a^{\alpha}_{\beta}\, DT^{\beta}$$

which is the analogy of (10-5.5). Accordingly, we have

$$'(DT^{\alpha}) = a^{\alpha}_{\beta}\, DT^{\beta} \qquad \text{for} \qquad 'T^{\alpha} = a^{\alpha}_{\beta} T^{\beta} \tag{10-5.16}$$

so that DT^{α} is indeed a covariant exterior derivative. What now remains is to substitute (10-5.11), (10-5.14), and (10-5.15) into (10-5.16):

$$\begin{aligned} a^{\alpha}_{\beta}\left(dT^{\beta} + \Gamma^{\beta}_{\sigma} \wedge T^{\sigma}\right) &= d\,'T^{\alpha} + '\Gamma^{\alpha}_{\beta} \wedge 'T^{\beta} \\ &= d\left(a^{\alpha}_{\beta} T^{\beta}\right) + '\Gamma^{\alpha}_{\beta} \wedge a^{\beta}_{\sigma} T^{\sigma}. \end{aligned}$$

A straightforward calculation gives

$$'\Gamma^{\alpha}_{\beta} a^{\beta}_{\sigma} \wedge T^{\sigma} = \left(a^{\alpha}_{\beta}\Gamma^{\beta}_{\sigma} - da^{\alpha}_{\sigma}\right) \wedge T^{\sigma}.$$

Thus for this to hold for all k-forms T^σ, we must have

$$'\Gamma^\alpha_\beta a^\beta_\sigma = a^\alpha_\beta \Gamma^\beta_\sigma - da^\alpha_\sigma. \tag{10-5.17}$$

Equation (10-5.17) is the transformation law for the Γ^α_β that we have been seeking. It is the group space analog of the transformation law $'\mathbf{T}\mathbf{A} = \mathbf{A}\boldsymbol{\Gamma} - d\mathbf{A}$ for the connection matrix $\boldsymbol{\Gamma}$ on $\boldsymbol{\Psi}$-space. The same purpose is served by this transformation law; namely, we can solve (10-5.17) for da^α_σ to obtain

$$da^\alpha_\sigma = a^\alpha_\beta \Gamma^\beta_\sigma - {}'\Gamma^\alpha_\beta a^\beta_\sigma \tag{10-5.18}$$

and thereby eliminate the quantities da^α_σ in any transformation formula. For example, if G_α has the transformation law $'G_\alpha = a^{-1\beta}_\alpha G_\beta$, then

$$\begin{aligned} d\,'G_\alpha &= d\left(a^{-1\beta}_\alpha\right) \wedge G_\beta + a^{-1\beta}_\alpha dG_\beta \\ &= -a^{-1\rho}_\sigma \left(da^\sigma_\rho\right) a^{-1\beta}_\alpha \wedge G_\beta + a^{-1\beta}_\alpha dG_\beta \\ &= a^{-1\beta}_\alpha dG_\beta - a^{-1\beta}_\sigma \left(a^\sigma_\eta \Gamma^\eta_\rho - {}'\Gamma^\sigma_\eta a^\eta_\rho\right) a^{-1\rho}_\alpha G_\beta \\ &= a^{-1\rho}_\alpha \left(dG_\rho - \Gamma^\beta_\rho \wedge G_\beta\right) + {}'\Gamma^\sigma_\alpha \wedge {}'G_\sigma. \end{aligned}$$

We therefore have the gauge covariant exterior derivative

$$DG_\alpha = dG_\alpha - \Gamma^\beta_\alpha \wedge G_\beta, \qquad '(DG_\alpha) = a^{-1\beta}_\alpha DG_\beta \tag{10-5.19}$$

for

$$'G_\alpha = a^{-1\beta}_\alpha G_\beta. \tag{10-5.20}$$

Since DG_α has the same transformation law as G_α, we can do the same thing again:

$$\begin{aligned} D\,DG_\alpha &= d\left(dG_\alpha - \Gamma^\sigma_\alpha \wedge G_\sigma\right) - \Gamma^\lambda_\alpha \wedge \left(dG_\lambda - \Gamma^\beta_\lambda \wedge G_\beta\right) \\ &= -\left(d\Gamma^\sigma_\alpha - \Gamma^\lambda_\alpha \wedge \Gamma^\sigma_\lambda\right) \wedge G_\sigma \\ &= -\left(d\Gamma^\sigma_\alpha + \Gamma^\sigma_\lambda \wedge \Gamma^\lambda_\alpha\right) \wedge G_\sigma. \end{aligned}$$

Now, Γ^α_β are the connection 1-forms on group space, and hence the curvature 2-forms on group space are given by

$$\Theta^\sigma_\alpha = d\Gamma^\sigma_\alpha + \Gamma^\sigma_\lambda \wedge \Gamma^\lambda_\alpha. \tag{10-5.21}$$

We accordingly have (remember that $\deg(\Theta_\alpha^\sigma) = 2$)

$$D\,DG_\alpha = -\Theta_\alpha^\sigma \wedge G_\sigma = -G_\sigma \wedge \Theta_\alpha^\sigma \tag{10-5.22}$$

in agreement with (10-4.26).

The relations (10-5.22) are of particular use since they allow immediate determination of integrability conditions (the relations (10-5.22) are the gauge covariant exterior derivative analog of the integrability identity $d\,d\omega = 0$ of the ordinary exterior derivative). For example, if G_α is to satisfy $DG_\alpha = P_\alpha$, then covariant exterior differentiation yields $DP_\alpha = D\,DG_\alpha = -G_\sigma \wedge \Theta_\alpha^\sigma$ as integrability conditions that the quantities P_α have to satisfy. It is therefore useful to obtain an explicit evaluation of the quantities Θ_β^α. Combination of the defining relations $\Gamma_\beta^\alpha = C_{\rho\beta}^\alpha W^\rho$, $\Theta_\beta^\alpha = d\Gamma_\beta^\alpha + \Gamma_\sigma^\alpha \wedge \Gamma_\beta^\sigma$ gives

$$\Theta_\beta^\alpha = C_{\sigma\beta}^\alpha\, dW^\sigma + C_{\rho\sigma}^\alpha C_{\lambda\beta}^\sigma W^\rho \wedge W^\lambda. \tag{10-5.23}$$

However,

$$\begin{aligned} C_{\rho\sigma}^\alpha C_{\lambda\beta}^\sigma W^\rho \wedge W^\lambda &= \tfrac{1}{2}\left(C_{\rho\sigma}^\alpha C_{\lambda\beta}^\sigma - C_{\lambda\sigma}^\alpha C_{\rho\beta}^\sigma\right) W^\rho \wedge W^\lambda \\ &= \tfrac{1}{2}\left(C_{\rho\sigma}^\alpha C_{\lambda\beta}^\sigma + C_{\lambda\sigma}^\alpha C_{\beta\rho}^\sigma\right) W^\rho \wedge W^\lambda \\ &= -\tfrac{1}{2} C_{\beta\sigma}^\alpha C_{\rho\lambda}^\sigma W^\rho \wedge W^\lambda = \tfrac{1}{2} C_{\sigma\beta}^\alpha C_{\rho\lambda}^\sigma W^\rho \wedge W^\lambda, \end{aligned}$$

where the first equality follows from the skew symmetry of $W^\rho \wedge W^\lambda$ in the indices (ρ, λ), the second equality obtains from the skew symmetry of $C_{\rho\beta}^\sigma$ in the indices (ρ, β), the third equality follows from the Jacobi identity, and the fourth equality is a consequence of the skew symmetry of $C_{\beta\sigma}^\alpha$ in the indices (β, σ). When this is put back into (10-5.23), we see that

$$\Theta_\beta^\alpha = C_{\sigma\beta}^\alpha\left(dW^\sigma + \tfrac{1}{2} C_{\rho\lambda}^\sigma W^\rho \wedge W^\lambda\right). \tag{10-5.24}$$

Finally, we use the definition of the Yang-Mills field tensors F^α given by (10-4.20) to obtain

$$\Theta_\beta^\alpha = C_{\sigma\beta}^\alpha F^\sigma \tag{10-5.25}$$

which is the analog of $\boldsymbol{\Theta} = F^\alpha \boldsymbol{\gamma}_\alpha$. The direct correspondences between Ψ-space and group space quantities thus obtain in the following direct and symmetric fashion:

$$\boldsymbol{\Gamma} = W^\alpha \boldsymbol{\gamma}_\alpha, \qquad \boldsymbol{\Theta} = F^\alpha \boldsymbol{\gamma}_\alpha,$$

$$\Gamma_\beta^\alpha = W^\sigma C_{\sigma\beta}^\alpha, \qquad \Theta_\beta^\alpha = F^\sigma C_{\sigma\beta}^\alpha.$$

In fact, it is easily seen that

$$\boldsymbol{\Theta}\boldsymbol{\gamma}_\alpha - \boldsymbol{\gamma}_\alpha\boldsymbol{\Theta} = F^\sigma(\boldsymbol{\gamma}_\sigma\boldsymbol{\gamma}_\alpha - \boldsymbol{\gamma}_\alpha\boldsymbol{\gamma}_\sigma) = F^\sigma C^\beta_{\sigma\alpha}\boldsymbol{\gamma}_\beta = \Theta^\beta_\alpha\boldsymbol{\gamma}_\beta,$$

and hence we have the correspondence

$$\boldsymbol{\Theta}\boldsymbol{\gamma}_\alpha - \boldsymbol{\gamma}_\alpha\boldsymbol{\Theta} = \Theta^\beta_\alpha\boldsymbol{\gamma}_\beta.$$

Equations (10-5.2) and (10-5.6) show that F^α transforms like T^α, $'F^\alpha = a^\alpha_\beta F^\beta$, and hence $DF^\alpha = dF^\alpha + C^\alpha_{\sigma\beta}W^\sigma \wedge F^\beta$. Thus (10-5.1) gives

$$D\boldsymbol{\Theta} = (DF^\alpha)\boldsymbol{\gamma}_\alpha.$$

However, $\boldsymbol{\Theta} = d\boldsymbol{\Gamma} + \boldsymbol{\Gamma} \wedge \boldsymbol{\Gamma}$ implies the identical satisfaction of $D\boldsymbol{\Theta} = \mathbf{0}$, and hence we have $DF^\alpha = 0$. On the other hand $\Theta^\alpha_\beta = d\Gamma^\alpha_\beta + \Gamma^\alpha_\rho \wedge \Gamma^\rho_\beta$ implies the identical satisfaction of $D\Theta^\alpha_\beta = 0$ provided Θ^α_β has the transformation law $'\Theta^\alpha_\beta = a^\alpha_\rho\Theta^\rho_\sigma a_\beta^{-1\sigma}$. Thus since $'F^\sigma = a^\sigma_\lambda F^\lambda$ and $\Theta^\alpha_\beta = F^\sigma C^\alpha_{\sigma\beta}$, the constants of structure must have the transformation law

(10-5.26) $$'C^\alpha_{\sigma\beta} = a^\alpha_\lambda C^\lambda_{\gamma\rho}a_\sigma^{-1\gamma}a_\beta^{-1\rho}, \qquad 'C_{\alpha\beta} = C_{\gamma\rho}a_\alpha^{-1\gamma}a_\sigma^{-1\rho}.$$

Although the quantities $C^\alpha_{\sigma\beta}$ and $C_{\alpha\beta}$ have the transformation laws (10-5.26), they retain their "constant" attribute: the transformation law and (10-5.12) imply

$$\begin{aligned} DC^\alpha_{\beta\gamma} &= dC^\alpha_{\beta\gamma} + \Gamma^\alpha_\lambda C^\lambda_{\beta\gamma} - \Gamma^\lambda_\beta C^\alpha_{\lambda\gamma} - \Gamma^\lambda_\gamma C^\alpha_{\beta\lambda} \\ &= W^\rho(C^\alpha_{\rho\lambda}C^\lambda_{\beta\gamma} - C^\lambda_{\rho\beta}C^\alpha_{\lambda\gamma} - C^\lambda_{\rho\gamma}C^\alpha_{\beta\lambda}) \end{aligned}$$

which vanish when the Jacobi identity is used. We thus have

$$DC^\alpha_{\beta\gamma} = 0, \qquad DC_{\alpha\beta} = 0,$$

and the quantities $C^\alpha_{\beta\gamma}$ and $C_{\alpha\beta}$ are gauge covariant constant.

The reason for these calculations can now be made evident. We have seen that $\Theta^\alpha_\beta = F^\sigma C^\alpha_{\sigma\beta}$, and hence these relations imply

$$D\Theta^\alpha_\beta = (DF^\sigma)C^\alpha_{\sigma\beta} + F^\sigma DC^\alpha_{\sigma\beta} = 0;$$

that is,

(10-5.27) $$DF^\alpha = 0 \Rightarrow D\Theta^\alpha_\beta = 0.$$

(Note that $D\Theta^\alpha_\beta = 0$, $DF^\alpha = 0$ would imply the constraints $F^\sigma \wedge DC^\alpha_{\sigma\beta} = 0$ without an independent verification of $DC^\alpha_{\sigma\beta} = 0$.) It is also useful to note that these same results follow from exterior covariant differentiation of $\boldsymbol{\Theta}\boldsymbol{\gamma}_\alpha - \boldsymbol{\gamma}_\alpha\boldsymbol{\Theta} = \Theta_\alpha^\beta\boldsymbol{\gamma}_\beta$.

There is, as yet, one problem still outstanding: the determination of the transformations and covariant exterior derivatives of the connecting quantities $\{\boldsymbol{\gamma}_\alpha\}$. If we go back to (10-5.6) and (10-5.10), we have $\mathbf{T} = T^\alpha\boldsymbol{\gamma}_\alpha$ and $'\mathbf{T} = {'T^\alpha}\boldsymbol{\gamma}_\alpha$ because we have agreed to keep the basis for the matrix Lie algebra of G_0 fixed. On the other hand, (10-5.8) and (10-5.9) give $\bar{\boldsymbol{\gamma}}_\alpha = \mathbf{A}\boldsymbol{\gamma}_\alpha\mathbf{A}^{-1} = a_\alpha^\beta\boldsymbol{\gamma}_\beta$, and hence we may combine them to obtain the transformation laws

$$'\boldsymbol{\gamma}_\alpha = a_\alpha^{-1\beta}\mathbf{A}\boldsymbol{\gamma}_\beta\mathbf{A}^{-1} = \boldsymbol{\gamma}_\alpha, \tag{10-5.28}$$

in which case $'\mathbf{T} = {'T^\alpha}{'\boldsymbol{\gamma}_\alpha} = {'T^\alpha}\boldsymbol{\gamma}_\alpha$. The transformation laws (10-5.28) thus make the connecting quantities $\boldsymbol{\gamma}_\alpha$ *absolute invariants* under the action of G; the gauge group does not change the basis for the matrix Lie algebra of G_0. Further, the transformation laws (10-5.28) serve to define the gauge covariant derivative of the connecting quantities $\boldsymbol{\gamma}_\alpha$ by

$$D\boldsymbol{\gamma}_\alpha = d\boldsymbol{\gamma}_\alpha + \boldsymbol{\Gamma}\boldsymbol{\gamma}_\alpha - \boldsymbol{\gamma}_\alpha\boldsymbol{\Gamma} - \Gamma_\alpha^\beta\boldsymbol{\gamma}_\beta.$$

We then need only substitute $\boldsymbol{\Gamma} = W^\sigma\boldsymbol{\gamma}_\sigma$, $\Gamma_\alpha^\beta = W^\sigma C_{\sigma\alpha}^\beta$ into this definition and note that $d\boldsymbol{\gamma}_\alpha = \mathbf{0}$ in order to obtain

$$\begin{aligned} D\boldsymbol{\gamma}_\alpha &= W^\sigma(\boldsymbol{\gamma}_\sigma\boldsymbol{\gamma}_\alpha - \boldsymbol{\gamma}_\alpha\boldsymbol{\gamma}_\sigma) - W^\sigma C_{\sigma\alpha}^\beta\boldsymbol{\gamma}_\beta \\ &= W^\sigma\{C_{\sigma\alpha}^\beta\boldsymbol{\gamma}_\beta - C_{\sigma\alpha}^\beta\boldsymbol{\gamma}_\beta\} \equiv \mathbf{0}. \end{aligned}$$

The connecting quantities $\{\boldsymbol{\gamma}_\alpha\}$ *are absolute gauge covariant constant matrices,*

$$D\boldsymbol{\gamma}_\alpha = \mathbf{0}, \qquad d\boldsymbol{\gamma}_\alpha = \mathbf{0}. \tag{10-5.29}$$

There is actually no real choice in this matter. We have seen that the definition of the gauge covariant exterior derivative of $\mathbf{T}$ gives $D\mathbf{T} = (DT^\alpha)\boldsymbol{\gamma}_\alpha$ for $\mathbf{T} = T^\alpha\boldsymbol{\gamma}_\alpha$ and $'\mathbf{T} = \mathbf{ATA}^{-1}$. On the other hand $D\mathbf{T} = (DT^\alpha)\boldsymbol{\gamma}_\alpha + (-1)^k T^\alpha \wedge D\boldsymbol{\gamma}_\alpha = (DT^\alpha)\boldsymbol{\gamma}_\alpha$ for all $\mathbf{T}$ only if $D\boldsymbol{\gamma}_\alpha = \mathbf{0}$. In fact, most discussions in the literature begin with the requirements $D\boldsymbol{\gamma}_\alpha = \mathbf{0}$.

10-6. DERIVATION OF THE FIELD EQUATIONS

The Lagrangian for the problem is given by

$$L = \mathscr{M}_\Gamma L_0 + s_1 L_1, \tag{10-6.1}$$

where L_1 is a gauge invariant function of the compensating fields W_i^α and their first derivatives and s_1 is a coupling constant. We thus have

$$L = L\left(x^j, \Psi^A, \partial_k \Psi^A + W_k^\alpha \gamma_{\alpha B}^A \Psi^B, W_k^\alpha, \partial_m W_k^\alpha\right).$$

In practice, things are not as general as this, for L_1 is usually assumed to be constructed solely from the components of the Yang-Mills field tensor

$$F_{ij}^\alpha = \partial_i W_j^\alpha - \partial_j W_i^\alpha + C_{\beta\rho}^\alpha W_i^\beta W_j^\rho. \tag{10-6.2}$$

Thus the only other place where the W's occur is in the arguments

$$y_k^A = D_k \Psi^A = \partial_k \Psi^A + W_k^\alpha \gamma_{\alpha B}^A \Psi^B. \tag{10-6.3}$$

We accordingly assume that L is given by

$$L = \hat{L}\left(x^j, \Psi^A, y_k^A, F_{ij}^\alpha\right). \tag{10-6.4}$$

Notice that (10-6.4) allows for dependence on the torsion 2-forms $\boldsymbol{\Sigma} = \boldsymbol{\Theta\Psi}$ since they may be expressed in terms of Ψ^A and F_{ij}^α. It is therefore useful to make use of the derivatives

$$Z_A = \partial \hat{L} / \partial \Psi^A, \tag{10-6.5}$$

$$M_A^k = \partial \hat{L} / \partial y_k^A, \tag{10-6.6}$$

$$G_\alpha^{ij} = \partial \hat{L} / \partial F_{ij}^\alpha, \tag{10-6.7}$$

where $\hat{L}$ is understood to be an explicit function of the variables x^j, Ψ^A, y_k^A and F_{ij}^α only.

The action for this problem is

$$A\left[\Psi^A, W_i^\alpha\right] = \int_{B_n} \hat{L}\left(x^j, \Psi^A, y_k^A, F_{ij}^\alpha\right)\mu \tag{10-6.8}$$

for the field quantities $\{\Psi^A, W_i^\alpha\}$. Let $\phi^A(x^j)$ be zero forms and let $\eta^\alpha(x^j)$ be 1-forms on E_n. The results established in Chapter Seven then show that $\{\phi^A, \eta_i^\alpha\}$ can be used to embed the variables Ψ^A, W_i^α in the 1-parameter family of variables by

$$\Psi^A \mapsto \Psi^A + s\delta\Psi^A + o(s), \qquad W_i^\alpha \mapsto W_i^\alpha + s\delta W_i^\alpha + o(s) \tag{10-6.9}$$

with

$$\delta\Psi^A = \phi^A, \qquad \delta W_i^\alpha = \eta_i^\alpha. \tag{10-6.10}$$

The definition of the variation of the action then gives

$$\delta A = \int_{B_n} \left(\frac{\partial \hat{L}}{\partial \Psi^A} \delta \Psi^A + \frac{\partial \hat{L}}{\partial y_k^A} \delta y_k^A + \frac{\partial \hat{L}}{\partial F_{ij}^{\alpha}} \delta F_{ij}^{\alpha} \right) \mu . \tag{10-6.11}$$

We therefore consider the n-form

$$\delta(\hat{L}\mu) = \left(\frac{\partial \hat{L}}{\partial \Psi^A} \delta \Psi^A + \frac{\partial \hat{L}}{\partial y_k^A} \delta y_k^A + \frac{\partial \hat{L}}{\partial F_{ij}^{\alpha}} \delta F_{ij}^{\alpha} \right) \mu \tag{10-6.12}$$

$$= \left(Z_A \delta \Psi^A + M_A^k \delta y_k^A + G_\alpha^{ij} \delta F_{ij}^{\alpha} \right) \mu .$$

It thus remains to determine the variations δy_i^A and δF_{ij}^{α} that are induced by the variations (10-6.9) and the defining relations (10-6.2), (10-6.3).

We first substitute (10-6.9) into (10-6.3). This gives

$$y_k^A \mapsto y_k^A + s\left(\partial_k \phi^A + W_k^{\alpha} \gamma_{\alpha B}^A \phi^B + \eta_k^{\alpha} \gamma_{\alpha \beta}^A \Psi^B \right) + o(s),$$

and hence

$$\delta y_k^A = \partial_k \phi^A + W_k^{\alpha} \gamma_{\alpha B}^A \phi^B + \eta_k^{\alpha} \gamma_{\alpha B}^A \Psi^B . \tag{10-6.13}$$

A similar calculation starting with (10-6.2) shows that

$$\delta F_{ij}^{\alpha} = \partial_i \eta_j^{\alpha} - \partial_j \eta_i^{\alpha} + C_{\beta\rho}^{\alpha}\left(W_i^{\beta} \eta_j^{\rho} - W_j^{\beta} \eta_i^{\rho} \right); \tag{10-6.14}$$

that is,

$$\delta F^{\alpha} = d\eta^{\alpha} + C_{\beta\rho}^{\alpha} W^{\beta} \wedge \eta^{\rho} = D\eta^{\alpha} . \tag{10-6.15}$$

When these evaluations are put back into (10-6.12), we obtain the explicit evaluation

$$\delta(\hat{L}\mu) = \left\{ Z_A \phi^A + M_A^k\left(\partial_k \phi^A + W_k^{\alpha} \gamma_{\alpha B}^A \phi^B + \eta_k^{\alpha} \gamma_{\alpha B}^A \Psi^B \right) \right.$$

$$\left. + G_\alpha^{ij}\left(\partial_i \eta_j^{\alpha} - \partial_j \eta_i^{\alpha} + C_{\beta\rho}^{\alpha}\left(W_i^{\beta} \eta_j^{\rho} - W_j^{\beta} \eta_i^{\rho} \right) \right) \right\} \mu ;$$

that is,

$$\delta(\hat{L}\mu) = \delta_{\Psi}(\hat{L}\mu) + \delta_W(\hat{L}\mu) \tag{10-6.16}$$

with

$$\text{(10-6.17)} \quad \delta_{\Psi}(\hat{L}\mu) = \left\{ Z_A \phi^A + M_A^k \left(\partial_k \phi^A + W_k^\alpha \gamma_{\alpha B}^A \phi^B \right) \right\} \mu,$$

$$\text{(10-6.18)} \quad \delta_W(\hat{L}\mu) = \left\{ M_A^k \eta_k^\alpha \gamma_{\alpha B}^A \Psi^B \right.$$

$$\left. + G_\alpha^{ij} \left(\partial_i \eta_j^\alpha - \partial_j \eta_i^\alpha + C_{\beta\rho}^\alpha \left(W_i^\beta \eta_j^\rho - W_j^\beta \eta_i^\rho \right) \right) \right\} \mu.$$

Noting that $\delta_\Psi(\hat{L}\mu)$ is linear and homogeneous in the variations $\phi^A = \delta\Psi^A$, and that $\delta_W(\hat{L}\mu)$ is linear and homogeneous in the variations $\eta_i^\alpha = \delta W_i^\alpha$, we may analyze (10-6.17) and (10-6.18) separately.

Let us first introduce the $(n-1)$-forms M_A by

$$\text{(10-6.19)} \qquad M_A = M_A^k \mu_k.$$

It is then a simple matter to see that (remember $dx^k \wedge \mu_i = \delta_i^k \mu$)

$$\text{(10-6.20)} \qquad d\phi^A \wedge M_A = M_A^k \partial_k \phi^A \mu.$$

Next, if we introduce the components Γ_B^A of the matrix of connection 1-forms by

$$\text{(10-6.21)} \qquad \Gamma_B^A = W_k^\alpha \gamma_{\alpha B}^A \, dx^k,$$

then

$$\text{(10-6.22)} \qquad \Gamma_B^A \wedge M_A = W_k^\alpha \gamma_{\alpha B}^A M_A^k \mu.$$

Comparison of (10-6.17), (10-6.20), and (10-6.22) then shows that

$$\text{(10-6.23)} \quad \delta_\Psi(\hat{L}\mu) = \phi^A Z_A \mu + d\phi^A \wedge M_A + \phi^B \Gamma_B^A \wedge M_A$$

$$= \phi^B \left\{ Z_B \mu - dM_B + \Gamma_B^A \wedge M_A \right\} + d\left(\phi^B M_B \right).$$

It is now simply a matter of noting that

$$\int_{B_n} d\left(\phi^B M_B \right) = \int_{\partial B_n} \phi^B M_B = 0$$

for variations that vanish on the boundary, ∂B_n, in order to see that *the field equations for the* Ψ *fields are*

$$\text{(10-6.24)} \qquad dM_B - \Gamma_B^A \wedge M_A = Z_B \mu, \qquad B = 1, \ldots, N,$$

with

(10-6.25) $$M_B = \frac{\partial \hat{L}}{\partial y_k^B} \mu_k, \qquad Z_B = \frac{\partial \hat{L}}{\partial \Psi^B},$$

(10-6.26) $$\Gamma_B^A = W_k^\alpha \gamma_{\alpha B}^A \, dx^k.$$

Next, we introduce the $(n-2)$-forms G_α by

(10-6.27) $$G_\alpha = \tfrac{1}{2} G_\alpha^{ij} \mu_{ij},$$

and put

(10-6.28) $$\eta^\alpha = \eta_i^\alpha \, dx^i, \qquad W^\alpha = W_i^\alpha \, dx^i.$$

Noting that $dx^k \wedge dx^m \wedge \mu_{ij} = (\delta_i^m \delta_j^k - \delta_j^m \delta_i^k)\mu$, direct calculations show that

(10-6.29) $$G_\alpha^{ij}(\partial_i \eta_j^\alpha - \partial_j \eta_i^\alpha)\mu = -2\, d\eta^\alpha \wedge G_\alpha,$$

(10-6.30) $$G_\alpha^{ij} C_{\beta\rho}^\alpha \left(W_i^\beta \eta_j^\rho - W_j^\beta \eta_i^\rho \right)\mu = 2\eta^\rho \wedge W^\beta C_{\beta\rho}^\alpha \wedge G_\alpha,$$

while (10-6.25) gives

(10-6.31) $$M_A^k \eta_k^\alpha \gamma_{\alpha B}^A \Psi^B \mu = \eta^\alpha \wedge M_A \gamma_{\alpha B}^A \Psi^B.$$

Thus if we introduce the $(n-1)$-forms J_α by

(10-6.32) $$J_\alpha = M_A \gamma_{\alpha B}^A \Psi^B \qquad \left(= \left. \frac{\partial L}{\partial W_i^\alpha} \right|_F \mu_i \right),$$

(10-6.17) and (10-6.29) through (10-6.32) yield

(10-6.33) $$\begin{aligned} \delta_W(\hat{L}\mu) &= \eta^\alpha \wedge J_\alpha - 2\, d\eta^\alpha \wedge G_\alpha + 2\eta^\rho \wedge W^\beta C_{\beta\rho}^\alpha \wedge G_\alpha \\ &= \eta^\rho \wedge \left\{ J_\rho - 2\, dG_\rho + 2 W^\beta C_{\beta\rho}^\alpha \wedge G_\alpha \right\} - 2d(\eta^\alpha \wedge G_\alpha). \end{aligned}$$

(Remember that η^α are 1-forms.) The same reasoning as that used in the case of δ_Ψ gives the field equations

$$dG_\rho - W^\beta C_{\beta\rho}^\alpha \wedge G_\alpha = \tfrac{1}{2} J_\rho.$$

It is now simply a matter of recalling (10-5.15) that define the group space connection 1-forms, $\Gamma_\rho^\alpha = C_{\beta\rho}^\alpha W^\beta$, in order to see that *the field equations for the*

W^α fields are

(10-6.34) $$dG_\rho - \Gamma^\alpha_\rho \wedge G_\alpha = \tfrac{1}{2} J_\rho,$$

with

(10-6.35) $$G_\alpha = \frac{1}{2}\frac{\partial \hat{L}}{\partial F^\alpha_{ij}}\mu_{ij}, \qquad J_\rho = \frac{\partial \hat{L}}{\partial y^A_k}\gamma^A_{\rho B}\Psi^B\mu_k = \left.\frac{\partial \hat{L}}{\partial W^\rho_k}\right|_F \mu_k,$$

(10-6.36) $$\Gamma^\alpha_\rho = C^\alpha_{\beta\rho}W^\beta.$$

10-7. TRANSFORMATION PROPERTIES AND GAUGE COVARIANCE

The first thing to be done is to obtain the transformation laws for the quantities Z_A, M^k_A, J_α, and G^{ij}_α that are induced by the gauge group. These follow directly from the invariance of the Lagrangian $\hat{L}$ under gauge transformations, as we now proceed to show. The arguments of $\hat{L}$ are $(x^j, \Psi^A, y^A_k = D_k\Psi^A, F^\alpha_{ij})$ that have the transformation laws

(10-7.1) $$'\Psi^A = A^A_B\Psi^B, \qquad 'y^A_k = '(D_k\Psi^A) = A^A_B y^B_k,$$
$$'F^\alpha_{ij} = a^\alpha_\beta F^\beta_{ij} \qquad (\text{i.e., } 'F^\alpha = a^\alpha_\beta F^\beta).$$

Invariance of the Lagrangian function $\hat{L}$ is then explicitly given by

(10-7.2) $$'\hat{L}(x^i, '\Psi^A, 'y^A_k, 'F^\alpha_{ij}) = '\hat{L}(x^i, A^A_B\Psi^B, A^A_B y^B_k, a^\alpha_\beta F^\beta_{ij})$$
$$= \hat{L}(x^j, \Psi^A, y^A_k, F^\alpha_{ij}).$$

The transformed quantities $'Z_A$, $'M^k_A$, $'G^{ij}_\alpha$ are defined by

(10-7.3) $$'Z_A = \partial'\hat{L}/\partial'\Psi^A, \qquad 'M^k_A = \partial'\hat{L}/\partial'y^A_k, \qquad 'G^{ij}_\alpha = \partial'\hat{L}/\partial'F^\alpha_{ij}.$$

Now, $Z_A = \partial\hat{L}/\partial\Psi^A$, hence

$$Z_A = \frac{\partial \hat{L}}{\partial \Psi^A} = \frac{\partial'\hat{L}}{\partial'\Psi^B}\frac{\partial'\Psi^B}{\partial\Psi^A} = 'Z_B A^B_A$$

by (10-7.1) through (10-7.3). Similarly,

$$M^k_A = \frac{\partial \hat{L}}{\partial y^A_k} = \frac{\partial'\hat{L}}{\partial'y^B_m}\frac{\partial'y^B_m}{\partial y^A_k} = 'M^k_B A^B_A,$$

$$G^{ij}_\alpha = \frac{\partial \hat{L}}{\partial F^\alpha_{ij}} = \frac{\partial'L}{\partial'F^\beta_{km}}\frac{\partial'F^\beta_{km}}{\partial F^\alpha_{ij}} = 'G^{ij}_\beta a^\beta_\alpha.$$

We therefore have

(10-7.4) $$'Z_A = A_A^{-1B} Z_B, \qquad 'M_A^k = A_A^{-1B} M_B^k$$

and

(10-7.5) $$'G_\alpha^{ij} = a_\alpha^{-1\beta} G_\beta^{ij}.$$

The exterior forms M_A and G_α thus have the transformation laws

(10-7.6) $$'M_A = A_A^{-1B} M_B,$$

(10-7.7) $$'G_\alpha = a_\alpha^{-1\beta} G_\beta.$$

Finally, (10-6.32) gives $J_\alpha = M_A \gamma_{\alpha B}^A \Psi^B$ and hence for a different gauge that is obtained by allowing $\mathbf{A} \in G$ to act, we have

(10-7.8) $$'J_\alpha = 'M_A \gamma_{\alpha B}^A {}' \Psi^B.$$

However,

$$J_\alpha = M_A \gamma_{\alpha B}^A \Psi^B = 'M_R A_A^R \gamma_{\alpha B}^A A_S^{-1B} {}' \Psi^S$$

$$= 'M_R a_\alpha^\beta \gamma_{\beta S}^R {}' \Psi^S = a_\alpha^\beta {}' J_\beta,$$

and hence we obtain the transformation law

(10-7.9) $$'J_\alpha = a_\alpha^{-1\beta} J_\beta.$$

Let $\mathbf{M}$ be the row matrix of $(n-1)$-forms with entries $\{M_1, M_2, \ldots, M_N\}$. The transformation law (10-7.6) then becomes $'\mathbf{M} = \mathbf{M}\mathbf{A}^{-1}$ so that $\mathbf{M}$ has the gauge covariant exterior derivative

$$D\mathbf{M} = d\mathbf{M} - (-1)^{n-1} \mathbf{M} \wedge \mathbf{\Gamma}$$

with the transformation law $'(D\mathbf{M}) = (D\mathbf{M})\mathbf{A}^{-1}$. When these are written back into component form, it follows that

(10-7.10) $$DM_A = dM_A - (-1)^{n-1} M_B \wedge \Gamma_A^B = dM_A - \Gamma_A^B \wedge M_B,$$

(10-7.11) $$'(DM_A) = (DM_B) A_A^{-1B} = A_A^{-1B} DM_B.$$

A comparison of (10-7.10) and (10-6.24) then shows that these field equations can be written in the equivalent form $DM_A = Z_A \mu$. Further, (10-7.11) and (10-7.4) give

$$'(DM_A) = A_A^{-1B} DM_B = A_A^{-1B} Z_B \mu = 'Z_A \mu$$

under action by the gauge group. *The field equations for the* Ψ *fields have the gauge covariant form*

$$DM_A = Z_A\mu; \tag{10-7.12}$$

they are form invariant under action by the gauge group.

The transformation law (10-7.7) for the G's makes all of the material of Section 10-5 available. We therefore have the gauge covariant exterior derivative

$$DG_\alpha = dG_\alpha - \Gamma^\beta_\alpha \wedge G_\beta \tag{10-7.13}$$

with the transformation law

$$'(DG_\alpha) = a_\alpha^{-1\beta}\, DG_\beta. \tag{10-7.14}$$

A direct comparison of (10-7.13) and (10-6.35) shows that the field equations for the W^α fields may be written in the equivalent form $DG_\alpha = \frac{1}{2}J_\alpha$. When these equations are combined with (10-7.14), we obtain the field equations

$$'(DG_\alpha) = a_\alpha^{-1\beta}\, dG_\beta = \tfrac{1}{2}a_\alpha^{-1\beta}J_\beta = \tfrac{1}{2}\,'J_\beta$$

that result under the action of the gauge group. *The field equations for the* W^α *fields have the gauge covariant form*

$$DG_\alpha = \tfrac{1}{2}J_\alpha. \tag{10-7.15}$$

10-8. INTEGRABILITY CONDITIONS AND CURRENT CONSERVATION

The gauge covariant field equations of the theory are

$$DM_A = Z_A\mu \tag{10-8.1}$$

for the Ψ fields and

$$DG_\alpha = \tfrac{1}{2}J_\alpha \tag{10-8.2}$$

for the W^α fields. The latter are supplemented by the first half of the Yang-Mills equations

$$DF^\alpha = 0 \tag{10-8.3}$$

which are identically satisfied by

(10-8.4) $$F^\alpha = dW^\alpha + \tfrac{1}{2}C^\alpha_{\beta\gamma}W^\beta \wedge W^\gamma.$$

Noting that the field equations involve gauge covariant exterior derivatives, they must entail satisfaction of the associated integrability conditions.

First, DM_A and $Z_A\mu$ are n-forms on E_n and hence further covariant exterior differentiation will only lead to identically satisfied statements $0 = 0$ (an $(n+1)$-form on E_n vanishes identically). Accordingly, the field equations (10-8.1) entail no further conditions of integrability.

Second, $\{F^\alpha\}$ are 2-forms on E_n with the transformation law $'F^\alpha = a^\alpha_\beta F^\beta$ and hence covariant exterior differentiation of (10-8.3) gives

$$0 = D\,DF^\alpha = d\left(dF^\alpha + \Gamma^\alpha_\beta \wedge F^\beta\right) + \Gamma^\alpha_\sigma \wedge \left(dF^\sigma + \Gamma^\sigma_\beta \wedge F^\beta\right)$$

$$= \left(d\Gamma^\alpha_\beta + \Gamma^\alpha_\sigma \wedge \Gamma^\sigma_\beta\right) \wedge F^\beta = \Theta^\alpha_\beta \wedge F^\beta.$$

However, use of (10-5.25) gives $\Theta^\alpha_\beta \wedge F^\beta = C^\alpha_{\rho\beta}F^\rho \wedge F^\beta$ which vanishes because of the skew symmetry of $C^\alpha_{\rho\beta}$ in the indices (ρ, β) and of the symmetry of $F^\rho \wedge F^\beta$ in the indices (ρ, β). (Recall that F^α is a 2-form for each value of α.) The integrability conditions for the field equations (10-8.3) are thus identically satisfied.

We now turn to the field equations (10-8.2) which are another matter altogether. Covariant exterior differentiation gives

(10-8.5) $$DJ_\alpha = 2\,DDG_\alpha = -2\Theta^\beta_\alpha \wedge G_\beta$$

when (10-5.22) is used. When the evaluation $\Theta^\beta_\alpha = C^\beta_{\sigma\alpha}F^\sigma$ given by (10-5-25) is used, it follows that

$$\Theta^\beta_\alpha \wedge G_\beta = C^\beta_{\sigma\alpha}F^\sigma \wedge G_\beta = -C^\beta_{\alpha\sigma}F^\sigma \wedge G_\beta.$$

The field equations (10-8.2) *entail the integrability conditions*

(10-8.6) $$DJ_\alpha = 2C^\beta_{\alpha\sigma}F^\sigma \wedge G_\beta.$$

For later convenience, we obtain an explicit evaluation of $2C^\beta_{\alpha\sigma}F^\sigma \wedge G_\beta$. This is simply a matter of recalling that $F^\sigma = \frac{1}{2}F^\sigma_{ij}\,dx^i \wedge dx^j$ and that $G_\beta = \frac{1}{2}(\partial\hat{L}/\partial F^\beta_{km})\mu_{km}$. Thus

$$2C^\beta_{\alpha\sigma}F^\sigma \wedge G_\beta = \frac{1}{2}C^\beta_{\alpha\sigma}F^\sigma_{ij}\frac{\partial\hat{L}}{\partial F^\beta_{km}}\,dx^i \wedge dx^j \wedge \mu_{km}$$

$$= \frac{1}{2}C^\beta_{\alpha\sigma}\left(F^\sigma_{ji} - F^\sigma_{ij}\right)\frac{\partial\hat{L}}{\partial F^\beta_{ij}}\mu.$$

Since $F_{ij}^{\sigma} = -F_{ji}^{\sigma}$, we obtain

(10-8.7) $$2C_{\alpha\sigma}^{\beta} F^{\sigma} \wedge G_{\beta} = -C_{\alpha\sigma}^{\beta} F_{ij}^{\sigma} \frac{\partial \hat{L}}{\partial F_{ij}^{\beta}} \mu.$$

The $(n-1)$-forms J_{α} have the evaluation (see (10-6.6), (10-6.19), and (10-6.32))

(10-8.8) $$J_{\alpha} = M_A \gamma_{\alpha B}^{A} \Psi^B = \frac{\partial \hat{L}}{\partial y_i^A} \gamma_{\alpha B}^{A} \Psi^B \mu_i = \frac{\partial \hat{L}}{\partial W_k^{\alpha}} \mu_k$$

and hence they are referred to as *current* $(n-1)$*-forms* of the Ψ fields. Accordingly, the integrability conditions (10-8.6) show that *the current* $(n-1)$*-forms are gauge covariantly conserved,*

(10-8.9) $$DJ_{\alpha} = 0,$$

if and only if

(10-8.10) $$C_{\alpha\sigma}^{\beta} F^{\sigma} \wedge G_{\beta} = 0.$$

The evaluation (10-8.7) then gives the following important result. *The current* $(n-1)$*-forms* J_{α} *are gauge covariantly conserved if and only if the Lagrangian function* $\hat{L}$ *depends on the arguments* $\{F_{ij}^{\alpha}\}$ *in such a way that*

(10-8.11) $$C_{\alpha\sigma}^{\beta} F_{ij}^{\sigma} \frac{\partial \hat{L}}{\partial F_{ij}^{\beta}} = 0, \qquad \alpha = 1, \ldots, r.$$

The usual situation in physics is that of Lagrangian functions that are quadratic in the derivatives of the field quantities (i.e., quadratic in the F_{ij}^{α}). When we recall the minimal coupling argument, the only place where the quantities F_{ij}^{α} enter is in the W_k^{α}-field Lagrangian L_1. It is therefore useful to consider the case where

(10-8.12) $$\hat{L}_1 = \tfrac{1}{2} F_{km}^{\gamma} h^{ki} S_{\gamma\beta} h^{mj} F_{ij}^{\beta},$$

where $(S_{\gamma\beta}, h^{ij})$ are symmetric in their respective indices. Since F^{α} transforms under the action of an element of G by $'F^{\alpha} = a_{\beta}^{\alpha} F^{\beta}$, gauge invariance of $\hat{L}_1$ is secured by the requirements that $S_{\alpha\beta}$ have the transformation law

(10-8.13) $$a_{\rho}^{\alpha} {}' S_{\alpha\beta} a_{\sigma}^{\beta} = S_{\rho\sigma}.$$

A substitution of $\hat{L}_1$ for $\hat{L}$ in (10-8.11) thus gives

$$C^{\beta}_{\alpha\sigma} F^{\sigma}_{ij} \frac{\partial \hat{L}_1}{\partial F^{\beta}_{ij}} = C^{\beta}_{\alpha\sigma} S_{\beta\gamma} F^{\sigma}_{ij} h^{im} h^{jn} F^{\gamma}_{mn}, \tag{10-8.14}$$

and hence the Lagrangian (10-8.12) satisfies the conditions (10-8.11) whenever

$$C^{\beta}_{\alpha\sigma} S_{\beta\gamma} = -C^{\beta}_{\alpha\gamma} S_{\beta\sigma}. \tag{10-8.15}$$

Most gauge theories that are studied in the literature start with field Lagrangian functions $\hat{L}_1$ that have the form given by (10-8.12) with $S_{\alpha\beta}$ taken as the components of the Cartan-Killing form. This choice satisfies the conditions (10-8.13) and (10-8.15), and hence gauge covariant conservation of the currents J_α are built in from the start. It is important to realize that gauge covariant conservation of the currents need not obtain for arbitrary gauge invariant choices of the field Lagrangian $\hat{L}_1$, for we have

$$DJ_\alpha = 2C^{\beta}_{\alpha\sigma} F^{\sigma} \wedge G_\beta$$

in the general case. Accordingly, (10-8.7) shows that there is a production of current

$$-C^{\beta}_{\alpha\sigma} F^{\sigma}_{ij} \frac{\partial \hat{L}}{\partial F^{\beta}_{ij}}$$

per unit of volume in the general case. On the other hand, since gauge covariant conservation of current may be taken to be one of the underlying postulates of gauge theory, *we are restricted to those Lagrangian functions* $\hat{L}$ *that satisfy the system of linear first order partial differential equations* (10-8.11).

There is an interesting and direct parallel that can be drawn between the W^α field equations

$$DF^\alpha = 0, \qquad DG_\alpha = \tfrac{1}{2} J_\alpha \tag{10-8.16}$$

and the field equations of electrodynamics with electric and magnetic currents. In order to see this, we write out the covariant exterior derivatives in (10-8.16) and use the fact that $\Gamma^{\alpha}_{\beta} = C^{\alpha}_{\sigma\beta} W^\sigma$:

$$dF^\alpha = -C^{\alpha}_{\beta\gamma} W^\beta \wedge F^\gamma, \tag{10-8.17}$$

$$DG_\alpha = \tfrac{1}{2} J_\alpha + C^{\beta}_{\sigma\alpha} W^\sigma \wedge G_\beta. \tag{10-8.18}$$

The first of these then implies

$$d\left(C^{\alpha}_{\beta\gamma} W^\beta \wedge F^\gamma\right) = 0, \tag{10-8.19}$$

while the second gives

(10-8.20) $$d\left(\tfrac{1}{2}J_\alpha + C^{\beta}_{\sigma\alpha}W^\sigma \wedge G_\beta\right) = 0.$$

The results of Chapter Nine can be used to identify $-C^{\alpha}_{\beta\gamma}W^\beta \wedge F^\gamma$ with r magnetic current $(n-1)$-forms and $\frac{1}{2}J_\alpha + C^{\beta}_{\sigma\alpha}W^\sigma \wedge G_\beta$ with r electric current $(n-1)$-forms (i.e., we identify (10-8.17) with the second of (9-2.7) for each value of α and (10-8.18) with the first of (9-2.7) for each value of α). The important thing to note here is that the magnetic current arises solely from the W^α fields. The electric current consists, on the other hand, of the sum of two currents; one is a pure field current $C^{\beta}_{\sigma\alpha}W^\sigma \wedge G_\beta$ while the other is the current $\frac{1}{2}J_\alpha$ that arises from the presence of $\boldsymbol{\Psi}$ fields. Accordingly (10-8.20) is a statement of the conservation of the total electric current. The reason why we do not use the conservation laws (10-8.19) and (10-8.20) directly in the theory is that the various quantities change drastically under gauge transformations. This is because the currents have parts that arise from the W^α fields and these fields do *not* transform homogeneously under gauge transformations. In fact, all that can be relied on in the context of gauge theories are those equations that are gauge covariant, namely

$$DF^\alpha = 0, \qquad DG_\alpha = \tfrac{1}{2}J_\alpha.$$

10-9. GAUGE GROUP ORBITS AND THE ANTIEXACT GAUGE

It is well known that the problem of solving Maxwell's equations is significantly simplified by an appropriate choice of gauge. Argument by analogy with electrodynamics leads directly to a similar expectation in the case of Yang-Mills fields. This expectation can be fully realized, and in a strictly algebraic way, as we now proceed to show.

Let G be an r-parameter matrix Lie group and, as before, let $\{\boldsymbol{\gamma}_\alpha, \alpha = 1,\ldots,r\}$ be a basis for the Lie algebra of G_0. We know that any matrix $\boldsymbol{\Gamma}$ of connection 1-forms of G takes values in the Lie algebra of G_0. Thus any such $\boldsymbol{\Gamma}$ belongs to the collection

(10-9.1) $$\mathcal{Y} = \left\{W^\alpha \boldsymbol{\gamma}_\alpha \mid W^\alpha \in \Lambda^1(E_n),\ \alpha = 1,\ldots,r\right\}.$$

In fact, at this point any element of $\mathcal{Y}$ could act as a matrix of connection 1-forms of G. Further, it follows from (10-4.5) that any element $\mathbf{A}$ of G generates the transformation

(10-9.2) $$'\boldsymbol{\Gamma} = \mathbf{A}\boldsymbol{\Gamma}\mathbf{A}^{-1} - d\mathbf{A}\,\mathbf{A}^{-1}$$

and hence G may be considered to act on the collection $\mathcal{Y}$ via (10-9.2):

(10-9.3) $$G_Y : \mathcal{Y} \to \mathcal{Y} \mid {}'\boldsymbol{\Gamma} = \mathbf{A}\boldsymbol{\Gamma}\mathbf{A}^{-1} - d\mathbf{A}\,\mathbf{A}^{-1}, \qquad \mathbf{A} \in G.$$

Since the Lagrangian function for the problem is invariant under the action of G (is gauge invariant), the Euler-Lagrange field equations for the problem are gauge covariant. Accordingly, the gauge group will map solutions of the field equations onto solutions, and hence the action of G_Y is an equivalence relation on $\mathscr{Y}$ (gauge equivalence). Under these circumstances, $\mathscr{Y}$ becomes a fiber space $\mathscr{Y}/G_Y$ under the identification of its equivalent elements. Thus it is only necessary that we characterize a cross section $\mathscr{K}$ of this fiber space in order to know all of $\mathscr{Y}$, for $\mathscr{Y}$ is then the orbit of $\mathscr{K}$ under the action of G_Y.

Pick a point P of E_n as center for the construction of the linear homotopy operation H, and let $\mathbf{\Gamma}$ denote a generic element of $\mathscr{Y}$. The matrix $\mathbf{A}_\Gamma$ that solves the Riemann-Graves integral equation

$$\mathbf{A}_\Gamma = \mathbf{I} + H(\mathbf{A}_\Gamma \mathbf{\Gamma}) \tag{10-9.4}$$

belongs to G because $\mathbf{\Gamma}$ is a matrix of infinitesimal generating 1-forms of G (Theorems 5-13.1 with the $\mathbf{A}$ of the theorem given by $\mathbf{A} = \mathbf{A}_\Gamma^{-1}$). Exterior differentiation of (10-9.4) yields $d\mathbf{A}_\Gamma = dH(\mathbf{A}_\Gamma \mathbf{\Gamma})$. However, H satisfies $dH + Hd =$ identity, and hence we have

$$d\mathbf{A}_\Gamma = \mathbf{A}_\Gamma \mathbf{\Gamma} - Hd(\mathbf{A}_\Gamma \mathbf{\Gamma}). \tag{10-9.5}$$

When this is substituted into the right-hand side of (10-9.2), we obtain

$$'\mathbf{\Gamma} = (Hd(\mathbf{A}_\Gamma \mathbf{\Gamma}))\mathbf{A}_\Gamma^{-1}.$$

$'\mathbf{\Gamma}$ thus belongs to the antiexact cross section

$$\mathscr{A}_Y = \{W^\alpha \gamma_\alpha | W^\alpha \in \mathscr{A}^1(E_n),\ \alpha = 1, \ldots, r\} \tag{10-9.6}$$

because $Hd(\mathbf{A}_\Gamma \mathbf{\Gamma})$ belongs to $\mathscr{A}_Y$ and the module property of antiexact forms. Thus *any element of $\mathscr{Y}$ can be mapped onto an element of $\mathscr{A}_Y$ by an appropriate choice of an element of G*; we have the map

$$\pi : \mathscr{Y} \to \mathscr{A}_Y | {'\mathbf{\Gamma}} = (\mathbf{A}_\Gamma \mathbf{\Gamma} - d\mathbf{A}_\Gamma)\mathbf{A}_\Gamma^{-1}, \tag{10-9.7}$$

$$\mathbf{A}_\Gamma = \mathbf{I} + H(\mathbf{A}_\Gamma \mathbf{\Gamma}).$$

Let $\mathbf{A}_1$ and $\mathbf{A}_2$ be the elements of G that map $\mathbf{\Gamma}_1$ and $\mathbf{\Gamma}_2$ onto the corresponding antiexact elements $'\mathbf{\Gamma}_1$ and $'\mathbf{\Gamma}_2$, respectively:

$$'\mathbf{\Gamma}_1 = (\mathbf{A}_1 \mathbf{\Gamma}_1 - d\mathbf{A}_1)\mathbf{A}_1^{-1}, \qquad '\mathbf{\Gamma}_2 = (\mathbf{A}_2 \mathbf{\Gamma}_2 - d\mathbf{A}_2)\mathbf{A}_2^{-1}.$$

We thus have

$$\mathbf{\Gamma}_1 = \mathbf{A}_1^{-1}({'\mathbf{\Gamma}_1}\mathbf{A}_1 + d\mathbf{A}_1), \qquad \mathbf{\Gamma}_2 = \mathbf{A}_2^{-1}({'\mathbf{\Gamma}_2}\mathbf{A}_2 + d\mathbf{A}_2). \tag{10-9.8}$$

Now $\boldsymbol{\Gamma}_1$ can be mapped onto $\boldsymbol{\Gamma}_2$ by an element $\mathbf{B}$ of G if and only if $\mathbf{B}$ satisfies $\boldsymbol{\Gamma}_2 = (\mathbf{B}\boldsymbol{\Gamma}_1 - d\mathbf{B})\mathbf{B}^{-1}$; that is, if and only if

$$d\mathbf{B} = \mathbf{B}\boldsymbol{\Gamma}_1 - \boldsymbol{\Gamma}_2\mathbf{B}. \tag{10-9.9}$$

If we make the substitution $\mathbf{B} = \mathbf{A}_2^{-1}\mathbf{C}\mathbf{A}_1$ and use (10-9.8) to eliminate $\boldsymbol{\Gamma}_1$ and $\boldsymbol{\Gamma}_2$ from (10-9.9), we obtain the requirement

$$d\mathbf{C} = \mathbf{C}\,{}'\mathbf{T}_1 - {}'\mathbf{T}_2\mathbf{C}. \tag{10-9.10}$$

Since ${}'\mathbf{T}_1$ and ${}'\mathbf{T}_2$ are antiexact, an integration of (10-9.10) by use of the homotopy operator H gives

$$\mathbf{C} = \mathbf{K} + H\left(\mathbf{C}\,{}'\mathbf{T}_1 - {}'\mathbf{T}_2\mathbf{C}\right) = \mathbf{K}, \qquad d\mathbf{K} = \mathbf{0}, \tag{10-9.11}$$

where $\mathbf{K}$ is a constant element of G (i.e., $\mathbf{K}$ belongs to G_0). When this result is put back into (10-9.10), we obtain the direct results

$$\mathbf{K}\,{}'\mathbf{T}_1 = {}'\mathbf{T}_2\mathbf{K}, \qquad \mathbf{B} = \mathbf{A}_2^{-1}\mathbf{K}\mathbf{A}_1, \qquad \mathbf{K} \in G_0. \tag{10-9.12}$$

Two elements of $\mathscr{Y}$ can be mapped onto each other by an element of G if and only if their images in $\mathscr{A}_Y$ are related through the adjoint action of an element of G_0. Thus $\mathscr{A}_Y$ is unique to within the adjoint action of the original group G_0.

When an element $\boldsymbol{\Gamma}$ of $\mathscr{Y}$ is mapped onto $\mathscr{A}_Y$ by an appropriate element of G, it becomes a Lie algebra-valued antiexact 1-form, $\boldsymbol{\Gamma}_a$. Accordingly, $\boldsymbol{\Gamma}_a$ satisfies the conditions

$$\mathscr{X} \lrcorner \boldsymbol{\Gamma}_a = \mathbf{0}, \qquad \boldsymbol{\Gamma}_a\left(x_0^i\right) = \mathbf{0}.$$

This process imposes a set of gauge conditions, *the antiexact gauge conditions*. The mapping π of $\mathscr{Y}$ onto $\mathscr{A}_Y$, that is given by (10-9.7), may thus be viewed as restriction to the *antiexact gauge* cross section of $\mathscr{Y}$. The result of the last paragraph shows that the antiexact gauge restriction is unique to within the adjoint action of the original group G_0. Accordingly, the antiexact gauge fixes $\boldsymbol{\Gamma}$ to within the adjoint action of the original homogeneous group G_0. This result should not be a surprise, for without it an assignment of gauge would break the action of the original homogeneous group G_0 of internal symmetries.

Up to this point we have assumed that the center of E_n has been fixed. Let H_1 and H_2 denote the linear homotopy operators that are constructed from two choices of the center of E_n, and let $\mathscr{A}_{Y1}$ and $\mathscr{A}_{Y2}$ be the corresponding antiexact cross sections of $\mathscr{Y}$ that are determined in the manner described above. If $\boldsymbol{\Gamma}$ is a generic element of $\mathscr{Y}$, then

$$\mathbf{A}_1 = \mathbf{I} + H_1(\mathbf{A}_1\boldsymbol{\Gamma}), \qquad \mathbf{A}_2 = \mathbf{I} + H_2(\mathbf{A}_2\boldsymbol{\Gamma}) \tag{10-9.13}$$

map Γ onto the corresponding elements

$$(10\text{-}9.14)\qquad {}'\Gamma_1 = (\mathbf{A}_1\Gamma - d\mathbf{A}_1)\mathbf{A}_1^{-1} \in \mathscr{A}_{Y1},$$

$$'\Gamma_2 = (\mathbf{A}_2\Gamma - d\mathbf{A}_2)\mathbf{A}_2^{-1} \in \mathscr{A}_{Y2},$$

respectively. An elimination of the common term Γ between these two equations then gives

$$(10\text{-}9.15)\qquad {}'\Gamma_2 = (\mathbf{B}\,{}'\Gamma_1 - d\mathbf{B})\mathbf{B}^{-1}, \qquad \mathbf{B} = \mathbf{A}_2\mathbf{A}_1^{-1} \in G.$$

Cross sections of $\mathscr{Y}$ that are determined by different choices of the center of E_n can be mapped one onto the other by the action of the gauge group G. Thus gauge covariance of the theory renders the theory insensitive to the choice of center!

This latter result is of particular importance in practice, for it allows us to choose the center of E_n in such a way as to give maximal simplification in actual calculations. An obvious choice is to take the origin of the coordinate cover as the center, in which case $x_0^i = 0$. The choice of center corresponds to the choice of a reference point for the Yang-Mills potential functions $W_i^\alpha(x^j)$, while $W_i^\alpha(x^j)dx^i \in \mathscr{A}^1(E_4)$ fixes the reference values at the center to be all zero (recall that $\omega \in \mathscr{A}^1(E_n)$ implies $\omega(x_0^i) = 0$).

The choice of the antiexact gauge is achieved by the requirements $W^\alpha \in \mathscr{A}(E_n)$. Accordingly, since $\Gamma_\beta^\alpha = C_{\sigma\beta}^\alpha W^\sigma$, *the antiexact gauge renders the connection* 1*-forms* Γ_β^α *on group space antiexact* 1*-forms.*

10-10. DIRECT INTEGRATION OF THE FIELD EQUATIONS

The fact that the elements of the canonical cross section $\mathscr{A}_Y$ are antiexact 1-form valued matrices leads to a remarkable simplification of the problem of solving the field equations for the W^α fields. Because it always occurs in practice, let us assume that the Lagrangian $\hat{L}$ is *quadratic* in the field quantities $\{F_{ij}^\alpha\}$ and such that gauge covariant conservation of the current $(n-1)$-forms J_α obtain (i.e., (10-8.11) is satisfied). We then have $\Theta_\alpha^\beta \wedge G_\beta = C_{\sigma\alpha}^\beta F^\sigma \wedge G_\beta = 0$, as previously noted. It is further assumed that the constitutive relations (10-6.7), $G_\alpha^{ij} = \partial\hat{L}/\partial F_{ij}^\alpha$, are invertible so that

$$(10\text{-}10.1)\qquad F_{ij}^\alpha = K_{ijmn}^{\alpha\beta}G_\beta^{mn}, \qquad K_{ijmn}^{\alpha\beta} \in \Lambda^0(E_n).$$

The problem is therefore that of solving the system of field equations

$$(10\text{-}10.2)\qquad F^\alpha = dW^\alpha + \tfrac{1}{2}C_{\beta\gamma}^\alpha W^\beta \wedge W^\gamma,$$

$$(10\text{-}10.3)\qquad DG_\alpha = dG_\alpha - C_{\sigma\alpha}^\beta W^\sigma \wedge G_\beta = \tfrac{1}{2}J_\alpha,$$

$$(10\text{-}10.4)\qquad DJ_\alpha = dJ_\alpha - C_{\sigma\alpha}^\beta W^\sigma \wedge J_\beta = 0$$

subject to the constitutive relations (10-10.1) and the gauge conditions $W^\alpha \in \mathscr{A}^1(E_n)$.

The simplest way of seeing what is involved is to solve the system (10-10.2) through (10-10.4) for the exterior derivatives of the various quantities. This gives

$$dW^\alpha = F^\alpha - \tfrac{1}{2}C^\alpha_{\beta\gamma}W^\beta \wedge W^\gamma, \tag{10-10.5}$$

$$dG_\alpha = \tfrac{1}{2}J_\alpha + C^\beta_{\sigma\alpha}W^\sigma \wedge G_\beta, \tag{10-10.6}$$

$$dJ_\alpha = C^\beta_{\sigma\alpha}W^\sigma \wedge J_\beta. \tag{10-10.7}$$

Now, (10-10.6) and (10-10.7) form a complete differential system under satisfaction of $\Theta^\alpha_\beta \wedge G_\beta = 0$ and (10-10.5). Accordingly, we may consider (10-10.6) and (10-10.7) as a system for the determination of $\{G_\alpha\}$ and $\{J_\alpha\}$ in terms of $\{W^\alpha\}$. This means that we take $\{W^\alpha\}$ as fixed elements of $\mathscr{A}^1(E_n)$ in (10-10.6), (10-10.7), in which case the integrability conditions

$$d\left(\tfrac{1}{2}J_\alpha + C^\beta_{\sigma\alpha}W^\sigma \wedge G_\beta\right) = 0, \tag{10-10.8}$$

$$d\left(C^\beta_{\sigma\alpha}W^\sigma \wedge J_\beta\right) = 0 \tag{10-10.9}$$

must be imposed. Once $\{G_\alpha\}$ and $\{J_\alpha\}$ have been found in terms of $\{W^\alpha\}$, we may substitute $\{G_\alpha\}$ into (10-10.1) and (10-10.5) in order to determine the required $\{W^\alpha\}$. This determination of $\{W^\alpha\}$ will then be such as to secure satisfaction of the integrability conditions (10-10.8), (10-10.9) provided the W^α thus determined are such as to secure the identical satisfaction of (10-10.2) with $\{F^\alpha\}$ determined by (10-10.1). The problem of solving the W^α field equations thus splits naturally into two parts.

The first part of the problem consists of finding the general solution of (10-10.6), (10-10.7) subject to the integrability conditions (10-10.8), (10-10.9) for $\{W^\alpha\}$ fixed elements of $\mathscr{A}^1(E_n)$. Since any $J_\alpha \in \Lambda^{n-1}(E_n)$ has the unique representation

$$J_\alpha = dHJ_\alpha + HdJ_\alpha,$$

any J_α that satisfy (10-10.7) has the unique representation

$$J_\alpha = d\eta_\alpha + C^\beta_{\sigma\alpha}H\left(W^\sigma \wedge J_\beta\right), \tag{10-10.10}$$

where we have set

$$\eta_\alpha = HJ_\alpha. \tag{10-10.11}$$

An iteration of the integral equations (10-10.10) then yields

$$J_\alpha = d\eta_\alpha + C^\beta_{\sigma\alpha} H(W^\sigma \wedge d\eta_\beta) \tag{10-10.12}$$

because $W^\alpha \in \mathscr{A}(E_n)$, $H(W^\sigma \wedge J_\beta) \in \mathscr{A}(E_n)$ imply that $W^\sigma \wedge H(W^\lambda \wedge J_\rho) \in \mathscr{A}(E_n)$ and hence belong to ker H. Conversely, if we take J_α to be given by (10-10.12) with η_α arbitrary elements of $\mathscr{A}^{n-2}(E_n)$, then application of H to both sides of (10-10.12) gives $HJ_\alpha = Hd\eta_\alpha = \eta_\alpha$ because H inverts d on $\mathscr{A}(E_n)$. We therefore recover (10-10.11). Exterior differentiation of (10-10.12), which is the same thing as (10-10.10), then gives

$$\begin{aligned} dJ_\alpha &= C^\beta_{\sigma\alpha} dH(W^\sigma \wedge J_\beta) \\ &= C^\beta_{\sigma\alpha} W^\sigma \wedge J_\beta - Hd(C^\beta_{\sigma\alpha} W^\sigma \wedge J_\beta) = C^\beta_{\sigma\alpha} W^\sigma \wedge J_\beta \end{aligned}$$

under satisfaction of the integrability conditions (10-10.9). Thus for $\{\eta_\alpha\}$ arbitrary antiexact $(n-2)$-forms every J_α given by (10-10.12) satisfies (10-10.7) whenever the integrability conditions (10-10.9) are satisfied.

Now that $\{J_\alpha\}$ have been determined, the unique representation

$$G_\alpha = dHG_\alpha + HdG_\alpha$$

can be used together with the field equation (10-10.6) to obtain

$$G_\alpha = d\omega_\alpha + H\left(\tfrac{1}{2}J_\alpha + C^\beta_{\sigma\alpha} W^\sigma \wedge G_\beta\right) \tag{10-10.13}$$

with

$$\omega_\alpha = HG_\alpha. \tag{10-10.14}$$

Substituting for $\{J_\alpha\}$ from (10-10.12) into (10-10.13) gives

$$G_\alpha = d\omega_\alpha + \tfrac{1}{2}\eta_\alpha + C^\beta_{\sigma\alpha} H(W^\sigma \wedge G_\beta)$$

because $\{\eta_\alpha\}$ is antiexact, and an iteration yields the final result

$$G_\alpha = d\omega_\alpha + \tfrac{1}{2}\eta_\alpha + C^\beta_{\sigma\alpha} H(W^\sigma \wedge d\omega_\beta). \tag{10-10.15}$$

The same reasoning as that given for $\{J_\alpha\}$ shows any $\{G_\alpha\}$ given by (10-10.15) will satisfy the field equations (10-10.6) for any choice of $\{\omega_\alpha\} \in \mathscr{A}^{n-3}(E_n)$ provided the integrability conditions (10-10.8) are satisfied.

The general integrals of the complete differential system of field equations (10-10.6) *through* (10-10.9) *are given by*

$$J_\alpha = d\eta_\alpha + C^\beta_{\sigma\alpha} H\big(W^\sigma \wedge d\eta_\beta\big), \tag{10-10.16}$$

$$G_\alpha = d\omega_\alpha + \tfrac{1}{2}\eta_\alpha + C^\beta_{\sigma\alpha} H\big(W^\sigma \wedge d\omega_\beta\big) \tag{10-10.17}$$

in terms of the antiexact forms W^α, η_α, *and* ω_α.

The second part of the problem is that of integrating (10-10.5). Since W^α are antiexact, $W^\alpha = HdW^\alpha$, and hence (10-10.5) gives

$$W^\alpha = HF^\alpha - \tfrac{1}{2}C^\alpha_{\beta\gamma} H\big(W^\beta \wedge W^\gamma\big) = HF^\alpha \tag{10-10.18}$$

because $W^\beta \wedge W^\gamma$ is antiexact. When the constitutive relations (10-10.1) are used, we have

$$F^\alpha = \tfrac{1}{2}F^\alpha_{ij}dx^i \wedge dx^j = \tfrac{1}{2}K^{\alpha\beta}_{ijmn}G^{mn}_\beta dx^i \wedge dx^j \tag{10-10.19}$$

and hence

$$W^\alpha = \tfrac{1}{2}H\big(K^{\alpha\beta}_{ijmn}G^{mn}_\beta dx^i \wedge dx^j\big). \tag{10-10.20}$$

Thus if we write (10-10.17) in the symbolic form

$$G^{mn}_\alpha = \big(d\omega_\alpha\big)^{mn} + \tfrac{1}{2}\eta^{mn}_\alpha + C^\beta_{\sigma\alpha} H\big(W^\sigma \wedge d\omega_\beta\big)^{mn},$$

(10-10.20) shows that W^α *satisfy the system of linear Riemann-Graves integral equations*

$$W^\alpha = \tfrac{1}{2}H\Big(K^{\alpha\beta}_{ijmn}\Big\{\big(d\omega_\beta\big)^{mn} + \tfrac{1}{2}\eta^{mn}_\beta + C^\rho_{\sigma\beta} H\big(W^\sigma \wedge d\omega_\rho\big)^{mn}\Big\} dx^i \wedge dx^j\Big). \tag{10-10.21}$$

A very strong note of caution must be sounded at this point. Although every solution of the W^α field equations satisfies the system of linear Riemann-Graves integral equations (10-10.21), not every solution of the system (10-10.21) will satisfy the W^α field equations. The reason for this is that we used the linear homotopy operator to integrate $dW^\alpha = F^\alpha - \frac{1}{2}C^\alpha_{\beta\gamma}W^\beta \wedge W^\gamma$ and the terms $C^\alpha_{\beta\gamma}W^\beta \wedge W^\gamma$ made no contribution since they belong to ker H by the module property of antiexact forms. It is therefore necessary that we secure satisfaction of

$$F^\alpha = dW^\alpha + \tfrac{1}{2}C^\alpha_{\beta\gamma}W^\beta \wedge W^\gamma \tag{10-10.22}$$

for $W^\alpha = HF^\alpha$. Noting that $F^\alpha = dHF^\alpha + HdF^\alpha$, substitution of $W^\alpha = HF^\alpha$ into (10-10.22) yields the conditions

$$\text{(10-10.23)} \qquad HdF^\alpha = \tfrac{1}{2}C^\alpha_{\beta\gamma}H(F^\beta) \wedge H(F^\gamma).$$

However, $DF^\alpha = dF^\alpha + C^\alpha_{\beta\gamma}W^\beta \wedge F^\gamma = 0$ as a consequence of (10-10.22), so that (10-10.23) gives

$$\text{(10-10.24)} \qquad C^\alpha_{\beta\gamma}\left(HF^\beta \wedge HF^\gamma - 2H(W^\beta \wedge F^\gamma)\right) = 0, \qquad \alpha = 1,\dots,r.$$

The r equations (10-10.24) are the nonlinear residual equations for the determination of the r antiexact $(n-3)$-forms ω_α that appear in (10-10.7) and (10-10.21).

10-11. MOMENTUM-ENERGY COMPLEXES AND INTERACTION FORCES

The minimal coupling construct gave the Lagrangian function

$$\text{(10-11.1)} \qquad \hat{L} = L_0 - s_1 L_1 = L_0\left(x^j, \Psi^A, y^A_k\right) + s_1 L_1\left(x^k, F^\alpha_{ij}\right),$$

where the form of L_1 is determined by the requirement

$$\hat{L} = \hat{L}\left(x^j, \Psi^A, y^A_k, F^\alpha_{ij}\right)$$

and

$$\text{(10-11.2)} \qquad y^A_k = \partial_k \Psi^A + W^\alpha_k \gamma^A_{\alpha B}\Psi^B, \qquad F^\alpha_{ij} = \partial_i W^\alpha_j - \partial_j W^\alpha_i + C^\alpha_{\beta\gamma}W^\beta_i W^\gamma_j.$$

We can accordingly write

$$\text{(10-11.3)} \qquad \hat{L} = \hat{L}_\Psi + \hat{L}_W$$

where

$$\text{(10-11.4)} \qquad \hat{L}_\Psi = L_0\left(x^j, \Psi^A, y^A_i\right)$$

is the Lagrangian function for the Ψ fields and

$$\text{(10-11.5)} \qquad \hat{L}_W = s_1 L_1\left(x^k, F^\alpha_{ij}\right)$$

is the Lagrangian function for the free W^α fields (i.e., W^α fields with $\Psi^A = 0$).

We therefore have

$$(10\text{-}11.6) \qquad Z_A = \frac{\partial \hat{L}}{\partial \Psi^A} = \frac{\partial \hat{L}_\Psi}{\partial \Psi^A}, \qquad M_A^k = \frac{\partial \hat{L}}{\partial y_k^A} = \frac{\partial \hat{L}_\Psi}{\partial y_k^A}$$

$$(10\text{-}11.7) \qquad G_\alpha^{ij} = \frac{\partial \hat{L}}{\partial F_{ij}^\alpha} = \frac{\partial \hat{L}_W}{\partial F_{ij}^\alpha}.$$

The components of the momentum-energy complex, $T_j^i(L)$, of a given Lagrangian function L were shown in Section 7-3 to be linear in the Lagrangian function,

$$T_j^i(L_1 + L_2) = T_j^i(L_1) + T_j^i(L_2).$$

We therefore have

$$(10\text{-}11.8) \qquad T_j^i(\hat{L}) = T_j^i(\hat{L}_\Psi) + T_j^i(\hat{L}_w)$$

when (10-11.3) is used. A straightforward computation then gives

$$(10\text{-}11.9) \qquad T_j^i(\hat{L}_\Psi) = \frac{\partial \hat{L}_\Psi}{\partial(\partial_i \Psi^A)} \partial_j \Psi^A - \delta_j^i \hat{L}_\Psi,$$

$$(10\text{-}11.10) \qquad T_j^i(\hat{L}_W) = \frac{\partial L_W}{\partial(\partial_i W_k^\alpha)} \partial_j W_k^\alpha - \delta_j^i \hat{L}_W.$$

It is then a simple matter to use (10-11.2), (10-11.6), and (10-11.7) to obtain the explicit evaluations

$$(10\text{-}11.11) \qquad T_j^i(\hat{L}_\Psi) = M_A^i \partial_j \Psi^A - \delta_j^i \hat{L}_\Psi,$$

$$(10\text{-}12.12) \qquad T_j^i(\hat{L}_W) = G_\alpha^{ik}(\partial_i W_j^\alpha - \partial_j W_i^\alpha) - \delta_j^i \hat{L}_W.$$

We now assume that $\hat{L}$ does not depend explicitly on the n variables $x^1, \ldots, x^n$; that is, $\hat{L} = \hat{L}(\Psi^A, y_k^A, F_{ij}^\alpha)$ and hence

$$(10\text{-}11.13) \qquad \hat{L}_\Psi = \hat{L}_\Psi(\Psi^A, y_k^A), \hat{L}_W = \hat{L}_W(F_{ij}^\alpha).$$

The relations (10-11.13) state that the functions $\hat{L}_\Psi$ and $\hat{L}_W$ are homogeneous with respect to the underlying space E_n, a situation that is the rule rather than the exception in practice. In effect, this assumption says that the Lagrangian functions $\hat{L}_\Psi$ and $\hat{L}_W$ do not distinguish between points in E_n. The upshot of this assumption is that any solution (Ψ^A, W_i^α) of the field equations will satisfy the identity

(10-11.14) $$\partial_i T_j^i(\hat{L}) = 0$$

by Theorem 7-3.2. (If $\hat{L}$ were to depend explicitly on the x's then the right-hand side of (10-11.14) would be replaced by $-\partial_j \hat{L}$ where the derivatives are taken for constant values of Ψ^A, y_k^A and F_{ij}^α.) It is now simply a matter of substituting (10-11.8) into (10-11.14) in order to obtain the fundamental result

(10-11.15) $$\partial_i T_j^i(\hat{L}_\Psi) + \partial_i T_j^i(\hat{L}_W) = 0.$$

Suppose that we could "switch off" the W_i^α fields. This would give us the original Lagrangian $\hat{L}_\Psi = L_0(\Psi^A, \partial_k \Psi^A)$ and (10-11.15) would reduce to $\partial_i T_j^i(L_0) = 0$. On the other hand, as soon as the W_i^α fields are switched on, we obtain the results $\partial_i T_j^i(L_\Psi) + \partial_i T_j^i(L_W) = 0$ instead of $\partial_i T_j^i(L_0) = 0$. This is reasonable, for we have seen that the field equations for the Ψ^A fields and for the W_i^α fields are coupled and hence there are specific interactions between the Ψ^A fields and the W_i^α fields. It is therefore natural to define the quantities

(10-11.16) $$f_{\Psi j} = \partial_i T_j^i(L_\Psi)$$

as the components of the *forces of interaction* that act on the Ψ^A fields as a consequence of the W_k^α fields, and

(10-11.17) $$f_{Wj} = \partial_i T_j^i(L_W)$$

as the components of the *forces of interaction* that act on the W_i^α fields as a consequence of the Ψ^A fields. The fundamental relations (10-11.15) then give

(10-11.18) $$f_{\Psi j} + f_{Wj} = 0;$$

that is, *the forces of interaction are self-equilibrating.* In view of this result we only need to evaluate one set of components of the forces of interaction.

A direct substitution from (10-11.11) gives

(10-11.19) $$f_{\Psi j} = (\partial_i M_A^i)\partial_j \Psi^A + M_A^i \partial_i \partial_j \Psi^A - \frac{\partial \hat{L}_\Psi}{\partial x^j}.$$

However,

$$\frac{\partial \hat{L}_\Psi}{\partial x^j} = \frac{\partial L_\Psi}{\partial \Psi^A}\partial_j \Psi^A + \frac{\partial \hat{L}_\Psi}{\partial y_i^A}\frac{\partial y_i^A}{\partial x^j}$$

$$= Z_A \partial_j \Psi^A + M_A^i \partial_j(\partial_i \Psi^A + \Gamma_{Bi}^A \Psi^B),$$

when (10-6.5), (10-6.6), and (10-11.2) are used. We thus have

$$f_{\Psi^j} = \left(\partial_i M_A^i\right)\partial_j \Psi^A + M_A^i \partial_i \partial_j \Psi^A - Z_A \partial_j \Psi^A$$

$$- M_A^i \partial_j \partial_i \Psi^A - M_A^i \Gamma_{Bi}^A \partial_j \Psi^B - M_A^i \Psi^B \partial_j \Gamma_{Bi}^A,$$

and hence the field equations $DM_A = Z_A \mu$ give the explicit evaluation

(10-11.20) $$f_{\Psi j} = -M_A^i \Psi^B \partial_j \Gamma_{Bi}^A.$$

As a last remark, we note that $J_\alpha = M_A \gamma_{\alpha B}^A \Psi^B = J_\alpha^i \mu_i$ and $\Gamma_{Bj}^A = W_j^\alpha \gamma_{\alpha B}^A$ give

(10-11.21) $$J_\alpha^i = M_A^i \gamma_{\alpha B}^A \Psi^B$$

and hence

(10-11.22) $$f_{\Psi j} = -J_\alpha^i \partial_j W_i^\alpha = W_i^\alpha \partial_j J_\alpha^i - \partial_j\left(J_\alpha^i W_i^\alpha\right).$$

Thus the components of the interaction force are determined by the components of the current and the derivatives of the W_i^α fields. In particular, *the interaction forces vanish with the currents.*

10-12. GENERAL SYMMETRY GROUPS OF THE ACTION *n*-FORM

The gauge theories considered in previous sections arise through recognition of a global *internal* symmetry group of the salient physical state variables. This recognition was achieved by noting that the group in question leaves the action *n*-form invariant. Further, the state space of the physical system was taken to be a representation space for the internal symmetry group, and hence the internal symmetry group acts linearly (i.e., $'\boldsymbol{\Psi} = \mathbf{A}\boldsymbol{\Psi}$).

There are two aspects of this construct that appear unduly restrictive. First, the group of symmetries of the action integral of a given physical system is usually much richer than just a linear internal symmetry group. It can be calculated without difficulty and its general properties are well known (Noetherian vector fields of the first kind). In particular, the group action can occur on both the physical state variables and the space-time labels, and the group action need not be linear. All that is required is that all quantities have well defined Lie derivatives with respect to vector fields on an appropriately structured space. The symmetry group is then obtained by exponentiation of a Lie algebra of Lie derivatives.

This brings us to the second aspect. The classic Yang-Mills minimal replacement construct introduces connection forms that take their values in the matrix Lie algebra of the linear internal symmetry group, while the general situation involves a group that is the exponentiation of a Lie algebra of Lie derivatives. It is then almost self-evident that the general case should involve connection forms that take their values in a Lie algebra of Lie derivatives; that is, we have to be able to deal with operator-valued connections. The approach to gauge theories for general symmetry groups through the use of operator-valued connections differs substantially from that presented in the current literature. The final results, however, agree with the more customary approach to gauge theories of gravity and related matters. In addition, the theory of operator-valued connections provides a natural vehicle for certain aspects of the exterior calculus that are often ignored.

There will be a number of different spaces involved in this discussion. For simplicity, $T(W)$ will be used to denote the tangent space of W and $\Lambda(W)$ denotes the exterior algebra of differential forms over W. If $S = R \times T$, then $\Lambda(R)$ and $\Lambda(T)$ will denote the exterior algebras of differential forms over R and T, respectively. If π_1 and π_2 are the projections onto the first and second factors, respectively, $\pi_1 : R \times T \to R$, $\pi_2 : R \times T \to T$, then $\Lambda(R)$ and $\Lambda(T)$ trivially lift to subspaces $(\pi_1)^*\Lambda(R)$ and $(\pi_2)^*\Lambda(T)$ of $\Lambda(R \times T)$, respectively.

Suppose that we are given a system of N quantities on E_n. In practice, these will be the state variables of a dynamical system on space-time. For the purposes of this discussion, let $\mathbb{R}^N$ be the range space of the given N quantities. The space K is defined by

$$K = E_n \times \mathbb{R}^N$$

(see Chapter Six for this construction). We assume that K is referred to a system of local coordinates $\{z^A | 1 \le A \le n + N\}$ for the time being. The general discussion will be carried out in this context. Only later, after identifying the state variables, will it be necessary to identify some of the z^A's with the x^i's.

Let G_r be a given r-parameter Lie group and let g_r be its Lie algebra. We assume that G_r acts on K as an r-parameter Lie group of point transformations,

$$(10\text{-}12.1) \qquad {}'z^A = \exp(u^a V_a) z^A,$$

where $\{u^a | 1 \le a \le r\}$ is a system of canonical parameters for G_r and $\{V_a \in T(K) | 1 \le a \le r\}$ is a basis for g_r in this representation. We thus have

$$(10\text{-}12.2) \qquad [V_a, V_b] = C^e_{ab} V_e,$$

where C^e_{ab} are the structure constants of G_r.

Since $V_a \in T(K)$, each V_a acts on the collection of C^∞ functions $\Lambda^0(K)$ by

$$V_a : \Lambda^0(K) \to \Lambda^0(K) | {}'f = V_a f.$$

The Lie algebra g_r may thus be realized in terms of the mappings $V_a : \Lambda^0(K) \to \Lambda^0(K)$. We denote this situation by $g_r(V_a; \Lambda^0(K))$. It is then a trivial matter to see that g_r may also be realized by $g_r(£_a; \Lambda(K))$ since

$$(10\text{-}12.3) \qquad [£_a, £_b] = £_a £_b - £_b £_a = C^e_{ab} £_e, \qquad £_a := £_{V_a},$$

and $\Lambda(K)$ is a domain for the Lie derivative. The r operators $\{£_a | 1 \le a \le r\}$ then form a basis for Lie algebra $g_r(£_a; \Lambda(K))$ and the group G_r acts on $\Lambda(K)$ by

$$(10\text{-}12.4) \qquad {}^{\backprime}\omega = \exp(u^a £_a)\omega, \qquad \omega \in \Lambda(K).$$

In view of these considerations, we can shift to the space

$$(10\text{-}12.5) \qquad \mathcal{G} = G_r \times K$$

with local coordinates $\{u^a;\ z^A | 1 \le a \le r,\ 1 \le A \le n + N\}$. It is now just one more step to consider the larger structure $\Lambda(\mathcal{G})$.

Let $\bar{d}$ denote the exterior derivative on $\mathcal{G}$. We then have

$$(10\text{-}12.6) \qquad \bar{d} = d + d_u, \qquad d = \bar{d}|_K, \qquad d_u = \bar{d}|_{G_r},$$

where $\bar{d}|_R$ denotes the restriction to R, and

$$(10\text{-}12.7) \qquad £_a u^b = 0, \qquad £_a \bar{d} u^b = 0$$

because $V_a \in T(K)$ and hence $\exp(u^a £_a)$ restricted to $\Lambda(G_r)$ is the identity. It is therefore consistent to allow G_r to act on $\Lambda(\mathcal{G})$ by

$$(10\text{-}12.8) \qquad {}^{\backprime}\Omega = \exp(u^a £_a)\Omega, \qquad \Omega \in \Lambda(\mathcal{G}).$$

In particular, we have $g_r(£_a; \Lambda(\mathcal{G}))$ and hence $\{£_a | 1 \le a \le r\}$ is an operator basis for $g_r(£_a; \Lambda(\mathcal{G}))$. It is then a simple matter to see that (10-12.8) applied to $(\pi_2)^*\Lambda(K)$ reproduces (10-12.4). The reason why we have to go to the larger structure $\Lambda(\mathcal{G})$ is because we will be interested in what happens when $\bar{d}$ is applied to the image of $\Lambda(K)$ under the action of G_r (${}^{\backprime}\omega = \exp(u^a £_a)\omega$) in order to prepare for an eventual dependence of the u^a's on the x^i's that comes about by a mapping from E_n into G_r.

10-13. OPERATOR-VALUED CONNECTION 1-FORMS

For any $\omega \in \Lambda(K)$, we have $d(\exp(u^a £_a)\omega) = \exp(u^a £_a) d\omega$ because d and $£_a$ commute and $d = \bar{d}|_K$. Thus the exterior derivative on $\Lambda(K)$ transforms covariantly under the action of G_r. When consideration is shifted to $\Lambda(\mathcal{G})$, situation involves a group that is the exponentiation of a Lie algebra of Lie derivatives. It is then almost self-evident that the general case should involve connection forms that take their values in a Lie algebra of Lie derivatives; that is, we have to be able to deal with operator-valued connections. The approach to gauge theories for general symmetry groups through the use of operator-valued connections differs substantially from that presented in the current literature. The final results, however, agree with the more customary approach to gauge theories of gravity and related matters. In addition, the theory of operator-valued connections provides a natural vehicle for certain aspects of the exterior calculus that are often ignored.

There will be a number of different spaces involved in this discussion. For simplicity, $T(W)$ will be used to denote the tangent space of W and $\Lambda(W)$ denotes the exterior algebra of differential forms over W. If $S = R \times T$, then $\Lambda(R)$ and $\Lambda(T)$ will denote the exterior algebras of differential forms over R and T, respectively. If π_1 and π_2 are the projections onto the first and second factors, respectively, $\pi_1 : R \times T \to R$, $\pi_2 : R \times T \to T$, then $\Lambda(R)$ and $\Lambda(T)$ trivially lift to subspaces $(\pi_1)^*\Lambda(R)$ and $(\pi_2)^*\Lambda(T)$ of $\Lambda(R \times T)$, respectively.

Suppose that we are given a system of N quantities on E_n. In practice, these will be the state variables of a dynamical system on space-time. For the purposes of this discussion, let $\mathbb{R}^N$ be the range space of the given N quantities. The space K is defined by

$$K = E_n \times \mathbb{R}^N$$

(see Chapter Six for this construction). We assume that K is referred to a system of local coordinates $\{z^A | 1 \leq A \leq n + N\}$ for the time being. The general discussion will be carried out in this context. Only later, after identifying the state variables, will it be necessary to identify some of the z^A's with the x^i's.

Let G_r be a given r-parameter Lie group and let g_r be its Lie algebra. We assume that G_r acts on K as an r-parameter Lie group of point transformations,

$$\text{(10-12.1)} \qquad {}'z^A = \exp(u^a V_a) z^A,$$

where $\{u^a | 1 \leq a \leq r\}$ is a system of canonical parameters for G_r and $\{V_a \in T(K) | 1 \leq a \leq r\}$ is a basis for g_r in this representation. We thus have

$$\text{(10-12.2)} \qquad [V_a, V_b] = C^e_{ab} V_e,$$

where C^e_{ab} are the structure constants of G_r.

Since $V_a \in T(K)$, each V_a acts on the collection of C^∞ functions $\Lambda^0(K)$ by

$$V_a : \Lambda^0(K) \to \Lambda^0(K) | {}'f = V_a f.$$

The Lie algebra g_r may thus be realized in terms of the mappings $V_a: \Lambda^0(K) \to \Lambda^0(K)$. We denote this situation by $g_r(V_a; \Lambda^0(K))$. It is then a trivial matter to see that g_r may also be realized by $g_r(\pounds_a; \Lambda(K))$ since

$$[\pounds_a, \pounds_b] = \pounds_a\pounds_b - \pounds_b\pounds_a = C^e_{ab}\pounds_e, \qquad \pounds_a := \pounds_{V_a}, \tag{10-12.3}$$

and $\Lambda(K)$ is a domain for the Lie derivative. The r operators $\{\pounds_a | 1 \le a \le r\}$ then form a basis for Lie algebra $g_r(\pounds_a; \Lambda(K))$ and the group G_r acts on $\Lambda(K)$ by

$$`\omega = \exp(u^a\pounds_a)\omega, \qquad \omega \in \Lambda(K). \tag{10-12.4}$$

In view of these considerations, we can shift to the space

$$\mathscr{G} = G_r \times K \tag{10-12.5}$$

with local coordinates $\{u^a;\ z^A | 1 \le a \le r,\ 1 \le A \le n + N\}$. It is now just one more step to consider the larger structure $\Lambda(\mathscr{G})$.

Let $\bar{d}$ denote the exterior derivative on $\mathscr{G}$. We then have

$$\bar{d} = d + d_u, \qquad d = \bar{d}|_K, \qquad d_u = \bar{d}|_{G_r}, \tag{10-12.6}$$

where $\bar{d}|_R$ denotes the restriction to R, and

$$\pounds_a u^b = 0, \qquad \pounds_a \bar{d}u^b = 0 \tag{10-12.7}$$

because $V_a \in T(K)$ and hence $\exp(u^a\pounds_a)$ restricted to $\Lambda(G_r)$ is the identity. It is therefore consistent to allow G_r to act on $\Lambda(\mathscr{G})$ by

$$`\Omega = \exp(u^a\pounds_a)\Omega, \qquad \Omega \in \Lambda(\mathscr{G}). \tag{10-12.8}$$

In particular, we have $g_r(\pounds_a; \Lambda(\mathscr{G}))$ and hence $\{\pounds_a | 1 \le a \le r\}$ is an operator basis for $g_r(\pounds_a; \Lambda(\mathscr{G}))$. It is then a simple matter to see that (10-12.8) applied to $(\pi_2)^*\Lambda(K)$ reproduces (10-12.4). The reason why we have to go to the larger structure $\Lambda(\mathscr{G})$ is because we will be interested in what happens when $\bar{d}$ is applied to the image of $\Lambda(K)$ under the action of G_r ($`\omega = \exp(u^a\pounds_a)\omega$) in order to prepare for an eventual dependence of the u^a's on the x^i's that comes about by a mapping from E_n into G_r.

10-13. OPERATOR-VALUED CONNECTION 1-FORMS

For any $\omega \in \Lambda(K)$, we have $d(\exp(u^a\pounds_a)\omega) = \exp(u^a\pounds_a)d\omega$ because d and $\pounds_a$ commute and $d = \bar{d}|_K$. Thus the exterior derivative on $\Lambda(K)$ transforms covariantly under the action of G_r. When consideration is shifted to $\Lambda(\mathscr{G})$, things are no longer so simple because

$$\bar{d}(\exp(u^a £_a)\omega) = d(\exp(u^a £_a)\omega) + d_u(\exp(u^a £_a)\omega), \tag{10-13.1}$$

by (10-12.6), and d_u does not commute with $\exp(u^a £_a)$. A direct way around this difficulty is to set

(10-13.2)

$$d_u(\exp(u^a £_a)\Omega) - \exp(u^a £_a) d_u \Omega = \exp(u^a £_a)(\Gamma \wedge \Omega) - {}^{`}\Gamma \wedge \exp(u^a £_a)\Omega,$$

as suggested by the Yang-Mills construction of gauge covariant derivatives. Noting that $g_r(£_a;\ \Lambda(\mathscr{G}))$ has the operators $\{£_a | 1 \le a \le r\}$ as a basis, we take Γ to be an element of $\Lambda^1(\mathscr{G})$ with values in $g_r(£_a;\ \Lambda(\mathscr{G}))$ and ${}^{`}\Gamma$ to be the image of Γ under the action of G_r that is defined by (10-13.2). We therefore have

$$\Gamma = W^a £_a, \qquad W^a \in \Lambda^1(\mathscr{G}), \tag{10-13.3}$$

and (10-13.1) and (10-13.2) combine to give

$$\bar{d}(\exp(u^a £_a)\Omega) + {}^{`}\Gamma \wedge \exp(u^a £_a)\Omega = \exp(u^a £_a)(\bar{d}\Omega + \Gamma \wedge \Omega). \tag{10-13.4}$$

This shows that

$$D\Omega = \bar{d}\Omega + \Gamma \wedge \Omega, \qquad {}^{`}D{}^{`}\Omega = \bar{d}{}^{`}\Omega + {}^{`}\Gamma \wedge {}^{`}\Omega \tag{10-13.5}$$

serve to define a *G_r-covariant exterior derivative* on $\Lambda(\mathscr{G})$:

$${}^{`}D{}^{`}\Omega = \exp(u^a £_a)(D\Omega), \qquad {}^{`}\Omega = \exp(u^a £_a)\Omega. \tag{10-13.6}$$

It is therefore consistent to refer to $\Gamma = W^a £_a$ as an *operator-valued connection* on $\Lambda(\mathscr{G})$.

The transformation law for the operator-valued connection Γ on $\Lambda(\mathscr{G})$ obtains directly from (10-13.2) and (10-13.3):

(10-13.7)

$${}^{`}\Gamma \wedge \exp(u^a £_a)\Omega = \exp(u^a £_a)(W^b \wedge £_b \Omega + d_u \Omega) - d_u(\exp(u^a £_a)\Omega).$$

For the moment, set

$$u^a £_a = £_R, \qquad R = u^a V_a. \tag{10-13.8}$$

We then have (Schouten, *Ricci-Calculus*. p. 109)

$$\exp(£_R) £_a \exp(-£_R)\alpha = G_a^b(u^e) £_b \alpha \tag{10-13.9}$$

for any linear geometric object field α, where the G's are functions of the u's only that are defined by

$$\exp(\pounds_R)V_a = G_a^b(u^e)V_b. \tag{10-13.10}$$

The relations (10-13.9) show that G_r acts on its Lie algebra $g_r(\pounds_a; \Lambda(\mathscr{G}))$ by the adjoint representation while (10-13.10) shows that the G's give the adjoint representation on the vector space $\text{span}(V_a | 1 \le a \le r)$. If we now set $\alpha = \exp(\pounds_R)\beta$ in (10-13.9), it follows that

$$\exp(\pounds_R)\pounds_a\beta = G_a^b\pounds_b\exp(\pounds_R)\beta \tag{10-13.11}$$

for any element β of $\Lambda(\mathscr{G})$. Next, we note that

$$\exp(\pounds_R)(a \wedge \beta) = (\exp(\pounds_R)\alpha) \wedge \exp(\pounds_R)\beta \tag{10-13.12}$$

for any $\alpha, \beta \in \Lambda(\mathscr{G})$ because $\exp(\pounds_R)(\alpha \wedge \beta) = T_R^*(\alpha \wedge \beta)$ and $T_R: K \to K$ is the automorphism of K that is generated by the flow of R with canonical orbital parameter equal to unity. A combination of (10-13.11) and (10-13.12) shows that

$$\begin{aligned}\exp(\pounds_R)(W^b \wedge \pounds_b\Omega) &= (\exp(\pounds_R)W^b) \wedge \exp(\pounds_R)\pounds_b\Omega \\ &= (\exp(\pounds_R)W^b) \wedge G_b^a\pounds_a\exp(\pounds_R)\Omega.\end{aligned}$$

When this is put back into (10-13.7), we have

$$\begin{aligned}{}^{`}\Gamma \wedge \exp(\pounds_R)\Omega = {}&(\exp(\pounds_R)W^b)G_b^a \wedge \pounds_a\exp(\pounds_R)\Omega \\ &+\exp(\pounds_R)d_u\Omega - d_u(\exp(\pounds_R)\Omega).\end{aligned} \tag{10-13.13}$$

The standard equations for the Lie group G_r and the fact that Ω is a differential form on $\mathscr{G}$ with coefficients from $\Lambda^0(\mathscr{G}) = \Lambda^0(G_r \times K)$ show that (see Rund, 1982)

$$\frac{\partial}{\partial u^a}(\exp(\pounds_R)\Omega) = \lambda_a^b\pounds_b\exp(\pounds_R)\Omega + \exp(\pounds_R)\frac{\partial}{\partial u^a}\Omega,$$

where the λ's are functions of the u's only for which

$$\frac{\partial}{\partial u^a}{}^{`}z^A = \lambda_a^b(u){}^{`}V_b\langle{}^{`}z^A\rangle,\ {}^{`}z^A = \exp(u^bV_b)z^A.$$

Accordingly, we obtain

$$d_u(\exp(\pounds_R)\Omega) = \exp(\pounds_R)d_u\Omega + \bar{d}u^a \wedge \lambda_a^b\pounds_b\exp(\pounds_R)\Omega. \tag{10-13.14}$$

When (10-13.14) is substituted into (10-13.13), it follows that

$$\text{(10-13.15)} \quad {}^{\backprime}\Gamma \wedge \exp(\pounds_R)\Omega = \left(G_a^b \exp(\pounds_R) W^a - \lambda_a^b \bar{d}u^a\right) \wedge \pounds_b \exp(\pounds_R)\Omega.$$

Since $\{\pounds_b | 1 \le b \le r\}$ is a basis for $g_r(\pounds_a;\ \Lambda(\mathscr{G}))$, this relation can be satisfied simultaneously for all $\Omega \in \Lambda(\mathscr{G})$ if and only if

$$\text{(10-13.16)} \qquad {}^{\backprime}\Gamma = {}^{\backprime}W^b \pounds_b,$$

$$\text{(10-13.17)} \qquad {}^{\backprime}W^b = G_a^b \exp(u^e \pounds_e) W^a - \lambda_a^b \bar{d}u^a.$$

Thus *the image of any operator-valued connection* $\Gamma = W^a \pounds_a$ *under the action of* G_r *is an operator-valued connection* ${}^{\backprime}\Gamma = {}^{\backprime}W^a \pounds_a$, *and the* 1*-forms* $W^a \in \Lambda^1(\mathscr{G})$ *transform under action of* G_r *by the generalized gauge transformations* (10-13.17). This latter result is not unexpected, for $\Gamma = W^a \pounds_a$ is an operator-valued connection and hence the W^a's should transform inhomogeneously under the action of G_r. The unusual aspect is the very complicated dependence, $\exp(u^e \pounds_e) W^a$, on the original 1-forms W^a.

10-14. PROPERTIES OF D AND OPERATOR-VALUED CURVATURE FORMS

Starting with (10-13.3) and (10-13.5),

$$\text{(10-14.1)} \qquad D\omega = \bar{d}\omega + W^a \wedge \pounds_a \omega$$

for any $W^a \in \Lambda(\mathscr{G})$, it is easily shown that

$$\text{(10-14.2)} \qquad D(\alpha + \beta) = D\alpha + D\beta,$$

$$\text{(10-14.3)} \qquad D(\alpha \wedge \beta) = (D\alpha) \wedge \beta + (-1)^{\deg(\alpha)} \alpha \wedge D\beta.$$

The G_r-covariant exterior derivative is thus an antiderivation on $\Lambda(\mathscr{G})$. The analogy with the exterior derivative stops here, however.

Since $D\omega$ belongs to $\Lambda(\mathscr{G})$, we have

$$\pounds_a D\omega = \pounds_a \bar{d}\omega + \pounds_a W^b \wedge \pounds_b \omega + W^b \wedge \pounds_a \pounds_b \omega.$$

On the other hand

$$D\pounds_a \omega = \bar{d}\pounds_a \omega + W^b \wedge \pounds_b \pounds_a \omega = \pounds_a \bar{d}\omega + W^b \wedge \pounds_b \pounds_a \omega$$

because $\bar{d}$ and $\pounds_a$ commute. When these are combined, the commutator of $\pounds_a$ and D is seen to have the evaluation

$$(\pounds_a D - D\pounds_a)\omega = \pounds_a W^b \wedge \pounds_b \omega + W^b \wedge (\pounds_a \pounds_b - \pounds_b \pounds_a)\omega.$$

Thus when (10-12.3) is used, we have

(10-14.4) $$(\pounds_a D - D\pounds_a)\omega = \rho_a \wedge \omega,$$

where

(10-14.5) $$\rho_a = \rho_a^b \pounds_b$$

is a system of operator-valued 1-forms and

(10-14.6) $$\rho_a^b = \pounds_a W^b + C_{ae}^b W^e$$

are 1-forms on $\mathscr{G}$.

If we start with an element α from $(\pi_2)^*\Lambda(K)$, $D\alpha$ belongs to $\Lambda(\mathscr{G})$ but not to $(\pi_2)^*\Lambda(K)$. It is for precisely this reason that the space $\mathscr{G}$ was introduced in the first place, for $D\alpha \in \Lambda(\mathscr{G})$ allows us to apply the operator D to this quantity again. In particular, since (10-13.5) holds, we have

(10-14.7) $$`D`D\exp(u^a\pounds_a)\omega = \exp(u^a\pounds_a)(DD\omega)$$

and hence $DD\omega$ is G_r-covariant. A direct substitution using $D = \bar{d} + W^a \wedge \pounds_a$ gives

(10-14.8) $$DD\omega = \Theta \wedge \omega$$

where

(10-14.9) $$\Theta = \theta^a \pounds_a, \qquad \theta^a = DW^a + \tfrac{1}{2}C_{be}^a W^b \wedge W^e.$$

Direct analogy with the results of differential geometry and gauge theory suggests that Θ be referred to as the *operator-valued curvature 2-form* associated with the operator-valued connection 1-form $\Gamma = W^a\pounds_a$ and that $\{\theta^a | 1 \leq a \leq r\}$ are the *curvature 2-forms* on $\Lambda(\mathscr{G})$ that arise from the connection 1-forms $\{W^a | 1 \leq a \leq r\}$. This is further borne out by noting that (10-14.8) is a G_r-covariant equation and hence (see (10-13.8))

$$\begin{aligned} `\theta^a \wedge \pounds_a`\Omega &= \exp(\pounds_R)(\theta^b \wedge \pounds_b\Omega) = \left(\exp(\pounds_R)\theta^b\right) \wedge \exp(\pounds_R)\pounds_b\Omega \\ &= \left(\exp(\pounds_R)\theta^b\right) \wedge G_b^a \pounds_a \exp(\pounds_R)\Omega \\ &= G_b^a\left(\exp(\pounds_R)\theta^b\right) \wedge \pounds_a`\Omega \end{aligned}$$

when (10-13.11) is used. Accordingly, the curvature 2-forms transform under the action of G_r by the homogeneous transformation law

(10-14.10) $$`\theta^a = G_b^a \exp(u^e\pounds_e)\theta^b.$$

We now look at the two expressions $D(DD\omega)$ and $DD(D\omega)$. Since

$$DD(D\omega) = \Theta \wedge D\omega = \theta^a \wedge \pounds_a D\omega$$

and

$$D(DD\omega) = D(\theta^a \wedge \pounds_a\omega) = D\theta^a \wedge \pounds_a\omega + \theta^a \wedge D\pounds_a\omega,$$

we have

$$D(DD\omega) - DD(D\omega) = D\theta^a \wedge \pounds_a\omega + \theta^a \wedge (D\pounds_a - \pounds_a D)\omega.$$

Accordingly, when (10-14.4) is used, we obtain

$$D(DD\omega) - DD(D\omega) = \beta \wedge \omega,$$

where

(10-14.11) $$\beta = \beta^a \pounds_a, \qquad \beta^a = D\theta^a - \rho_e^a \wedge \theta^e.$$

An elementary calculation based upon (10-14.6), (10-14.9) and the Jacobi identity shows that $\beta^a \equiv 0$ on $\mathcal{G}$ for any choice of $W^a \in \Lambda^1(\mathcal{G})$. We accordingly have the desired result,

(10-14.12) $$D(DD\omega) \equiv DD(D\omega)$$

and the corresponding Bianchi identities $\beta^a \equiv 0$; that is,

(10-14.13) $$D\theta^a = \rho_e^a \wedge \theta^e.$$

There is an interesting point that should be observed here. Slight rearrangements of (10-14.9) and (10-14.13) give

(10-14.14) $$DW^a = \theta^a - \tfrac{1}{2}C_{be}^a W^b \wedge W^e, \qquad D\theta^a = \rho_e^a \wedge \theta^e.$$

Accordingly, the ideal

(10-14.15) $$\overline{W} = I\{W^1, \ldots, W^r, \theta^1, \ldots, \theta^r\}$$

of $\Lambda(\mathcal{G})$ is G_r-covariant differentially closed; that is,

(10-14.16) $$D\overline{W} \subset \overline{W}.$$

An exceptional aspect of the G_r-covariant exterior derivative operator D is that *any constant element of* $\Lambda^0(\mathcal{G})$ *is* G_r*-covariant constant*, as follows directly from $Dk = \bar{d}k + W^a \wedge \pounds_a k = 0$. We therefore have

(10-14.17) $$DC_{be}^a = 0$$

since the structure constants of G_r are constant functions on $\mathcal{G}$. It is useful to compare this one line derivation of (10-14.17) for operator-valued connections with the more customary approach in which derivation of the same result usually covers pages.

The fields of 1-forms W^a that occur in the operator-valued connection $\Gamma = W^a £_a$ are often referred to in the literature as compensating fields; that is, fields that compensate for changes that arise from the local action of the group G_r. This same interpretation obtains here. To see this, suppose that ω is a G_r-invariant form, $\omega = \exp(u^a £_a)\omega$. We then have $£_a \omega = 0$ and hence $D\omega = \bar{d}\omega + W^a \wedge £_a \omega = \bar{d}\omega$. *The G_r-covariant exterior derivative of a G_r-invariant form ω reduces to the exterior derivative of ω*;

$$(10\text{-}14.18) \qquad £_a \omega = \omega, \qquad 1 \le a \le r \Rightarrow D\omega = \bar{d}\omega.$$

The G_r-covariant exterior derivative thus differs from the exterior derivative only if the action of G_r changes things. This result is a special case of a general situation that will be of importance later. Let I be an ideal of $\Lambda(\mathcal{G})$. It then follows directly from $D\omega = \bar{d}\omega + W^a \wedge £_a \Omega$ that

$$(10\text{-}14.19) \qquad £_a \omega \equiv 0 \bmod I \Rightarrow D\omega \equiv \bar{d}\omega \bmod I.$$

10-15. LIE CONNECTIONS

Up to this point, the 1-forms W^a have been arbitrary elements of $\Lambda^1(\mathcal{G})$. We now specialize to the important case where the W^a are invariant under transport along the orbits of G_r; that is,

$$(10\text{-}15.1) \qquad £_b W^a = 0$$

for all values of a and b in the range 1 through r. Operator-valued connections $\Gamma = W^a £_a$ with the W^a's satisfying (10-15.1) will be referred to as *Lie connections*.

The constraints (10-15.1) are not as severe as might appear on first reading, for W^a are 1-forms on $\mathcal{G}$ rather than on K and $£_a u^b = 0$ are identically satisfied on $\mathcal{G}$. The full scope of this can be seen by setting $W^a = T^a + w^a_b \bar{d}u^b$ with $£_b T^a = 0$. In this event, (10-15.1) is satisfied provided $V_b\langle w^a_e \rangle = 0$ hold. However, $V_b\langle w^a_e \rangle = 0$ is a complete system of linear first-order partial differential equations because $[V_a, V_b] = C^e_{ab} V_e$. Thus since $\{u^a | 1 \le a \le r\}$ are r primitive integrals of $V_a\langle f \rangle = 0$, we have $w^a_e = \Psi^a_e(u^b; \eta^\sigma)$ where the Ψ's are arbitrary C^1 functions of their arguments and $\{\eta^\sigma | 1 \le \sigma \le n + N - r\}$ together with $\{u^a | 1 \le a \le r\}$ constitute a complete system of primitive integrals of $V_a\langle f \rangle = 0$. Here of course, it is assumed that the set $\{\eta^\sigma\}$ is vacuous if $n + N \le r$.

If $\Gamma = W^a £_a$ is a Lie connection, we have

$$DW^a = \bar{d}W^a + W^b \wedge £_b W^a = \bar{d}W^a. \tag{10-15.2}$$

Equations (10-14.9) then show that the associated *Lie curvature* 2-forms $\{\theta^a\}$ are given by

$$\theta^a = \bar{d}W^a + \tfrac{1}{2}C^a_{be}W^b \wedge W^e, \tag{10-15.3}$$

which constitute the familiar representation for curvature 2-forms in gauge theory.

Satisfaction of (10-15.1) implies $\exp(u^e £_e)W^a = W^a$. Thus if $\Gamma = W^a £_a$ is a Lie connection, (10-13.15) and (10-13.16) give

$$`\Gamma = `W^a £_a, \tag{10-15.4}$$

$$`W^a = G^a_b W^b - \lambda^a_b \bar{d}u^b. \tag{10-15.5}$$

Thus if G_r is restricted to a constant section ($\bar{d}u^b = 0$), the W^a's transform by the adjoint representation. In the general case, it is useful to write

$$`W^a = G^a_b W^b - \lambda^a \tag{10-15.6}$$

where

$$\lambda^a = \lambda^a_b \bar{d}u^b \tag{10-15.7}$$

are 1-forms on G_r that satisfy the Maurer equations

$$\bar{d}\lambda^a = \tfrac{1}{2}C^a_{be}\lambda^b \wedge \lambda^e. \tag{10-15.8}$$

Noting that G^a_b and λ^a_b are functions of the u's only, it follows that $£_m G^a_b = 0$, $£_m \lambda^a_b \bar{d}u^b = 0$. Accordingly, (10-15.5) gives

$$£_m `W^a = G^a_b £_m W^b = 0, \tag{10-15.9}$$

and hence *the action of G_r takes Lie connections into Lie connections*. The collection of all Lie connections is thus closed under the action of G_r.

If $\{\theta^a\}$ are the Lie curvature 2-forms of a Lie connection $\Gamma = W^a £_a$, (10-15.3) shows that

$$£_m \theta^a = \bar{d}£_m W^a + \tfrac{1}{2}C^a_{be}\left(£_m W^b \wedge W^e + W^b \wedge £_m W^e\right) = 0. \tag{10-15.10}$$

We thus have $\exp(u^e £_e)\theta^a = \theta^a$, and hence (10-14.10) gives

$$`\theta^a = G^a_b \theta^b; \tag{10-15.11}$$

Lie curvature 2-forms transform under action of G_r by the adjoint representation. This result lies at the heart of later matters since it provides the means whereby a G_r-invariant 4-form may be constructed. The coefficients $\{G^a_b(u)\}$ of the adjoint action of any element of G_r satisfy (see Rund, 1982)

$$G^a_e(u)C^e_{br} = C^a_{fm}G^f_b(u)G^m_r(u) \tag{10-15.12}$$

and hence

$$C_{ab} = C_{fm}G^f_a(u)G^m_b(u), \tag{10-15.13}$$

where

$$C_{ab} = C^e_{am}C^m_{be} = C_{ba} \tag{10-15.14}$$

are the components of the Cartan-Killing form on G_r. It is then a simple matter to see that

$$\rho = C_{ab}\theta^a \wedge \theta^b \tag{10-15.15}$$

is a G_r-invariant 4-form on $\mathcal{G}$ for any Lie connection $\Gamma = W^a £_a$ (i.e., $C_{ab}{}^\backprime\theta^a \wedge {}^\backprime\theta^b = C_{ab}G^a_eG^b_f\theta^e \wedge \theta^f = C_{ef}\theta^e \wedge \theta^f$). We note as a matter of consistency that (10-15.5),

$${}^\backprime\theta^a = \bar{d}{}^\backprime W^a + \tfrac{1}{2}C^a_{be}{}^\backprime W^b \wedge {}^\backprime W^e, \tag{10-15.16}$$

and the group equations (see Rund, 1982)

$$\bar{d}\lambda^a = \tfrac{1}{2}C^a_{be}\lambda^b \wedge \lambda^e, \qquad \bar{d}G^a_b = C^a_{ef}G^f_b\lambda^e \tag{10-15.17}$$

lead directly to the transformation law (10-15.11). Conversely, (10-15.3), (10-15.5), (10-15.11) and (10-15.17) lead to the determination (10-15.16) for ${}^\backprime\theta^a$: *evaluations of θ^a in terms of W^a are G_r-invariant.*

10-16. THE MINIMAL REPLACEMENT CONSTRUCT

We assume that there are m physical state variables and that the physics is described by means of an action functional $A[\Phi]$ with Lagrangian function $L(x^i, q^\alpha, y_i^\alpha)$ as in Chapter Seven. Here the y's are correlated with the q's through the contact 1-forms $C^\alpha = dq^\alpha - y_i^\alpha dx^i$ and Φ is such that $\Phi^*\mu \neq 0$, $\Phi^*C^\alpha = 0$.

Let $N_1(L)$ be the collection of all Noetherian vector fields of the first kind for the Lagrangian L. The action $A[\Phi]$ is then invariant under the global action of any $g_r \subset N_1(L)$ for any regular map Φ and we have ${}^\backprime(L\mu) \equiv$

Let the replacement operator

$$\mathscr{M}: \bar{d} \mapsto D \tag{10-16.3}$$

be defined by

$$\mathscr{M}(\alpha + \beta) = \mathscr{M}\alpha + \mathscr{M}\beta, \qquad \mathscr{M}(\alpha \wedge \beta) = (\mathscr{M}\alpha) \wedge (\mathscr{M}\beta),$$

$$\mathscr{M}(\bar{d}z^A) = Dz^A, \qquad \mathscr{M}(\bar{d}u^a) = Du^a = \bar{d}u^a, \qquad \mathscr{M}f = f \quad \forall f \in \Lambda^0(\mathscr{G}).$$

Thus since $\Lambda(\mathscr{G})$ is a module over $\Lambda^0(\mathscr{G})$ that is generated from the basis $(1, \bar{d}z^A, \bar{d}u^b)$, $\mathscr{M}$ is well defined on $\Lambda(\mathscr{G})$. Some care must be exercised here, for $\mathscr{M}(d\alpha) \neq D(\mathscr{M}\alpha)$. Simply observe that $\mathscr{M}(d(z^A dz^B)) = \mathscr{M}(dz^A \wedge dz^B) = Dz^A \wedge Dz^B$, while $D\mathscr{M}(z^A dz^B) = D(z^A Dz^B) = Dz^A \wedge Dz^B + z^A DDz^B$ and $DDz^B = \theta^a £_a z^B \neq 0$. On the other hand,

$$\mathscr{M}(\bar{d}f) = \frac{\partial f}{\partial z^A} Dz^A + \frac{\partial f}{\partial u^a} \bar{d}u^a$$

for any $f \in \Lambda^0(\mathscr{G})$. We thus have

$$\mathscr{M}C^\alpha = Dq^\alpha - y_i^\alpha Dx^i = C^\alpha + W^a(V_a \lrcorner C^\alpha), \tag{10-16.4}$$

$$\mathscr{M}(L\mu) = \mathscr{M}(L) Dx^1 \wedge Dx^2 \wedge Dx^3 \wedge Dx^4, \tag{10-16.5}$$

while (10-16.2) shows that

$$(\mathscr{M}C^\alpha)|_k = C^\alpha, \qquad (\mathscr{M}(L\mu))|_k = L\mu. \tag{10-16.6}$$

Further ${}^{\backprime}(D\alpha) = \exp(u^a £_a) D\alpha = {}^{\backprime}D(\exp(u^a £_a)\alpha) = {}^{\backprime}D{}^{\backprime}\alpha$, and $Du^a = \bar{d}u^a$, ${}^{\backprime}u^a = u^a$ because $£_b u^a = 0$. We therefore have ${}^{\backprime}(\mathscr{M}\beta) = \exp(u^a £_a)\mathscr{M}\beta = \mathscr{M}(\exp(u^a £_a)\beta) = \mathscr{M}({}^{\backprime}\beta)$ where $\mathscr{M}(\bar{d}{}^{\backprime}z^A) = {}^{\backprime}D{}^{\backprime}z^A$:

$${}^{\backprime}(\mathscr{M}\beta) = \mathscr{M}({}^{\backprime}\beta) \qquad \forall \beta \in \Lambda(\mathscr{G}). \tag{10-16.7}$$

Accordingly,

$${}^{\backprime}(\mathscr{M}(L\mu)) = \mathscr{M}({}^{\backprime}(L\mu)) \tag{10-16.8}$$

under the action of the group G_r.

There is quite a bit more here, however, for $(\mathscr{M}(L\mu))|_k = L\mu$ and also $(\mathscr{M}(L\mu) + \eta)|_k = L\mu$ for any $\eta \in \Lambda^4(\mathscr{G})$ that vanishes on G_r-constant sections of $\mathscr{G}$. Further, (10-16.8) gives

$${}^{\backprime}(\mathscr{M}(L\mu) + \eta) = \mathscr{M}({}^{\backprime}(L\mu)) + \eta \tag{10-16.9}$$

$L\mu \bmod C$. On the other hand, gauge theory arises by allowing
elements of G_r to act at different points of E_4 while preserving the i
of the action. There are clearly two parts to this problem. The first i
the canonical parameters to vary from one point to another over E_4
second is to retain the invariance of the action $A[\Phi]$ under the resul
action of the group G_r.

The simplest way of accomplishing these tasks is to lift considerat
K to the larger space $\mathscr{G} = G_r \times K$ with local coordinates $\{u^a;\ x^i, q^\alpha,$
is, we make the identification

$$\{z^A | 1 \leq A \leq 4 + 5m\} = \{x^i, q^\alpha, y_i^\alpha | 1 \leq i \leq 4, 1 \leq \alpha \leq m$$

$n = 4$ and $N = 5m$. For this general setting, the group space coordin
are independent quantities that may vary in any way we please. On
have been analyzed in $\mathscr{G}$, we will be able to consider sections S:
without difficulty. There is actually no real choice in the ma
$S^*(\exp(u^a £_a)\alpha)$ is quite different from $\exp(S^*(u^a)£_a)\alpha$ for $\alpha \in \Lambda$
differently, position dependent action of G_r means that different elem
act at different positions, that is, $S^*(\exp(u^a £_a)\alpha)$, not $\exp(S^*(u^a$
fact, $\exp(S^*(u^a)£_a)z^B$ will belong to G_r only if $S: E_4 \to G_r$ defines
stant section $\{u^a = k^a | 1 \leq a \leq r\}$.

Let $\Gamma = W^a £_a$ be a Lie connection for a group G_r with $g_r \subset N$
assume that the W^a's have the form

$$W^a = W_b^a \bar{d} u^b \tag{10-16.1}$$

with $£_a \langle W_c^b \rangle = 0$. If $\alpha|_k$ denotes the restriction of any exterior form o
constant section of G_r (i.e., $u^a = \bar{u}^a = \text{constant}, 1 \leq a \leq r$), we have
and hence

$$(D\alpha)|_k = d(\alpha|_k), \qquad (\bar{d}\alpha)|_k = d(\alpha|_k). \tag{10-16.2}$$

This, however, is exactly the case in which the group G_r acts globall
To remove the restriction to G_r-constant sections of $\mathscr{G}$ we note tha

$$`(dz^A) = \exp(u^a £_a)dz^A \neq \bar{d}(\exp(u^a £_a)z^A) = \bar{d}`z^A$$

because the u's can change, but

$$`(Dz^A) = \exp(u^a £_a)Dz^A = `D(\exp(u^a £_a)z^A) = `D`z^A$$

by (10-13.6). Further (10-16.2) shows that Dz^A restricted to any G_r-
section of $\mathscr{G}$ agrees with dz^A. Thus if we simply replace the exterior d
by the G_r-covariant exterior derivative in all statements, these sta
become G_r-covariant statements on $\mathscr{G}$.

provided η is a G_r-invariant 4-form on $\mathcal{G}$ ($\exp(u^a £_a)\eta = \eta$). Thus *the transition*

$$L\mu \mapsto \mathcal{M}(L\mu) + \eta, \tag{10-16.10}$$

for any $\eta \in \Lambda^4(\mathcal{G})$ *such that*

$$\eta|_k = 0, \qquad `\eta = \exp(u^a £_a)\eta = \eta, \tag{10-16.11}$$

lifts $L\mu \in \Lambda^4(K)$ *up to an element of* $\Lambda^4(\mathcal{G})$ *for which*

$$(\mathcal{M}(L\mu) + \eta)|_k = L\mu, \tag{10-16.12}$$

$$`(\mathcal{M}(L\mu) + \eta) = \mathcal{M}(`(L\mu)) + \eta. \tag{10-16.13}$$

The partial transition $L\mu \mapsto \mathcal{M}(L\mu)$ will turn out to be the Yang-Mills *minimal replacement* while $\mathcal{M}(L\mu) \mapsto \mathcal{M}(L\mu) + \eta$ is the basis for the Yang-Mills *minimal coupling* construct.

The construct arrived at in this way is more general than actually needed, for we are interested only in what happens when the canonical parameters $\{u^a | 1 \le a \le r\}$ vary over the space-time manifold E_4. It is therefore necessary to cut things down by introducing mappings

$$S : E_4 \mapsto G_r | u^a = s^a(x^j). \tag{10-16.14}$$

When S acts, (10-16.1) gives

$$S^*W^a = (S^*W_b^a)\frac{\partial s^b}{\partial x^i}dx^i,$$

and hence we may write

$$S^*W^a = W_i^a(x^j)\,dx^i \tag{10-16.15}$$

where $\{W_i^a(x^j) | 1 \le a \le r, 1 \le i \le 4\}$ is taken to be a system of $4r$ *new fields* that compensate for the local space-time action $u^a = s^a(x^j)$ of the group G_r. Now, $S^*\bar{d}\alpha = dS^*\alpha$ and hence

$$S^*D\alpha = dS^*\alpha + W_i^a dx^i \wedge S^*£_a\alpha. \tag{10-16.16}$$

If α depends on the u^a's in any way, $S^*£_a\alpha \ne £_a S^*\alpha$. On the other hand, if $\beta \in \Lambda(K)$ then $S^*\beta = \beta$ and we have

$$S^*D\beta = d\beta + W_i^a dx^i \wedge £_a\beta, \tag{10-16.17}$$

which we will simply write as $\overset{*}{D}\beta$ for $\beta \in \Lambda(K)$ from now on. The G_r-covariant exterior derivative $\overset{*}{D}$ thus induces the G_r-covariant derivative $\overset{*}{D}_i$, where

$$\overset{*}{D}_i\gamma = \partial_i\gamma + W_i^a £_a\gamma$$

for any linear geometric object field γ on M_4.

A combination of the two operations S^* and $\mathscr{M}$ gives what is usually called the *minimal replacement*

$$\mathscr{M}^* = S^*\mathscr{M}. \tag{10-16.18}$$

In particular, we have

$$\mathscr{M}^*dx^i = \overset{*}{D}x^i = dx^i + W_j^a dx^j £_a x^i = \left(\delta_j^i + W_j^a £_a x^i\right)dx^j, \tag{10-16.19}$$

$$\mathscr{M}^*dq^\alpha = \overset{*}{D}q^\alpha = dq^\alpha + W_j^a £_a q^\alpha dx^j, \tag{10-16.20}$$

and hence (10-16.4) and (10-16.5) give

$$\mathscr{M}^*C^\alpha = C^\alpha + (V_a \lrcorner C^\alpha)W_i^a dx^i, \tag{10-16.21}$$

$$\mathscr{M}^*(L\mu) = \mathscr{M}^*(L)\det\left(\delta_j^i + W_j^a £_a x^i\right)\mu. \tag{10-16.22}$$

The $4r$ quantities $\{W_j^a(x^i) | 1 \le a \le r, 1 \le j \le 4\}$ constitute a system of new fields that compensate for the space-time dependence of the action of the group G_r. Therefore, it must be clearly noted that we have gone from the system of $r + r^2$ quantities $\{u^a, W_b^a\}$ to the system of $4r$ quantities $\{W_i^a\}$ since the individual u^a's become lost among the other x^i-dependences once the map $S: M_4 \to G_r$ has been effected. This, however, is the standard situation in gauge theory, for a specific mapping $S: M_4 \to G_r$ is not obtained, only the compensating fields $\{W_i^a(x)\}$. Accordingly, we must adjoin the $4r$ quantities $\{W_i^a\}$ to the list $\{q^\alpha, y_i^\alpha\}$ as a system of new state variables.

There is now an important question that must be resolved; namely, what is the image of a quantity $S^*\alpha$ under the action of the group G_r? The considerations given at the beginning of this section concerning the nature of the map S^* show that *the action of G_r must be computed in $\mathscr{G}$ and only afterward cut down by sectioning with S*. This means that the image of $S^*\alpha$ can be defined only by

$$`(S^*\alpha) = S^*(\exp(u^a £_a)\alpha) = S^*(`\alpha). \tag{10-16.23}$$

Thus since $\mathscr{M}^* = S^*\mathscr{M}$, (10-16.23) and (10-16.7) give $`(\mathscr{M}^*\alpha) = `(S^*\mathscr{M}\alpha) = S^*(`(\mathscr{M}\alpha)) = S^*\mathscr{M}(`\alpha)$; that is,

$$`(\mathscr{M}^*\alpha) = \mathscr{M}^*(`\alpha). \tag{10-16.24}$$

In view of the transition $C^\alpha \to \mathscr{M}^*C^\alpha$, the induced transition of the contact ideal C is

$$C \to C^* = \mathscr{M}^*C = I\{\mathscr{M}^*C^1, \dots, \mathscr{M}^*C^m\}. \tag{10-16.25}$$

Now, $g_r \subset N_1(L)$ so that $\grave{}(L\mu) \equiv L\mu \bmod C$. We thus have

$$\mathscr{M}^*(\grave{}(L\mu)) \equiv \mathscr{M}^*(L\mu) \bmod C^*$$

and hence (10-16.24) gives

$$\grave{}\mathscr{M}^*(L\mu) \equiv \mathscr{M}^*(L\mu) \bmod C^*. \tag{10-16.26}$$

For $S: E_4 \to G_r$, the transition (10-16.10) becomes

$$L\mu \to \mathscr{M}^*(L\mu) + S^*\eta. \tag{10-16.27}$$

The new action integral is thus given by

$$\bar{A}[\Phi] = \int_{M_4} \Phi^*(\mathscr{M}^*(L\mu) + S^*\eta) \tag{10-16.28}$$

for any regular map

$$\Phi: E_4 \to K \times \mathbb{R}^{4r} | x^i = x^i, \qquad q^\alpha = \phi^\alpha(x^j), \qquad W_i^a = W_i^a(x^j), \tag{10-16.29}$$

$$\Phi^*\mathscr{M}^*\mu \neq 0, \qquad \Phi^*C^* = 0.$$

(Recall that minimal replacement induces the transition $\mu \to \mathscr{M}^*\mu$ and that the quantities $\{W_i^a\}$ are to be included as new field variables.) We now simply observe that (10-16.11) and (10-16.23) give $\grave{}(S^*\eta) = S^*\eta$ and hence

$$\grave{}(\mathscr{M}^*(L\mu) + S^*\eta) \equiv \mathscr{M}^*(L\mu) + S^*\eta \bmod C^* \tag{10-16.30}$$

by (10-16.26). Accordingly, (10-16.28) through (10-16.30) show that *the new action integral, $\bar{A}[\Phi]$, is invariant under the local action of the Lie group G_r.*

It should be noted that we started in K where each V_a that generates G_r is an isovector of the contact ideal,

$$\pounds_b C^\alpha = A^\alpha_{b\beta} C^\beta, \tag{10-16.31}$$

for which $\pounds_b(L\mu) \equiv 0 \bmod C$. Under minimal replacement, C^α is replaced by $\mathscr{M}^*C^\alpha = C^\alpha + W^a(V_a \lrcorner C^\alpha)$. Thus when (10-16.31) is used, we have

$$\pounds_b \mathscr{M}^*C^\alpha = (V_e \lrcorner C^\alpha)(\pounds_b W^e + W^a C^e_{ba}) \bmod C^*. \tag{10-16.32}$$

The generators of G_r fail to be isovectors of C^*. Thus although the new action $\bar{A}[\Phi]$ is G_r-invariant, the contact forms are only G_r-covariant:

$$(10\text{-}16.33)\qquad `(\mathscr{M}^*C^\alpha) = \mathscr{M}^*(`C^\alpha) = `D`q^\alpha - `y_i^\alpha`D`x^i.$$

What this means is that the r conserved currents that arise from global action of G_r go over into r balanced currents that are integrability conditions on the field equations for the compensating fields of the local action of G_r, as we shall see in the next section.

10-17. VARIATIONS AND THE FIELD EQUATIONS

The problem at hand is that of obtaining the governing Euler-Lagrange field equations. These obtain from rendering the action

$$(10\text{-}17.1)\qquad \bar{A}[\Phi] = \int_{M_4} \Phi^*(\mathscr{M}^*(L\mu) + S^*\eta)$$

stationary in value subject to the constraints

$$(10\text{-}17.2)\qquad \Phi^*C^* = 0, \qquad \Phi^*\mathscr{M}^*\mu \neq 0$$

where

$$(10\text{-}17.3)\qquad \Phi : E_4 \to K \times \mathbb{R}^{4r} | x^i = x^i, \qquad q^\alpha = \phi^\alpha(x^j), \qquad W_i^a = W_i^a(x^j),$$

$$\Phi^*\mu \neq 0, \qquad \Phi^*\mathscr{M}^*C^\alpha = 0,$$

and η is a G_r-invariant 4-form on $\mathscr{G}$ that vanishes on G_r-constant sections of G_r. Now,

$$(10\text{-}17.4)\qquad \mathscr{M}^*dx^i = \overset{*}{D}x^i = T_j^i dx^j,$$

where

$$(10\text{-}17.5)\qquad T_j^i = \delta_j^i + W_j^a \pounds_a x^i,$$

and hence

$$\mathscr{M}^*\mu = \det\left(T_j^i\right)\mu.$$

We therefore have $\mathscr{M}^*(L\mu) = \mathscr{M}^*(L)\det(T_j^i)\mu$. Now, $L \in \Lambda^0(K)$, $L = L(x^i, q^\alpha, y_i^\alpha)$, and hence $\mathscr{M}^*(L) = L$; that is,

$$(10\text{-}17.6)\qquad \mathscr{M}^*(L\mu) = L\det\left(T_j^i\right)\mu.$$

The reader accustomed to the standard minimal replacement construct might expect to see the y_i^α's change in L. This is not the case here, for the y_i^α's are independent quantities in the space K. We shall see, however, that $\Phi^* y_i^\alpha$ will be drastically different as a consequence of satisfaction of the constraints $\Phi^* \mathscr{M}^* C^\alpha = 0$ rather than $\Phi^* C^\alpha = 0$. Minimal replacement does have its expected effect on the derivatives of the field variables, but these effects come about only after application of Φ^*.

The exact nature of the 4-form η, and hence $S^*\eta$, is somewhat arbitrary at this point, although $S^*\eta$ must account for the presence of G_r-curvature terms. Accordingly, we shall deal with the problem in the general form

(10-17.7) $$\mathscr{M}^*(L\mu) + S^*\eta = \bar{L}\mu,$$

with

(10-17.8) $$\bar{L} = \bar{L}\left(x^j, q^\alpha, y_i^\alpha, W_i^a, \Theta_{ij}^a\right).$$

Here, we have set

(10-17.9) $$S^*\theta^a = \Theta^a = \tfrac{1}{2}\Theta_{ij}^a dx^i \wedge dx^j, \qquad \Theta_{ij}^a = -\Theta_{ji}^a,$$

$$\Theta^a = dW^a + \tfrac{1}{2}C_{be}^a W^b \wedge W^e, \qquad W^a = W_i^a dx^i.$$

The easiest way of dealing with this variational problem is to shift directly to the space R with local coordinates (x^i, q^α, W_i^a). A vector field on R has the form

$$U = U^i \partial_i + U^\alpha \partial_\alpha + U_i^a \partial_a^i,$$

where we have set $\partial_a^i = \partial/\partial W_i^a$. The classic variational process requires increments of the field variables that are functions on E_4, while the points of E_4 itself are unchanged by the variation process. It is therefore sufficient to our purposes to take $U^i = 0$ and all of the remaining U's to be functions of the x^i's only; that is,

(10-17.10) $$U = U^\alpha(x^j)\partial_\alpha + U_i^a(x^j)\partial_a^i.$$

We may then use Lie differentiation with respect to U to compute the variations in y_i^α and Θ_{ij}^a that arise from the variations (U^α, U_i^a) in the basic fields (q^α, W_i^a), respectively.

The induced variations in the y_i^α's are obtained through satisfaction of the conditions

(10-17.11) $$\pounds_U \mathscr{M}^* C^\alpha = 0;$$

that is, the variations preserve the constraints (10-17.2). Now, a combination of

(10-16.9) through (10-16.21) and (10-17.5) yields

$$\mathscr{M}^*C^\alpha = dq^\alpha + W_i^a £_a q^\alpha dx^i - y_i^\alpha T_j^i dx^j. \tag{10-17.12}$$

We therefore have

$$\Phi^*\mathscr{M}^*C^\alpha = \left(\partial_j\phi^\alpha + W_j^a\Phi^*(£_a q^\alpha) - \Phi^* y_i^\alpha \Phi^* T_j^i\right)dx^j,$$

and hence satisfaction of the constraints (10-17.2) demands that

$$\Phi^*(y_i^\alpha)\Phi^*\left(T_j^i\right) = \partial_j\phi^\alpha + W_j^a\Phi^*(£_a q^\alpha)$$

for any map Φ of the form (10-17.3). These are the relations that determine the place holders y_i^α in $\bar{L}$ when we come down to actual evaluations in terms of the fields $\{\phi^\alpha(x^j), W_i^a(x^j)\}$.

Noting that (10-17.10) gives

$$£_U £_a x^i = U^\beta \partial_\beta £_a x^i, \qquad £_U £_a q^\alpha = U^\beta \partial_\beta £_a q^\alpha,$$

(10-17.11) and (10-17.12) lead to

$$\begin{aligned}(£_U y_i^\alpha)T_j^i = \partial_j U^\alpha + U_j^a £_a q^\alpha + U^\beta W_j^a \partial_\beta £_a q^\alpha \\ - y_i^\alpha\left(U_j^a £_a x^i + U^\beta W_j^a \partial_\beta £_a x^i\right).\end{aligned} \tag{10-17.13}$$

It is clear from (10-17.5) and $\Phi^*\mathscr{M}^*\mu \neq 0$ that we must require

$$\det\left(T_j^i\right) \neq 0, \tag{10-17.14}$$

and hence we may introduce the quantities t_j^i by

$$T_j^i t_k^j = \delta_k^i. \tag{10-17.15}$$

Thus (10-17.13) yields the desired specific evaluation

$$\begin{aligned}£_U y_i^\alpha = \Big(\partial_j U^\alpha + U_j^a £_a q^\alpha + U^\beta W_j^a \partial_\beta £_a q^\alpha \\ - y_k^\alpha\left(U_j^a £_a x^k + U^\beta W_j^a \partial_\beta £_a x^k\right)\Big)t_i^j.\end{aligned} \tag{10-17.16}$$

Computation of the variations that are induced in Θ_{ij}^a are most easily accomplished by noting that (10-17.10) yields

$$£_U W^a = U_i^a dx^i. \tag{10-17.17}$$

Accordingly, (10-16.15) gives

(10-17.18) $$\pounds_U \Theta^a = d\pounds_U W^a + \pounds_U W^b \wedge C^a_{bc} W^c.$$

Further expansion is unnecessary, as we shall see presently.

We now have all of the results needed to proceed with the final calculations. Since $\pounds_U(\bar{L}\mu) = (\pounds_U \bar{L})\mu$ because $\pounds_U \mu = 0$ (recall that $U^i = 0$), we need only compute $\pounds_U \bar{L}$. Thus, introducing the notation

(10-17.19) $$L_\alpha = \partial \bar{L}/\partial q^\alpha, \qquad L^i_\alpha = \partial \bar{L}/\partial y^\alpha_i,$$

(10-17.20) $$G^{ij}_a = \partial \bar{L}/\partial \Theta^a_{ij}, \qquad \sigma^i_a = \partial \bar{L}/\partial W^a_i|_\Theta,$$

we have

(10-17.21) $$\pounds_U(\bar{L}\mu) = \left(L_\alpha U^\alpha + L^i_\alpha \pounds_U y^\alpha_i + \sigma^i_a U^a_i + G^{ij}_a \pounds_U \Theta^a_{ij}\right)\mu.$$

It is now simply a matter of substituting (10-17.16) and (10-17.18) into (10-17.21) and then discarding all divergences and/or exact 4-forms in order to obtain the Euler-Lagrange field equations.

The field equations for the q^α's (for the ϕ^α's) come from collecting together all terms that involve the U^α's and their derivatives that appear in (10-17.21)

$$\partial_j\left(t^j_i L^i_\gamma U^\gamma\right)\mu + U^\gamma\left\{L_\gamma - \partial_j\left(t^j_i L^i_\gamma\right) + t^j_i L^i_\alpha W^a_j\left(\partial_\gamma \pounds_a q^\alpha - y^\alpha_k \partial_\gamma \pounds_a x^k\right\}\mu.$$

Standard practices of the calculus of variations thus give *the Euler-Lagrange equations for the q^α-fields*:

(10-17.22) $$\Phi^*\left\{\partial_j\left(t^j_i L^i_\alpha\right)\right\} = \Phi^*\left\{L_\alpha + t^j_i L^i_\beta W^a_j\left(\partial_\alpha \pounds_a q^\beta - y^\beta_k \partial_\alpha \pounds_a x^k\right)\right\}.$$

The terms in (10-17.12) that involve the variations U^a_i in the W^a-fields are given by

(10-17.23) $$\xi = \left[U^a_j\left\{\sigma^j_a + t^j_i L^i_\alpha\left(\pounds_a q^\alpha - y^\alpha_k \pounds_a x^k\right)\right\} + G^{ij}_a \pounds_U \Theta^a_{ij}\right]\mu.$$

If we set

(10-17.24) $$J_a = \left\{\sigma^j_a + t^j_i L^i_\alpha\left(\pounds_a q^\alpha - y^\alpha_k \pounds_a x^k\right)\right\}\mu_j \in \Lambda^3,$$

(10-17.25) $$G_a = \tfrac{1}{2} G^{ij}_a \mu_{ij} \in \Lambda^2,$$

where

(10-17.26) $$\mu_{ij} = \partial_i \lrcorner \mu_j, \qquad dx^k \wedge \mu_{ij} = \delta^k_i \mu_j - \delta^k_j \mu_i,$$

then (10-17.9), (10-17.17) and (10-17.23) through (10-17.26) give the particularly simple evaluation

$$\xi = £_U W^a \wedge J_a - 2£_U \Theta^a \wedge G_a. \tag{10-17.27}$$

When (10-17.18) is used, an elementary rearrangement gives

$$£_U \Theta^a \wedge G_a = d(£_U W^a \wedge G_a) + £_U W^a \wedge (dG_a + C^b_{ac} W^c \wedge G_b).$$

We therefore have

$$\xi = £_U W^a \wedge \{J_a - 2dG_a - 2C^b_{ac} W^c \wedge G_b\} - 2d(£_U W^a \wedge G_a). \tag{10-17.28}$$

Standard practices of the calculus of variations thus give *the Euler-Lagrange equations for the* W_i^a*-fields*:

$$\Phi^*\{dG_a + C^b_{ac} W^c \wedge G_b\} = \tfrac{1}{2}\Phi^* J_a. \tag{10-17.29}$$

The field equations (10-17.29) obviously entail integrability conditions. If we write (10-17.29) in the equivalent form ($G_a^* = \Phi^* G_a$, $J_a^* = \Phi^* J_a$)

$$dG_a^* = \tfrac{1}{2} J_a^* - C^b_{ac} W^c \wedge G_b^*, \tag{10-17.30}$$

then exterior differentiation gives

$$\tfrac{1}{2} dJ_a^* = C^b_{ac}(dW^c \wedge G_b^* - W^c \wedge dG_b^*). \tag{10-17.31}$$

When (10-17.30) and $\Theta^c = dW^c + \frac{1}{2} C^c_{ef} W^e \wedge W^f$ are used to eliminate dG_b^* and dW^c from the right-hand side of (10-17.31) and the Jacobi identity is applied, *the integrability conditions for the* W_i^a*-field equations are*

$$dJ_a^* - C^b_{ac} W^c \wedge J_b^* = 2C^b_{ac} \Theta^c \wedge G_b^*. \tag{10-17.32}$$

If the dependence of the Lagrangian $\bar{L}$ on W_i^a and Θ^a_{ij} is such that

$$C^b_{ac} \Theta^c \wedge G_b^* = 0, \tag{10-17.33}$$

which would appear to be the case as a consequence of G_r-invariance, we obtain the G_r-covariant current conservation laws

$$dJ_a^* + C^b_{ac} W^c \wedge J_b^* = 0. \tag{10-17.34}$$

In order to see that these results are consistent with those obtained in Section 10-6, we consider the situation in which G_r is a linear internal symmetry group. We then have

$$V_a = \gamma^{\alpha}_{a\beta} q^{\beta} \partial_{\alpha},$$

where $\boldsymbol{\gamma}_a = ((\gamma^{\alpha}_{a\beta}))$ $a = 1, \ldots, r$, is a basis for the matrix Lie algebra of G_r. In this event,

$$\pounds_a x^i = 0, \qquad \pounds_a q^{\alpha} = \gamma^{\alpha}_{a\beta} q^{\beta}$$

hence (10-17.12) gives

$$\mathscr{M}^* C^{\alpha} = dq^{\alpha} + W^a_i \gamma^{\alpha}_{a\beta} q^{\beta} dx^i - y^{\alpha}_i dx^i = \overset{*}{D} q^{\alpha} - y^{\alpha}_i dx^i$$

while (10-17.5) and (10-17.15) show that

$$T^i_j = \delta^i_j, \qquad t^i_j = \delta^i_j.$$

Accordingly, $\Phi^* \mathscr{M}^* C^{\alpha} = 0$ gives

$$\Phi^* y^{\alpha}_i = \partial_i \phi^{\alpha} + W^a_i \gamma^{\alpha}_{a\beta} \phi^{\beta} = \overset{*}{D}_i \phi^{\alpha},$$

which is just (10-6.3) in the alternative notation $\Psi^A \mapsto \phi^{\alpha}$. Further the Euler-Lagrange equations for the q^{α}-fields, (10-17.22), become

$$\Phi^* \left\{ \partial_j L^j_{\alpha} \right\} = \Phi^* \left\{ L_{\alpha} + L^i_{\beta} W^a_i \gamma^{\beta}_{a\alpha} \right\};$$

that is,

$$\Phi^* \left\{ \partial_j L^i_{\alpha} - W^a_j \gamma^{\beta}_{a\alpha} L^j_{\beta} \right\} = \Phi^* \{ L_{\alpha} \}.$$

These, however, are the same as (10-6.24) with the notation change $\Psi^A \mapsto \phi^{\alpha}$ and (10-6.25), (10-6.26). Finally, (10-17.24) and (10-17.20) give

$$J_a = \left\{ \sigma^j_a + L^j_{\alpha} \gamma^{\alpha}_{a\beta} q^{\beta} \right\} \mu_j, \qquad \sigma^i_a = \partial \bar{L} / \partial W^a_i |_{\theta^b}.$$

However (10-6.8) gives $\hat{L} = \hat{L}(x^j, \psi^A, y^A_k, F^{\alpha}_{ij}) = \bar{L}(x^j, q^{\alpha}, y^{\alpha}_k, \Theta^a_{ij})$ for the altered notation, and hence correspondence with the assumptions of Section 10-6 give $\sigma^j_a = 0$; that is,

$$J_a = L^j_{\alpha} \gamma^{\alpha}_{a\beta} q^{\beta} \mu_j,$$

which agrees with (10-6.32) and (10-6.25). It is then an easy matter to see that (10-17.30) agrees with (10-6.35) through (10-6.37).

These considerations point up a very important fact: gauge theories based on operator-valued Lie connections are adaptive. Any group generated by Noetherian vector fields of the first kind may be gauged by this method because the occurrence of $\pounds_a x^i, \pounds_a q^\alpha$ adapt the theory to the group in question. A further example of this is given in the next section where we consider the 10-parameter Poincaré group.

10-18. GAUGE THEORY FOR THE POINCARÉ GROUP

Most variational principles of current interest in physics are manifestly invariant under the 10-parameter Poincaré group, $P_{10}(\mathbb{R}) = L(4, \mathbb{R}) \rhd T(4)$, where $L(4, \mathbb{R})$ is the Lorentz group, $T(4)$ is the 4-parameter translation group, and $\rhd$ denotes the semidirect product. In addition, the flat space-time manifold E_4 carries a *Lorentz structure*

(10-18.1) $$ds^2 = h_{ij} dx^i \otimes dx^j, \qquad ((h_{ij})) = \text{diag}(1, 1, 1, -1)$$

for which $P_{10}(\mathbb{R})$ is the maximal group of isometries.

In view of the semidirect product structure of P_{10}, it is natural that we decompose the canonical parameters $\{u^a | 1 \le a \le 10\}$ into two sets by

$$\{u^a | 1 \le a \le 10\} = \{u^r; u^i | 1 \le r \le 6, 1 \le i \le 4\}.$$

If $\{v_a | 1 \le a \le 10\}$ is a basis for the Lie algebra of P_{10} realized as a group of automorphisms of E_4, we have

(10-18.2) $$V = u^a v_a = u^r l^i_{rj} x^j \partial_i + u^i \partial_i,$$

where the $\{l^i_{rj}\}$ is a basis for the matrix Lie algebra of $L(4, \mathbb{R})$;

(10-18.3) $$h_{ik} l^k_{rj} + h_{jk} l^k_{ri} = 0.$$

Now,

(10-18.4) $$[v_a, v_b] = C^e_{ab} v_e,$$

where the C's are the structure constants of P_{10}, while (10-18.2) show that $[\partial_i, \partial_j] = 0$. The Cartan-Killing form $\{C_{ab}\}$ thus has rank equal to six. The 6-by-6 form $\{C_{rs} | 1 \le r, s \le 6\}$ is nonsingular if we identify the first six u^a's with the six u^r's. We therefore set

(10-18.5) $$u^a = \delta^a_r u^r + \delta^a_{6+i} u^i.$$

The statement that the action is invariant under P_{10} means that P_{10} must be lifted to an isomorphic global group G_{10} of Noetherian vector fields on kinematic space K. Now, K is a $(4 + 5m)$-dimensional space with local coordinates $\{x^i, q^\alpha, y_i^\alpha\}$, so we must say something about how the state variables $\{q^\alpha\}$ behave when E_4 is subjected to the action of P_{10}. It is reasonable to assume that $\{\Phi^* q^\alpha\}$ transform under the global action of P_{10} as linear differential geometric object fields (as combinations of scalars, vectors, tensors, etc.). As such, the global translation part, $T(4)$, of P_{10} will have no effect and we may write

$$u^a V_a = u^i \partial_i + u^r l^i_{rj} x^j \partial_i + u^r M^\alpha_{r\beta} q^\beta \partial_\alpha + u^a Z_i (V_a \lrcorner C^\alpha) \partial^i_\alpha. \tag{10-18.6}$$

Here $\partial_\alpha = \partial / \partial q^\alpha$, $\partial^i_\alpha = \partial / \partial y^\alpha_i$, and the M's are constants that are determined by the transformation properties of the q^α's and are such that

$$[V_a, V_b] = C^e_{ab} V_e; \tag{10-18.7}$$

that is,

$$M^\alpha_{r\beta} M^\gamma_{s\alpha} - M^\alpha_{s\beta} M^\gamma_{r\alpha} = c^t_{rs} M^\gamma_{t\beta}, \qquad 1 \le r, s, t \le 6. \tag{10-18.8}$$

Here the lower-case C's are the structure constants of $L(4, \mathbb{R})$. For example, if 4 of the q's, say $\{T^i\}$, constitute the components of a vector field on E_4 when pulled back by any regular map $\Phi : E_4 \to K$, we will have

$$\begin{aligned} {}'T^i = T^j \partial' x^i / \partial x^j &= T^j \big(\delta^i_j + u^r l^i_{rj} + o(u^r) \big) \\ &= T^i + u^r l^i_{rj} T^j + o(u^r). \end{aligned}$$

Hence the corresponding terms in (10-18.6) will be given by $u^r l^i_{rj} T^j (\partial / \partial T^i)$. These clearly satisfy (10-18.8).

The minimal replacement construct for P_{10} may now be obtained without further ado; simply apply the results obtained in Section 10-16. To this end, we set

$$W^a = \delta^a_r W^r + \delta^a_{6+i} W^i \tag{10-18.9}$$

in conformity with (10-18.5), where

$$W^r = W^r_i dx^i, \qquad 1 \le r \le 6 \tag{10-18.10}$$

are the compensating fields for $L(4, \mathbb{R})$ and

$$W^i = W^i_j dx^j, \qquad 1 \le i \le 4 \tag{10-18.11}$$

are the compensating fields for $T(4)$. Thus since $\overset{*}{D}x^i = dx^i + W^a \pounds_a x^i$, $\overset{*}{D}q^\alpha = dq^\alpha + W^a \pounds_a q^\alpha$, we have

$$\overset{*}{D}x^i = \left(\delta^i_j + W^i_j + W_j{}^r l^i_{rk} x^k\right) dx^j, \tag{10-18.12}$$

$$\overset{*}{D}q^\alpha = dq^\alpha + W_j{}^r M^\alpha_{r\beta} q^\beta dx^j. \tag{10-18.13}$$

Accordingly, (10-17.12) and (10-17.6) yield

$$\mathscr{M}^* C^\alpha = dq^\alpha + W_j{}^r M^\alpha_{r\beta} q^\beta dx^j - y^\alpha_i \left(\delta^i_j + W^i_j + W_j{}^r l^i_{rk} x^k\right) dx^j, \tag{10-18.14}$$

$$\mathscr{M}^*(L\mu) = L \det\left(\delta^i_j + W^i_j + W_j{}^r l^i_{rk} x^k\right)\mu. \tag{10-18.15}$$

An interesting and characteristic result now obtains. If we apply the minimal replacement operator to the Lorentz structure (10-18.1), we have

$$dS^2 = \mathscr{M}^*(ds^2) = T^i_k h_{ij} T^j_l dx^k \otimes dx^l, \tag{10-18.16}$$

where we have set (see (10-17.5))

$$T^i_k = \delta^i_k + W^i_k + W_k{}^r l^i_{rj} x^j. \tag{10-18.17}$$

Accordingly, we may write

$$dS^2 = g_{ij} dx^i \otimes dx^j \tag{10-18.18}$$

where

$$g_{ij} = T_i{}^k h_{kl} T^l_j = g_{ji}. \tag{10-18.19}$$

Minimal replacement may thus be viewed as a construct that replaces the Lorentz structure ds^2 on E_4 by the more complicated pseudo-Riemannian structure dS^2 through the transition process $h_{ij} \mapsto g_{ij}$. It is assumed that the minimal replacement construct is *regular* in the sense that

$$\det\left(T_j{}^i\right) \neq 0, \tag{10-18.20}$$

which is clearly necessary in view of (10-18.15). (10-18.19) then shows that g_{ij} and h_{ij} both have the same signature, namely 2, and that dS^2 defines a proper pseudo-Riemannian structure on E_4. Further, it follows directly from (10-18.19) that

$$\det(g_{ij}) = \det(h_{ij}) \det\left(T_j{}^i\right)^2 = -\det\left(T_j{}^i\right)^2. \tag{10-18.21}$$

Thus if we set

(10-18.22) $$g = \det(g_{ij}),$$

which is necessarily negative, (10-18.21) gives

(10-18.23) $$\det\left(T_j^{\,i}\right) = \sqrt{-g}\,.$$

Accordingly, when (10-18.17) is used, (10-18.15) may be rewritten in the equivalent form

(10-18.24) $$\mathscr{M}^*(L\mu) = L\sqrt{-g}\,\mu,$$

while (10-18.14) becomes

(10-18.25) $$\mathscr{M}^*C^\alpha = dq^\alpha + \left(W_j^{\,r}M_{r\beta}^{\alpha}q^\beta - y_i^{\,\alpha}T_j^{\,i}\right)dx^j.$$

The form $L\sqrt{-g}\,\mu$ given by (10-18.24) is immediately recognized as the standard form of an action 4-form on a pseudo-Riemannian space-time with fundamental metric form $g_{ij}dx^i \otimes dx^j$.

The total action functional in this new context is given by

(10-18.26) $$\bar{A}[\Phi] = \int_{E_4} \Phi^*\left(L\sqrt{-g}\,\mu + S^*\eta\right)$$

where η is a G_{10}-invariant 4-form on G_{10} and

(10-18.27)
$$\Phi: E_4 \to K \times \mathbb{R}^{40} | x^i = x^i, \qquad q^\alpha = \phi^\alpha(x^j), \qquad W_i^a = W_i^a(x^j),$$
$$\Phi^*\mu \neq 0, \qquad \Phi^*\mathscr{M}^*C^\alpha = 0,$$

see (10-17.1) and (10-17.3).

In order that we may determine possible forms for η, we first calculate the forms of the curvature 2-forms, Θ^a, for G_{10}. Since $\Theta^a = dW^a + \frac{1}{2}C_{bc}^{a}W^bW^c$, (10-18.5) induces the decomposition

(10-18.28) $$\Theta^a = \delta_r^{\,a}\Theta^r + \delta_{6+i}^{a}\Theta^i.$$

Noting that G_{10} and P_{10} have the same structure constants, we have

(10-18.29) $$C_{rs}^{i} = 0, \qquad C_{ij}^{a} = 0, \qquad C_{ri}^{s} = 0$$

with $1 \leq i, j \leq 4$, $1 \leq r, s \leq 6$. It is then a simple matter to see that

$$\Theta^r = dW^r + \tfrac{1}{2}C^r_{st}W^s \wedge W^t, \qquad 1 \leq r, s, t \leq 6, \tag{10-18.30}$$

and

$$\Theta^k = dW^k + C^k_{si}W^s \wedge W^i, \qquad i \leq i, k \leq 4, \qquad 1 \leq s \leq 6. \tag{10-18.31}$$

The 2-forms $\{\Theta^r | 1 \leq r \leq 6\}$ are the curvature 2-forms associated with the local action of $L(4, \mathbb{R})$, while $\{\Theta^k | 1 \leq k \leq 4\}$ are the curvature 2-forms associated with the semidirect product action of $T(4)$, as evidenced by the coupling terms $C^k_{si}W^s \wedge W^i$. Now, $((C_{rs}))$ is a nonsingular 6-by-6 matrix and hence applying S^* shows that we have the G_{10}-invariant scalar

$$\alpha = \Theta^r_{ij} h^{ik} h^{jl} \Theta^s_{kl} C_{rs} \tag{10-18.32}$$

where $2\Theta^r = \Theta^r_{ij} dx^i \wedge dx^j$. Here we have used the fact that G_{10} is a group of isometrics of E_4 and hence ${}'h_{ij} = h_{ij}$.

An inspection of (10-18.32) shows that α is independent of the curvature coefficients $\{\Theta^k_{ij}\}$ associated with $T(4)$. This is a direct consequence of the fact that G_{10} is *not* a semisimple group and has been a source of certain difficulties in the past; it is necessary that we go back to $\Lambda(\mathscr{G})$ in order to determine other G_{10}-invariant quantities. Another possibility is afforded by the scalar invariants that can be formed from the Riemannian curvature tensor based on the metric tensor $g_{ij} = T_i^k h_{kl} T_j^l$. Such invariants have the right properties, for (10-18.17) shows that $g_{ij} = h_{ij}$ for G_{10}-constant sections of $\mathscr{G}$ ($W^a = 0$) and the Riemannian curvature tensor formed from h_{ij} vanishes throughout E_4. In any event, even after further invariants are found, there is still the question of selecting an appropriate representation for $S^*\eta$. We leave this aspect of the problem for future study and simply take

$$\mathscr{M}^*(L\mu) + S^*\eta = \left(L\sqrt{-g} + f(\alpha, \cdots)\right)\mu = \bar{L}\mu \tag{10-18.33}$$

with

$$\bar{L} = \bar{L}\left(x^j, q^\alpha, y^\alpha_i, W^i_j, W^r_j, \Theta^i_{jk}, \Theta^r_{jk}\right) \tag{10-18.34}$$

in conformity with (10-17.8). The field equations for the gauge theory of the Poincaré group then follow directly from the results given in Section 10-17:

$$\Phi^*\partial_j\left(t^j_i L^i_\alpha\right) = \Phi^*\left\{L_\alpha + t^j_i L^i_\beta W^a_j \partial_\alpha \pounds_a q^\beta\right\} \tag{10-18.35}$$

$$dG^*_i + C^b_{ic} W^c \wedge G^*_b = \tfrac{1}{2}J^*_i, \qquad 1 \leq i \leq 4,$$

$$dG^*_r + C^b_{rc} W^c \wedge G^*_b = \tfrac{1}{2}J^*_r, \qquad 1 \leq r \leq 6. \tag{10-18.36}$$

The reader should compare the results of this section with those given in the References for Further Study.

REFERENCES FOR FURTHER STUDY

Chapter 8

Bataille, J. and J. Kestin, General Forms of the Dissipation Inequality. *J. Non-Equilib. Thermodyn.* **1** (1976), 25–31.

de Groot, S. R. and P. Mazur, *Non-Equilibrium Thermodynamics*. North-Holland, Amsterdam, 1969.

Edelen, D. G. B., The Thermodynamics of Evolving Chemical Systems and the Approach to Equilibrium. *Adv. Chem. Phys.* **33** (1975), 399–442.

Edelen, D. G. B., A Thermodynamics with Internal Degrees of Freedom and Nonconservative Forces. *Int. J. Engng. Sci.* **14** (1976), 1013–1032.

Gyarmati, I., *Non-Equilibrium Thermodynamics*. Springer-Verlag, Berlin, 1970.

Kestin, J., *A Course in Thermodynamics*. Blaisdell, New York, 1968.

Lamprecht, I. and A. I. Zotin, eds. *Thermodynamics and Kinetics of Biological Processes*. Walter de Gruyter, Berlin, 1983.

Meixner, J., Processes in Simple Thermodynamic Materials. *Arch. Rational Mech. Anal.* **33**, 33–53.

Prigogine, I., *Introduction to Thermodynamics of Irreversible Processes*. John Wiley, New York, 1967.

Truesdell, C. and R. A. Toupin, The Classical Field Theories. In *Handbuch der Physik* 111/1. Springer-Verlag, Berlin, 1960.

Chapter 9

Cabibbo, N. and E. Farrari, Quantum Electrodynamics with Dirac Monopoles. *Il Nuovo Cimento* **23** (1962), 1147–1154.

Dirac, P. A. M., Quantized Singularities in the Electromagnetic Field. *Proc. Roy. Soc.* **A 133** (1931), 60–72.

Edelen, D. G. B., A Metric Free Electrodynamics with Electric and Magnetic Charges. *Ann. Phys.* (*N.Y.*). **112** (1978), 366–400.

Han, M. Y. and L. C. Biedenharn, Manifest Dyality Invariance and the Cabibbo–Ferrari Theory of Magnetic Monopoles. *Il Nuove Cimento* **2A** (1971), 544–556.

Mignani, R., Symmetries of Electrodynamics with Magnetic Monopoles and the Hertz Tensor. *Phys. Rev.* **D. 13** (1976), 2437–2440.

Schwinger, J., Magnetic Charge and the Charge Quantization Condition. *Phys. Rev.* **D. 12** (1975), 3105–3111.

Stratton, J. A., *Electromagnetic Theory*. McGraw-Hill, New York, 1941.

van Dantzig, D., Electromagnetism, Independent of Metrical Geometry. *Akad. Wetensch. Amsterdam* **37** (1934), 521–525, 526–531, 643–652, 825–836.

Wu, T. T. and C. N. Yang, Dirac's Monopole without Strings: Classical Lagrangian Theory. *Phys. Rev.* **D 14** (1976), 437–445.

Chapter 10

Actor, A., Classical Solutions of SU (2) Yang-Mills Theories. *Rev. Mod. Phys.* **51** (1979), 461–525.

Bernstein, J., Spontaneous Symmetry Breaking, Gauge Theories, The Higgs Mechanism and All That. *Rev. Mod. Phys.* **46** (1974), 7–48.

Corrigan, E., et al., Magnetic Monopoles and SU(3) Gauge Theories. *Nuclear Phys.* **B106** (1976), 475–492.

Daniel, M. and C. M. Viallet, The Geometrical Setting for Gauge Theories of the Yang-Mills Type. *Rev. Mod. Phys.* **52** (1980), 175–197.

Drechsler, W. and M. E. Mayer, *Figer Bundle Techniques in Gauge Theories*. Lecture Notes in Physics No. 67. Springer-Verlag, Berlin, 1977.

Edelen, D. G. B., On the Intrinsic Structure of Yang-Mills Fields. *Ann. Phys.* (*N.Y.*) **133** (1981), 286–303.

Kadić, Aida and D. G. B. Edelen, *A Gauge Theory of Dislocations and Disclinations*. Lecture Notes in Physics No. 174. Springer-Verlag, Berlin, 1983.

Kikkawa, K., N. Nakanishi and H. Nariai, eds. *Gauge Theory and Gravitation*. Lecture Notes in Physics No. 176. Springer-Verlag, Berlin, 1983.

Prasad, M. K. and C. M. Sommerfield, Solutions of Classical Gauge Field Theories with Spin and Internal Symmetry. *Nuclear Phys.* **B110** (1976), 153–172.

Rund, H., Differential-Geometric and Variational Background of Classical Gauge Field Theories. *Aeq. Math.* **24** (1982), 121–174.

Schouten, J. A., *Ricci-Calculus*, Second Edition. Springer-Verlag, Berlin, 1954.

Weinberg, S., Recent Progress in Gauge Theories of the Weak, Electromagnetic and Strong Interactions. *Rev. Mod. Phys.* **48** (1974), 255–277.

Wu, T. T. and C. N. Yang, Concept of Nonintegrable Phase Factors and the Global Formulation of Gauge Fields. *Phys. Rev.* **D. 12** (1975), 3845–3857.

Yang, C. N. and R. L. Mills, Conservation of Isotopic Spin and Isotopic Gauge Invariance. *Phys. Rev.* **96** (1954), 191–195.

APPENDIX

The first thing that needs doing is to prove the fixed point theorem that is stated at the beginning of Section 1–3.

Let T be a map of a complete metric space, S, into itself that satisfies

$$\text{(A-1)} \qquad \rho(T\phi, T\psi) < k\rho(\phi, \psi), \qquad k < 1,$$

where ρ is the distance function on S. Then there is a *unique fixed point of T; that is,* $T\phi_0 = \phi_0$ *for one and only one* $\phi_0 \in S$. *Further,* $\phi_0 = \lim_{n\to\infty} T^n\phi$ *for any* $\phi \in S$, where $T^n\phi = TT^{n-1}\phi$.

Proof. We first prove uniqueness. Suppose that ϕ_0 and ψ_0 are two fixed points with $\rho(\phi_0, \psi_0) \neq 0$. We then have $T\phi_0 = \phi_0$, $T\psi_0 = \psi_0$, and hence $\rho(T\phi_0, T\psi_0) = \rho(\phi_0, \psi_0)$ which contradicts (A-1). A simple calculation shows that

$$\rho(T^{n+1}\phi, T^n\phi) = \rho(TT^n\phi, TT^{n-1}\phi) < k\rho(T^n\phi, T^{n-1}\phi)$$
$$< k^n\rho(T\phi, \phi).$$

Thus since $k < 1$, the limit as n tends to infinity of $T^n\phi$ exists. Further since S is complete, $\phi_0 = \lim_{n\to\infty} T^n\phi$ belongs to S. Finally, $\rho(T\phi_0, \phi_0) = \lim_{n\to\infty}(T^{n+1}\phi, T^n\phi) < \rho(T\phi, \phi)\lim_{n\to\infty}(k^n) = 0$, so that $T\phi_0 = \phi_0$. □

Proof of the Implicit Function Theorem. We use the notation given in the statement of the Implicit Function Theorem in Section 1–3. Let $((B^i_j))$ be the inverse of the matrix $((\partial F^i/\partial y^j))$ when the latter is evaluated at the origin.

We can then replace the equations (1-3.5) by the equivalent system of equations

$$B^i_j F^j = G^i = 0. \tag{A-2}$$

This change has the useful effect that $((\partial G^i/\partial y^j))$ is equal to the identity matrix at the origin. Accordingly, we can write

$$G^i = y^i - f^i \tag{A-3}$$

with $\partial f^i/\partial y^j = 0$ at the origin for all values of i and j. Equations (1-3.5) now become

$$\phi^i(x^m) = f^i(\phi^1(x^m), \ldots, \phi^k(x^m);\ x^1, \ldots, x^n), \qquad i = 1, \ldots, k. \tag{A-4}$$

For any neighborhood U of the origin in E_n, let $\boldsymbol{\Phi}(U)$ be the metric space of all k-tuples of continuous functions $\boldsymbol{\phi} = (\phi^1, \ldots, \phi^k)$ with metric

$$\rho(\boldsymbol{\phi}, \boldsymbol{\psi}) = \sum_{i=1}^{k} \sup_{x \in U} |\phi^i(x^m) - \psi^i(x^m)|. \tag{A-5}$$

We then have $\|\boldsymbol{\phi}\| = \rho(\boldsymbol{\phi}, \mathbf{0})$. $\boldsymbol{\Phi}(U)$ is then a complete metric space since (A-5) and $\|\boldsymbol{\phi}\|$ give $\boldsymbol{\Phi}(U)$ the norm of uniform convergence (i.e., a uniformly convergent sequence of continuous functions has a continuous limit function). Let T denote the mapping of a subset of $\boldsymbol{\Phi}(U)$ into itself that is given by $T\boldsymbol{\phi} = \boldsymbol{\psi}$ where

$$\psi^i(x^m) = f^i(\phi^1(x^m), \ldots, \phi^k(x^m);\ x^1, \ldots, x^n).$$

Since the functions f^i are continuous with continuous derivatives, as a consequence of the continuity assumptions satisfied by the functions F^i, the map $T\boldsymbol{\phi}$ is defined for all $\boldsymbol{\phi}$ with $\|\boldsymbol{\phi}\|$ sufficiently small. Now, $f^i(0, \ldots, 0; 0, \ldots, 0) = 0$, so that we may assume that $|f^i(y^1, \ldots, y^k;\ x^1, \ldots, x^n)| < \epsilon$ for $|x^m| < \delta$, and $|y^i| < \delta$. Also, since $\partial f^i/\partial y^j = 0$ at the origin, we can assume that $|\partial f^i/\partial y^j| < \epsilon$ and $|\partial f^i/\partial x^r| < K$ for $|x^m| < \delta$, $|y^k| < \delta$. The mean value theorem then implies that

$$|f^i(\phi^1(x^m), \ldots, \phi^k(x^m);\ x^1, \ldots, x^n)| \leq \frac{\partial f^i}{\partial y^j}\phi^j + \frac{\partial f^i}{\partial x^r}x^r.$$

We therefore have $\|T\boldsymbol{\phi}\| < k(\epsilon\|\boldsymbol{\phi}\| + nK\delta)$. Suppose that T is defined on the set of those $\boldsymbol{\phi} \in \boldsymbol{\Phi}(U)$ with $\|\boldsymbol{\phi}\| < \eta$. We then choose U sufficiently small so that δ and ϵ satisfy

$$k(\epsilon\eta + nK\delta) < \eta. \tag{A-6}$$

The map T then maps the set $\|\phi\| < \eta$ into itself. Further the mean value theorem also shows that $\|T\phi - T\psi\| < k\epsilon\|\phi - \psi\|$ and hence $\rho(T\phi, T\psi) < k\epsilon\rho(\phi, \psi)$. It is then a simple matter to see that we can satisfy $k\epsilon < 1$ and (A-6) simultaneously; for example, take $k\epsilon < \frac{1}{2}$, $\delta < \eta/2knK$. All of the conditions of the contraction mapping theorem are thus met from which we conclude existence and uniqueness of solutions of (A-4). The continuity properties of the solutions then follow from the assumed continuity properties of the functions $F^i(y^j; x^m)$. □

Proof of the Existence Theorem for Autonomous Systems of First Order Differential Equations. The notation used here will be the same as that given in Section 1–4. It is a simple matter to see that the differential equations (1-4.1) and the initial data (1-4.2) are equivalent to the system of integral equations

$$\phi^i(t; x_0^j) = x_0^i + \int_0^t f^i(\phi^k(\tau))\, d\tau. \tag{A-7}$$

Let us assume that the functions $f^i(x^k)$ are defined for $|x^i| < \eta$ and that $|f^i(x^k)| < M$ for $|x^i| < \eta$. Let $C(a, \eta)$ be the set of all n-tuples of functions $\phi = (\phi^1(t), \ldots, \phi^n(t))$ that are defined for $|t| < a$ and that satisfy $|\phi^i(t)| < \eta$ for $|t| < b$. If $|x_0^i| < \eta/2$, and $|t| < 1/2M$, $\phi \in C(1/2M, \eta)$ is such that $T\phi = \psi = (\psi^1, \ldots, \psi^n)$, with

$$\psi^i(t; x_0^i) = x_0^i + \int_0^t f^i(\phi^k(\tau))\, d\tau, \tag{A-8}$$

is well defined and lies in $C(1/2M, \eta)$. Thus if we set

$$\|\phi\| = \sum_{i=1}^{n} \left(\sup_{|t| < 1/2M} |\phi^i(t)| \right),$$

the mean value theorem gives

$$\|T\psi - T\phi\| < n\bar{a}K\|\psi - \phi\|$$

for ϕ, $\psi \in C(\bar{a}, \eta)$; simply take $K = \sup_{i,j,x}|\partial f^i/\partial x^j|$. The choice of $\bar{a}$ such that $\bar{a} < 1/nK$ then makes T a contraction mapping of $C(\bar{a}, \eta)$ into itself. Application of the contraction mapping theorem then gives existence and uniqueness. Continuity and differentiability in t then follow directly from (A-7). Continuity and differentiability properties with respect to the variables (x_0^i) follow from similar estimates that obtain from

$$\phi^i(t; x_0^j) = x_0^i + \int_0^t f^i(\phi^j(\tau; x_0^k))\, d\tau. \qquad \square$$

INDEX